Local Fields and Their Extensions

Translations of

MATHEMATICAL MONOGRAPHS

Volume 121

Local Fields and Their Extensions

Second Edition

I. B. Fesenko
S. V. Vostokov

American Mathematical Society
Providence, Rhode Island

2000 *Mathematics Subject Classification*. Primary 11Sxx, 11S15, 11S20, 11S31, 11S70, 11S85, 11S99.

Library of Congress Cataloging-in-Publication Data

Fesenko, I. B. (Ivan B.)

Local fields and their extensions / I. B. Fesenko, S. V. Vostokov.—2nd ed.

p. cm. — (Translations of mathematical monographs, ISSN 0065-9282 ; v. 121)

Includes bibliographical references and index.

ISBN 0-8218-3259-X (alk. paper)

1. Algebraic fields. 2. Field extensions (Mathematics) I. Vostokov, S. V. II. Title. III. Series.

QA247.F47 2002

512′.74—dc21 2002066561

∞ The paper used in this book is acid-free and falls within the guidelines
established to ensure permanence and durability.
Visit the AMS home page at URL: http://www.ams.org/

10 9 8 7 6 5 4 3 2 1 07 06 05 04 03 02

Contents

Introduction to the Second Edition

The class of discrete valuation fields appears to be next in significance and order of complexity to the class of finite fields. Among discrete valuation fields a highly important place, both for themselves and in connection with other theories, is occupied by complete discrete valuation fields.

This book is devoted to local fields, i.e. complete discrete valuation fields with perfect residue field.

The time distance between the second edition of "Local Fields and Their Extensions" and its first edition is ten years. During this period, according to Math Reviews, almost one thousand papers on local fields have been published. Some of them have further developed and clarified various topics described in the first edition of this book. On the other hand, the authors of this book have received a variety of useful suggestions and remarks from several dozen readers of the first edition.

All these have naturally led to the second edition of the book.

This book is aimed to serve as an easy exposition of the arithmetical properties of local fields by using explicit and constructive tools and methods. Almost everywhere it does not require more prerequisites than a standard course in Galois theory and a first course in number theory which includes p-adic numbers.

The book consists of nine chapters which form the following groups:

group 1: elementary properties of local fields (Chapter I–III)

group 2: class field theory for various types of local fields and generalizations (Chapter IV-V)

group 3: explicit formulas for the Hilbert pairing (Chapter VI-VIII)

group 4: Milnor K-groups of local fields (Chapter IX).

Chapters of the third group were mainly written by S. V. Vostokov and the rest was written by I. B. Fesenko.

The first page of each chapter provides a detailed description of its contents, so here we just emphasize the most important issues and also indicate changes with respect to the first edition.

Chapter I describes the most elementary properties of local fields when one does not look at connections between them, but concentrates on a single field.

Chapter II deals with extensions of discrete valuation fields and already section 1 and 2 introduce a very important class of Henselian fields and describe relations between Henselian and complete fields. We have included more information than in the first edition on ramification subgroups in section 4.

The main object of study in Chapter III is the norm map acting on the multiplicative group and its arithmetical properties. In section 1 we describe its behaviour for cyclic extensions of prime degree. Section 2 shows that almost all cyclic extensions of degree equal to the characteristic of the perfect residue field are generated by roots of Artin–Schreier polynomials. In section 3 we introduce a function which takes into account certain properties of the norm map acting on higher principal units. Our approach to the definition of the Hasse–Herbrand function is different from the approach in other textbooks (where the definition involves ramification groups). Sections 3 and 4 in the second edition now include more applications of our treatment of the Hasse–Herbrand function. Section 5 is devoted to the Fontaine–Wintenberger theory of fields of norms for arithmetically profinite extensions of local fields. This theory links certain infinite extensions of local fields of characteristic zero or p with local fields of characteristic p. Now the section contains more details on applications of this theory, some of which have been published since 1993.

Chapter IV is on class field theory of local fields with finite residue fields. For this edition we have chosen a slightly different approach from the first edition: for totally ramified extensions we work simultaneously with both the Neukirch map and Hazewinkel homomorphism (which are almost inverse to each other). We hope that this method explains more fully on what is going on behind definitions, constructions and calculations and therefore gives the reader more chances to appreciate the theory. This method is also very useful for applications. Section 1 contains new subsections (1.6)–(1.9) which are required for the study of the reciprocity maps. Sections 2–4 differs significantly from the corresponding parts of the first edition. After proving the main results of local class field theory we review all other approaches to it in the new section 7. The new section 8 presents to the reader a recent noncommutative reciprocity map, which is not a homomorphism but a Galois 1-cycle. This theory is based a generalization of the approach to (abelian) class field theory in this book. We also review results on the absolute Galois group of a local field.

Chapter V studies abelian extensions of local fields with infinite residue field. In the same way as in the first edition, the first three sections discuss in detail class field theory of local fields with quasi-finite residue field. In the new section 4 we describe recent theory of abelian totally ramified p-extensions of a local field with perfect residue fields of characteristic p which can be viewed as the largest possible generalization of class field theory of Chapter IV. If a complete discrete valuation field has imperfect residue field, then its class field theory becomes much more difficult. Still, some results on abelian totally ramified p-extensions of such fields and their norm groups can be established in the framework of this book; we explain some features in the new section 5. The latter also includes a class field theory interpretation of results on some abelian varieties over local fields.

Chapter VI serves as a prerequisite for Chapters VII and VIII. For a finite extension of the field of p-adic numbers it presents a very useful formal power series method for the study of elements of the fields. The Artin–Hasse–Shafarevich exponential map is described in section 2 and the Shafarevich basis of the group of principal units in section 5. This Chapter contains many technical results, especially in section 3 and 4, which are of use in Chapter VI.

The aim of Chapter VII is to explain to the reader explicit formulas for the Hilbert symbol. The method is to introduce at first an independent pairing on formal power series and to show that it is well defined and satisfies the Steinberg property (subsection (2.1)). Then a pairing on the multiplicative group of the field induced by the previous pairing is defined. Its properties (independence of a power series presentation and invariance with respect to the choice of a prime element) help one easily show its equality with the Hilbert pairing. The second edition contains many simplifications of the first edition and it also includes more material on interpretations of the explicit formulas and their applications.

Chapter VIII is an exposition of a generalization of the method of Chapter VII to formal groups. The simplest among the groups are Lubin–Tate groups which are introduced in section 1; exercises let the reader see the well known applications of them to local class field theory. Explicit formulas for the generalized Hilbert pairing associated to a Lubin–Tate formal group are presented in section 2. The new section 3 describes a recent generalization to Honda formal groups.

Chapter IX describes the Milnor K-groups of fields. Calculations of the Milnor K-groups of local fields in section 4 shed a new light on the Hilbert symbol of Chapter IV.

The bibliography includes comments on introductory texts on various applications of local fields.

Numerous remarks and exercises often indicate further important results and theories left outside this introductory book. The most challenging exercises are marked by $(\diamond)$.

Those readers who prefer to start with class field theory of local fields with finite residue fields are recommended to read sections 1–7 of Chapter IV and follow the references to the previous Chapters if necessary.

One of more advanced theories closely related to the material of this book and its presentation is higher local class field theory; for an introduction to higher local fields see [FK].

A reference in Chapter n to an assertion in Chapter m does not state the number m explicitly if and only if $m = n$. Briefly on notations: For a field F an algebraic closure of F is denoted by F^{alg} and the separable closure of F in F^{alg} is denoted by F^{sep}. Separable and algebraic closures of fields are assumed suitably chosen where it is necessary to make such conventions. $G_F = \text{Gal}(F^{\text{sep}}/F)$ stands for the absolute Galois group of F, μ_n denotes the group of all nth roots of unity in F^{sep}.

The text is typed using AMSTeX and a modified version of osudeG style (written by W. Neumann and L. Siebenmann and available from the public domain of Department of Mathematics of Ohio State University, pub/osutex).

March 2002 I. B. Fesenko S. V. Vostokov

Foreword to the First Edition

A. Weil was undoubtedly rightwhen he asserted, in the preface to the Russian edition of his book on number theory, that since class field theory pertains to the foundation of mathematics, every mathematician should be as familiar with it as with Galois theory. Moreover, just like Galois theory before it, class field theory was reputed to be very complicated and accessible only to specialists.

Here, however, the parallels between these two theories come to an end. A mathematician who has decided to become acquainted with Galois theory is not confronted with the problem of choosing a suitable exposition: all expositions of it are essentially equivalent, differing only in didactic details. For class field theory, on the other hand, there is a wide range of essentially different expositions, so that it is not immediately obvious even whether the subject is the same.

In the 1960s, it seemed that a universal Galois cohomology approach to class field theory had been found. What is more, the role of homological algebra as a common language unifying various branches of mathematics was becoming clear. Homological algebra could be likened to medieval Latin that served as the means of communication within educated circles. However, just as Latin could not effectively stand up against the originality of individual national languages, so Galois cohomology theory no longer offers the "only reasonable" understanding of class field theory. The goal of the cohomological method was the formation of class fields in which both number and local fields and their arithmetic properties disappear, the whole theory being formalized as a system of axioms. But other expositions of class field theory reveal remarkable properties of number and local fields, that are ignored in the cohomological approach. It has become evident that class field theory is not just an application of cohomology groups, but that it is also closely related with other profound theories such as the theory of formal groups, K-theory, etc.

The exposition of this book does not use homological algebra. It presents specific realities of local fields as clear as possible. Despite its limited volume, the book contains a vast amount of information on local fields. It offers the reader the possibility to see the beauty and diversity of this subject.

30 June 1992, Moscow — I. R. Shafarevich

CHAPTER 1

Complete Discrete Valuation Fields

This chapter introduces local fields as complete discrete valuation fields with perfect residue field. The material of sections 1–4, 7–8 is standard. Section 5 describes raising to the pth power on the group of principal units and section 6 treats the group of principal units as a multiplicative $\mathbb{Z}_p$-module in terms of convergent power series. Section 9 introduces various modifications of the logarithm map for local fields; those are important for Chapters VI–VIII. The reader is supposed to have some preliminary knowledge on p-adic numbers, e.g., to the extent supplied by the first chapters of [Gou] or any other elementary book on p-adic numbers.

1. Ultrametric Absolute Values

We start with a classical characterization of absolute values on the field of rational numbers which demonstrates that the p-adic norms and p-adic numbers are as important as the better known absolute value and real numbers.

(1.1). A map $\|\cdot\|\colon \mathbb{Q} \to \mathbb{R}$ is said to be an *absolute value* if the following three properties are satisfied:

$$\begin{aligned} &\|\alpha\| > 0 \quad \text{if} \quad \alpha \neq 0, \quad \|0\| = 0, \\ &\|\alpha\beta\| = \|\alpha\| \, \|\beta\|, \\ &\|\alpha + \beta\| \leqslant \|\alpha\| + \|\beta\| \qquad \text{(triangle inequality)}. \end{aligned}$$

Obviously, the usual absolute value $|\cdot|$ of $\mathbb{Q}$ induced from $\mathbb{C}$ satisfies these conditions, and we will also denote it by $\|\cdot\|_\infty$. The absolute value $\|\cdot\|$ on $\mathbb{Q}$ such that $\|\mathbb{Q}^*\| = 1$ is called trivial.

For a prime p and a non-zero integer m let $k = v_p(m)$ be the maximal integer such that p^k divides m. Extend v_p to rational numbers putting $v_p(m/n) = v_p(m) - v_p(n)$; $v_p(0) = +\infty$.

Define the *p-adic norm* of a rational number α:

$$\|\alpha\|_p = p^{-v_p(\alpha)}.$$

A complete description of absolute values on $\mathbb{Q}$ is supplied by the following result.

THEOREM (OSTROWSKI). *An absolute value $\|\cdot\|$ on $\mathbb{Q}$ either coincides with $\|\cdot\|_\infty^c$ for some real $c \geqslant 0$, or with $\|\cdot\|_p^c$ for some prime p and real c.*

Proof. For an integer $a > 1$ and an integer $b > 0$ write

$$b = b_n a^n + b_{n-1}a^{n-1} + \dots + b_0, \qquad 0 \leqslant b_i < a, a^n \leqslant b.$$

Then

$$\|b\| \leqslant (\|b_n\| + \|b_{n-1}\| + \dots + \|b_0\|) \max(1, \|a\|^n)$$

and

$$\|b\| \leqslant (\log_a b + 1) d \max(1, \|a\|^{\log_a b}),$$

with $d = \max(\|0\|, \|1\|, \dots, \|a-1\|)$. Substituting b^s instead of b in the last inequality, we get

$$\|b\| \leqslant (s \log_a b + 1)^{1/s} d^{1/s} \max(1, \|a\|^{\log_a b}).$$

When $s \to +\infty$ we deduce

$$\|b\| \leqslant \max(1, \|a\|^{\log_a b}).$$

There are two cases to consider for the nontrivial absolute value $\|\cdot\|$.

(1) Suppose that $\|b\| > 1$ for some natural b. Then

$$1 < \|b\| \leqslant \max(1, \|a\|^{\log_a b}),$$

and $\|a\| > 1$, $\|b\| = \|a\|^{\log_a b}$ for any integer $a > 1$. It follows that $\|a\| = \|a\|_\infty^c$, with real $c > 0$ satisfying the equation $\|b\| = \|b\|_\infty^c$.

(2) Suppose that $\|a\| \leqslant 1$ for each integer a. Let a_0 be the minimal positive integer, such that $\|a_0\| < 1$. If $a_0 = a_1 a_2$ with positive integers a_1, a_2, then $\|a_1\| \, \|a_2\| < 1$ and either $a_1 = 1$ or $a_2 = 1$. This means that $a_0 = p$ is a prime. If $q \notin p\mathbb{Z}$, then $pp_1 + qq_1 = 1$ with some integers p_1, q_1 and hence $1 = \|1\| \leqslant \|p\| \, \|p_1\| + \|q\| \, \|q_1\| \leqslant \|p\| + \|q\|$. Writing q^s instead of q we get $\|q\|^s \geqslant 1 - \|p\| > 0$. When s is sufficiently large we obtain $\|q\| = 1$. Therefore, $\|\alpha\| = \|p\|^{v_p(\alpha)}$, which was to be proved. □

We are naturally led to look more closely at absolute values of the type indicated in case (2). As $v_p(\alpha + \beta) \geqslant \min(v_p(\alpha), v_p(\beta))$, for such absolute values we get that

$$\|\alpha + \beta\| \leqslant \max(\|\alpha\|, \|\beta\|) \quad \text{(ultrametric inequality)}.$$

Such absolute values are said to be *ultrametric*.

(1.2). One can generalize the notions discussed above. Call a map $\|\cdot\| \colon F \to \mathbb{R}$ for a field F an *absolute value* if it satisfies the three conditions formulated in (1.1). An absolute value is called trivial if $\|F^*\| = 1$. Similarly one can introduce the notion of an ultrametric absolute value on F.

Note that for an ultrametric absolute value $\|\cdot\|$ on F, if $\|\alpha\| < \|\beta\|$, then

$$\|\alpha+\beta\| \leqslant \max(\|\alpha\|, \|\beta\|) = \|\beta\| = \|\alpha+\beta-\alpha\| \leqslant \max(\|\alpha+\beta\|, \|\alpha\|).$$

Therefore, $\|\alpha+\beta\| = \|\beta\|$. This means that any triangle has two equal sides with respect to the ultrametric absolute value $\|\cdot\|$.

Let $F = K(X)$ and let $\|\cdot\|$ be a nontrivial absolute value on F such that $\|K^*\| = 1$. If $\alpha, \beta \in F$, then

$$\|(\alpha+\beta)\|^n \leqslant \|\alpha\|^n + \|\alpha\|^{n-1}\|\beta\| + \cdots + \|\beta\|^n \leqslant (n+1)\max(\|\alpha\|^n, \|\beta\|^n).$$

Taking the n th root of both sides in the last inequality, and letting n tend to $+\infty$, we obtain that $\|\cdot\|$ is ultrametric.

We consider two cases.

(1) $\|X\| > 1$. Put $\deg(f(X)/g(X)) = \deg f(X) - \deg g(X)$, if $f(X)$, $g(X) \in K[X]$. Hence

$$\|\alpha\| = \|X^{-1}\|^{-\deg\alpha}.$$

Put $v_\infty(\alpha) = -\deg\alpha$, $v_\infty(0) = +\infty$. Note that $v_\infty(\frac{1}{X}) = 1$.

(2) $\|X\| \leqslant 1$. Then $\|\alpha\| \leqslant 1$ for $\alpha \in K[X]$. Let $p(X) \in K[X]$ be a monic polynomial of minimal positive degree satisfying the condition $\|p(X)\| < 1$. One shows similarly to case (2) in (1.1) that $p(X)$ is irreducible and

$$\|\alpha\| = \|p(X)\|^{v_{p(X)}(\alpha)},$$

where $v_{p(X)}(f(X))$ is the largest integer k such that $p(X)^k$ divides polynomial $f(X)$, and $v_{p(X)}(f/g) = v_{p(X)}(f) - v_{p(X)}(g)$ for polynomials f, g, $v_{p(X)}(0) = +\infty$.

Thus, nontrivial absolute values on $F = K(X)$, which are trivial on K, are in one-to-one correspondence (up to raising to a positive real power) with irreducible monic polynomials of positive degree in $K[X]$ and $\frac{1}{X}$.

Exercises.

1. Show that $\big|\|\alpha\| - \|\beta\|\big| \leqslant \|\alpha+\beta\| \leqslant \|\alpha\| + \|\beta\|$ for $\alpha, \beta \in F$, where $\|\cdot\|$ is an absolute value on F.
2. Show that every absolute value on a finite field is trivial.
3. Let A be a subring of F generated by 1 of F. Show that an absolute value $\|\cdot\|$ on F is ultrametric if and only if there exists $c > 0$ such that $\|a\| \leqslant c$ for all $a \in A$.
4. Show that every absolute value on a field of positive characteristic is ultrametric.
5. Show that an absolute value $\|\cdot\|$ on a field F is ultrametric if and only if $\|\cdot\|^c$ is an absolute value on F for all real $c > 0$.
6. Find the set of real $c > 0$, such that $\|\cdot\|_\infty^c$ is not an absolute value on $\mathbb{Q}$.
7. Let S be the set of all positive primes in $\mathbb{Z}$, $S' = S \cup \{\infty\}$. Show that if $\alpha \in \mathbb{Q}^*$, then $\|\alpha\|_i = 1$ for almost all $i \in S$ and

$$\prod_{i\in S'} \|\alpha\|_i = 1.$$

8. Let $0 < d < 1$, let I be the set of all irreducible monic polynomials of positive degree over K, and $I' = I \cup \{\infty\}$. Let

$$\|\alpha\|_\infty = d^{-\deg \alpha}, \qquad \|\alpha\|_{p(X)} = d^{\deg p(X) v_{p(X)}(\alpha)} \quad \text{for } \alpha \in K(X)^*.$$

Show that $\|\alpha\|_i = 1$ for almost all $i \in I$ and

$$\prod_{i \in I'} \|\alpha\|_i = 1 \quad \text{for} \quad \alpha \in K(X)^*.$$

2. Valuations and Valuation Fields

In this section we initiate the study of valuations.

(2.1). One can generalize the properties of v_p of (1.1) and $v_{p(X)}$ of (1.2) and proceed to the concept of valuation. Let Γ be an additively written totally ordered abelian group. Add to Γ a formal element $+\infty$ with the properties $a \leqslant +\infty$, $+\infty \leqslant +\infty$, $a + (+\infty) = +\infty$, $(+\infty) + (+\infty) = +\infty$, for each $a \in \Gamma$; denote $\Gamma' = \Gamma \cup \{+\infty\}$.

A map $v \colon F \to \Gamma'$ with the properties

$$v(\alpha) = +\infty \Leftrightarrow \alpha = 0$$
$$v(\alpha\beta) = v(\alpha) + v(\beta)$$
$$v(\alpha + \beta) \geqslant \min(v(\alpha), v(\beta))$$

is said to be a *valuation* on F; in this case F is said to be a valuation field. The map v induces a homomorphism of F^* to Γ and its value group $v(F^*)$ is a totally ordered subgroup of Γ. If $v(F^*) = \{0\}$, then v is called the *trivial valuation*. Similarly to (1.2) it is easy to show that if $v(\alpha) \neq v(\beta)$, then $v(\alpha + \beta) = \min(v(\alpha), v(\beta))$.

(2.2). Let $\mathcal{O}_v = \{\alpha \in F : v(\alpha) \geqslant 0\}$, $\mathcal{M}_v = \{\alpha \in F : v(\alpha) > 0\}$. Then $\mathcal{M}_v$ coincides with the set of non-invertible elements of $\mathcal{O}_v$. Therefore, $\mathcal{O}_v$ is a local ring with the unique *maximal ideal* $\mathcal{M}_v$; $\mathcal{O}_v$ is called the *ring of integers* (with respect to v), and the field $\overline{F}_v = \mathcal{O}_v / \mathcal{M}_v$ is called the *residue field*, or residue class field. The image of an element $\alpha \in \mathcal{O}_v$ in $\overline{F}_v$ is denoted by $\overline{\alpha}$, it is called the *residue* of α in $\overline{F}_v$. The set of invertible elements of $\mathcal{O}_v$ is a multiplicative group $U_v = \mathcal{O}_v - \mathcal{M}_v$, it is called the *group of units*.

(2.3). EXAMPLES OF VALUATIONS AND VALUATION FIELDS.

1. A valuation v on F is said to be *discrete* if the totally ordered group $v(F^*)$ is isomorphic to the naturally ordered group $\mathbb{Z}$.

The map v_p of (1.1) is a discrete valuation with the ring of integers

$$\mathcal{O}_{v_p} = \left\{ \frac{m}{n} : m, n \in \mathbb{Z}, \quad n \text{ is relatively prime to } p \right\}.$$

The residue field $\overline{\mathbb{Q}}_{v_p}$ is a finite field of order p. The map v_∞ of (1.2) is a discrete valuation with the residue field K. The map $v_{p(X)}$ of (1.2) is a discrete valuation with the ring of integers

$$\mathcal{O}_{v_{p(X)}} = \left\{ \frac{f(X)}{g(X)} : f(X), g(X) \in K[X], g(X) \text{ is relatively prime to } p(X) \right\}$$

and the residue field is $K[X]/p(X)K[X]$.

2. Let $\Gamma_1, \ldots \Gamma_n$ be totally ordered abelian groups. One can order the group $\Gamma_1 \times \cdots \times \Gamma_n$ lexicographically, namely setting $(a_1, \ldots, a_n) < (b_1, \ldots, b_n)$ if and only if $a_1 = b_1, \ldots, a_{i-1} = b_{i-1}$, $a_i < b_i$ for some $1 \leqslant i \leqslant n$. A valuation v on F is said to be *discrete of rank* n if the value group $v(F^*)$ is isomorphic to the lexicographically ordered group $(\mathbb{Z})^n = \underbrace{\mathbb{Z} \times \cdots \times \mathbb{Z}}_{n \text{ times}}$. Note that the first component v_1 of a discrete valuation $v = (v_1, \ldots, v_n)$ of rank n is a discrete valuation (of rank 1) on the field F.

3. Let F be a field with a valuation v. For $f(X) = \sum_{i=m}^k \alpha_i X^i \in F[X]$ with $\alpha_m \neq 0$, $m \leqslant k$, put

$$v^*(f(X)) = (m, v(\alpha_m)) \in \mathbb{Z} \times v(F^*).$$

One can naturally extend v^* to $F(X)$. If we order the group $\mathbb{Z} \times v(F^*)$ lexicographically, we obtain the valuation v^* on $F(X)$ with the residue field $\overline{F}_v$. Similarly, it is easy to define a valuation on $F(X_1)\ldots(X_n)$ with the value group $(\mathbb{Z})^{n-1} \times v(F^*)$ ordered lexicographically. In particular, for $F = \mathbb{Q}$, $v = v_p$ we get a discrete valuation of rank n on $\mathbb{Q}(X_1)\ldots(X_{n-1})$ and for $F = K(X)$, $v = v_{p(X)}$ we get a discrete valuation of rank n on $K(X)(X_1)\ldots(X_{n-1})$.

4. Let F, v be as in Example 3. Fix an integer c. For $f(X) = \sum_{i=m}^k \alpha_i X^i \in F[X]$ with $\alpha_m \neq 0$, $m \leqslant k$ put

$$w_c(f(X)) = \min_{m \leqslant i \leqslant k} v(\alpha_i) + ic.$$

Extending w_c to $F(X)$ we obtain the discrete valuation w_c with residue field $\overline{F}_v(X)$ (make substitution $X = Y\beta$ with $v(\beta) = c$ to reduce to the case $c = 0$).

5. Let F, v be as in Example 3. For $f(X) = \sum_{i=m}^k \alpha_i X^i \in F[X]$, $\alpha_m \neq 0$, $m \leqslant k$ put

$$v_*(f(X)) = \min_{m \leqslant i \leqslant k} (v(\alpha_i), i) \in v(F^*) \times \mathbb{Z}, \quad v_*(0) = (+\infty, +\infty)$$

for $v(F^*) \times \mathbb{Z}$ ordered lexicographically. Extending v_* to $F(X)$, we obtain the valuation v_*. The residue field of v_* is $\overline{F}_v$.

For a general valuation theory see [Bou], [Rib], [E].

Exercises.

1. Find the ring of integers, the group of units and the maximal ideal of the ring of integers for the preceding examples.
2. Show that $\bigcap_{p\in S} \mathcal{O}_{v_p} = \mathbb{Z}$ for $S = S' - \{\infty\}$ (see Exercise 7 section 1) and $\bigcap_{p(X)\in I} \mathcal{O}_{v_{p(X)}} = K[X]$ for $I = I' - \{\infty\}$ (see Exercise 8 section 1).
3. Let $F = K(X)$, $F_m = F(X^{\frac{1}{m}})$ for a natural $m \geqslant 1$ and $L = \cup F_m$. For $f = \sum_{a\in\mathbb{Q}} \alpha_a X^a \in L$, $\alpha_a \in K$, put $v(f) = \min\{a \in \mathbb{Q} : \alpha_a \neq 0\}$. Show that v is a valuation on L with the residue field K and the value group $\mathbb{Q}$.
4. A subring $\mathcal{O}$ of a field F is said to be a valuation ring if $\alpha \in \mathcal{O}$ or $\alpha^{-1} \in \mathcal{O}$ for every nonzero element $\alpha \in F$. Show that the ring of integers of a valuation on F is a valuation ring. Conversely, for a valuation ring $\mathcal{O}$ in F one can order the group $F^*/\mathcal{O}^*$ as follows: $\alpha\mathcal{O}^* \leqslant \beta\mathcal{O}^*$ if and only if $\beta\alpha^{-1} \in \mathcal{O}$. Show that the canonical map $F \to (F^*/\mathcal{O}^*)'$ (see (2.1)) is a valuation with the ring of integers $\mathcal{O}$.
5. Let $\mathcal{O}$ be a valuation ring of F and $\mathcal{O}_1$ a subring of F containing $\mathcal{O}$. Show that $\mathcal{O}_1$ is a valuation ring of F with the maximal ideal $\mathcal{M}_1$, which is a prime ideal of $\mathcal{O}$. Conversely, show that for a prime ideal P of $\mathcal{O}$ the ring of fractions $\mathcal{O}_P = \left\{\frac{\alpha}{\beta} : \alpha, \beta \in \mathcal{O}, \beta \notin P\right\}$ is a valuation ring of F.
6. A valuation v on F is said to be a p-valuation of rank d for a prime integer p if $\operatorname{char}(F) = 0$, $\operatorname{char}(\overline{F}_v) = p$, and $\mathcal{O}_v/p\mathcal{O}_v$ is of order p^d. Show that

$$\min\{v(\alpha) > 0 : \alpha \in F^*\} = \frac{v(p)}{e}$$

and $d = ef$, where $p^f = |\overline{F}_v|$, for some natural e.
A field F is said to be a formally p-adic field if it admits at least one nontrivial p-valuation. (For the theory of formally p-adic fields see [PR], [Po]).

3. Discrete Valuation Fields

Now we concentrate on discrete valuations.

(3.1). A field F is said to be a *discrete valuation field* if it admits a nontrivial discrete valuation v (see Example 1 in (2.3)). An element $\pi \in \mathcal{O}_v$ is said to be a *prime element* (*uniformizing element*) if $v(\pi)$ generates the group $v(F^*)$. Without loss of generality we shall often assume that the homomorphism

$$v\colon F^* \to \mathbb{Z}$$

is *surjective.*

(3.2). LEMMA. *Assume that* $\operatorname{char}(F) \neq \operatorname{char}(\overline{F}_v)$. *Then* $\operatorname{char}(F) = 0$ *and* $\operatorname{char}(\overline{F}_v) = p > 0$.

Proof. Suppose that $\operatorname{char}(F) = p \neq 0$. Then $p = 0$ in F and therefore in $\overline{F}_v$. Hence $p = \operatorname{char}(\overline{F}_v)$. □

(3.3). LEMMA. *Let F be a discrete valuation field, and π be a prime element. Then the ring of integers $\mathcal{O}_v$ is a principal ideal ring, and every proper ideal of $\mathcal{O}_v$ can be written as $\pi^n \mathcal{O}_v$ for some $n > 0$. In particular, $\mathcal{M}_v = \pi \mathcal{O}_v$. The intersection of all proper ideals of $\mathcal{O}_v$ is the zero ideal.*

Proof. Let I be a proper ideal of $\mathcal{O}_v$. Then there exists $n = \min\{v(\alpha) : \alpha \in I\}$ and hence $\pi^n \varepsilon \in I$ for some unit ε. It follows that $\pi^n \mathcal{O}_v \subset I \subset \pi^n \mathcal{O}_v$ and $I = \pi^n \mathcal{O}_v$. If α belongs to the intersection of all proper ideals $\pi^n \mathcal{O}_v$ in $\mathcal{O}_v$, then $v(\alpha) = +\infty$, i.e., $\alpha = 0$. □

Further characterization of discrete valuation fields via commutative algebra can be found in [Se3] and [Bou].

(3.4). LEMMA. *Any element $\alpha \in F^*$ can be uniquely written as $\pi^n \varepsilon$ for some $n \in \mathbb{Z}$ and $\varepsilon \in U_v$.*

Proof. Let $n = v(\alpha)$. Then $\alpha \pi^{-n} \in U_v$ and $\alpha = \pi^n \varepsilon$ for $\varepsilon \in U_v$. If $\pi^n \varepsilon_1 = \pi^m \varepsilon_2$, then $n + v(\varepsilon_1) = m + v(\varepsilon_2)$. As $\varepsilon_1, \varepsilon_2 \in U_v$, we deduce $n = m$, $\varepsilon_1 = \varepsilon_2$. □

(3.5). Let v be a discrete valuation on F, $0 < d < 1$. The mapping $d_v \colon F \times F \to \mathbb{R}$ defined by $d_v(\alpha, \beta) = d^{v(\alpha - \beta)}$ is a metric on F. Therefore, it induces a Hausdorff topology on F. For every $\alpha \in F$ the sets $\alpha + \pi^n \mathcal{O}_v$, $n \in \mathbb{Z}$, form a basis of open neighborhoods of α. This topology on F and the induced topology on U_v and $1 + \mathcal{M}_v$ is called the *discrete valuation topology*.

LEMMA. *The field F with the above-defined topology is a topological field.*

Proof. As

$$v((\alpha - \beta) - (\alpha_0 - \beta_0)) \geqslant \min(v(\alpha - \alpha_0), v(\beta - \beta_0)),$$
$$v(\alpha\beta - \alpha_0\beta_0) \geqslant \min(v(\alpha - \alpha_0) + v(\beta), v(\beta - \beta_0) + v(\alpha_0)),$$
$$v(\alpha^{-1} - \alpha_0^{-1}) = v(\alpha - \alpha_0) - v(\alpha) - v(\alpha_0),$$

we obtain the continuity of subtraction, multiplication and division. □

(3.6). LEMMA. *The topologies on F defined by two discrete valuations v_1, v_2 coincide if and only if $v_1 = v_2$ (recall that $v_1(F^*) = v_2(F^*) = \mathbb{Z}$).*

Proof. Let the topologies induced by v_1, v_2 coincide. Observe that α^n tends to 0 when n tends to $+\infty$ in the topology defined by a discrete valuation v if and only if $v(\alpha) > 0$. Therefore, $v_1(\alpha) > 0$ if and only if $v_2(\alpha) > 0$. Let π_1, π_2 be prime elements with respect to v_1 and v_2. Then we conclude that $v_2(\pi_1) \geqslant 1$ and $v_1(\pi_2) \geqslant 1$. If $v_2(\pi_1) > 1$ then $v_2(\pi_1\pi_2^{-1}) > 0$. Consequently, $v_1(\pi_1\pi_2^{-1}) > 0$, i.e., $v_1(\pi_2) < 1$. Thus, $v_2(\pi_1) = 1$ and this equality holds for all prime elements π_1 with respect to v_1. This shows the equality $v_1 = v_2$. □

(3.7). PROPOSITION (APPROXIMATION THEOREM). *Let $v_1, \dots, v_n$ be distinct discrete valuations on F. Then for every $\alpha_1, \dots, \alpha_n \in F$, $c \in \mathbb{Z}$, there exists $\alpha \in F$ such that $v_i(\alpha_i - \alpha) > c$ for $1 \leqslant i \leqslant n$.*

Proof. It is easy to show that if $v(\alpha) > 0$ then $v(\alpha^m(1+\alpha^m)^{-1}) \to +\infty$ as $m \to +\infty$, and if $v(\alpha) < 0$ then $v(\alpha^m(1+\alpha^m)^{-1} - 1) \to +\infty$ as $m \to +\infty$. We proceed by induction to deduce that there exists an element $\beta_1 \in F$ such that $v_1(\beta_1) < 0$, $v_i(\beta_1) > 0$ for $2 \leqslant i \leqslant n$. Indeed, one can first verify that there is an element $\gamma_1 \in F$ such that $v_1(\gamma_1) \geqslant 0$, $v_2(\gamma_1) < 0$. Using the proof of Lemma (3.6), take elements $\pi_1, \pi_2 \in F$ with $v_2(\pi_1) \neq 1 = v_1(\pi_1)$, $v_1(\pi_2) \neq 1 = v_2(\pi_2)$. If $v_2(\pi_1) < 0$ put $\gamma_1 = \pi_1$. If $v_2(\pi_1) \geqslant 0$, then $v_2(\rho) \neq 0 = v_1(\rho)$ for $\rho = \pi_2\pi_1^{-v_1(\pi_2)}$. Put $\gamma_1 = \rho$ or $\gamma_1 = \rho^{-1}$. Now let $\gamma_2 \in F$ be such that $v_2(\gamma_2) \geqslant 0$, $v_1(\gamma_2) < 0$. Then $\beta_1 = \gamma_1^{-1}\gamma_2$ is the desired element for $n = 2$.

Let $n > 2$. Then, by the induction assumption, there exists $\delta_1 \in F$ with $v_1(\delta_1) < 0$, $v_i(\delta_1) > 0$ for $2 \leqslant i \leqslant n-1$ and $\delta_2 \in F$ with $v_1(\delta_2) < 0$, $v_n(\delta_2) > 0$. One can put $\beta_2 = \delta_1$ if $v_n(\delta_1) > 0$, $\beta_2 = \delta_1^m\delta_2$ if $v_n(\delta_1) = 0$, and $\beta_2 = \delta_1\delta_2^m(1+\delta_2^m)^{-1}$ if $v_n(\delta_1) < 0$ for a sufficiently large m.

To complete the proof we take $\beta_1, \dots, \beta_n \in F$ with $v_i(\beta_i) < 0$, $v_i(\beta_j) > 0$ for $i \neq j$. Put $\alpha = \sum_{i=1}^n \alpha_i\beta_i^m(1+\beta_i^m)^{-1}$. Then α is the desired element for a sufficiently large m. □

Exercises.

1. Show that every interior point of an open ball in the topology induced by a discrete valuation is a center of the ball.
2. Do Lemmas (3.3) and (3.4) hold for a discrete valuation of rank n ?
3. Let v be a discrete valuation on F. Show that the map $\|\cdot\|\colon F^* \to \mathbb{R}^*$ defined as $\|\alpha\| = d^{v(\alpha)}$ for some real d, $0 < d < 1$, is an absolute value on F and $\|F^*\|$ is a discrete subgroup of $\mathbb{R}^*$.
4. Let $\|\cdot\|$ be an absolute value on F. As a basis of neighborhoods of $\alpha \in F$ one can take the sets $U_\varepsilon(\alpha) = \{\beta \in F : \|\alpha - \beta\| < \varepsilon\}$. The topology defined in this way is said to be induced by $\|\cdot\|$.
   a) Show that for the ultrametric absolute value related to a discrete valuation, this topology coincides with the above-defined topology induced by the valuation.
   b) Two absolute values are said to be equivalent if the induced topologies coincide. Show that $\|\cdot\|_1$ and $\|\cdot\|_2$ are equivalent if and only if $\|\cdot\|_2 = \|\cdot\|_1^c$ for some real $c > 0$.

4. Completion

Completion of a discrete valuation field is an object which is easier to understand than the original field. The central object of this book, local fields, is defined in (4.6).

(4.1). Let F be a field with a discrete valuation v (as usual, $v(F^*) = \mathbb{Z}$). As F is a metric topological space one can introduce the notion of a Cauchy sequence. A sequence $(\alpha_n)_{n\geqslant 0}$ of elements of F is called a Cauchy sequence if for every real c there is $n_0 \geqslant 0$ such that $v(\alpha_n - \alpha_m) \geqslant c$ for $m, n \geqslant n_0$. In view of the properties of valuations, for such a sequence there exists $\lim v(\alpha_n)$.

LEMMA. *The set A of all Cauchy sequences forms a ring with respect to componentwise addition and multiplication. The set of all Cauchy sequences $(\alpha_n)_{n\geqslant 0}$ with $\alpha_n \to 0$ as $n \to +\infty$ forms a maximal ideal M of A. The field A/M is a discrete valuation field with its discrete valuation $\widehat{v}$ defined by $\widehat{v}((\alpha_n)) = \lim v(\alpha_n)$ for a Cauchy sequence $(\alpha_n)_{n\geqslant 0}$.*

Proof. A sketch of the proof is as follows. It suffices to show that M is a maximal ideal of A. Let $(\alpha_n)_{n\geqslant 0}$ be a Cauchy sequence with $\alpha_n \nrightarrow 0$ as $n \to +\infty$. Hence, there is an $n_0 \geqslant 0$ such that $\alpha_n \neq 0$ for $n \geqslant n_0$. Put $\beta_n = 0$ for $n < n_0$ and $\beta_n = \alpha_n^{-1}$ for $n \geqslant n_0$. Then $(\beta_n)_{n\geqslant 0}$ is a Cauchy sequence and $(\alpha_n)(\beta_n) \in (1) + M$. Therefore, M is maximal. □

(4.2). A discrete valuation field F is called a *complete discrete valuation field* if every Cauchy sequence $(\alpha_n)_{n\geqslant 0}$ is convergent, i.e., there exists $\alpha = \lim \alpha_n \in F$ with respect to v. A field $\widehat{F}$ with a discrete valuation $\widehat{v}$ is called a *completion* of F if it is complete, $\widehat{v}|_F = v$, and F is a dense subfield in $\widehat{F}$ with respect to $\widehat{v}$.

PROPOSITION. *Every discrete valuation field has a completion which is unique up to an isomorphism over F.*

Proof. We verify that the field A/M with the valuation $\widehat{v}$ is a completion of F. F is embedded in A/M by the formula $\alpha \to (\alpha) \bmod M$. For a Cauchy sequence $(\alpha_n)_{n\geqslant 0}$ and real c, let $n_0 \geqslant 0$ be such that $v(\alpha_n - \alpha_m) \geqslant c$ for all $m, n \geqslant n_0$. Hence, for $\alpha_{n_0} \in F$ we have $\widehat{v}((\alpha_{n_0}) - (\alpha_n)_{n\geqslant 0}) \geqslant c$, which shows that F is dense in A/M. Let $((\alpha_n^{(m)})_n)_m$ be a Cauchy sequence in A/M with respect to $\widehat{v}$. Let $n(0)$, $n(1)$, ... be an increasing sequence of integers such that $v(\alpha_{n_2}^{(m)} - \alpha_{n_1}^{(m)}) \geqslant m$ for n_1, $n_2 \geqslant n(m)$. Then $(\alpha_{n(m)}^{(m)})_m$ is a Cauchy sequence in F and the limit of $((\alpha_n^{(m)})_n)_m$ with respect to $\widehat{v}$ in A/M. Thus, we obtain the existence of the completion A/M, $\widehat{v}$.

If there are two completions $\widehat{F}_1$, $\widehat{v}_1$ and $\widehat{F}_2$, $\widehat{v}_2$, then we put $f(\alpha) = \alpha$ for $\alpha \in F$ and extend this homomorphism by continuity from F, as a dense subfield in $\widehat{F}_1$, to $\widehat{F}_1$. It is easy to verify that the extension $\widehat{f}\colon \widehat{F}_1 \to \widehat{F}_2$ is an isomorphism and $\widehat{v}_2 \circ \widehat{f} = \widehat{v}_1$. □

We shall denote the completion of the field F with respect to v by $\widehat{F}_v$ or simply $\widehat{F}$.

(4.3). LEMMA. *Let F be a field with a discrete valuation v and $\widehat{F}$ its completion with the discrete valuation $\widehat{v}$. Then the ring of integers $\mathcal{O}_v$ is dense in $\mathcal{O}_{\hat{v}}$, the maximal ideal $\mathcal{M}_v$ is dense in $\mathcal{M}_{\hat{v}}$, and the residue field $\overline{F}_v$ coincides with the residue field of $\widehat{F}$ with respect to $\widehat{v}$.*

Proof. It follows immediately from the construction of A/M in (4.1) and Proposition (4.2). □

(4.4). Although we have considered the completion of discrete valuation fields, such a construction can be realized for any valuation field using the notion of filter. As a basis of neighborhoods of 0 one uses the sets $\{\alpha \in F : v(\alpha) > c\}$ where $c \in v(F^*)$. Assertions, similar to (4.2) and (4.3), hold in general (see [Bou, sect. 5 Ch. VI]).

(4.5). EXAMPLES OF COMPLETE VALUATION FIELDS.

1. The completion of $\mathbb{Q}$ with respect to v_p of (1.1) is denoted by $\mathbb{Q}_p$ and is called the p-adic field. Certainly, the completion of $\mathbb{Q}$ with respect to the absolute value $\|\cdot\|_\infty$ of (1.1) is $\mathbb{R}$. Embeddings of $\mathbb{Q}$ in $\mathbb{Q}_p$ for all prime p and in $\mathbb{R}$ is a tool to solve various problems over $\mathbb{Q}$. An example is the Minkowski–Hasse Theorem (see [BSh, Ch. 1]): an equation $\sum a_{ij}X_iX_j = 0$ for $a_{ij} \in \mathbb{Q}$ has a nontrivial solution in $\mathbb{Q}$ if and only if it admits a nontrivial solution in $\mathbb{R}$ and in $\mathbb{Q}_p$ for all prime p. A generalization of this result is the so-called Hasse local-global principle which is of great importance in algebraic number theory. It is interesting that, from the standpoint of model theory, the complex field $\mathbb{C}$ is locally equivalent to the algebraic closure of $\mathbb{Q}_p$ for each prime p (see [Roq2]).

The ring of integers of $\mathbb{Q}_p$ is denoted by $\mathbb{Z}_p$ and is called the ring of p-adic integers. The residue field of $\mathbb{Q}_p$ is the finite field $\mathbb{F}_p$ consisting of p elements.

2. The completion of $K(X)$ with respect to v_X is the formal power series field $K((X))$ of all formal series $\sum_{-\infty}^{+\infty} \alpha_n X^n$ with $\alpha_n \in K$ and $\alpha_n = 0$ for almost all negative n. The ring of integers with respect to v_X is $K[[X]]$, that is, the set of all formal series $\sum_0^{+\infty} \alpha_n X^n$, $\alpha_n \in K$. Its residue field may be identified with K.

3. Let F be a field with a discrete valuation v, and $\widehat{F}$ its completion. Then the valuation v^* on $F(X)$ defined in Example 3 of (2.3) can be naturally extended to $\widehat{F}((X))$. For $f(X) = \sum_{n \geqslant m} \alpha_n X^n$, $\alpha_n \in \widehat{F}$, $\alpha_m \neq 0$, put

$$v^*(f(X)) = (m, \widehat{v}(\alpha_m)).$$

The ring of integers of v^* on $\widehat{F}((X))$ is $\mathcal{O}_{\hat{v}} + X\widehat{F}[[X]]$.

4. Let F be the same as in Example 3. Then the valuation v_* on $F(X)$ defined in Example 5 of (2.3) can be naturally extended to the field

$$\widehat{F}\{\{X\}\} = \Big\{\sum_{-\infty}^{+\infty} \alpha_n X^n : \alpha_n \in \widehat{F}, \inf_n\{\widehat{v}(\alpha_n)\} > -\infty,\ \widehat{v}(\alpha_n) \to +\infty \text{ as } n \to -\infty\Big\}.$$

For $f(X) = \sum_{-\infty}^{+\infty} \alpha_n X^n \in \widehat{F}\{\{X\}\}$ put

$$v_*(f(X)) = \min_n(\widehat{v}(\alpha_n), n).$$

The ring of integers of v_* is $\mathcal{O}_{\widehat{v}}\{\{X\}\} = \big\{\sum_{-\infty}^{+\infty} \alpha_n X^n : \alpha_n \in \mathcal{O}_{\widehat{v}}\big\}$ and the residue field is $\overline{F}_v$.

(4.6). DEFINITIONS.

1. A complete discrete valuation field with perfect residue field is called a *local fields*. For example, $\mathbb{Q}_p$ and $F((X))$ are local fields where F is a perfect field (of positive or zero characteristic). Local fields with finite residue field are sometimes called *local number fields* if they are of characteristic zero and *local functional fields* if they are of positive characteristic.

2. Local fields are sometimes called 1-dimensional local fields. An *n-dimensional local field* ($n \geqslant 2$) is a complete discrete valuation fields whose residue field is an $(n-1)$-dimensional local field. For example, $\mathbb{Q}_p((X_2))\dots((X_n))$, $F((X_1))\dots((X_n))$ (F is a perfect field), $K\{\{X_1\}\}\dots\{\{X_{n-1}\}\}$ (K is a 1-dimensional local field of characteristic zero) are n-dimensional local fields. See [FK] for an introduction to n-dimensional local fields.

Exercises.

1. Let F be a complete discrete valuation field.
   a) Show that a series $\sum_{n\geqslant 0} \alpha_n$ converges in F if and only if $v(\alpha_n) \to +\infty$ as $n \to +\infty$.
   b) Prove that F is an uncountable set.
2. Show that if the residue field is finite then the ring of integers $\mathcal{O}_v$ of a complete discrete valuation field is isomorphic and homeomorphic with the projective limit $\varprojlim \mathcal{O}_v/\pi^n\mathcal{O}_v$, where the topology of $\mathcal{O}_v/\pi^n\mathcal{O}_v$ is discrete.
3. Let $f\colon \mathbb{Q}_p \to \mathbb{Q}_q$ be an isomorphism and homeomorphism. Show that $p = q$ (see also Exercise 5e in section 1 Ch. 2).
4. Let L be a field with a valuation v and let $\mathcal{M} = \mathcal{M}_v$ be the maximal ideal; $\mathcal{M}$-adic topology on L is defined as follows: the sets $\alpha + \mathcal{M}^n$, $n \geqslant 0$, are taken as open neighbourhoods of $\alpha \in L$. Show that for the case of a discrete valuation v the completion of L with respect to the $\mathcal{M}$-adic topology coincides with $\widehat{L}$. Does the completion of $L = F(X)$, where F is as in Examples 3 and 4, with respect to the $\mathcal{M}$-adic topology coincide with $\widehat{F}((X))$, $\widehat{F}\{\{X\}\}$? Does the completion of $L = F(X)$ with respect to the filter (see (4.4)) coincide with $\widehat{F}((X))$, $\widehat{F}\{\{X\}\}$?
5. Find the maximal ideal and the group of units in the examples in (4.5).

6. Show that the fields $\widehat{F}((X))$, $\widehat{F}\{\{X\}\}$ in (4.5) are complete discrete valuation fields with respect to the first component of v^*, v_* (see Example 2 in (2.3)), and find their residue fields .
7. Find the completion of $F(X)$ with respect to w_c (see Example 4 in (2.3)).
8. Define a completion of a field with respect to an absolute value. Then
   a) Using (1.1) show that if $\|\cdot\|$ is a nontrivial non-ultrametric absolute value on $\mathbb{R}$ then $\|\cdot\|$ coincides, up to an automorphism of $\mathbb{R}$, with $\|\cdot\|_\infty^c$ for some real $c > 0$.
   b) Prove that if $\|\cdot\|$ is a nontrivial non-ultrametric absolute value on $\mathbb{C}$, then $\|\cdot\|$ coincides, up to an automorphism of $\mathbb{C}$, with $\|\cdot\|_\infty^c$ for some real $c > 0$, where $\|\cdot\|_\infty$ is the usual absolute value.

   A Theorem of *A. Ostrowski* asserts that every complete field F with respect to a nontrivial non-ultrametric absolute value is isomorphic to ($\mathbb{R}$, $\|\cdot\|_\infty$) or ($\mathbb{C}$, $\|\cdot\|_\infty$) (see [Cas, Ch. 3], [Wes], [Bah]).

5. Filtrations of Discrete Valuation Fields

In this section we study natural filtrations on the multiplicative group of a discrete valuation field F; in particular, its behaviour with respect to raising to the pth power. For simplicity, we will often omit the index v in notations U_v, $\mathcal{O}_v$, $\mathcal{M}_v$, $\overline{F}_v$. We fix a prime element π of F.

(5.1). A set R is said to be a *set of representatives* for a valuation field F if $R \subset \mathcal{O}$, $0 \in R$ and R is mapped bijectively on $\overline{F}$ under the canonical map $\mathcal{O} \to \mathcal{O}/\mathcal{M} = \overline{F}$. Denote by rep: $\overline{F} \to R$ the inverse bijective map. For a set S denote by $(S)_n^{+\infty}$ the set of all sequences $(a_i)_{i \geqslant n}$, $a_i \in S$. Let $(S)_{-\infty}^{+\infty}$ denote the union of increasing sets $(S)_n^{+\infty}$ where $n \to -\infty$.

(5.2). The additive group F has a natural filtration

$$\dots \supset \pi^i \mathcal{O} \supset \pi^{i+1} \mathcal{O} \supset \dots .$$

The factor filtration of this filtration is easy to calculate: $\pi^i \mathcal{O}/\pi^{i+1}\mathcal{O} \xrightarrow{\sim} \overline{F}$.

PROPOSITION. *Let F be a complete field with respect to a discrete valuation v. Let $\pi_i \in F$ for each $i \in \mathbb{Z}$ be an element of F with $v(\pi_i) = i$. Then the map*

$$\text{Rep}: (\overline{F})_{-\infty}^{+\infty} \to F, \quad (a_i)_{i \in \mathbb{Z}} \mapsto \sum_{-\infty}^{+\infty} \text{rep}(a_i)\pi_i$$

is a bijection. Moreover, if $(a_i)_{i\in\mathbb{Z}} \neq (0)_{i\in\mathbb{Z}}$ then $v(\text{Rep}(a_i)) = \min\{i : a_i \neq 0\}$.

Proof. The map Rep is well defined, because for almost all $i < 0$ we get $\text{rep}(a_i) = 0$ and the series $\sum \text{rep}(a_i)\pi_i$ converges in F. If $(a_i)_{i\in\mathbb{Z}} \neq (b_i)_{i\in\mathbb{Z}}$ and

$$n = \min\{i \in \mathbb{Z} : a_i \neq b_i\},$$

then $v(a_n\pi_n - b_n\pi_n) = n$. Since $v(a_i\pi_i - b_i\pi_i) > n$ for $i > n$, we deduce that

$$v(\mathrm{Rep}(a_i) - \mathrm{Rep}(b_i)) = n.$$

Therefore Rep is injective.

In particular, $v(\mathrm{Rep}(a_i)) = \min\{i : a_i \neq 0\}$. Further, let $\alpha \in F$. Then $\alpha = \pi^n\varepsilon$ with $n \in \mathbb{Z}$, $\varepsilon \in U$. We also get $\alpha = \pi_n\varepsilon'$ for some $\varepsilon' \in U$. Let a_n be the image of ε' in $\overline{F}$; then $a_n \neq 0$ and $\alpha_1 = \alpha - \mathrm{rep}(a_n)\pi_n \in \pi^{n+1}\mathcal{O}$. Continuing in this way for α_1, we obtain a convergent series $\alpha = \sum \mathrm{rep}(a_i)\pi_i$. Therefore, Rep is surjective. □

COROLLARY. *We often take* $\pi_n = \pi^n$. *Therefore, by the preceding Proposition, every element* $\alpha \in F$ *can be uniquely expanded as*

$$\alpha = \sum_{-\infty}^{+\infty} \theta_i\pi^i, \qquad \theta_i \in R \quad \textit{and} \quad \theta_i = 0 \quad \textit{for almost all } i < 0.$$

We shall discuss the choice of the set of representatives in section 7.

DEFINITION. If $\alpha - \beta \in \pi^n\mathcal{O}$, we write $\alpha \equiv \beta \mod \pi^n$.

(5.3). DEFINITION. The group $1 + \pi\mathcal{O}$ is called the *group of principal units* U_1 and its elements are called *principal units*. Introduce also *higher groups of units* as follows: $U_i = 1 + \pi^i\mathcal{O}$ for $i \geqslant 1$.

(5.4). The multiplicative group F^* has a natural filtration $F^* \supset U \supset U_1 \supset U_2 \supset \dots$. We describe the factor filtration of the introduced filtration on F^*.

PROPOSITION. *Let* F *be a discrete valuation field. Then*

(1) *The choice of a prime element* π $(1 \in \mathbb{Z} \to \pi \in F^*)$ *splits the exact sequence* $1 \to U \to F^* \xrightarrow{v} \mathbb{Z} \to 0$. *The group* F^* *is isomorphic to* $U \times \mathbb{Z}$.

(2) *The canonical map* $\mathcal{O} \to \mathcal{O}/\mathcal{M} = \overline{F}$ *induces the surjective homomorphism*

$$\lambda_0 : U \to \overline{F}^*, \qquad \varepsilon \mapsto \overline{\varepsilon};$$

λ_0 *maps* U/U_1 *isomorphically onto* $\overline{F}^*$.

(3) *The map*

$$\lambda_i : U_i \to \overline{F}, \qquad 1 + \alpha\pi^i \mapsto \overline{\alpha}$$

for $\alpha \in \mathcal{O}$ *induces the isomorphism* λ_i *of* U_i/U_{i+1} *onto* $\overline{F}$ *for* $i \geqslant 1$.

Proof. The statement (1) follows for example from Lemma (3.4).

(2) The kernel of λ_0 coincides with U_1 and λ_0 is surjective.

(3) The induced map $U_i/U_{i+1} \to \overline{F}$ is a homomorphism, since

$$(1 + \alpha_1\pi^i)(1 + \alpha_2\pi^i) = 1 + (\alpha_1 + \alpha_2)\pi^i + \alpha_1\alpha_2\pi^{2i}.$$

This homomorphism is bijective, since $\lambda_i(1 + \mathrm{rep}(\overline{\alpha})\pi^i) = \overline{\alpha}$. □

(5.5). COROLLARY. *Let l be not divisible by* $\operatorname{char}(\overline{F})$. *Raising to the l th power induces an automorphism of U_i/U_{i+1} for $i \geqslant 1$.*

If F is complete, then the group U_i for $i \geqslant 1$ is uniquely l-divisible.

Proof. If $\varepsilon = 1 + \alpha\pi^i$ with $\alpha \in \mathcal{O}$, then $\varepsilon^l \equiv 1 + l\alpha\pi^i \mod \pi^{i+1}$. Absence of nontrivial l-torsion in the additive group $\overline{F}$ implies the first property. It also shows that U_i has no nontrivial l-torsion.

For an element $\eta = 1 + \beta\pi^i$ with $\beta \in \mathcal{O}^*$ we have $\eta = (1 + l^{-1}\beta\pi^i)^l\eta_1$ with $\eta_1 \in U_{i+1}$. Applying the same argument to η_1 and so on, we get an l th root of η in F in the case of complete F. □

(5.6). Let $\operatorname{char}(\overline{F}) = p > 0$. Lemma (3.2) implies that either $\operatorname{char}(F) = p$ or $\operatorname{char}(F) = 0$. We shall study the operation of raising to the pth power. Denote this homomorphism by

$$\uparrow p\colon \alpha \to \alpha^p.$$

The first and simplest case is $\operatorname{char}(F) = p$.

PROPOSITION. *Let* $\operatorname{char}(F) = \operatorname{char}(\overline{F}) = p > 0$. *Then the homomorphism $\uparrow p$ maps U_i injectively into U_{pi} for $i \geqslant 1$. For $i \geqslant 1$ it induces the commutative diagram*

$$\begin{array}{ccc} U_i/U_{i+1} & \xrightarrow{\uparrow p} & U_{pi}/U_{pi+1} \\ {\scriptstyle\lambda_i}\downarrow & & {\scriptstyle\lambda_{pi}}\downarrow \\ \overline{F} & \xrightarrow{\uparrow p} & \overline{F} \end{array}$$

Proof. Since $(1 + \varepsilon\pi^i)^p = 1 + \varepsilon^p\pi^{pi}$ and there is no nontrivial p-torsion in $\overline{F}^*$ and F^*, the assertion follows. □

COROLLARY. *Let F be a field of characteristic $p > 0$ and let $\overline{F}$ be perfect, i.e $\overline{F} = \overline{F}^p$. Then $\uparrow p$ maps the quotient group U_i/U_{i+1} isomorphically onto the quotient group U_{pi}/U_{pi+1} for $i \geqslant 1$.*

(5.7). We now consider the case of $\operatorname{char}(F) = 0$, $\operatorname{char}(\overline{F}) = p > 0$. As $p = 0$ in the residue field $\overline{F}$, we conclude that $p \in \mathcal{M}$ and, therefore, for the surjective discrete valuation v of F we get $v(p) = e \geqslant 1$.

DEFINITION. The number $e = e(F) = v(p)$ is called *the absolute ramification index of F*.

Let π be a prime element in F. Let R be a set of representatives, and let $\overline{\theta}_0 \in \overline{F}$ be the element of $\overline{F}$ uniquely determined by the relation $p - \operatorname{rep}(\overline{\theta}_0)\pi^e \in \pi^{e+1}\mathcal{O}$ (see Corollary (5.2)).

PROPOSITION. *Let F be a discrete valuation field of characteristic zero with residue field of positive characteristic p. Then the homomorphism $\uparrow p$ maps U_i to U_{pi} for $i \leqslant e/(p-1)$, and U_i to U_{i+e} for $i > e/(p-1)$. This homomorphism induces the following commutative diagrams*

(1) *if $i < e/(p-1)$,*

$$\begin{array}{ccc} U_i/U_{i+1} & \xrightarrow{\uparrow p} & U_{pi}/U_{pi+1} \\ {\scriptstyle\lambda_i}\downarrow & & \downarrow{\scriptstyle\lambda_{pi}} \\ \overline{F} & \xrightarrow{\overline{\alpha}\mapsto\overline{\alpha}^p} & \overline{F} \end{array}$$

(2) *if $i = e/(p-1)$ is an integer,*

$$\begin{array}{ccc} U_i/U_{i+1} & \xrightarrow{\uparrow p} & U_{pi}/U_{pi+1} \\ {\scriptstyle\lambda_i}\downarrow & & \downarrow{\scriptstyle\lambda_{pi}} \\ \overline{F} & \xrightarrow{\overline{\alpha}\mapsto\overline{\alpha}^p+\overline{\theta}_0\overline{\alpha}} & \overline{F} \end{array}$$

(3) *if $i > e/(p-1)$,*

$$\begin{array}{ccc} U_i/U_{i+1} & \xrightarrow{\uparrow p} & U_{i+e}/U_{i+e+1} \\ {\scriptstyle\lambda_i}\downarrow & & \downarrow{\scriptstyle\lambda_{i+e}} \\ \overline{F} & \xrightarrow{\overline{\alpha}\mapsto\overline{\theta}_0\overline{\alpha}} & \overline{F} \end{array}$$

The horizontal homomorphisms are injective in cases (1), (3) *and surjective in case* (3).

If a primitive pth root ζ_p of unity is contained in F, then $v(1-\zeta_p) = e/(p-1)$ and the kernel of the horizontal homomorphisms in case (2) *is of order p.*

If $e/(p-1) \in \mathbb{Z}$, $U_{pe/(p-1)+1} \subset U^p_{e/(p-1)+1}$ and there is no nontrivial p-torsion in F^*, then the homomorphism is injective in case (2).

Proof. Let $1+\alpha \in U_i$. Writing

$$(1+\alpha)^p = 1 + p\alpha + \frac{p(p-1)}{2}\alpha^2 + \cdots + p\alpha^{p-1} + \alpha^p$$

and calculating $v(p\alpha) = e+i$, $v\left(\frac{p(p-1)}{2}\alpha^2\right) = e+2i, \ \dots, \ v(p\alpha^{p-1}) = e+(p-1)i$, $v(\alpha^p) = pi$, we get

$$\begin{aligned} v((1+\alpha)^p - 1) &= v(\alpha^p + p\alpha), && \text{if} \quad v(\alpha^p) \neq v(p\alpha), \\ v((1+\alpha)^p - 1) &\geqslant v(\alpha^p + p\alpha), && \text{otherwise.} \end{aligned}$$

These formulas reveal the behavior of $\uparrow p$ acting on the filtration in U_1, because $v(\alpha^p) \leqslant v(p\alpha)$ if and only if $i \leqslant e/(p-1)$. Moreover, for a unit α we obtain

$$
\begin{aligned}
(1+\alpha\pi^i)^p &\equiv 1+\alpha^p\pi^{pi} \mod \pi^{pi+1}, && \text{if } i < e/(p-1),\\
(1+\alpha\pi^i)^p &\equiv 1+\operatorname{rep}(\overline{\theta}_0)\alpha\pi^{i+e} \mod \pi^{i+e+1}, && \text{if } i > e/(p-1),\\
(1+\alpha\pi^i)^p &\equiv 1+(\alpha^p+\operatorname{rep}(\overline{\theta}_0)\alpha)\pi^{pi} \mod \pi^{pi+1}, && \text{if } i = e/(p-1) \in \mathbb{Z}.
\end{aligned}
$$

Thus, we conclude that the diagrams in the Proposition are commutative. Further, the homomorphism $\uparrow p$ is an isomorphism in case (3) and injective in case (1).

Assume that $\zeta_p \in F$. The assertions obtained above imply that $v(1-\zeta_p) = e/(p-1)$ and $e/(p-1) \in \mathbb{Z}$. Therefore, the homomorphism $\overline{\alpha} \mapsto \overline{\alpha}^p + \overline{\theta}_0\overline{\alpha}$ is not injective. Its kernel $\sqrt[p-1]{-\overline{\theta}_0}\,\mathbb{F}_p$ in this case is of order p.

Now let $e/(p-1)$ be an integer and let $U_{pe/(p-1)+1} \subset U^p_{e/(p-1)+1}$. Assume that the horizontal homomorphism in case (2) is not injective. Let $\overline{\alpha}_0 \in \overline{F}$ satisfy the equation $\overline{\alpha}_0^p + \overline{\theta}_0\overline{\alpha}_0 = 0$. Then $(1+\operatorname{rep}(\overline{\alpha}_0)\pi^{e/(p-1)})^p \in U_j$ for some $j > pe/(p-1)$. Therefore $(1+\operatorname{rep}(\overline{\alpha}_0)\pi^{e/(p-1)})^p = \varepsilon_1^p$ for some $\varepsilon_1 \in U_{e/(p-1)+1}$. Thus, $(1+\operatorname{rep}(\overline{\alpha}_0)\pi^{e/(p-1)})\varepsilon_1^{-1} \in U_{e/(p-1)}$ is a primitive pth root of unity. □

(5.8). COROLLARY 1. *Let* $\operatorname{char}(F) = 0$ *and let* $\overline{F}$ *be a perfect field of characteristic* $p > 0$. *Then* $\uparrow p$ *maps the quotient group* U_i/U_{i+1} *isomorphically onto* U_{pi}/U_{pi+1} *for* $1 \leqslant i < e/(p-1)$ *and isomorphically onto* U_{i+e}/U_{i+e+1} *for* $i > e/(p-1)$.

COROLLARY 2. *Let* F *be a complete field. Let* $i > pe/(p-1)$. *Then* $U_i \subset U^p_{i-e}$. *Therefore, if* F^* *has no nontrivial* p*-torsion then the homomorphism is injective in case* (2).

In addition, if the residue field of F *is finite and* F *contains no nontrivial* p*th roots of unity, then* $U_i \subset U^p_{i-e}$ *for* $i \geqslant pe/(p-1)$

Proof. Use the completeness of F. Due to surjectivity of the homomorphisms in case (3) we get $U_i \subset U_{i+1}U^p_{i-e} \subset U_{i+2}U^p_{i-e} \subset \cdots \subset U^p_{i-e}$.

If the residue field of F is finite, then the injectivity of the homomorphism in case (2) implies its surjectivity. □

(5.9). PROPOSITION. *Let* F *be a complete discrete valuation field.*

If $\operatorname{char}(F) = 0$, *then* F^{*n} *is an open subgroup in* F^* *for* $n \geqslant 1$. *If* $\operatorname{char}(F) = p > 0$, *then* F^{*n} *is an open subgroup in* F^* *if and only if* n *is relatively prime to* p.

Proof. If $\operatorname{char}(\overline{F}) = 0$, then by Corollary (5.5) we get $U_1 \subset F^{*n}$ for $n \geqslant 1$. It means that F^{*n} is open. If $\operatorname{char}(\overline{F}) = p$, then by Corollary (5.5) $U_1 \subset F^{*n}$ for $(n,p) = 1$ and F^{*n} is open. In this case, if $\operatorname{char}(F) = p$, then by Proposition (5.6) $1+\pi^i \notin F^{*p}$ for $(i,p) = 1$. Then F^{*p} is not open. If $\operatorname{char}(F) = 0$, then using Corollary 2 of (5.8) we obtain $U_i \subset F^{*p^m}$ when $i > pe/(p-1) + (m-1)e$. Therefore F^{*n} is open for $n \geqslant 1$. □

This Proposition demonstrates that topological properties are closely connected with the algebraic ones for complete discrete valuation fields of characteristic 0 with residue field of characteristic p. This is not the case when $\operatorname{char}(F) = p$.

(5.10). Finally, we deduce a multiplicative analog of the expansion in Proposition (5.2).

PROPOSITION (HENSEL). *Let F be a complete discrete valuation field. Let R be a set of representatives and let π_i be as in (5.2). Then for $\alpha \in F^*$ there exist uniquely determined $n \in \mathbb{Z}$, $\theta_i \in R$ for $i \geqslant 0$, such that α can be expanded in the convergent product*

$$\alpha = \pi^n \theta_0 \prod_{i \geqslant 1} (1 + \theta_i \pi_i).$$

Proof. The existence and uniqueness of n and θ_0 immediately follow from Proposition (5.4). Assume that $\varepsilon \in U_m$, then, using Proposition (5.2), one can find $\theta_m \in R$ with $\varepsilon(1 + \theta_m \pi_m)^{-1} \in U_{m+1}$. Proceeding by induction, we obtain an expansion of α in a convergent product. If there are two such expansions $\prod(1 + \theta_i \pi_i) = \prod(1 + \theta_i' \pi_i)$, then the residues $\overline{\theta}_i$, $\overline{\theta_i'}$ coincide in $\overline{F}$. Thus, $\theta_i = \theta_i'$. □

Exercise.

1. Keeping the hypotheses and notations of (5.7), assume that a primitive pth root ζ_p of unity is contained in F^* and $\zeta_p = 1 + \operatorname{rep}(\overline{\theta}_1)\pi^{e/(p-1)} + \dots$ for some $\overline{\theta}_1 \in \overline{F}$. Show that $\overline{\theta}_0 = -\overline{\theta}_1^{\,p-1}$.

6. The Group of Principal Units as a $\mathbb{Z}_p$-module

We study $\mathbb{Z}_p$-structure of the group of principal units of a complete discrete valuation field F with residue field $\overline{F}$ of characteristic $p > 0$ by using convergent series and results of the previous section. Everywhere in this section F is a complete discrete valuation field with residue field of positive characteristic p.

(6.1). Propositions (5.6), (5.7) imply that $\varepsilon^{p^n} \to 1$ as $n \to +\infty$ for $\varepsilon \in U_1$. This enables us to write

$$\varepsilon^a = \lim_{n\to\infty} \varepsilon^{a_n} \qquad \text{if} \qquad \lim_{n\to\infty} a_n = a \in \mathbb{Z}_p, \quad a_n \in \mathbb{Z}.$$

LEMMA. *Let $\varepsilon \in U_1$, $a \in \mathbb{Z}_p$. Then $\varepsilon^a \in U_1$ is well defined and $\varepsilon^{a+b} = \varepsilon^a \varepsilon^b$, $\varepsilon^{ab} = (\varepsilon^a)^b$, $(\varepsilon\eta)^a = \varepsilon^a \eta^a$ for $\varepsilon, \eta \in U_1$, $a, b \in \mathbb{Z}_p$. The multiplicative group U_1 is a $\mathbb{Z}_p$-module under the operation of raising to a power. Moreover, the structure of the $\mathbb{Z}_p$-module U_1 is compatible with the topologies of $\mathbb{Z}_p$ and U_1.*

Proof. Assume that $\lim a_n = \lim b_n$; hence $a_n - b_n \to 0$ as $n \to +\infty$ and $\lim \varepsilon^{a_n - b_n} = 1$. Propositions (5.6), (5.7) show that a map $\mathbb{Z}_p \times U_1 \to U_1$ ($(a, \varepsilon) \to \varepsilon^a$) is continuous with respect to the p-adic topology on $\mathbb{Z}_p$ and the discrete valuation topology on U_1. This argument can be applied to verify the other assertions of the Lemma. □

(6.2). PROPOSITION. *Let F be of characteristic p with perfect residue field. Let R be a set of representatives, and let R_0 be a subset of it such that the residues of its elements in $\overline{F}$ form a basis of $\overline{F}$ as a vector space over $\mathbb{F}_p$. Let an index-set J numerate the elements of R_0. Assume that π_i are as in (5.2). Let v_p be the p-adic valuation.*

Then every element $\alpha \in U_1$ can be uniquely represented as a convergent product

$$\alpha = \prod_{\substack{(i,p)=1 \\ i>0}} \prod_{j \in J} (1 + \theta_j \pi_i)^{a_{ij}}$$

where $\theta_j \in R_0$, $a_{ij} \in \mathbb{Z}_p$ and the sets $J_{i,c} = \{j \in J : v_p(a_{ij}) \leqslant c\}$ are finite for all $c \geqslant 0$, $(i, p) = 1$.

Proof. We first show that the element α can be written modulo U_n for $n \geqslant 1$ in the desired form with $a_{ij} \in \mathbb{Z}$. Proceeding by induction, it will suffice to consider an element $\varepsilon \in U_n$ modulo U_{n+1}. Let $\varepsilon \equiv 1 + \theta\pi_n \mod U_{n+1}$, $\theta \in R$. If $(n, p) = 1$, then one can find $\theta_1, \dots, \theta_m \in R_0$ and $b_1, \dots, b_m \in \mathbb{Z}$ such that $1 + \theta\pi_n \equiv \prod_{k=1}^m (1 + \theta_k \pi_n)^{b_k} \mod U_{n+1}$ for some m. If $n = p^s n'$ with an integer n', $(n', p) = 1$, then using the Corollary of (5.6), one can find $\theta_1, \dots, \theta_m \in R_0$ and $b_1, \dots, b_m \in \mathbb{Z}$ such that $1 + \theta\pi_n \equiv \prod_{k=1}^m (1 + \theta_k \pi_{n'})^{p^s b_k} \mod U_{n+1}$ for some m. Now due to the continuity we get the desired expression for $\alpha \in U_1$ with the above conditions on the sets $J_{i,c}$.

Assume that there is a convergent product for 1 with θ_j, a_{ij}. Let $(i_0, p) = 1$ and $j_0 \in J$ be such that $n = p^{v_p(a_{i_0 j_0})} i_0 \leqslant p^{v_p(a_{ij})} i$ for all $(i, p) = 1$, $j \in J$. Then the choice of R_0 and (5.5), (5.6) imply $\prod (1 + \theta_j \pi_i)^{a_{ij}} \notin U_{n+1}$, which concludes the proof. □

COROLLARY. *The group U_1 has a free topological basis $1 + \theta_j \pi_i$ where where $\theta_j \in R_0$, $(i, p) = 1$ (for the definition of a topological basis see Exercise 2).*

(6.3). For subsequent consideration, we return to the horizontal homomorphism

$$\psi\colon \overline{F} \to \overline{F}, \quad \overline{\alpha} \mapsto \overline{\alpha}^p + \overline{\theta}_0\overline{\alpha}$$

of case (2) in Proposition (5.7). Suppose that a primitive pth root of unity ζ_p belongs to F and $\zeta_p \equiv 1 + \operatorname{rep}(\overline{\theta}_1)\pi^{e/(p-1)} \mod \pi^{e/(p-1)+1}$ ($v(\zeta_p - 1) = e/(p-1)$ according to Proposition (5.7)). As $\overline{\theta}_1 \in \ker\psi$, we conclude that $\psi(\overline{\alpha}) = \overline{\theta}_1^p(\eta^p - \eta)$ where $\eta = \overline{\alpha}\overline{\theta}_1^{-1}$. The homomorphism $\eta \mapsto \eta^p - \eta$ is usually denoted by $\wp$. In this terminology we get $\psi(\overline{F}) = \overline{\theta}_1^p\wp(\overline{F})$. Note that the theory of Artin–Schreier extensions sets a correspondence between abelian extensions of exponent p and subgroups of $\overline{F}/\wp(\overline{F})$ (see Exercise 6 section 5 Ch. V and [La1, Ch. VIII]). In particular, if $\overline{F}$ is finite, then the cardinalities of the kernel of ψ and of the cokernel of ψ coincide. In this simple case $\psi(\overline{F}) = \overline{F}$ if and only if there is no nontrivial p-torsion in F^*, and $\psi(\overline{F})$ is of index p if and only if $\zeta_p \in F^*$ (see (5.7)). The homomorphism $\wp$ will play an important role in class field theory.

More generally, if instead of π^n we use π_n as in (5.2), then we can describe raising to the pth power in a similar way. Suppose that $e/(p-1) \in \mathbb{Z}$. Let $\pi_{e/(p-1)}^p = \eta_1\pi_{pe/(p-1)}$ with $\eta_1 \in \mathcal{O}$. Then raising to the pth power in case (2) is described by

$$\psi\colon \overline{\alpha} \mapsto \overline{\eta}_1\overline{\theta}_1^p\wp(\overline{\alpha}\overline{\theta}_1^{-1}).$$

(6.4). PROPOSITION. *Let F be of characteristic 0 with perfect residue field of characteristic p. Let π_i be as in* (5.2). *If $e = v(p)$ is divisible by $p-1$, let $\psi\colon \overline{F} \to \overline{F}$ be the map introduced in* (6.3).

Let R be a set of representatives and let R_0 (resp. R_0') be a subset of it such that the residues of its elements in $\overline{F}$ form a basis of $\overline{F}$ as a vector space over $\mathbb{F}_p$ (resp. are generators of $\overline{F}/\psi(\overline{F})$). Let the index-set J (resp. J') numerate the elements of R_0 (resp. R_0'). Let

$$I = \{i : i \in \mathbb{Z}, 1 \leqslant i < pe/(p-1), (i,p) = 1\}.$$

Let v_p be the p-adic valuation.

Then every element $\alpha \in U_1$ can be represented as a convergent product

$$\alpha = \prod_{i\in I}\prod_{j\in J}(1 + \theta_j\pi_i)^{a_{ij}} \prod_{j\in J'}(1 + \eta_j\pi_{pe/(p-1)})^{a_j}$$

where $\theta_j \in R_0$, $\eta_j \in R_0'$, $a_{ij}, a_j \in \mathbb{Z}_p$ (the second product occurs when $e/(p-1)$ is an integer) and the sets

$$J_{i,c} = \{j \in J : v_p(a_{ij}) \leqslant c\}, \quad J_c' = \{j \in J' : v_p(a_j) \leqslant c\}$$

are finite for all $c \geqslant 0$, $i \in I$.

Proof. We shall show how to obtain the required form for $\varepsilon \in U_n$ modulo U_{n+1}. Put $\pi_n = \pi^n$ for $n = pe/(p-1)$. Let $\varepsilon = 1 + \theta\pi_n \mod U_{n+1}$, $\theta \in R$. There are four cases to consider:

(1) $n \in I$. One can find $\theta_1, \dots, \theta_m \in R_0$ and $b_1, \dots, b_m \in \mathbb{Z}$ satisfying the congruence $1 + \theta\pi_n \equiv \prod_{k=1}^m (1 + \theta_k \pi_n)^{b_k} \mod U_{n+1}$ for some m.

(2) $n < pe/(p-1)$, $n = p^s n'$ with $n' \in I$. Corollary 1 in (5.8) and (5.5) show that there exist $\theta_1, \dots, \theta_m \in R_0$, $b_1, \dots, b_m \in \mathbb{Z}$ such that

$$1 + \theta\pi_n \equiv \prod_{k=1}^{m} (1 + \theta_k \pi_{n'})^{p^s b_k} \mod U_{n+1} \quad \text{for some } m.$$

(3) $e/(p-1) \in \mathbb{Z}$, $n = pe/(p-1)$. Proposition (5.7) and (5.5) and the definition of R_0' imply that if $n = p^s n'$ with $n' \in I$, then there exist $\theta_1, \dots, \theta_m \in R_0$, $\eta_1, \dots, \eta_r \in R_0'$, $b_1, \dots, b_m, c_1, \dots, c_r \in \mathbb{Z}$ such that

$$1 + \theta\pi_n \equiv \prod_{k=1}^{m} (1 + \theta_k \pi_{n'})^{p^s b_k} \prod_{l=1}^{r} (1 + \eta_l \pi_n)^{c_l} \mod U_{n+1} \quad \text{for some } m, r.$$

(4) $n > pe/(p-1)$. Proposition (5.7) and Corollary 1 in (5.8) imply that if $d = \min\{d : n - de \leqslant pe/(p-1)\}$ and $n' = n - de$, then

$$1 + \theta\pi_n \equiv (1 + \theta' \pi_{n'})^{p^d} \mod U_{n+1} \quad \text{for some } \theta' \in R.$$

Now applying the arguments of the preceding cases to $1 + \theta'\pi_{n'}$, we can write $1 + \theta\pi_n$ mod U_{n+1} in the required form. □

(6.5). From Proposition (5.7) we deduce that F contains finitely many roots of unity of order a power of p.

COROLLARY. *Let F be of characteristic 0 with perfect residue field of characteristic p.*

(1) *If F does not contain nontrivial pth roots of unity then the representation in Proposition* (6.4) *is unique. Therefore the elements of Proposition* (6.4) *form a topological basis of $U_{1,F}$.*

(2) *If F contains a nontrivial pth root of unity let r be the maximal integer such that F contains a primitive p^r th root of unity. Then the numbers a_{ij}, a_j of Proposition* (6.4) *are determined uniquely modulo p^r. Therefore the elements of Proposition* (6.4) *form a topological basis of $U_{1,F}/U_{1,F}^{p^r}$.*

(3) *If the residue field of F is finite then U_1 is the direct sum of a free $\mathbb{Z}_p$-module of rank ef and the torsion part.*

Proof. (1) All horizontal homomorphisms of the diagrams in Proposition (5.7) are injective when $\zeta_p \notin F$. Repeating the arguments for uniqueness from the proof of Proposition (6.2), we get the first assertion of the Corollary.

(2) We can argue by induction on r and explain the induction step. Write a primitive p^r th root ζ_{p^r} in the form of Proposition (6.4)

$$\zeta_{p^r} = \prod_{i \in I} \prod_{j \in J} (1 + \theta_j \pi_i)^{c_{ij}} \prod_{j \in J'} (1 + \eta_j \pi_{pe/(p-1)})^{c_j}$$

and raise the expression to the p^r th power which demonstrates the non-uniqueness of the expansion in Proposition (6.4).

Now if

$$1 = \prod_{i\in I}\prod_{j\in J}(1+\theta_j\pi_i)^{a_{ij}} \prod_{j\in J'}(1+\eta_j\pi_{pe/(p-1)})^{a_j}$$

then by the same argument as in the proof of Proposition (6.2) we deduce that $a_{ij} = pb_{ij}, a_j = pb_j$ with p-adic integers b_{ij}, b_j. Then

$$\prod_{i\in I}\prod_{j\in J}(1+\theta_j\pi_i)^{b_{ij}} \prod_{j\in J'}(1+\eta_j\pi_{pe/(p-1)})^{b_j}$$

is a pth root of unity, and so is equal to

$$\Big(\prod_{i\in I}\prod_{j\in J}(1+\theta_j\pi_i)^{c_{ij}} \prod_{j\in J'}(1+\eta_j\pi_{pe/(p-1)})^{c_j}\Big)^{p^{r-1}c}$$

for some integer c. Now by the induction assumption all $b_{ij} - p^{r-1}cc_{ij}, b_j - p^{r-1}cc_j$ are divisible by p^{r-1}. Thus, all a_{ij}, a_j are divisible by p^r.

(3) If the residue field of F is finite then U_1 is a module of finite type over the principal ideal domain $\mathbb{Z}_p$. Note that the group $\wp\left(\overline{F}\right)$ is of index p in $\overline{F}$ because $\overline{F}$ is finite (see (6.3)).

Finally the cardinality of I is equal to $e = [pe/(p-1)] - [[pe/(p-1)]/p]$. □

Exercises.

1. Let B be a set of elements of U_1 such that the subset $B \cap (U_n \setminus U_{n+1})$ is finite for every n. Show that the product $\prod_{\alpha\in B}\alpha$ converges.
2. Let A be a $\mathbb{Z}_p$-module endowed with a topology compatible with the structure of the module and the p-adic topology of $\mathbb{Z}_p$. A set $\{a_i\}_{i\in I}$ of elements of A is called a set of topological generators of A if every element of A is a limit of a convergent sequence of elements of the submodule of A generated by this set. A set of topological generators is called a topological basis if for every $j \in I$ and every non-zero $c \in \mathbb{Z}_p$ the element ca_j is not a limit of a convergent sequence of elements of the submodule of A generated by $\{a_i : i \neq j\}$.
   Show that the elements indicated in Proposition (6.2) and (6.4) form a set of topological generators of U_1 with respect to the topology induced by the discrete valuation. Show that if the p-torsion of F^* consists of one element then those elements form a topological basis of U_1.
3. Assume that a primitive p^r th root ζ_{p^r} of unity belongs to a set of topological generators a_i of U_1 as in Proposition (6.4) by replacing one of appropriate elements of the set of generators indicated there, if necessary. By studying the unit $(1+\theta\pi^i)^{p^m l}$ with l relatively prime to p show that U_1 is a direct sum of its torsion part and the submodule topologically generated by $\{a_i : a_i \neq \zeta_{p^r}\}$ and these elements form a topological basis of the latter submodule.

7. Set of Multiplicative Representatives

We maintain the notations and hypotheses of section 5; F is a complete discrete valuation field. We shall introduce a special set $\mathcal{R}$ of multiplicative representatives which is closed with respect to multiplication. We will describe coefficients of the sum and product of convergent power series with multiplicative representatives.

(7.1). Assume that $\operatorname{char}(\overline{F}) = p > 0$.

Let $a \in \overline{F}$. An element $\alpha \in \mathcal{O}$ is said to be a *multiplicative representative* (*Teichmüller representative*) of a if $\overline{\alpha} = a$ and $\alpha \in \bigcap_{m\geqslant 0} F^{p^m}$. This definition is justified by the following Proposition.

PROPOSITION. *An element $a \in \overline{F}$ has a multiplicative representative if and only if $a \in \bigcap_{m\geqslant 0} \overline{F}^{p^m}$. A multiplicative representative for such a is unique. If a and b have the multiplicative representatives α and β, then $\alpha\beta$ is the multiplicative representative of ab.*

Proof. We need the following Lemma.

(7.2). LEMMA. *Let $\alpha, \beta \in \mathcal{O}$ and $v(\alpha-\beta) \geqslant m$, $m > 0$. Then $v(\alpha^{p^n}-\beta^{p^n}) \geqslant n+m$.*

Proof. Put $\alpha = \beta+\pi^m\gamma$; then $\alpha^p = \beta^p+p\beta^{p-1}\pi^m\gamma+\dots+p\beta(\pi^m\gamma)^{p-1}+\pi^{pm}\gamma^p$, and as $v(p) \geqslant 1$ (recall $\operatorname{char}(\overline{F}) = p$), we have $v(p\beta^{p-1}\pi^m\gamma) \geqslant m+1, \dots, v(\pi^{pm}\gamma^p) \geqslant m+1$, and $\alpha^p - \beta^p \in \pi^{m+1}\mathcal{O}$. Now the required assertion follows by induction. □

To prove the first assertion of the Proposition, suppose that $a \in \bigcap_{m\geqslant 0} \overline{F}^{p^m}$. Since $\overline{F}$ has no nontrivial p-torsion, there exist unique elements $a_m \in \overline{F}$ satisfying the equations $a_m^{p^m} = a$. Let $\beta_m \in \mathcal{O}$ be such that $\overline{\beta}_m = a_m$. Then $\overline{\beta_{m+1}^p} = \overline{\beta}_m$ and $v(\beta_{m+1}^p - \beta_m) \geqslant 1$. Lemma (7.2) implies $v(\beta_{m+1}^{p^{n+1}} - \beta_m^{p^n}) \geqslant n+1$. Hence, the sequence $(\beta_m^{p^{m-n}})_{m\geqslant n}$ is Cauchy. It has the limit $\alpha_n = \lim \beta_m^{p^{m-n}} \in \mathcal{O}$. We see that $\alpha_n^{p^n} = \alpha_0$ for $n \geqslant 0$ and $\overline{\alpha}_0 = a$, i.e., α_0 is a multiplicative representative of a. Conversely, if $a \in \overline{F}$ has a multiplicative representative α, then $\overline{\alpha} \in \bigcap_{m\geqslant 0} \overline{F}^{p^m}$.

Furthermore, if α and β are multiplicative representatives of $a \in \overline{F}$, then writing $\alpha = \alpha_m^{p^m}, \beta = \beta_m^{p^m}$ for some $\alpha_m, \beta_m \in \mathcal{O}$, we have $\overline{\alpha}_m^{p^m} = \overline{\beta}_m^{p^m}$ and $\overline{\alpha}_m = \overline{\beta}_m$ because of the injectivity of $\uparrow p^m$ in $\overline{F}$. Now Lemma (7.2) implies $v(\alpha-\beta) \geqslant m+1$, hence $\alpha = \beta$.

Finally, if α and β are the multiplicative representatives of a and b, then $\overline{\alpha\beta} = ab$ and $\alpha\beta \in \bigcap_{m\geqslant 0} F^{p^m}$. Therefore, $\alpha\beta$ is the multiplicative representative of ab. □

(7.3). Denote the set of multiplicative representatives in $\mathcal{O}$ by $\mathcal{R}$.

COROLLARY 1. *If $\overline{F}$ is perfect (i.e. F is a local field) then every element of $\overline{F}$ has its multiplicative representative in $\mathcal{R}$. The map $r\colon \overline{F} \to \mathcal{R}$ induces an isomorphism $\overline{F}^* \xrightarrow{\sim} \mathcal{R} \setminus \{0\}$. The correspondence $r\colon \overline{F} \to \mathcal{R}$ is called the Teichmüller map.*

If $\overline{F}$ is finite then $\mathcal{R} \setminus \{0\}$ is a cyclic group of order equal to $|\overline{F}| - 1$.

COROLLARY 2. *Let $\operatorname{char}(F) = p$. If α, β are the multiplicative representatives of $a, b \in \overline{F}$, then $\alpha + \beta$ is the multiplicative representative of $a + b$.*

Proof. Let $\alpha = \alpha_m^{p^m}, \beta = \beta_m^{p^m}$. Then $\alpha+\beta = (\alpha_m + \beta_m)^{p^m}$, hence $\alpha+\beta \in \bigcap_{m\geqslant 0} F^{p^m}$ and $\overline{\alpha + \beta} = a + b$. □

(7.4). Now we focus our attention exclusively on the case where $\operatorname{char}(F) = 0$ and $\operatorname{char}(\overline{F}) = p$. Suppose that we have two elements $\alpha, \beta \in \mathcal{O}$, and ($\pi$ is a prime element)

$$\alpha = \sum_{i\geqslant 0} \theta_i \pi^i, \qquad \beta = \sum_{i\geqslant 0} \eta_i \pi^i,$$

with $\theta_i, \eta_i \in \mathcal{R}$. Suppose also that $\alpha + \beta$ and $\alpha\beta$ are written in the form

$$\alpha + \beta = \sum_{i\geqslant 0} \rho_i^{(+)} \pi^i, \qquad \alpha\beta = \sum_{i\geqslant 0} \rho_i^{(\times)} \pi^i,$$

and $\rho_i{}^{(+)}, \rho_i{}^{(\times)} \in \mathcal{R}$.

Corollary (5.2) implies that $\rho_i^{(+)}, \rho_i^{(\times)}$ are uniquely determined by θ_i, η_i. Our intention is to reveal the dependence of $\rho_n^{(+)}, \rho_n^{(\times)}$ on θ_i, η_i, $i \leqslant n$. In order to obtain a polynomial relation we introduce elements $\theta_i = \varepsilon_i^{p^{n-i}}$, $\eta_i = \xi_i^{p^{n-i}}$, $\rho_i^{(*)} = \lambda_i^{(*)p^{n-i}}$ for ε_i, ξ_i, $\lambda_i^{(*)} \in \mathcal{R}$ and $* = +$ or $* = \times$, $i \geqslant 0$.

Then we deduce that

$$\Big(\sum_{i=0}^{n} \pi^i \varepsilon_i^{p^{n-i}}\Big) * \Big(\sum_{i=0}^{n} \pi^i \xi_i^{p^{n-i}}\Big) \equiv \Big(\sum_{i=0}^{n} \pi^i \lambda_i^{(*)p^{n-i}}\Big) \mod \pi^{n+1}, \tag{$*$}$$

for $* = +$ or $* = \times$. We see that if the residues $\overline{\varepsilon}_i, \overline{\xi}_i$ for $0 \leqslant i \leqslant n$ and $\overline{\lambda_i^{(*)}}$ for $0 \leqslant i \leqslant n-1$ are known, then by using Lemma (7.2) we can calculate $\pi^i \varepsilon_i^{p^{n-i}}$, $\pi^i \xi_i^{p^{n-i}}$, $\pi^i \lambda_i^{p^{n-i}} \mod \pi^{n+1}$. Hence, $\overline{\lambda_n^{(*)}}$ are uniquely determined from $(*)$.

(7.5). Let $A = \mathbb{Z}[X_0, X_1, \ldots, Y_0, Y_1, \ldots]$ be the ring of polynomials in variables $X_0, X_1, \ldots,\ Y_0, Y_1, \ldots$ with coefficients from $\mathbb{Z}$. Introduce polynomials

$$W_n(X_0, \ldots, X_n) = \sum_{i=0}^{n} p^i X_i^{p^{n-i}}, \qquad n \geqslant 0.$$

In particular, $W_0(X_0) = X_0$, $W_1(X_0, X_1) = X_0^p + pX_1$.

PROPOSITION. *There exist unique polynomials*

$$\omega_n^{(*)}(X_0, \dots, X_n, Y_0, \dots, Y_n) \in A,\ n \geqslant 0$$

satisfying the equations

$$W_n(X_0, \dots, X_n) * W_n(Y_0, \dots, Y_n) = W_n(\omega_0^{(*)}, \dots, \omega_n^{(*)})$$

for $n \geqslant 0$, *where* $* = +$ *or* $* = \times$. *Moreover, the polynomial*

$$\omega_n^{(*)}(X_0, \dots, X_n, Y_0, \dots, Y_n)^p - \omega_n^{(*)}(X_0^p, \dots, X_n^p, Y_0^p, \dots, Y_n^p)$$

belongs to pA.

Proof. We get

$$\omega_0^{(+)} = X_0 + Y_0, \quad \omega_1^{(+)} = X_1 + Y_1 + (X_0^p + Y_0^p - (X_0 + Y_0)^p)/p,$$
$$\omega_0^{(\times)} = X_0 Y_0, \quad \omega_1^{(\times)} = X_1 Y_0^p + Y_1 X_0^p + pX_1 Y_1,$$
$$\dots .$$

Assume now that $\omega_i^{(*)} \in A$ for $0 \leqslant i \leqslant n-1$ and proceed by induction. Then for a suitable polynomial $f \in A$ we get

$$\begin{aligned}\pi^n \omega_n^{(*)} &= W_{n-1}(X_0^p, \dots, X_{n-1}^p) * W_{n-1}(Y_0^p, \dots, Y_{n-1}^p)\\ &\quad - W_{n-1}(\omega_0^{(*)p}, \dots, \omega_{n-1}^{(*)p}) + p^n f\end{aligned} \qquad (**)$$

We get $g(X_0, Y_0, \dots)^p - g(X_0^p, Y_0^p, \dots) \in pA$ for $g \in A$. We also deduce that for $m \geqslant 0$

$$g(X_0, Y_0, \dots)^{p^m} - g(X_0^p, Y_0^p, \dots)^{p^{m-1}} \in p^m A.$$

One obtains immediately that

$$W_{n-1}(\omega_0^{(*)p}, \dots, \omega_{n-1}^{(*)p}) - W_{n-1}(\omega_0^{(*)}(X_0^p, Y_0^p), \dots, \omega_{n-1}^{(*)}(X_0^p, \dots, Y_0^p, \dots)) \in p^n A.$$

From

$$\begin{aligned}&W_{n-1}(X_0^p, \dots, X_n^p) * W_{n-1}(Y_0^p, \dots, Y_{n-1}^p)\\ &= W_{n-1}(\omega_0^{(*)}(X_0^p, Y_0^p), \dots, \omega_{n-1}^{(*)}(X_0^p, \dots, Y_0^p))\end{aligned}$$

using $(**)$ we conclude that $\omega_n^{(*)} \in A$. The last assertion of the Proposition is evident. □

(7.6). We now return to the original problem to find an expression for $\rho_i^{(*)}$ in the partial case of $\pi = p$ (the general case can be handled in a similar way).

PROPOSITION. *Let* $\left(\sum \theta_i p^i\right) * \left(\sum \eta_i p^i\right) = \sum \rho_i^{(*)} p^i$ *with* $\theta_i, \eta_i, \rho_i^{(*)} \in \mathcal{R}$ *and* $* = +$ *or* $* = \times$. *Then*

$$\rho_i^{(*)} \equiv \omega_i^{(*)}(\theta_0^{p^{-i}}, \theta_1^{p^{-i+1}}, \dots, \theta_i, \eta_0^{p^{-i}}, \eta_1^{p^{-i+1}}, \dots, \eta_i) \mod p, \quad i \geqslant 0,$$

where $\omega_i^{(*)}$ *are defined in* (7.5).

Proof. Assume that the assertion of the Proposition holds for $i \leqslant n-1$. Using notations of (7.4) this means that

$$\lambda_i^{(*)p^{n-i}} \equiv \omega_i^{(*)}(\varepsilon_0^{p^{n-i}}, \dots, \varepsilon_i^{p^{n-i}}, \xi_0^{p^{n-i}}, \dots, \xi_i^{p^{n-i}}) \mod p, \quad i \leqslant n-1.$$

From Proposition (7.5) we obtain that for $i \leqslant n-1$

$$\omega_i^{(*)}(\varepsilon_0^{p^{n-i}}, \dots, \varepsilon_i^{p^{n-i}}, \xi_0^{p^{n-i}}, \dots, \xi_i^{p^{n-i}}) \equiv \omega_i^{(*)}(\varepsilon_0, \dots, \varepsilon_i, \xi_0, \dots, \xi_i)^{p^{n-i}} \mod p.$$

Hence

$$\lambda_i^{(*)} \equiv \omega_i^{(*)}(\varepsilon_0, \dots, \varepsilon_i, \xi_0, \dots, \xi_i) \mod p, \quad i \leqslant n-1.$$

By Lemma (7.2) we have

$$p^i \lambda_i^{(*)p^{n-i}} \equiv p^i \omega_i^{(*)}(\varepsilon_0, \dots, \varepsilon_i, \xi_0, \dots, \xi_i)^{p^{n-i}} \mod p^{n+1}, \quad i \leqslant n-1.$$

The congruence $(*)$ for $\pi = p$ can be rewritten as

$$W_n(\lambda_0^{(*)}, \dots, \lambda_n^{(*)}) \equiv W_n(\varepsilon_0, \dots, \varepsilon_n) * W_n(\xi_0, \dots, \xi_n) \mod p^{n+1}.$$

We conclude that

$$p^n \lambda_n^{(*)} \equiv p^n \omega_n^{(*)}(\varepsilon_0, \dots, \varepsilon_n, \xi_0, \dots, \xi_n) \mod p^{n+1}$$

which implies the assertion. □

COROLLARY 1. *Let* $\left(\sum \theta_i^{p^{-i}} p^i\right) * \left(\sum \eta_i^{p^{-i}} p^i\right) = \sum \rho_i^{(*)p^{-i}} p^i$ *with* $\theta_i, \eta_i, \rho_i^{(*)} \in \mathcal{R}$, $* = +$ *or* $* = \times$. *Then*

$$\rho_i^{(*)} \equiv \omega_i^{(*)}(\theta_0, \dots, \theta_i, \eta_0, \dots, \eta_i) \mod p.$$

Proof. In fact, this has already been shown in the proof of the Proposition. □

COROLLARY 2. $\left(\sum \theta_i^p p^i\right) * \left(\sum \eta_i^p p^i\right) = \sum \rho_i^{(*)p} p^i$.

Proof. This follows immediately from the Proposition and the last assertion of Proposition (7.5). □

Exercises.

1. Let $W_n(X_1, \dots, X_n) = \sum_{m|n} m X_m^{n/m}$. Show that the polynomials

$$\Omega_n^{(*)} \in \mathbb{Q}[X_1, \dots, X_n, Y_1, \dots, Y_n],$$

which are defined via W_n in the same manner as the $\omega_n^{(*)}$ are defined via the W_n in the text above, have integer coefficients.

2. Let F be a complete discrete valuation field with perfect residue field, $\mathrm{char}(F) = 0$, $\mathrm{char}(\overline{F}) = p$. Let π be a prime element in F. In the notation of (6.4) show that
   a) If $e < i < pe/(p-1)$, then $\theta \in \mathcal{R}$ there exists $\theta' \in \mathcal{R}$ for satisfying the congruence
   $$1 + \theta\pi^i \equiv 1 + \theta'\pi^{p(i-e)} \mod U_{i+1}U_1^p$$
   b) Proposition (6.4) holds if the set I is replaced by the set $I' = \{i : i \in \mathbb{Z}, 1 \leqslant i \leqslant e\}$, $R = \mathcal{R}$ and π_i by π^i.
   c) Proposition (6.4) does not hold in general if I is replaced by I' but π_i are not replaced by π^i.

8. The Witt Ring

Closely related to the constructions of the previous section is the notion of Witt vectors. Witt vectors over a perfect field K of positive characteristic form the ring of integers of a local field with prime element p and residue field K.

(8.1). Let B be an arbitrary commutative ring with unity. Let the polynomials

$$W_n(X_0, \dots, X_n) = \sum_{i=0}^{n} p^i X_i^{p^{n-i}}, \qquad n \geqslant 0$$

over B be the images of the polynomials $W_n \in \mathbb{Z}[X_0, \dots, X_n]$ defined in (7.5) under the natural homomorphism $\mathbb{Z} \to B$. For $(a_i)_{i\geqslant 0}$, put

$$(a^{(i)}) = (W_0(a_0), W_1(a_0, a_1), \dots) \in (A)_0^{+\infty};$$

see (5.1). The sequences $(a_i) \in (B)_0^{+\infty}$ are called *Witt vectors* (or, more generally, p-Witt vectors), and the $a^{(i)}$ for $i \geqslant 0$ are called the ghost components of the Witt vector (a_i).

The map $(a_i) \mapsto (a^{(i)})$ is a bijection of $(B)_0^{+\infty}$ onto $(B)_0^{+\infty}$ if p is invertible in B.

Transfer the ring structure of $(a^{(i)}) \in (B)_0^{+\infty}$ under the natural componentwise addition and multiplication on $(a_i) \in (B)_0^{+\infty}$. Then for $(a_i), (b_i) \in (B)_0^{+\infty}$ we get

$$(a_i) * (b_i) = (\omega_0^{(*)}(a_0, b_0), \omega_1^{(*)}(a_0, a_1, b_0, b_1), \dots)$$

for $* = +$ or $* = \times$, where the polynomial $\omega_i^{(*)}$ is the image of the polynomial $\omega_i^{(*)} \in \mathbb{Z}[X_0, X_1, \dots, Y_0, Y_1, \dots]$ under the canonical homomorphism $\mathbb{Z} \to B$.

If p is invertible in B, then the set of Witt vectors is clearly a commutative ring under the operations defined above. In the general case, when p is not invertible in B, the property of the set $(B)_0^{+\infty}$ of being a commutative ring under the operations $+, \times$ defined above can be expressed via certain equations for the coefficients of the polynomials $\omega_i^{(*)} \in B[X_0, X_1, \dots, Y_0, Y_1, \dots]$. This implies that if a ring B satisfies these conditions, then the same is true for a subring, quotient ring and the polynominal ring. Since every ring can be obtained in this way from a ring $\mathcal{B}$ in which p is invertible, one deduces that under the image in B of the above defined operations for $\mathcal{B}$ the set $(B)_0^{+\infty}$ is a commutative ring with the unity $(1, 0, 0, \dots)$. This ring is called the *Witt ring* of B and is denoted by $W(B)$. It is easy to verify that if B is an integer domain, then $W(B)$ is an integer domain as well.

(8.2). Assume from now on that $p = 0$ in B.

LEMMA. *Define the maps $r_0 \colon B \to W(B)$, $\mathbf{V} \colon W(B) \to W(B)$ (the "Verschiebung" map), $\mathbf{F} \colon W(B) \to W(B)$ (the "Frobenius" map) by the formulas*

$$r_0(a) = (a, 0, 0, \dots) \in W(B),$$
$$\mathbf{V}(a_0, a_1, \dots) = (0, a_0, a_1, \dots),$$
$$\mathbf{F}(a_0, a_1, \dots) = (a_0^p, a_1^p, \dots).$$

Then

$$r_0(ab) = r_0(a) r_0(b),$$
$$\mathbf{F}(\alpha + \beta) = \mathbf{F}(\alpha) + \mathbf{F}(\beta), \mathbf{F}(\alpha\beta) = \mathbf{F}(\alpha)\mathbf{F}(\beta),$$
$$\mathbf{V}(\alpha + \beta) = \mathbf{V}(\alpha) + \mathbf{V}(\beta), \quad \mathbf{VF}(\alpha) = \mathbf{FV}(\alpha) = p\alpha$$

for $\alpha, \beta \in W(B)$.

Proof. All these properties can be deduced from properties of $\omega_i^{(*)}$. □

The map $\mathbf{F} - \mathrm{id}$ is often denoted by $\wp \colon W(B) \to W(B)$.

Put $W_n(B) = W(B)/\mathbf{V}^n W(B)$. This is a ring consisting of finite sequences $(a_0, \dots, a_{n-1})$.

(8.3). The following assertion is of great importance, since it provides a construction of a local field of characteristic zero with prime element p and given perfect residue field K.

PROPOSITION. *Let K be a perfect field of characteristic p. For a Witt vector $\alpha = (a_0, a_1, \dots) \in W(K)$ put*

$$v(\alpha) = \min\{i : a_i \neq 0\} \quad \text{if} \quad \alpha \neq 0, \qquad v(0) = +\infty.$$

Let F_0 be the field of fractions of $W(K)$ and $v \colon F_0^* \to \mathbb{Z}$ the extension of v from $W(K)$ ($v(\alpha\beta^{-1}) = v(\alpha) - v(\beta)$).

Then v is a discrete valuation on F_0 and F_0 is a complete discrete valuation field of characteristic 0 with ring of integers $W(K)$ and residue field isomorphic to K. The set of multiplicative representatives in F_0 coincides with $r_0(K)$ and the map r_0 with the Teichmüller map $K \to W(K)$.

Proof. If $\alpha = (\underbrace{0, \dots, 0}_{m \text{ times}}, \dots)$, $\beta = (\underbrace{0, \dots, 0}_{n \text{ times}}, \dots)$, then using the properties of the polynomials $\omega_i^{(*)}$, we get

$$\alpha + \beta = (\underbrace{0, \dots, 0}_{l \text{ times}}, \dots), \quad \alpha\beta = (\underbrace{0, \dots, 0}_{n+m \text{ times}}, \dots)$$

with $l \geqslant \min(m, n)$. Hence, the extension of v to F_0 is a discrete valuation.

Note that $p = (0, 1, 0, \dots) \in W(K)$ and $p^n \to 0$ as $n \to +\infty$ with respect to v. Since K is perfect, by Lemma (8.2) one can write an element $\alpha = (a_0, a_1, \dots) \in W(K)$ as the convergent sum

$$\alpha = (a_0, 0, 0, \dots) + (0, a_1, 0, \dots) + \dots = \sum_{i=0}^{\infty} r_0(a_i^{p^{-i}})p^i \qquad (*)$$

Moreover, such expressions for Witt vectors are compatible with addition and multiplication in $W(K)$.

We also obtain that $W(K)$ is complete with respect to v, and if $v(\alpha) = 0$ for $\alpha \in W(K)$, then $\alpha^{-1} \in W(K)$. Consequently, $v(\alpha) \geqslant v(\beta)$ for α, $\beta \in W(K)$ implies $\alpha\beta^{-1} \in W(K)$, i.e., the ring of integers coincides with $W(K)$ and F_0 is complete. The maximal ideal of $W(K)$ is $\mathbf{V}W(K)$ and the residue field is isomorphic to K.

Finally, $r_0(K) = \bigcap_{n \geqslant 0} F_0^{p^n}$, and hence, using Proposition (7.1), we complete the proof. □

REMARK. The notion of Witt vectors and Proposition (8.3) can be generalized to ramified Witt vectors by replacing p with π (see [Dr2], [Haz4]).

Exercises.

1. Can the maps $\mathbf{V}, \mathbf{F}, r_0$ (with properties similar to (8.2)) be defined for a ring with $p \neq 0$?
2. Show that $\mathbf{V}$ and $\mathbf{F}$ are injective in $W(K)$ if K is a field of characteristic p.
3. Show that $\mathbf{F}$ is an automorphism of $W(K)$ if K is a perfect field of characteristic p.
4. Show that $W(\mathbb{F}_p) \simeq \mathbb{Z}_p$, $W_n(\mathbb{F}_p) \simeq \mathbb{Z}/p^n\mathbb{Z}$.
5. Show that $r_0(a)(b_0, b_1, \dots) = (ab_0, a^p b_1, \dots)$.
6. Let K be a field of characteristic p. Show that $\wp\colon W(K) \to W(K)$ is a ring homomorphism and $\ker(\wp) = W(\mathbb{F}_p)$.
7. a) Let $\Omega_n^{(*)}$ be the polynomials defined in Exercise 1 of section 7. Show that one can introduce a big Witt ring $W_b(B)$ by these polynomials.

   b) Show that the canonical map

$$(a_1, a_2, \dots) \in W_b(B) \to (a_1, a_p, a_{p^2}, \dots) \in W(B)$$

   is a surjective ring homomorphism.

9. Artin–Hasse Maps

This section introduces several Artin–Hasse maps which can be viewed as a generalization of the exponential map; for a more advanced generalization see section 2 Ch. VI. In section (9.1) we define an Artin–Hasse function which is not additive; in section (9.2) we introduce its modification which is a group homomorphism from Witt vectors over B to formal power series in $1 + XB[[X]]$; for a local field F section (9.3) presents another modification which is a group homomorphism from $W(\overline{F})$ to $1 + X\mathcal{O}[[X]]$.

(9.1). The exponential map relates the additive and multiplicative structures. In the case of a complete discrete valuation field of characteristic zero $\exp\colon \mathcal{M}^n \to 1 + \mathcal{M}^n$ is an isomorphism for large n (see (1.4) of Ch. VI). We are interested in modifications of $\exp$ so that the new map is defined on the whole $\mathcal{M}$.

Introduce the formal power series

$$E(X) = \exp\left(\sum_{i\geq 0} \frac{X^{p^i}}{p^i}\right),$$

called the *Artin–Hasse function* (in fact, *E. Artin* and *H. Hasse* worked with $1/E(X)$, see [AH2]). Considering $\mathbb{Z}$ as a subring of $\mathbb{Z}_p$, we use the notation

$$\mathbb{Z}_{(p)} = \mathbb{Q} \cap \mathbb{Z}_p = \left\{\frac{m}{n} : m, n \in \mathbb{Z},\ (n,p) = 1\right\}.$$

Lemma. $E(X) = \prod_{(i,p)=1}(1 - X^i)^{-\mu(i)/i} \in \mathbb{Z}_{(p)}[[X]]$, *and* $E(X) \equiv 1 + X \mod X^2$, *where* μ *is the Möbius function* ($-\mu(i)/i$ *is viewed as an element of* $\mathbb{Z}_p$, *see* (6.1)).

Proof. Put $\lambda(X) = \sum_{i\geqslant 0} X^{p^i}/p^i \in 1 + X\mathbb{Q}[[X]]$. Then it is easy to verify that

$$\log(1 - X) = -\sum_{(i,p)=1} \frac{1}{i}\lambda(X^i).$$

The properties of the function μ imply

$$\lambda(X) = -\sum_{(i,p)=1} \frac{\mu(i)}{i}\log(1 - X^i)$$

and thereby

$$E(X) = \exp(\lambda(X)) = \prod_{(i,p)=1} (1 - X^i)^{-\mu(i)/i}.$$

□

Remark. For a generalization of $E(X)$ using formal groups see Exercise 4 in section 1 Ch. VIII.

(9.2). Let B be an arbitrary commutative ring in which all integers relatively prime to p are invertible. We shall denote also by $E(X)$ the image of $E(X)$ in $1+XB[[X]]$ under the canonical homomorphism $\mathbb{Z}_{(p)} \to B$.

The ring $B[[X]]$ of formal power series over a commutative ring B has the natural X-adic topology with $X^n B[[X]]$ as a basis of open neighborhoods of 0.

For $\alpha = (a_0, a_1, \dots) \in W(B)$ and $u(X) \in XB[[X]]$, define

$$E(\alpha, u(X)) = \prod_{i\geqslant 0} E(a_i u(X)^{p^i})$$

(the product converges in $1 + XB[[X]]$, since $u(X)^n \to 0$ as $n \to +\infty$).

LEMMA. *Let p be invertible in B. Then*

$$E(\alpha, u(X)) = \exp\left(\sum_{i\geqslant 0} \frac{a^{(i)} u(X)^{p^i}}{p^i}\right),$$

where $a^{(i)} = \sum_{j=0}^{j=i} p^j a_j^{p^{i-j}} \in B$ *are the ghost components of* α *defined in* (8.1).

Proof. This follows directly from

$$E(a_i u(X)^{p^i}) = \exp\left(\sum_{j\geqslant 0} \frac{a_i^{p^j} u(X)^{p^{i+j}}}{p^j}\right).$$

□

PROPOSITION. *Let B be a commutative ring in which all integers relatively prime to p are invertible. Then*

(1) $E(\alpha - \beta, u(X)) = E(\alpha, u(X)) E(\beta, u(X))^{-1}$ *for every* $\alpha, \beta \in W(B)$.

(2) $E(\mathbf{V}\alpha, u(X)) = E(\alpha, u(X)^p)$.

(3) $E(\alpha, u(X)) \equiv 1 + a_0 u(X) \mod X^{2n}$ *if* $u(X) \in X^n B[[X]]$.

The map $E(\cdot\,, u(X)) \colon W(B) \to 1 + XB[[X]]$ *is a continuous homomorphism of the additive group of* $W(B)$ *(with the topology given by* $\mathbf{V}^i W(B)$*) to the multiplicative group* $1 + XB[[X]]$.

Proof. If p is invertible in B then (1) follows from the previous Lemma and the definition of the Witt ring. In the general case property (1) can be reformulated as certain conditions imposed on the coefficients of the polynominals $\omega_i^{(+)}$. Repeating now the arguments of (8.1), we deduce that property (1) holds.

Further,

$$E(\mathbf{V}\alpha, u(X)) = \prod_{i\geqslant 0} E(a_i u(X)^{p^{i+1}}) = E(\alpha, u(X)^p).$$

The congruence $E(X) \equiv 1 + X \mod X^2$ implies property (3). Finally, one can deduce by induction that

$$E(\mathbf{V}^i W(B), u(X)) \subset 1 + X^m B[[X]] \qquad \text{for } m \leqslant p^i n,$$

provided $u(X) \in X^n B[[X]]$. This shows the continuity of $E\big(\cdot\,, u(X)\big)$ and completes the proof. □

COROLLARY. *The map* $E\big(\cdot\,, u(X)\big): W(B) \to 1 + XB[[X]]$ *is a continuous injective homomorphism for* $u(X) \in XB[[X]]$.

Proof. These assertions can be verified by induction, starting with the following: if $E(\alpha, u(X)) = 1$, then, by property (3) of Proposition, $a_0 = 0$; hence $\alpha = \mathbf{V}\beta$ and $E(\beta, u(X)^p) = 1$ by property (2). □

(9.3). Now we assume that B is the residue field $\overline{F}$ of a complete discrete valuation field F and that $\overline{F}$ is a perfect field of characteristic p. Let $\mathcal{O}$ be the ring of integers of F.

We endow the group $1 + \mathcal{O}[[X]]$ with the topology having a basis $1 + \pi^m \mathcal{O}[[X]] + X^n \mathcal{O}[[X]]$ of open neighborhoods of 1, where π is a prime element in F.

Let $\alpha \in W(\overline{F})$; then the relation $(*)$ in (8.3) allows us to write

$$\alpha = \sum_{i \geqslant 0} r_0(c_i) p^i \qquad \text{with } c_i \in \overline{F},$$

where r_0 is the Teichmüller map $\overline{F} \to W(\overline{F})$ (see Proposition (8.3)). Note that c_i are uniquely determined by α. We also have the Teichmüller map $r: \overline{F} \to \mathcal{O}$ (see Corollary 1 in (7.3)). Put

$$\mathcal{E}(\alpha, u(X)) = \prod_{i \geqslant 0} E(r(c_i) u(X))^{p^i} \qquad \text{with } u(X) \in X\mathcal{O}[[X]]$$

(the product converges, since $u(X)^n \to 0$ as $n \to +\infty$). The map

$$\mathcal{E}(\cdot, X): W(\overline{F}) \to 1 + X\mathcal{O}[[X]]$$

is called the Artin–Hasse map (*H. Hasse* employed it for a field F of characteristic 0, see [Has9], [Sha2]).

PROPOSITION.
(1) $\mathcal{E}(\alpha - \beta, u(X)) = \mathcal{E}(\alpha, u(X)) \mathcal{E}(\beta, u(X))^{-1}$ *for* $\alpha, \beta \in W(\overline{F})$.
(2) $\mathcal{E}(\mathbf{V}\alpha, u(X)) = \mathcal{E}(\alpha, u(X))^p$.
(3) $\mathcal{E}(\alpha, u(X)) \equiv 1 + r(c_0) u(X) \mod p u(X) \mathcal{O}[[X]] + u(X)^2 \mathcal{O}[[X]]$ *if* $\alpha = \sum_{i \geqslant 0} r_0(c_i) p^i$.

The map $\mathcal{E}(\cdot, u(X)): W(\overline{F}) \to 1 + X\mathcal{O}[[X]]$ *for* $u(X) \neq 0$ *is an injective continuous homomorphism of the additive group of* $W(\overline{F})$ *into the multiplicative group* $1 + X\mathcal{O}[[X]]$.

Proof. Assume first that $\operatorname{char}(F) = 0$. Then

$$\mathcal{E}(\alpha, u(X)) = \exp\Bigl(\sum_{j\geqslant 0}\Bigl(\sum_{i\geqslant 0} r(c_i)^{p^j} p^i\Bigr) u(X)^{p^j} p^{-j}\Bigr).$$

Let $\beta = \sum_{i\geqslant 0} r_0(d_i)p^i$ and $\alpha + \beta = \sum_{i\geqslant 0} r_0(e_i)p^i$. In this case property (1) will follow if we show that

$$\sum_{i\geqslant 0} r(e_i)^{p^j} p^i = \sum_{i\geqslant 0} r(c_i)^{p^j} p^i + \sum_{i\geqslant 0} r(d_i)^{p^j} p^i \quad \text{for } j \geqslant 0.$$

By Corollary 2 in (7.6) and section 8 it suffices to verify the last relation for $j = 0$.

Applying Proposition (7.6) for $\theta_i = r(c_i), \eta_i = r(d_i), \rho_i = r(e_i)$, we deduce that it should be shown that

$$e_i = \omega_i^{(+)}(c_0^{p^{-i}}, c_1^{p^{-i+1}}, \dots, c_i, d_0^{p^{-i}}, d_1^{p^{-i+1}}, \dots, d_i).$$

But by the same Proposition, these relations are equivalent to

$$\sum_{i\geqslant 0} r_0(c_i)p^i + \sum_{i\geqslant 0} r_0(d_i)p^i = \sum_{i\geqslant 0} r_0(e_i)p^i;$$

thus we have proved property (1) in the case of $\operatorname{char}(F) = 0$.

Since property (1) can be reformulated as certain conditions on the coefficients of the polynomials $\omega_i^{(+)}$ we obtain this property in the general case.

Properties (2) and (3) follow from the definition of $\mathcal{E}$. □

Exercises.

1. Let $E(X) = \sum_{n\geqslant 0} d_n X^n, d_n \in \mathbb{Q}$. Show that $d_0 = 1$, and

$$d_n = \frac{1}{n} \sum_{0\leqslant i\leqslant v_p(n)} d_{n-p^i}.$$

2. Show that

$$1 - X = \prod_{\substack{i\geqslant 1\\ (i,p)=1}} E(-i^{-1}, X^i) = \prod_{\substack{i\geqslant 1\\ (i,p)=1}} E(X^i)^{-1/i}.$$

3. *B. Dwork* introduced a function $F(\alpha, X)$ for $\alpha \in W(B)$ by the formula

$$F(\alpha, X) = \exp\Bigl(\sum_{i\geqslant 0} \sum_{\substack{m\geqslant 1\\ (m,p)=1}} \frac{\alpha^{p^i}}{mp^i} X^{mp^i}\Bigr).$$

Show that

$$F(\alpha, X) = \prod_{\substack{(m,p)=1\\ m\geqslant 1}} E(\alpha X^m)^{1/m}, \quad E(\alpha X) = \prod_{\substack{(m,p)=1\\ m\geqslant 1}} F(\alpha, X^m)^{\mu(m)/m}.$$

4. Let K be a field of characteristic p and let the map $P\colon K[[X]] \to K[[X]]$ be defined as follows:

$$P\Big(\sum_{i\geqslant 0} a_i X^i\Big) = \sum_{i\geqslant 0} a_i^p X^i.$$

Show that

a) $E(d_0\alpha, u(X)) = E(\alpha, u(X))^{d_0}$ for $d_0 \in W(\mathbb{F}_p) \simeq \mathbb{Z}_p$ (see Exercise 4 of section 8), $\alpha \in W(K), u(X) \in XK[[X]]$,

b) $E(r_0(a)\alpha, u(X)) = E(\alpha, au(X))$ for $a \in K, \alpha \in W(K), u(X) \in XK[[X]]$,

c) $E(\mathbf{F}\alpha, Pu(X)) = PE(\alpha, u(X))$ for $\alpha \in W(K), u(X) \in XK[[X]]$.

d) $\mathcal{E}(\mathbf{F}\alpha, Pu(X)) = P\mathcal{E}(\alpha, u(X))$ for $\alpha \in W(K)$ and $u(X) \in 1 + X\mathcal{O}[[X]]$.

5. Let K be as in Exercise 4. Show that

$$1 + XK[[X]] = \prod_{(i,p)=1} E(W(K), X^i).$$

6. ($\diamond$) (*K. Kanesaka* and *K. Sekiguchi*)

a) Let K be as above and for $m \geqslant 2$, let $B_m(K)$ denote the set

$$\{B \in M_m(K) : B = \begin{pmatrix} 1 & b_1 & b_2 & \cdots & b_{m-1} \\ & \ddots & & & \vdots \\ & & \ddots & & b_2 \\ & 0 & & \ddots & b_1 \\ & & & & 1 \end{pmatrix} = [1, b_1, \dots, b_{m-1}]\}$$

Show that $B_m(K)$ is a subgroup of $GL_m(K)$.

b) Let $h\colon B_m(K) \to 1 + XK[[X]]/(1 + X^m K[[X]])$ be defined as

$$h([1, b_1, \dots, b_{m-1}]) = 1 + b_1 X + \dots + b_{m-1}X^{m-1} \mod 1 + X^m K[[X]].$$

Show that h is an isomorphism.

c) Let $g_m\colon 1 + XK[[X]] \to B_m(K)$ be the surjective homomorphism induced by h and let $f_n\colon W(K) \to W_n(K)$ be canonical projection. Show that if $p^{n-1} + 1 \leqslant m \leqslant p^n$, then there exists an injective homomorphism $E_n\colon W_n(K) \to B_m(K)$ such that $E_n \circ f_n = g_m \circ E(\cdot, X)$. (This homomorphism allows one to connect Witt theory of abelian extensions of K of exponent p^n and *Inaba*'s theory of finite extensions of K, see[KnS]).

7. Let $W_b(B)$ be the big Witt ring (see Exercise 7 of section 8). Show that the map

$$(a_1, a_2, \dots) \to \prod_{i\geqslant 1}(1 - a_i X^i) \in 1 + XB[[X]]$$

is an isomorphism of the additive group of $W_b(B)$ onto the multiplicative group of $1 + XB[[X]]$.

8. ($\diamond$) Let K be a perfect field of characteristic p and $\mathcal{O} = W(K)$. Using Exercise 7 of section 8, Exercise 7 and Proposition (9.3) define the composition

$$\mathcal{E}\colon W(K) \xrightarrow{\mathcal{E}(\cdot, X)} 1 + XW(K)[[X]] \xrightarrow{\sim} W_b(W(K)) \to W(W(K)),$$

which is called the *Artin–Hasse exponential*.

Define

$$\varphi_n: W(W(K)) \to W(K), \quad (\alpha_0, \alpha_1, \dots) \mapsto \alpha_0^{p^n} + p\alpha_1^{p^{n-1}} + \dots + p^n \alpha_n \in W(K).$$

Show that $\varphi_n \circ \mathcal{E} = \mathbf{F}^n$ for $n \geqslant 1$ (the Artin–Hasse exponential $\mathcal{E}$ can be generalized to arbitrary rings and ramified Witt vectors, see [Haz4]).

CHAPTER 2

Extensions of Discrete Valuation Fields

This chapter studies discrete valuation fields in relation to each other. The first section introduces the class of Henselian fields which are quite similar to complete fields; the key property of the former is given by the Hensel Lemma. The long section 2 deals with the problem of extensions of valuations from a field to its algebraic extension. Section 3 describes first properties of unramified and totally ramified extensions. In the case of Galois extensions ramification subgroups are introduced in section 4. Structural results on complete discrete valuation fields are proved in section 5.

1. The Hensel Lemma and Henselian Fields

Complete fields are not countable (see Exercise 1 section 4 Ch. I) and therefore are relatively huge; algebraic extensions of complete fields are not necessarily complete with respect to any natural extension of the valuation. One of the most important features of complete fields is the Hensel Lemma (1.2). Fields satisfying this lemma are called Henselian. They can be relatively small; and, as we shall see later, an algebraic extension of a Henselian field is a Henselian field.

Let F be a valuation field with the ring of integers $\mathcal{O}$, the maximal ideal $\mathcal{M}$, and the residue field $\overline{F}$. For a polynomial $f(X) = a_nX^n + \dots + a_0 \in \mathcal{O}[X]$ we will denote the polynomial $\overline{a}_nX^n + \dots + \overline{a}_0$ by $\overline{f}(X) \in \overline{F}[X]$. We will write

$$f(X) \equiv g(X) \mod \mathcal{M}^m$$

if $f(X) - g(X) \in \mathcal{M}^m[X]$.

(1.1). We assume that the reader is familiar with the notion of resultant $R(f,g)$ of two polynomials f,g (see, e.g., [La1, Ch. V]). We shall use now the following of its properties: for an $f(X) \in \mathcal{O}[X]$ the map $R(f,\cdot)\colon \mathcal{O}[X] \to \mathcal{O}$ transforms the ideal $\mathcal{M}^m[X]$ to $\mathcal{M}^m$ and induces the map $R(f,\cdot)\colon \mathcal{O}/\mathcal{M}^m[X] \to \mathcal{O}/\mathcal{M}^m$; if resultant $R(f_1,f_2)$ of $f_1,f_2 \in \mathcal{O}[X]$ does not belong to the maximal ideal $\mathcal{M}$, then the ideal generated by $f_1(X), f_2(X)$ in $\mathcal{O}[X]$ coincides with $\mathcal{O}[X]$.

PROPOSITION. *Let F be a complete discrete valuation field with the ring of integers $\mathcal{O}$ and the maximal ideal $\mathcal{M}$. Let $g_0(X), h_0(X), f(X)$ be polynomials over $\mathcal{O}$ such that*

$\deg f(X) = \deg g_0(X) + \deg h_0(X)$ *and the leading coefficient of* $f(X)$ *coincides with that of* $g_0(X)h_0(X)$. *Let* $R(g_0, h_0) \notin \mathcal{M}^{s+1}$ *and* $f(X) \equiv g_0(X)h_0(X) \mod \mathcal{M}^{2s+1}$ *for an integer* $s \geqslant 0$.

Then there exist polynomials $g(X), h(X)$ *such that*

$$\begin{aligned} &f(X) = g(X)h(X), \\ &\deg g(X) = \deg g_0(X), \quad g(X) \equiv g_0(X) \mod \mathcal{M}^{s+1}, \\ &\deg h(X) = \deg h_0(X), \quad h(X) \equiv h_0(X) \mod \mathcal{M}^{s+1}. \end{aligned}$$

Proof. We first construct polynomials $g_i(X), h_i(X) \in \mathcal{O}[X]$ with the following properties: $\deg(g_i - g_0) < \deg g_0$, $\deg(h_i - h_0) < \deg h_0$

$$g_i(X) \equiv g_{i-1}(X) \mod \mathcal{M}^{i+s}, \quad h_i(X) \equiv h_{i-1}(X) \mod \mathcal{M}^{i+s},$$
$$f(X) \equiv g_i(X)h_i(X) \mod \mathcal{M}^{i+2s+1}.$$

Proceeding by induction, we can assume that the polynomials $g_j(X), h_j(X)$, for $j \leqslant i-1$, have been constructed. For a prime element π put

$$g_i(X) = g_{i-1}(X) + \pi^{i+s}G_i(X), \quad h_i(X) = h_{i-1}(X) + \pi^{i+s}H_i(X)$$

with $G_i(X), H_i(X) \in \mathcal{O}[X]$, $\deg G_i(X) < \deg g_0(X)$, $\deg H_i(X) < \deg h_0(X)$. Then

$$\begin{aligned} &g_i(X)h_i(X) - g_{i-1}(X)h_{i-1}(X) \\ &\equiv \pi^{i+s}\big(g_{i-1}(X)H_i(X) + h_{i-1}(X)G_i(X)\big) \mod \mathcal{M}^{i+2s+1}. \end{aligned}$$

Since by the induction assumption $f(X) - g_{i-1}(X)h_{i-1}(X) = \pi^{i+2s}f_1(X)$ for a suitable $f_1(X) \in \mathcal{O}[X]$, we deduce that it suffices for $G_i(X)$, $H_i(X)$ to satisfy the congruence

$$\pi^s f_1(X) \equiv g_{i-1}(X)H_i(X) + h_{i-1}(X)G_i(X) \mod \mathcal{M}^{s+1}.$$

However, $R(g_{i-1}(X), h_{i-1}(X)) \equiv R(g_0(X), h_0(X)) \not\equiv 0 \mod \mathcal{M}^{s+1}$. Then the properties of the resultant imply the existence of the required polynomials $G_i(X)$, $H_i(X)$.

Now put $g(X) = \lim g_i(X), h(X) = \lim h_i(X)$ and get $f(X) = g(X)h(X)$. □

(1.2). COROLLARY 1 (HENSEL LEMMA). *Let* F *be as in the Proposition and* $\overline{F}$ *the residue field of* F. *Let* $f(X), g_0(X), h_0(X)$ *be monic polynomials with coefficients in* $\mathcal{O}$ *and* $\overline{f}(X) = \overline{g}_0(X)\overline{h}_0(X)$. *Suppose that* $\overline{g}_0(X), \overline{h}_0(X)$ *are relatively prime in* $\overline{F}[X]$. *Then there exist monic polynomials* $g(X), h(X)$ *with coefficients in* $\mathcal{O}$, *such that*

$$f(X) = g(X)h(X), \ \overline{g}(X) = \overline{g}_0(X), \ \overline{h}(X) = \overline{h}_0(X).$$

Proof. We have $R(f_0(X), g_0(X)) \notin \mathcal{M}$ and we can apply the previous Proposition for $s = 0$. The polynomials $g(X)$ and $h(X)$ may be assumed to be monic, as it follows from the proof of the Proposition. □

Valuation fields satisfying the assertion of Corollary 1 are said to be *Henselian*. Corollary 1 demonstrates that complete discrete valuation fields are Henselian.

COROLLARY 2. *Let F be a Henselian field and $f(X)$ a monic polynomial with coefficients in $\mathcal{O}$. Let $\overline{f}(X) \in \overline{F}[X]$ have a simple root β in $\overline{F}$. Then $f(X)$ has a simple root $\alpha \in \mathcal{O}$ such that $\overline{\alpha} = \beta$.*

Proof. Let $\gamma \in \mathcal{O}$ be such that $\overline{\gamma} = \beta$. Put $g_0(X) = X - \gamma$ in Corollary 1. □

(1.3). COROLLARY 3. *Let F be a complete discrete valuation field. Let $f(X)$ be a monic polynomial with coefficients in $\mathcal{O}$. Let $f(\alpha_0) \in \mathcal{M}^{2s+1}$, $f'(\alpha_0) \notin \mathcal{M}^{s+1}$ for some $\alpha_0 \in \mathcal{O}$ and integer $s \geqslant 0$. Then there exists $\alpha \in \mathcal{O}$ such that $\alpha - \alpha_0 \in \mathcal{M}^{s+1}$ and $f(\alpha) = 0$.*

Proof. Put $g_0(X) = X - \alpha_0$ and write $f(X) = f_1(X)(X - \alpha_0) + \delta$ with $\delta \in \mathcal{O}$. Then $\delta \in \mathcal{M}^{2s+1}$. Put $h_0(X) = f_1(X) \in \mathcal{O}[X]$. Hence $f(X) \equiv g_0(X)h_0(X) \mod \mathcal{M}^{2s+1}$ and $f'(\alpha_0) = h_0(\alpha_0) \notin \mathcal{M}^{s+1}$. This means that $R(g_0(X), h_0(X)) \notin \mathcal{M}^{s+1}$, and the Proposition implies the existence of polynomials $g(X), h(X) \in \mathcal{O}[X]$ such that $g(X) = X - \alpha, \alpha \equiv \alpha_0 \mod \mathcal{M}^{s+1}$, and $f(X) = g(X)h(X)$. □

A direct proof of Corollary 3 can be found in Exercises.

COROLLARY 4. *Let F be a complete discrete valuation field. For every positive integer m there is n such that $1 + \mathcal{M}^n \subset F^{*m}$.*

Proof. Put $f_a(X) = X^m - a$ with $a \in 1 + \mathcal{M}^n$. Let $m \in \mathcal{M}^s \setminus \mathcal{M}^{s+1}$. Then $f'_a(1) \in \mathcal{M}^s \setminus \mathcal{M}^{s+1}$. Therefore for every $a \in 1 + \mathcal{M}^{2s+1}$ due to Corollary 3 the polynomial $f_a(X)$ has a root $\alpha \equiv 1 \mod \mathcal{M}^{s+1}$. □

(1.4). The following assertion will be used in the next section.

LEMMA. *Let F be a complete discrete valuation field and let*

$$f(X) = X^n + \alpha_{n-1}X^{n-1} + \dots + \alpha_0$$

be an irreducible polynomial with coefficients in F. Then the condition $v(\alpha_0) \geqslant 0$ implies $v(\alpha_i) \geqslant 0$ for $0 \leqslant i \leqslant n - 1$.

Proof. Assume that $\alpha_0 \in \mathcal{O}$ and that j is the maximal integer such that $v(\alpha_j) = \min_{0 \leqslant i \leqslant n-1} v(\alpha_i)$. If $\alpha_j \notin \mathcal{O}$, then put

$$\begin{aligned} f_1(X) &= \alpha_j^{-1} f(X), \\ g_0(X) &= X^j + \alpha_j^{-1}\alpha_{j-1}X^{j-1} + \dots + \alpha_j^{-1}\alpha_0, \\ h_0(X) &= \alpha_j^{-1}X^{n-j} + 1 \end{aligned}$$

We have $\overline{f}_1(X) = \overline{g}_0(X)\overline{h}_0(X)$, and $\overline{g}_0(X), \overline{h}_0(X)$ are relatively prime. Therefore, by Proposition (1.1), $f_1(X)$ and $f(X)$ are not irreducible. □

REMARK. Later in (2.9) we show that all the assertions of this section hold for Henselian discrete valuation fields.

Exercises.

1. Let F be a complete discrete valuation field, and $f(X)$ a monic polynomial with coefficients in $\mathcal{O}$. Let $\alpha_0 \in \mathcal{O}$ be such that $f(\alpha_0) \in \mathcal{M}^{2s+1}$ and $f'(\alpha_0) \notin \mathcal{M}^{s+1}$. Show that the sequence $\{\alpha_m\}, \alpha_m = \alpha_{m-1} - \dfrac{f(\alpha_{m-1})}{f'(\alpha_{m-1})}$, is convergent and $\alpha = \lim \alpha_m$ is a root of $f(X)$.
2. Let F be a complete discrete valuation field and $f(X) = X^n + \alpha_{n-1}X^{n-1} + \dots + \alpha_0$ an irreducible polynomial over F.
   a) Show that $v(\alpha_0) > 0$ implies $v(\alpha_i) > 0$ for $1 \leqslant i \leqslant n-1$.
   b) Show that if $v(\alpha_0) \leqslant 0$, then $v(\alpha_0) = \min\limits_{0 \leqslant i \leqslant n-1} v(\alpha_i)$.
3. a) Let F be a field with a valuation v and the maximal ideal $\mathcal{M}_v$. Assume that F is complete with respect to the $\mathcal{M}_v$-adic topology (see Exercise 4 in section 4 Ch. 1). Show that F is Henselian, by modifying the proof of Proposition (1.1) for $s = 0$ and using appropriate $\pi_k \in \mathcal{M}_v^k$ instead of π^k.
   b) Show that the fields of Examples 3, 4 in section 4 Ch. I are Henselian.
4. Let F be a Henselian field and $f(X) \in \mathcal{O}[X]$ an irreducible monic polynomial. Show that $\overline{f}(X)$ is a power of some irreducible polynomial in $\overline{F}[X]$.
5. Let F be a Henselian field with the residue field $\overline{F}$.
   a) Show that the group μ of all the roots in F (of order relatively prime with $\mathrm{char}(\overline{F})$, if $\mathrm{char}(\overline{F}) \neq 0$), is isomorphic with the group of all roots of unity in $\overline{F}$.
   b) Let n be any integer (relatively prime to $\mathrm{char}(\overline{F})$, if $\mathrm{char}(\overline{F}) \neq 0$). Show that raising to the n th power is an automorphism of $1 + \mathcal{M}$.
   c) Let F be a Henselian discrete valuation field, and σ an isomorphism of F onto a subfield of F. Show that $\sigma(\mathcal{M}) \subset \mathcal{O}, \sigma(U) \subset U$.
   d) Let $F = \mathbb{Q}_p$. Show that every isomorphism of F onto a subfield of F is continuous.
   e) Show that if $p \neq q$, then $\mathbb{Q}_p$ is not isomorphic to $\mathbb{Q}_q$.

2. Extensions of Valuation Fields

In this rather lengthy section we study extensions of discrete valuations. In Theorem (2.5) we show that if a field F is complete with respect to a discrete valuation, then there is exactly one extension of the valuation to a finite extension of F. The non-complete case will be described in Theorem (2.6). In Theorem (2.8) we give three new equivalent definitions of a Henselian discrete valuation field.

(2.1). Let F be a field and L an extension of F with a valuation $w: L \to \Gamma'$. Then w induces the valuation $w_0 = w|_F: F \to \Gamma'$ on F. In this context L/F is said to be an *extension of valuation fields*. The group $w_0(F^*)$ is a totally ordered subgroup of $w(L^*)$ and the index of $w_0(F^*)$ in $w(L^*)$ is called the *ramification index* $e(L/F, w)$. The ring of integers $\mathcal{O}_{w_0}$ is a subring of the ring of integers $\mathcal{O}_w$ and the maximal ideal $\mathcal{M}_{w_0}$ coincides with $\mathcal{M}_w \cap \mathcal{O}_{w_0}$. Hence, the residue field $\overline{F}_{w_0}$ can be considered as a subfield of the residue field $\overline{L}_w$. Therefore, if α is an element of $\mathcal{O}_{w_0}$, then its residue in the field $\overline{F}_{w_0}$ can be identified with the image of α as an element of $\mathcal{O}_w$ in the field $\overline{L}_w$. We shall denote this image of α by $\overline{\alpha}$. The degree of the extension $\overline{L}_w/\overline{F}_{w_0}$ is called the *residue degree* $f(L/F, w)$. An immediate consequence is the following Lemma.

LEMMA. *Let L be an extension of F and let w be a valuation on L. Let $L \supset M \supset F$ and let w_0 be the induced valuation on M. Then*

$$e(L/F, w) = e(L/M, w)e(M/F, w_0),$$
$$f(L/F, w) = f(L/M, w)f(M/F, w_0).$$

(2.2). Assume that L/F is a finite extension and w_0 is a discrete valuation. Let the elements $\alpha_1, \dots, \alpha_e \in L^*$ for natural $e \leqslant e(L/F, w)$ be such that $w(\alpha_1) + w(F^*), \dots, w(\alpha_e) + w(F^*)$ are distinct in $w(L^*)/w(F^*)$. If $\sum_{i=1}^{e} c_i\alpha_i = 0$ holds with $c_i \in F$, then, as $w(c_i\alpha_i)$ are all distinct, by (2.1) Ch. I we get

$$w\Big(\sum_{i=1}^{e} c_i\alpha_i\Big) = \min_{1 \leqslant i \leqslant e} w(c_i\alpha_i) \quad \text{and} \quad c_i = 0 \quad \text{for } 1 \leqslant i \leqslant e.$$

This shows that $\alpha_1, \dots, \alpha_e$ are linearly independent over F and hence $e(L/F, w)$ is finite. Let π be a prime element with respect to w_0. Then we deduce that there are only a finite number of positive elements in $w(L^*)$ which are $\leqslant w(\pi)$. Consider the smallest positive element in $w(L^*)$. It generates the group $w(L^*)$, and we conclude that w is a discrete valuation. Thus, we have proved the following result.

LEMMA. *Let L/F be a finite extension and w_0 discrete for a valuation w on L. Then w is discrete.*

(2.3). Hereafter we shall consider discrete valuations. Let F and L be fields with discrete valuations v and w respectively and $F \subset L$. The valuation w is said to be an *extension of the valuation* v, if the topology defined by w_0 is equivalent to the topology defined by v. We shall write $w|v$ and use the notations $e(w|v), f(w|v)$ instead of $e(L/F, w), f(L/F, w)$. If $\alpha \in F$ then $w(\alpha) = e(w|v)v(\alpha)$.

LEMMA. *Let L be a finite extension of F of degree n; then*

$$e(w|v)f(w|v) \leqslant n.$$

Proof. Let $e = e(w|v)$ and let f be a positive integer such that $f \leqslant f(w|v)$. Let $\theta_1, \dots, \theta_f$ be elements of $\mathcal{O}_w$ such that their residues in $\overline{L}_w$ are linearly independent over $\overline{F}_v$. It suffices to show that $\{\theta_i \pi_w^j\}$ are linearly independent over F for $1 \leqslant i \leqslant f, 0 \leqslant j \leqslant e-1$. Assume that

$$\sum_{i,j} c_{ij}\theta_i\pi_w^j = 0$$

for $c_{ij} \in F$ and not all $c_{ij} = 0$.

Multiplying the coefficients c_{ij} by a suitable power of π_v, we may assume that $c_{ij} \in \mathcal{O}_v$ and not all $c_{ij} \in \mathcal{M}_v$. Note that if $\sum_i c_{ij}\theta_i \in \mathcal{M}_w$, then $\sum_i \overline{c}_{ij}\overline{\theta}_i = 0$ and $c_{ij} \in \mathcal{M}_v$. Therefore, there exists an index j such that $\sum_i c_{ij}\theta_i \notin \mathcal{M}_w$. Let j_0 be the minimal such index. Then $j_0 = w(\sum c_{ij}\theta_i\pi_w^j)$, which is impossible. We conclude that all $c_{ij} = 0$. Hence, $ef \leqslant n$ and $e(w|v)f(w|v) \leqslant n$. □

For instance, let $\widehat{F}$ be the completion of F with the discrete valuation $\widehat{v}$ (see section 4 Ch. 1). Then $e(\widehat{v}|v) = 1, f(\widehat{v}|v) = 1$. Note that if F is not complete, then $|\widehat{F} : F| \neq e(\widehat{v}|v)f(\widehat{v}|v)$. On the contrary, in the case of complete discrete valuation fields we have

(2.4). PROPOSITION. *Let L be an extension of F and let F, L be complete with respect to discrete valuations v, w. Let $w|v, f = f(w|v)$ and $e = e(w|v) < \infty$. Let $\pi_w \in L$ be a prime element with respect to w and $\theta_1, \dots, \theta_f$ elements of $\mathcal{O}_w$ such that their residues form a basis of $\overline{F}_w$ over $\overline{F}_v$. Then $\{\theta_i\pi_w^j\}$ is a basis of the F-space L and of the $\mathcal{O}_v$-module $\mathcal{O}_w$, with $1 \leqslant i \leqslant f, 0 \leqslant j \leqslant e-1$. If $f < \infty$, then L/F is a finite extension of degree $n = ef$.*

Proof. Let R be a set of representatives for F (see (5.1) Chapter I). Then the set

$$R' = \Big\{\sum_{i=1}^{f} a_i\theta_i : a_i \in R \text{ and almost all } a_i = 0\Big\}$$

is the set of representatives for L. For a prime element π_v with respect to v put $\pi_m = \pi_v^k\pi_w^j$, where $m = ek + j, 0 \leqslant j < e$. Using Proposition (5.2) Ch. I we obtain that an element $\alpha \in L$ can be expressed as a convergent series

$$\alpha = \sum_m \eta_m \pi_m \qquad \text{with} \qquad \eta_m \in R'.$$

Writing

$$\eta_m = \sum_{i=1}^{f} \eta_{m,i}\theta_i \qquad \text{with} \qquad \eta_{m,i} \in R,$$

we get

$$\alpha = \sum_{i,j} \Big(\sum_k \eta_{ek+j,i}\pi_v^k\Big)\theta_i \pi_w^j.$$

Thus, α can be expressed as $\sum \rho_{i,j}\theta_i\pi_w^j$ with

$$\rho_{i,j} = \sum_k \eta_{ek+j,i}\pi_v^k \in F, \quad 1 \leqslant i \leqslant f, 0 \leqslant j \leqslant e-1.$$

By the proof of the previous Lemma this expression for α is unique. We conclude that $\{\theta_i\pi_w^j\}$ form a basis of L over F and of $\mathcal{O}_w$ over $\mathcal{O}_v$. □

(2.5). Further we shall assume that $v(F^*) = \mathbb{Z}$ for a discrete valuation v. Then $e(w|v) = |\mathbb{Z} : w(F^*)|$ for an extension w of v.

Theorem. *Let F be a complete field with respect to a discrete valuation v and L a finite extension of F. Then there is precisely one extension w on L of the valuation v and $w = \dfrac{1}{f} v \circ N_{L/F}$ with $f = f(w|v)$. The field L is complete with respect to w.*

Proof. Let $w' = v \circ N_{L/F}$. First we verify that w' is a valuation on L. It is clear that $w'(\alpha) = +\infty$ if and only if $\alpha = 0$ and $w'(\alpha\beta) = w'(\alpha) + w'(\beta)$. Assume that $w'(\alpha) \geqslant w'(\beta)$ for $\alpha, \beta \in L^*$, then

$$w'(\alpha + \beta) = w'(\beta) + w'\Big(1 + \frac{\alpha}{\beta}\Big)$$

and it suffices to show that if $w'(\gamma) \geqslant 0$, then $w'(1+\gamma) \geqslant 0$. Let

$$f(X) = X^m + a_{m-1}X^{m-1} + \cdots + a_0$$

be the monic irreducible polynomial of γ over F. Then we get $(-1)^m a_0 = N_{F(\gamma)/F}(\gamma)$ and if $s = |L : F(\gamma)|$, then $((-1)^m a_0)^s = N_{L/F}(\gamma)$. We deduce that $v(a_0) \geqslant 0$, and making use of (1.4), we get $v(a_i) \geqslant 0$ for $0 \leqslant i \leqslant m-1$. However,

$$(-1)^m N_{F(\gamma)/F}(1+\gamma) = f(-1) = (-1)^m + a_{m-1}(-1)^{m-1} + \cdots + a_0,$$

hence

$$v\left(N_{F(\gamma)/F}(1+\gamma)\right) \geqslant 0 \quad \text{and} \quad v\left(N_{L/F}(1+\gamma)\right) \geqslant 0,$$

i.e., $w'(1+\gamma) \geqslant 0$. Thus, we have shown that w' is a valuation on L.

Let $n = |L : F|$; then $w'(\alpha) = nv(\alpha)$ for $\alpha \in F^*$. Hence, the valuation $(1/n)w'$ is an extension of v to L (note that $(1/n)w'(L^*) \neq \mathbb{Z}$ in general). Let $e = e(L/F, (1/n)w')$. By Lemma (2.3) e is finite. Put $w = (e/n)w' \colon L^* \to \mathbb{Q}$, hence $w(L^*) = w(\pi_w)\mathbb{Z} = \mathbb{Z}$ with a prime element π_w with respect to w. Therefore, $w = (e/n)v \circ N_{L/F}$ is at once a discrete valuation on L and an extension of v.

Furthermore, let $\widehat{L}$ be the completion of L with respect to w and let $\widehat{w}$ be the discrete valuation on $\widehat{L}$. Then $e(\widehat{w}|v) = e(\widehat{w}|w) = e$, $f(\widehat{w}|v) = f(\widehat{w}|w) = f$ by Lemma (2.1). Proposition (2.4) implies $|\widehat{L} : F| = ef$. However, Lemma (2.3) shows that $ef \leqslant n = |L : F|$. It follows that $L = \widehat{L}$ is complete, and $w = (1/f)v \circ N_{L/F}$.

Let $\widetilde{w}$ be an extension of v to L with $\widetilde{w}(L^*) = \mathbb{Z}$. Then $\widetilde{w}$ is a discrete valuation by Lemma (2.2). Taking the same basis $\{\theta_i \pi_w^j\}$ of L over F as in Proposition (2.4), one can write an element $\alpha \in L$ as

$$\alpha = \sum_{i,j} \rho_{ij}(\alpha)\theta_i \pi_w^j \qquad \text{with} \qquad \rho_{ij}(\alpha) \in F.$$

Denote by M the field extension of F which is generated by elements θ_i. Let π_v be a prime element of F with respect to v. Let $\min \widetilde{w}(\theta_i) = c$. Then, according to Proposition 2.4 every element β of M is a convergent power series $\sum_{n \geqslant n_0} \pi_v^n (\sum_i \theta_{i,n}\theta_i)$ with $\theta_{i,n}$ in the set of multiplicative representatives in F. We deduce that $w(\beta) \geqslant n_0$ implies $\widetilde{w}(\beta) \geqslant c + n_0$. Therefore, for every integer m the sequence $(\pi_v \theta_i^m)^n$ tends to 0 with respect to $\widetilde{w}$ when n tends to $+\infty$. Then $\widetilde{w}(\theta_i) = 0$ and thus the restriction w_0 of w on M coincides with the restriction of $\widetilde{w}$ on M.

Now, according to (2.1) Ch. I,

$$w(\alpha) = \min_j \Big(w_0\big(\sum_i \rho_{ij}(\alpha)\theta_i\big) + j \Big),$$

we deduce that the condition $\alpha_k \to 0$ as $k \to +\infty$ with respect to w implies $\sum_i \rho_{ij}(\alpha_k)\theta_i \to 0$ as $k \to +\infty$ with respect to w_0 for $0 \leqslant j \leqslant e-1$. Since

$$\widetilde{w}(\alpha) \geqslant \min_j \Big(\widetilde{w}_0\big(\sum_i \rho_{ij}(\alpha)\theta_i\big) + j\widetilde{w}(\pi_w) \Big),$$

we obtain that in this case $\alpha_k \to 0$ as $k \to +\infty$ with respect to $\widetilde{w}$. Putting $\alpha_k = \pi_w^k$ we deduce that $\alpha \in \mathcal{M}_w$ implies $\alpha \in \mathcal{M}_{\widetilde{w}}$ and $\mathcal{M}_w^i \subset \mathcal{M}_{\widetilde{w}}^i$. If $w(\pi_{\widetilde{w}}) \leqslant 0$, then $\pi_w \pi_{\widetilde{w}}^k \in \mathcal{M}_w$ for any negative integer k. Hence, $\pi_w \pi_{\widetilde{w}}^k \in \mathcal{M}_{\widetilde{w}}$, a contradiction. Therefore, $\pi_{\widetilde{w}} \in \mathcal{M}_w$ and $\mathcal{M}_{\widetilde{w}}^i \subset \mathcal{M}_w^i$. This shows that the topologies induced by w and $\widetilde{w}$ coincide. Lemma (3.6) Ch. 1 shows that $w = \widetilde{w}$, and the proof is complete. □

(2.6). Now we treat extensions of discrete valuations in the general case.

THEOREM. *Let F be a field with a discrete valuation v. Let $\widehat{F}$ be the completion of F, and $\widehat{v}$ the discrete valuation of $\widehat{F}$. Suppose that $L = F(\alpha)$ is a finite extension of F and $f(X)$ the monic irreducible polynomial of α over F. Let $f(X) = \prod_{i=1}^{k} g_i(X)^{e_i}$ be the decomposition of the polynomial $f(X)$ into irreducible monic factors in $\widehat{F}[X]$. For a root α_i of the polynomial $g_i(X)$ $(\alpha_1 = \alpha)$ put $L_i = \widehat{F}(\alpha_i)$. Let $\widehat{w}_i$ be the discrete valuation on L_i, the unique extension of $\widehat{v}$.*

Then L is embedded as a dense subfield in the complete discrete valuation field L_i under $F \hookrightarrow \widehat{F}$, $\alpha \to \alpha_i$, and the restriction w_i of $\widehat{w}_i$ on L is a discrete valuation on L which extends v. The valuations w_i are distinct and every discrete valuation which is an extension of v to L coincides with some w_i for $1 \leqslant i \leqslant k$.

Proof. First let w be a discrete valuation on L which extends v. Let $\widehat{L}_w$ be the completion of L with respect to w. By Proposition (4.2) Ch. I there exists an embedding $\sigma\colon \widehat{F} \to \widehat{L}_w$ over F. As $\alpha \in \widehat{L}_w$, we get $\sigma(\widehat{F})(\alpha) \subset \widehat{L}_w$. Since $\sigma(\widehat{F})(\alpha)$ is a finite extension of $\sigma(\widehat{F})$, Theorem (2.5) shows that $\sigma(\widehat{F})(\alpha)$ is complete. Therefore, $\widehat{L}_w \subset \sigma(\widehat{F})(\alpha)$ and, moreover, $\widehat{L}_w = \sigma(\widehat{F})(\alpha)$. Let $g(X) \in \widehat{F}[X]$ be the monic irreducible polynomial of α over $\sigma\widehat{F}$. Then $\sigma^{-1}g(X)$ divides $f(X)$ and $\sigma^{-1}g(X) = g_i(X)$ for some $1 \leqslant i \leqslant k$, $w = w_i$.

Conversely, assume that $g(X) = g_i(X)$ and $\widehat{w}_i$ is the unique discrete valuation on $L_i = \widehat{F}(\alpha_i)$ which extends $\widehat{v}$. Since F is dense in $\widehat{F}$, we deduce that the image of L is dense in L_i and w_i extends v.

If $w_i = w_j$ for $i \neq j$ then there is an isomorphism between $\widehat{F}(\alpha_i)$ and $\widehat{F}(\alpha_j)$ over $\widehat{F}$ which sends α_i to α_j, but this is impossible. □

COROLLARY. *Let L/F be a purely inseparable finite extension. Then there is precisely one extension to L of the discrete valuation v of F.*

Proof. Assume $L = F(\alpha)$. Then $f(X)$ is decomposed as $(X - \alpha)^{p^m}$ in the fixed algebraic closure F^{alg} of F. Therefore, $k = 1$ and there is precisely one extension of v to L. If there were two distinct extensions w_1, w_2 of v to L in the general case of a purely inseparable extension L/F, we would find $\alpha \in L$ such that $w_1(\alpha) \neq w_2(\alpha)$, and hence the restriction of w_1 and w_2 on $F(\alpha)$ would be distinct. This leads to contradiction. □

(2.7). REMARKS.

1. More precisely, Theorem (2.6) should be formulated as follows.

The tensor product $L \otimes_F \widehat{F}$ may be treated as an L-module and $\widehat{F}$-algebra. Then the quotient of $L \otimes_F \widehat{F}$ by its radical decomposes into the direct sum of complete fields which correspond to the discrete valuations on L that are extensions of v.

Under the conditions of Theorem $L \otimes_F \widehat{F} = \widehat{F}[X]/f(X)$, and we have the surjective homomorphism

$$L \otimes_F \widehat{F} = \widehat{F}[X]/f(X) \longrightarrow \bigoplus_{i=1}^{k} \widehat{F}[X]/g_i(X) \xrightarrow{\sim} \bigoplus_{i=1}^{k} \widehat{F}(\alpha_i) = \bigoplus_{w_i|v} \widehat{L}_{w_i}$$

with the kernel $\left(\prod_{i=1}^k g_i(X)\right)\widehat{F}[X]/f(X)$, where $\widehat{L}_{w_i} = \widehat{F}(\alpha_i)$. Note that this kernel coincides with the radical of $L \otimes_F \widehat{F}$. Under the conditions of the previous Theorem, if L/F is separable, then all e_i are equal to 1.

2. Assume that L/F is as in the Theorem and, in addition, L/F is Galois. Then $\widehat{F}(\alpha_i)/\widehat{F}$ is Galois. Let $G = \mathrm{Gal}(L/F)$. Note that if w is a valuation on L, then $w \circ \sigma$ is a valuation on L for $\sigma \in G$. Put

$$H_i = \{\sigma \in G : w_1 \circ \sigma = w_i\} \quad \text{for } 1 \leqslant i \leqslant k.$$

Then it is easy to show that G is a disjoint union of the H_i and $H_i = H_1\sigma_i$ for $\sigma_i \in H_i$. Theorem (2.6) implies that H_i coincides with $\{\sigma \in G : \sigma g_i(X) = g_1(X)\}$, whence $\{\sigma \in G : \sigma g_i(X) = g_i(X)\} = \sigma_i^{-1} H_1 \sigma_i$. Then $\deg g_i(X) = \deg g_1(X)$, $e_i = 1$. The subgroup H_1 is said to be the *decomposition group* of w_1 over F. The fixed field $M = L^{H_1}$ is said to be the decomposition field of w_1 over F. Note that the field M is obtained from F by adjoining coefficients of the polynomial $g_1(X)$. We get $L = M(\alpha_1)$, and $g_1(X) \in M[X]$ is irreducible over $\widehat{F} = \widehat{M}$. Theorem (2.6) shows that w_1 is the unique extension to L of $w_1|_M$; there are k distinct discrete valuations on M which extend v.

EXAMPLE. Let $E = F(X)$. Recall that the discrete valuations on E which are trivial on F are in one-to-one correspondence with irreducible monic polynomials $p(X)$ over F: $p(X) \to v_{p(X)}$, $v \to p_v(X)$ and there is the valuation v_∞ with a prime element $\frac{1}{X}$ (see (1.2) Ch. I). If a_n is the leading coefficient of $f(X)$, then

$$f(X) = a_n \prod_{v \neq v_\infty} p_v(X)^{v(f(X))}.$$

Let F_1 be an extension of F. Then a discrete valuation on $E_1 = F_1(X)$, trivial on F_1, is an extension of some discrete valuation on $E = F(X)$, trivial on F. Let $p(X) = p_v(X)$ be an irreducible monic polynomial over F. Let $p(X)$ be decomposed into irreducible monic factors over $F_1 : p(X) = \prod_{i=1}^k p_i(X)^{e_i}$. Then one immediately deduces that the $w_i = w_{p_i(X)}$, $1 \leqslant i \leqslant k$, are all discrete valuations, trivial on F_1, which extend the valuation $v_{p(X)}$. We also have $e\left(w_{p_i(X)}|v_{p(X)}\right) = e_i$. There is precisely one extension w_∞ of v_∞. Thus, for every v

$$p_v(X) = \prod_{w_i|v} p_{w_i}(X)^{e(w_i|v)}$$

and we have the surjective homomorphism $F(\alpha)\otimes_F F_1 \to \bigoplus F_1(\alpha_i)$, where α is a root of $p(X)$ and α_i is a root of $p_i(X)$. Here the kernel of this homomorphism also coincides with the radical of $F(\alpha)\otimes_F F_1$.

(2.8). Finally we treat extensions of Henselian discrete valuation fields.

LEMMA (GAUSS). *Let F be a discrete valuation field, $\mathcal{O}$ its ring of integers. Then if a polynomial $f(X)\in\mathcal{O}[X]$ is not irreducible in $F[X]$, it is not irreducible in $\mathcal{O}[X]$.*

Proof. Assume that $f(X)=g(X)h(X)$ with $g(X),h(X)\in F[X]$. Let

$$g(X)=\sum_{i=0}^{n} b_iX^i,\quad h(X)=\sum_{i=0}^{m} c_iX^i,\quad f(X)=\sum_{i=0}^{n+m} a_iX^i.$$

Let

$$j_1=\min\Big\{i: v(b_i)=\min_{0\leqslant k\leqslant n} v(b_k)\Big\},\quad j_2=\min\Big\{i: v(c_i)=\min_{0\leqslant k\leqslant m} v(c_k)\Big\}.$$

Then $v\left(b_ic_{j_1+j_2-i}\right) > v\left(b_{j_1}c_{j_2}\right)$ for $i\neq j_1$; hence $v\left(a_{j_1+j_2}\right)=v\left(b_{j_1}\right)+v\left(c_{j_2}\right)$. If $c=v\left(b_{j_1}\right)<0$, then we obtain $v\left(c_{j_2}\right)\geqslant -v\left(b_{j_1}\right)$, and one can write $f(X)=\left(\pi^{-c}g(X)\right)\left(\pi^{c}h(X)\right)$, as desired. □

THEOREM. *Let v be a discrete valuation on F. The following conditions are equivalent:*

(1) *F is a Henselian field with respect to v.*
(2) *The discrete valuation v has a unique extension to every finite algebraic extension L of F.*
(3) *If L is a finite separable extension of F of degree n, then*

$$n=e(w|v)f(w|v),$$

where w is an extension of v on L.
(4) *F is separably closed in $\widehat{F}$.*

Proof.

(1)⇒(2). Using Corollary (2.6), we can assume that L/F is separable. Moreover, it suffices to verify (2) for the case of a Galois extension. Let $L=F(\alpha)$ be Galois, $f(X)$ be the irreducible polynomial of α over F. Let $f(X)=g_1(X)\dots g_k(X)$ be the decomposition of $f(X)$ over $\widehat{F}$ as in remark 2 of (2.7). Let H_1 and $M=L^{H_1}$ be as in remark 2. Put $w_i'=w_i|_M$ for $1\leqslant i\leqslant k$ and suppose that $k\geqslant 2$. Then, w_i' for $1\leqslant i\leqslant k$ induce distinct topologies on M. w_i' $w_2',\dots,w_l'$. We get $w_i'=w_1\circ\sigma_i|_M$ for $\sigma_1,\dots,\sigma_l\in G,\sigma_1=1$. Taking into account Proposition (3.7) Ch. I, one can find an element $\beta\in M$ such that

$$-c=w_1'(\beta)<0,\quad w_2'(\beta)>c,\ \dots,\quad w_k'(\beta)>c.$$

Let $\tau_1, \dots, \tau_r$ ($\tau_1 = 1$) be the maximal set of elements of $G = \mathrm{Gal}(L/F)$ for which the elements $\beta, \tau_2(\beta), \dots, \tau_r(\beta)$ are distinct. Then $\tau_2, \dots, \tau_r \notin H_1$, and $w_1(\beta) = -c$, $w_1(\tau_i(\beta)) > c$ for $2 \leqslant i \leqslant r$.

Let $h(X) = X^r + b_{r-1}X^{r-1} + \dots + b_0$ be the irreducible monic polynomial of β over F. Then

$$w_1(b_0) = \sum_{i=1}^{r} w_1(\tau_i(\beta)) > 0$$

and, similarly, $w_1(b_i) > 0$ for $i < r - 1$. We also obtain that

$$w_1(b_{r-1}) = \min_{1 \leqslant i \leqslant r} w_1(\tau_i(\beta)) = -c < 0.$$

Hence, $v(b_i) > 0$ for $0 \leqslant i < r - 1$ and $v(b_{r-1}) < 0$. Put $h_1(X) = b_{r-1}^{-r} h(b_{r-1}X)$. Then $h_1(X)$ is a monic polynomial with integer coefficients. Since $\overline{h}_1(X) = (X + 1)X^{r-1}$, by the Hensel Lemma (1.2), we obtain that $h_1(X)$ is not irreducible, implying the same for $h(X)$, and we arrive at a contradiction. Thus, $k = 1$, and the discrete valuation v is uniquely extended on L.

(2) $\Rightarrow$ (3). Let $L = F(\alpha)$ be a finite separable extension of F and let L/F be of degree n. Since v can be uniquely extended to L, we deduce from Theorem (2.6) that $f(X) = g_1(X)$ is the decomposition of the irreducible monic polynomial $f(X)$ of α over F in $\widehat{F}[X]$. Therefore, the extension $\widehat{F}(\alpha)/\widehat{F}$ is of degree n. We have also $e(w|v) = e(\widehat{w}|\widehat{v})$, $f(w|v) = f(\widehat{w}|\widehat{v})$, because $e(\widehat{w}|w) = 1$, $f(\widehat{w}|w) = 1$, $e(\widehat{v}|v) = 1$, $f(\widehat{v}|v) = 1$; see (2.3). Now Proposition (2.4) shows that $n = e(\widehat{w}|\widehat{v}) f(\widehat{w}|\widehat{v})$ and hence $n = e(w|v) f(w|v)$.

(3) $\Rightarrow$ (4). Let $\alpha \in \widehat{F}$ be separable over F. Put $L = F(\alpha)$ and $n = |L : F|$. Let w be the discrete valuation on L which induces the same topology on L as $\widehat{v}|_L$. Then $e(w|v) = f(w|v) = 1$, and hence $n = 1, \alpha \in F$.

(4) $\Rightarrow$ (1). Let $f(X), g_0(X), h_0(X)$ be monic polynomials with coefficients in $\mathcal{O}$. Let $\overline{f}(X) = \overline{g}_0(X)\overline{h}_0(X)$ and $\overline{g}_0(X), \overline{h}_0(X)$ be relatively prime in $\overline{F}_v[X]$; $\widehat{F}$ is Henselian according to (1.1). Then there exist monic polynomials $g(X)$, $h(X)$ over the ring of integers $\widehat{\mathcal{O}}$ in $\widehat{F}$, such that $f(X) = g(X)h(X)$ and $\overline{g}(X) = \overline{g}_0(X), \overline{h}(X) = \overline{h}_0(X)$. The polynomials $g_0(X), h_0(X)$ are relatively prime in $\mathcal{O}[X]$ because their residues possess this property. Consequently, they are relatively prime in $F[X]$ by the previous Lemma. The roots of the polynomial $f(X)$ are algebraic over F, hence the roots of the polynomials $g(X), h(X)$ are algebraic over F and the coefficients of $g(X), h(X)$ are algebraic over F. Since F is separably closed in $\widehat{F}$, we obtain that $g(X)^{p^m}, h(X)^{p^m} \in F[X]$ for some $m \geqslant 0$. Then $f(X)^{p^m}$ is the product of two relatively prime polynomials in $F[X]$. We conclude that $g(X)^{p^m} = g_1(X)^{p^m}$ and $h(X)^{p^m} = h_1(X)^{p^m}$ for some polynomials $g_1(X), h_1(X) \in F[X]$ and, finally, the

polynomial $g(X)$ coincides with $g_1(X) \in \mathcal{O}[X]$, the polynomial $h(X)$ coincides with $h_1(X) \in \mathcal{O}[X]$. □

REMARK. The equality $e(w|v)f(w|v) = n$ does not hold in general for algebraic extensions of Henselian fields; see Exercise 3.

(2.9). COROLLARY 1. *Let F be a Henselian discrete valuation field and L an algebraic extension of F. Then there is precisely one valuation $w: L^* \to \mathbb{Q}$ (not necessarily discrete), such that the restriction $w|_F$ coincides with the discrete valuation v on F. Moreover, L is Henselian with respect to w.*

Proof. Let M/F be a finite subextension of L/F, and let, in accordance with the previous Theorem, $w_M: M^* \to \mathbb{Q}$ be the unique valuation on M for which $w_M|_F = v$. For $\alpha \in L^*$ we put $w(\alpha) = w_M(\alpha)$ with $M = F(\alpha)$. It is a straightforward Exercise to verify that w is a valuation on L and that $w|_F = v$. If there were another valuation w' on L with the property $w'|_F = v$, we would find $\alpha \in L$ with $w(\alpha) \neq w'(\alpha)$, and hence $w|_{F(\alpha)}$ and $w'|_{F(\alpha)}$ would be two distinct valuations on $F(\alpha)$ with the property $w|_F = w'|_F = v$. Therefore, there exists exactly one valuation w on L for which $w|_F = v$. To show that L is Henselian we note that polynomials $f(X) \in \mathcal{O}_w[X], g_0(X) \in \mathcal{O}_w[X], h_0(X) \in \mathcal{O}_w[X]$ belong in fact to $\mathcal{O}_1[X]$, where $\mathcal{O}_1$ is the ring of integers for some finite subextension M/F in L/F. Clearly, the polynomials $\overline{g}_0(X), \overline{h}_0(X)$ are relatively prime in $\overline{M}_{w_M}[X]$, hence there exist polynomials $g(X), h(X) \in \mathcal{O}_1[X]$, such that $f(X) = g(X)h(X)$, $\overline{g}(X) = \overline{g}_0(X)$ and $\overline{h}(X) = \overline{h}_0(X)$. □

COROLLARY 2. *Let F be a Henselian discrete valuation field, and let L/F be a finite separable extension. Let v be the valuation on F and w the extension of v to L. Let $e, f, \pi_w, \theta_1, \dots, \theta_f$ be as in Proposition (2.4). Then $\theta_i\pi_w^j$ is a basis of the F-space L and of the $\mathcal{O}_v$-module $\mathcal{O}_w$, with $1 \leqslant i \leqslant f, 0 \leqslant j \leqslant e-1$. In particular, if $e = 1$, then*

$$\mathcal{O}_w = \mathcal{O}_v\left[\{\theta_i\}\right], \quad L = F\left(\{\theta_i\}\right),$$

and if $f = 1$, then

$$\mathcal{O}_w = \mathcal{O}_v\left[\pi_w\right], \quad L = F\left(\pi_w\right).$$

Proof. One can show, similarly to the proof of Lemma (2.3), that the elements $\theta_i\pi_w^j$ for $1 \leqslant i \leqslant f, 0 \leqslant j \leqslant e-1$ are linearly independent over F. As $n = ef$, these elements form a basis of $\mathcal{O}_w$ over $\mathcal{O}_v$ and of L over F. □

COROLLARY 3. *Let F be a Henselian discrete valuation field, and L/F a finite separable extension. Let w be the discrete valuation on L and $\sigma: L \to F^{\mathrm{alg}}$ an* *embedding over* F. *Then* $w \circ \sigma^{-1}$ *is the discrete valuation on* σL *and* $\mathcal{M}_{\sigma L} = \sigma\mathcal{M}_L, \mathcal{O}_{\sigma L} = \sigma\mathcal{O}_L$.

COROLLARY 4. *If* F *is a Henselian discrete valuation field, then Proposition* (1.1), *Corollary 3 and 4 of* (1.3), *and Lemma* (1.4) *hold for* F.

Proof. In terms of Proposition (1.1) we obtain that there exist polynomials $g, h \in \widehat{\mathcal{O}}[X]$ (where $\widehat{\mathcal{O}}$ is the ring of integers of $\widehat{F}$), such that $f = gh$, $g \equiv g_0 \mod \widehat{\mathcal{M}}^{s+1}$, $h \equiv h_0 \mod \widehat{\mathcal{M}}^{s+1}$, $\deg g = \deg g_0$, $\deg h = \deg h_0$ (where $\widehat{\mathcal{M}}$ is the maximal ideal of $\widehat{\mathcal{O}}$). Proceeding now analogously to the part (4)$\Rightarrow$(1) of the proof of Theorem (2.8), we conclude that g^{p^m} and h^{p^m} belong to $\mathcal{O}[X]$ for some $m \geqslant 0$. As $g_0(X), h_0(X)$ are relatively prime in $F[X]$ because $R(g_0(X), h_0(X)) \neq 0$, we obtain that $g(X) = g_0(X), h(X) = h_0(X)$ and Proposition (1.1) holds for F. Corollary 3 of (1.3) and Lemma (1.4) for F are formally deduced from the latter. □

REMARK. Corollary 1 does not hold if the word "Henselian" is replaced by "complete". For instance, if the maximal unramified extension of a complete discrete valuation field is of infinite degree over the field, then it is not complete (see Exercise 1 in the next section). However, it is always Henselian. The assertions in (2.8) show that many properties of complete discrete valuation fields are retained for Henselian valuation fields. For the valuation theory with more commutative algebra flavour see [Bou], [Rib], [E], [Ra].

The separable closure of F in $\widehat{F}$ is called the *Henselization* of F (this is a least Henselian field containing F). For example, the separable closure of $\mathbb{Q}$ in $\mathbb{Q}_p$ is a Henselian countable field with respect to the p-adic valuation.

Exercises.

1. a) In terms of Theorem (2.6) and remark 2 show that if $\widehat{F}$ is separable over F or L is separable over F then $L \otimes_F \widehat{F} \simeq \oplus_{w_i|v} \widehat{L}_{w_i}$ with $\widehat{L}_{w_i} = \widehat{F}(\alpha_i)$.
   b) Let L be separable over F. Show that
$$|L : F| = \sum_{i=1}^{k} e(w_i|v) f(w_i|v).$$
   c) Show that $|L : F| = p^m e(w|v) f(w|v)$ for some $m \geqslant 0$ if L is a finite extension of a Henselian discrete valuation field F.
2. (*E. Artin*) Let α_i be elements of $\mathbb{Q}_2^{\text{alg}}$ such that $\alpha_1^2 = 2$, $\alpha_{i+1}^2 = \alpha_i$ for $i \geqslant 1$. Put $F = \mathbb{Q}_2(\alpha_1, \alpha_2, \dots)$. Then the discrete 2-adic valuation is uniquely extended to F. Let $\widehat{F}$ be its completion. Show that $\widehat{F}(\sqrt{-1})/\widehat{F}$ is of degree 2 and if w is the valuation on $\widehat{F}(\sqrt{-1})$, then
$$w\left(\sqrt{-1} - 1 - 2(\alpha_1^{-1} + \dots + \alpha_m^{-1})\right) = \left(1 - 2^{-m-1}\right) w(2).$$

Then the index of ramification and the residue degree of $\widehat{F}(\sqrt{-1})/\widehat{F}$ are equal to 1.

3. a) Using Exercise 1 section 4 Ch. I show that there exists an element $\alpha = \sum_{i\geqslant 0} a_i X^i \in \mathbb{F}_p((X))$ which is not algebraic over $\mathbb{F}_p(X)$.

   b) Let $\beta = \alpha^p$ and let F be the separable algebraic closure of $\mathbb{F}_p(X)(\beta)$ in $\mathbb{F}_p((X))$. Show that F is dense in $\mathbb{F}_p((X))$ and Henselian. Let $L = F(\alpha)$. Show that L/F is of degree p, and that the index of ramification and the residue degree of L/F are equal to 1.

4. Let F be a field with a discrete valuation v. Show that the following conditions are equivalent:

   (1) F is a Henselian discrete valuation field.

   (2) If $f(X) = X^n + \alpha_{n-1}X^{n-1} + \dots + \alpha_0$ is an irreducible polynomial over F and $\alpha_0 \in \mathcal{O}$, then $\alpha_i \in \mathcal{O}$ for $0 \leqslant i \leqslant n-1$.

   (3) If $f(X) = X^n + \alpha_{n-1}X^{n-1} + \dots + \alpha_0$ is an irreducible polynomial over $F, n \geqslant 1, \alpha_{n-2}, \dots, \alpha_0 \in \mathcal{O}$, then $\alpha_{n-1} \in \mathcal{O}$.

   (4) If $f(X) = X^n + \alpha_{n-1}X^{n-1} + \dots + \alpha_0$ is an irreducible polynomial over $F, n \geqslant 1, \alpha_{n-2}, \dots, \alpha_0 \in \mathcal{M}, \alpha_{n-1} \in \mathcal{O}$, then $\alpha_{n-1} \in \mathcal{M}$.

   (5) If $f(X)$ is a monic polynomial with coefficients in $\mathcal{O}$ and $\overline{f}(X) \in \overline{F}[X]$ has a simple root $\theta \in \overline{F}$, then there exists $\alpha \in \mathcal{O}$ such that $f(\alpha) = 0$ and $\overline{\alpha} = \theta$.

5. Let v be a discrete valuation on F. Let $w_c = w_c(v)$ be the discrete valuation on $F(X)$ defined in Example 4 (2.3) Ch. I. Suppose that F is Henselian with respect to v. Show that for an irreducible separable polynomial $f(X) \in F[X]$ there exists an integer d, such that if $g(X) \in F[X]$, $\deg g(X) = \deg f(X)$ and $w_c(f(X) - g(X)) > d$, then $g(X)$ is irreducible. In this case for every root α of $f(X)$ there is a root β of $g(X)$ with $F(\alpha) = F(\beta)$.

6. ([Rim]) Let F be a Henselian field with respect to nontrivial valuations $v, v'\colon F \to \mathbb{Q}$. Assume the topologies induced by v and v' are not equivalent (see (4.4) Ch. I).

   a) Show that if v is discrete, then v' is not.

   b) By using an analogue of approximation Theorem (3.7) Ch. I show that if $f(X)$ is an irreducible separable polynomial in $F[X]$ of degree $n > 1$, then for positive integers c_1, c_2 there exists a polynomial $g(X) \in F[X]$ with the property

$$w_0(f(X) - g(X)) > c_1, \quad w_0'\left(X^{n-1}(X-1) - g(X)\right) > c_2,$$

   where $w_0 = w_0(v)$ and $w_0' = w_0(v')$ as in Exercise 5.

   c) Show that F is separably closed.

3. Unramified and Ramified Extensions

In this section we look at two types of finite extensions of a Henselian discrete valuation field F: unramified and totally ramified.

In view of Exercise 6 in the previous section the field F has the unique surjective discrete valuation $F^* \to \mathbb{Z}$ with respect to which it is Henselian; we shall denote it from now on by v_F.

Let L/F be an algebraic extension. If v_L is the unique discrete valuation on L which extends the valuation $v = v_F$ on F, then we shall write $e(L|F), f(L|F)$ instead of $e(v_L|v_F)$, $f(v_F|v_F)$. We shall write $\mathcal{O}$ or $\mathcal{O}_F, \mathcal{M}$ or $\mathcal{M}_F, U$ or U_F, π or $\pi_F, \overline{F}$ for the ring of integers $\mathcal{O}_v$, the maximal ideal $\mathcal{M}_v$, the group of units U_v, a prime element π_v with respect to v, and the residue field $\overline{F}_v$, respectively.

(3.1). LEMMA. *Let L/F be a finite extension. Let $\alpha \in \mathcal{O}_L$ and let $f(X)$ be the monic irreducible polynomial of α over F. Then $f(X) \in \mathcal{O}_F[X]$. Conversely, let $f(X)$ be a monic polynomial with coefficients in $\mathcal{O}_F$. If $\alpha \in L$ is a root of $f(X)$, then $\alpha \in \mathcal{O}_L$.*

Proof. It is well known that $\beta = \alpha^{p^m}$ is separable over F for some $m \geqslant 0$ (see [La1, sect. 4 Ch. VII]). Let M be a finite Galois extension of F with $\beta \in M$. Then, in fact, $\beta \in \mathcal{O}_M$ and the monic irreducible polynomial $g(X)$ of β over F can be written as

$$g(X) = \prod_{i=1}^{r}(X - \sigma_i\beta), \qquad \sigma_i \in \mathrm{Gal}(M/F),\ \sigma_1 = 1.$$

Since $\beta \in \mathcal{O}_M$ we get $\sigma_i\beta \in \mathcal{O}_M$ using Corollary 3 of (2.9). Hence we obtain $g(X) \in \mathcal{O}_F[X]$ and $f(X) = g\left(X^{p^m}\right) \in \mathcal{O}_F[X]$. If $\alpha \in L$ is a root of the polynomial $f(X) = X^n + a_{n-1}X^{n-1} + \dots + a_0 \in \mathcal{O}_F[X]$ and $\alpha \notin \mathcal{O}_L$, then $1 = -a_{n-1}\alpha^{-1} - \dots - a_0\alpha^{-n} \in \mathcal{M}_L$, contradiction. Thus, $\alpha \in \mathcal{O}_L$. □

A finite extension L of a Henselian discrete valuation field F is called *unramified* if $\overline{L}/\overline{F}$ is a separable extension of the same degree as L/F. A finite extension L/F is called *totally ramified* if $f(L|F) = 1$. A finite extension L/F is called *tamely ramified* if $\overline{L}/\overline{F}$ is a separable extension and $p \nmid e(L|F)$ when $p = \mathrm{char}(\overline{F}) > 0$.

We deduce by Lemma (2.3) that $e(L|F) = 1, f(L|F) = |L : F|$ if L/F is unramified.

(3.2). First we treat the case of unramified extensions.

PROPOSITION.

(1) *Let L/F be an unramified extension, and $\overline{L} = \overline{F}(\theta)$ for some $\theta \in \overline{L}$. Let $\alpha \in \mathcal{O}_L$ be such that $\overline{\alpha} = \theta$. Then $L = F(\alpha)$, and L is separable over F, $\mathcal{O}_L = \mathcal{O}_F[\alpha]$; θ is a simple root of the polynomial $\overline{f}(X)$ irreducible over $\overline{F}$, where $f(X)$ is the monic irreducible polynomial of α over F.*

(2) *Let $f(X)$ be a monic polynomial over $\mathcal{O}_F$, such that its residue is a monic separable polynomial over $\overline{F}$; and $\overline{f}(X) = g(X)$. Let α be a root of $f(X)$ in F^{alg}, and let $L = F(\alpha)$. Then the extension L/F is unramified and $\overline{L} = \overline{F}(\theta)$ for $\theta = \overline{\alpha}$.*

Proof. (1) By the preceding Lemma $f(X) \in \mathcal{O}_F[X]$. We have $f(\alpha) = 0$ and $\overline{f}(\overline{\alpha}) = 0$, $\deg f(X) = \deg \overline{f}(X)$. Furthermore,

$$|L : F| \geqslant |F(\alpha) : F| = \deg f(X) = \deg \overline{f}(X) \geqslant |\overline{F}(\theta) : \overline{F}| = |L : F|.$$

It follows that $L = F(\alpha)$ and θ is a simple root of the irreducible polynomial $\overline{f}(X)$. Therefore, $\overline{f}'(\theta) \neq 0$ and $f'(\alpha) \neq 0$, i.e., α is separable over F. It remains to use Corollary 2 of (2.9) to obtain $\mathcal{O}_L = \mathcal{O}_F[\alpha]$.

(2) Let $f(X) = \prod_{i=1}^n f_i(X)$ be the decomposition of $f(X)$ into irreducible monic factors in $F[X]$. Lemma (2.8) shows that $f_i(X) \in \mathcal{O}_F[X]$. Suppose that α is a root of $f_1(X)$. Then $g_1(X) = \overline{f}_1(X)$ is a monic separable polynomial over $\overline{F}$. The Henselian property of F implies that $g_1(X)$ is irreducible over $\overline{F}$. We get $\alpha \in \mathcal{O}_L$ by Lemma (3.1). Since $\theta = \overline{\alpha} \in \overline{L}$, we obtain $\overline{L} \supset \overline{F}(\theta)$ and

$$\deg f_1(X) = |L : F| \geqslant |\overline{L} : \overline{F}| \geqslant |\overline{F}(\theta) : \overline{F}| = \deg g_1(X) = \deg f_1(X).$$

Thus, $\overline{L} = \overline{F}(\theta)$, and L/F is unramified. □

COROLLARY.

(1) *If $L/F, M/L$ are unramified, then M/F is unramified.*

(2) *If L/F is unramified, M is an algebraic extension of F and M is the discrete valuation field with respect to the extension of the valuation of F, then ML/M is unramified.*

(3) *If $L_1/F, L_2/F$ are unramified, then L_1L_2/F is unramified.*

Proof. (1) follows from Lemma (2.1).

To verify (2) let $L = F(\alpha)$ with $\alpha \in \mathcal{O}_L$, $f(X) \in \mathcal{O}_F[X]$ as in the first part of the Proposition. Then $\alpha \notin \mathcal{M}_L$ because $\overline{L} = \overline{F}(\overline{\alpha})$. Observing that $ML = M(\alpha)$, we denote the irreducible monic polynomial of α over M by $f_1(X)$. By the Henselian property of M we obtain that $\overline{f}_1(X)$ is a power of an irreducible polynomial over $\overline{M}$. However, $\overline{f}_1(X)$ divides $\overline{f}(X)$, hence $\overline{f}_1(X)$ is irreducible separable over $\overline{M}$. Applying the second part of the Proposition, we conclude that ML/M is unramified.

(3) follows from (1) and (2). □

An algebraic extension L of a Henselian discrete valuation field F is called *unramified* if $L/F, \overline{L}/\overline{F}$ are separable extensions and $e(w|v) = 1$, where v is the discrete valuation on F, and w is the unique extension of v on L.

The third assertion of the Corollary shows that the compositum of all finite unramified extensions of F in a fixed algebraic closure F^{alg} of F is unramified. This extension is a Henselian discrete valuation field (it is not complete in the general case, see Exercise 1). It is called the *maximal unramified extension* F^{ur} of F. Its maximality implies $\sigma F^{\mathrm{ur}} = F^{\mathrm{ur}}$ for any automorphism of the separable closure F^{sep} over F. Thus, F^{ur}/F is Galois.

(3.3). PROPOSITION.

(1) *Let* L/F *be an unramified extension and let* $\overline{L}/\overline{F}$ *be a Galois extension. Then* L/F *is Galois.*

(2) *Let* L/F *be an unramified Galois extension. Then* $\overline{L}/\overline{F}$ *is Galois. For an automorphism* $\sigma \in \mathrm{Gal}(L/F)$ *let* $\overline{\sigma}$ *be the automorphism in* $\mathrm{Gal}(\overline{L}/\overline{F})$ *satisfying the relation* $\bar{\sigma}\bar{\alpha} = \overline{\sigma\alpha}$ *for every* $\alpha \in \mathcal{O}_L$. *Then the map* $\sigma \to \overline{\sigma}$ *induces an isomorphism of* $\mathrm{Gal}(L/F)$ *onto* $\mathrm{Gal}(\overline{L}/\overline{F})$.

Proof. (1) It suffices to verify the first assertion for a finite unramified extension L/F. Let $\overline{L} = \overline{F}(\theta)$ and let $g(X)$ be the irreducible monic polynomial of θ over $\overline{F}$. Then

$$g(X) = \prod_{i=1}^{n}(X - \theta_i),$$

with $\theta_i \in \overline{L}, \theta_1 = \theta$. Let $f(X)$ be a monic polynomial over $\mathcal{O}_F$ of the same degree as $g(X)$ and $\overline{f}(X) = g(X)$. The Henselian property $\big($Corollary 2 in (1.2)$\big)$ implies

$$f(X) = \prod_{i=1}^{n}(X - \alpha_i),$$

with $\alpha_i \in \mathcal{O}_L, \overline{\alpha}_i = \theta_i$. Proposition (3.2) shows that $L = F(\alpha_1)$, and we deduce that L/F is Galois.

(2) Note that the automorphism $\overline{\sigma}$ is well defined. Indeed, if $\beta \in \mathcal{O}_L$ with $\overline{\beta} = \overline{\alpha}$, then $\sigma(\alpha - \beta) \in \mathcal{M}_L$ by Corollary 3 in (2.9) and $\overline{\sigma\alpha} = \overline{\sigma\beta}$. It suffices to verify the second assertion for a finite unramified Galois extension L/F. Let $\alpha, \theta, f(X)$ be as in the first part of Proposition (3.2). Since all roots of $f(X)$ belong to L, we obtain that all roots of $\overline{f}(X)$ belong to $\overline{L}$ and $\overline{L}/\overline{F}$ is Galois. The homomorphism $\mathrm{Gal}(L/F) \to \mathrm{Gal}(\overline{L}/\overline{F})$ defined by $\sigma \to \overline{\sigma}$ is surjective because the condition $\overline{\sigma}\theta = \theta_i$ implies $\sigma\alpha = \alpha_i$ for the root α_i of $f(X)$ with $\overline{\alpha}_i = \theta_i$. Since $\mathrm{Gal}(L/F)$, $\mathrm{Gal}(\overline{L}/\overline{F})$ are of the same order, we conclude that $\mathrm{Gal}(L/F)$ is isomorphic to $\mathrm{Gal}(\overline{L}/\overline{F})$. □

COROLLARY. *The residue field of* F^{ur} *coincides with the separable closure* $\overline{F}^{\mathrm{sep}}$ *of* $\overline{F}$ *and* $\mathrm{Gal}(F^{\mathrm{ur}}/F) \simeq \mathrm{Gal}(\overline{F}^{\mathrm{sep}}/\overline{F})$.

Proof. Let $\theta \in \overline{F}^{\mathrm{sep}}$, let $g(X)$ be the monic irreducible polynomial of θ over $\overline{F}$, and $f(X)$ as in the second part of Proposition (3.2). Let $\{\alpha_i\}$ be all the roots of $f(X)$ and $L = F(\{\alpha_i\})$. Then $L \subset F^{\mathrm{ur}}$ and $\theta = \overline{\alpha}_i \in \overline{F^{\mathrm{ur}}}$ for a suitable i. Hence, $\overline{F^{\mathrm{ur}}} = \overline{F}^{\mathrm{sep}}$. □

(3.4). Let L be an algebraic extension of F, and let L be a discrete valuation field. We will assume that $F^{\text{alg}} = L^{\text{alg}}$ in this case.

PROPOSITION. *Let L be an algebraic extension of F and let L be a discrete valuation field. Then $L^{\text{ur}} = LF^{\text{ur}}$, and $L_0 = L \cap F^{\text{ur}}$ is the maximal unramified subextension of F which is contained in L. Moreover, $\overline{L}/\overline{L_0}$ is a purely inseparable extension.*

Proof. The second part of Corollary (3.2) implies $L^{\text{ur}} \supset LF^{\text{ur}}$. Since the residue field of LF^{ur} contains the compositum of the fields $\overline{L}$ and $\overline{F}^{\text{sep}}$, which coincides with $\overline{L}^{\text{sep}}$ because $\overline{L}/\overline{F}$ is algebraic, we deduce $L^{\text{ur}} = LF^{\text{ur}}$. An unramified subextension of F in L is contained in L_0, and L_0/F is unramified. Let $\theta \in \overline{L}$ be separable over $\overline{F}$, and let $g(X)$ be the monic irreducible polynomial of θ over $\overline{F}$. Let $f(X)$ be a monic polynomial with coefficients in $\mathcal{O}_F$ of the same degree as $g(X)$, and $\overline{f}(X) = g(X)$. Then there exists a root $\alpha \in \mathcal{O}_L$ of the polynomial $f(X)$ with $\overline{\alpha} = \theta$ because of the Henselian property. Proposition (3.2) shows that $F(\alpha)/F$ is unramified, and hence $\theta \in \overline{L_0}$. □

COROLLARY. *Let L be a finite separable (resp. finite) extension of a Henselian (resp. complete) discrete valuation field F, and let $\overline{L}/\overline{F}$ be separable. Then L is a totally ramified extension of L_0, L^{ur} is a totally ramified extension of F^{ur}, and $|L : L_0| = |L^{\text{ur}} : F^{\text{ur}}|$.*

Proof. Theorem (2.8) and Proposition (2.4) show that $f(L|L_0) = 1$, and $e(L|L_0) = |L : L_0|$. Lemma (2.1) implies $e(L^{\text{ur}}|F^{\text{ur}}) = e(L^{\text{ur}}|F) = e(L|L_0)$. Since $|L : L_0| \geqslant |L^{\text{ur}} : F^{\text{ur}}|$, we obtain that $|L : L_0| = |L^{\text{ur}} : F^{\text{ur}}|$, $e(L^{\text{ur}}|F^{\text{ur}}) = |L^{\text{ur}} : F^{\text{ur}}|$, and $f(L^{\text{ur}}|F^{\text{ur}}) = 1$. □

(3.5). We treat the case of tamely ramified extensions.

PROPOSITION.

(1) *Let L be a finite separable (resp. finite) tamely ramified extension of a Henselian (resp. complete) discrete valuation field and let L_0/F be the maximal unramified subextension in L/F. Then $L = L_0(\pi)$ and $\mathcal{O}_L = \mathcal{O}_{L_0}[\pi]$ with a prime element π in L satisfying the equation $X^e - \pi_0 = 0$ for some prime element π_0 in L_0, where $e = e(L|F)$.*

(2) *Let L_0/F be a finite unramified extension, $L = L_0(\alpha)$ with $\alpha^e = \beta \in L_0$. Let $p \nmid e$ if $p = \operatorname{char}(\overline{F}) > 0$. Then L/F is separable tamely ramified.*

Proof. (1) The Corollary of Proposition (3.4) shows that L/L_0 is totally ramified. Let π_1 be a prime element in L_0, then $\pi_1 = \pi_L^e \varepsilon$ for a prime element π_L in L and $\varepsilon \in U_L$ according to (2.3). Since $\overline{L} = \overline{L_0}$, there exists $\eta \in \mathcal{O}_{L_0}$ such that $\overline{\eta} = \overline{\varepsilon}$. Hence $\pi_1\eta^{-1} = \pi_L^e\rho$ for the principal unit $\rho = \varepsilon\eta^{-1} \in \mathcal{O}_L$. For the polynomial $f(X) = X^e - \rho$ we have $f(1) \in \mathcal{M}_L$, $f'(1) = e$. Now Corollary 2 of (1.2) shows the existence of an element $\nu \in \mathcal{O}_L$ with $\nu^e = \rho$, $\overline{\nu} = 1$. Therefore, $\pi = \pi_1\eta^{-1}$, $\pi_0 = \pi_L\nu$ are the elements desired for the first part of the Proposition. It remains to use Corollary 2 of (2.9).

(2) Let $\beta = \pi_1^a\varepsilon$ for a prime element π_1 in L_0 and a unit $\varepsilon \in U_{L_0}$. The polynomial $g(X) = X^e - \overline{\varepsilon}$ is separable in $\overline{L}_0[X]$ and we can apply Proposition (3.2) to $f(X) = X^e - \varepsilon$ and a root $\eta \in F^{\text{sep}}$ of $f(X)$. We deduce that $L_0(\eta)/L_0$ is unramified and hence it suffices to verify that M/M_0 for $M = L(\eta), M_0 = L_0(\eta)$, is tamely ramified. We get $M = M_0(\alpha_1)$ with $\alpha_1 = \alpha\eta^{-1}$, $\alpha_1^e = \pi_1^a$. Put $d = \text{g.c.d.}(e, a)$. Then $M \subset M_0(\alpha_2, \zeta)$ with $\alpha_2^{e/d} = \pi_1^{a/d}$ and a primitive eth root ζ of unity. Since the extension $M_0(\zeta)/M_0$ is unramified (this can be verified by the same arguments as above), π_1 is a prime element in $M_0(\zeta)$. Let v be the discrete valuation on $M_0(\alpha_2, \zeta)$. Then $(a/d)v(\pi_1) \in (e/d)\mathbb{Z}$ and $v(\pi_1) \in (e/d)\mathbb{Z}$, because a/d and e/d are relatively prime. This shows that $e\big(M_0(\alpha_2, \zeta) \mid M_0(\zeta)\big) \geqslant e/d$. However, $|M_0(\zeta, \alpha_2) : M_0(\zeta)| \leqslant e/d$, and we conclude that $M_0(\zeta, \alpha_2)/M_0(\zeta)$ is tamely and totally ramified. Thus, $M_0(\zeta, \alpha_2)/M_0$ and M/M_0 are tamely ramified extensions. □

COROLLARY.

(1) *If $L/F, M/L$ are separable tamely ramified, then M/F is separable tamely ramified.*

(2) *If L/F is separable tamely ramified, M/F is an algebraic extension, and M is discrete, then ML/M is separable tamely ramified.*

(3) *If $L_1/F, L_2/F$ are separable tamely ramified, then L_1L_2/F is separable tamely ramified.*

If F is complete, then all the assertions hold without the assumption of separability.

Proof. It is carried out similarly to the proof of Corollary (3.2). To verify (2) one can find the maximal unramified subextension L_0/F in L/F. Then it remains to show that ML/ML_0 is tamely ramified. Put $L = L_0(\pi)$ with $\pi^e = \pi_0$. Then we get $ML = ML_0(\pi)$, and the second part of the Proposition yields the required assertion. □

(3.6). Finally we treat the case of totally ramified extensions. Let F be a Henselian discrete valuation field. A polynomial

$$f(X) = X^n + a_{n-1}X^{n-1} + \dots + a_0 \quad \text{over } \mathcal{O}$$

is called an *Eisenstein polynomial* if

$$a_0, \dots, a_{n-1} \in \mathcal{M}, \quad a_0 \notin \mathcal{M}^2.$$

PROPOSITION.

(1) *The Eisenstein polynomial $f(X)$ is irreducible over F. If α is a root of $f(X)$, then $F(\alpha)/F$ is a totally ramified extension of degree n, and α is a prime element in $F(\alpha)$, $\mathcal{O}_{F(\alpha)} = \mathcal{O}_F[\alpha]$.*

(2) *Let L/F be a separable totally ramified extension of degree n, and let π be a prime element in L. Then π is a root of an Eisenstein polynomial over F of degree n.*

Proof. (1) Let α be a root of $f(X)$, $L = F(\alpha)$, $e = e(L|F)$. Then

$$nv_L(\alpha) = v_L\left(\sum_{i=0}^{n-1} a_i\alpha^i\right) \geqslant \min_{0\leqslant i\leqslant n-1} (ev_F(a_i) + iv_L(\alpha)),$$

where v_F and v_L are the discrete valuations on F and L. It follows that $v_L(\alpha) > 0$. Since $ev_F(a_0) < ev_F(a_i) + iv_L(\alpha)$ for $i > 0$, one has $nv_L(\alpha) = ev_F(a_0) = e$. Lemma (2.3) implies $v_L(\alpha) = 1, n = e, f = 1$, and $\mathcal{O}_L = \mathcal{O}_F[\alpha]$ similarly to Corollary 2 of (2.9).
(2) Let π be a prime element in L. Then $L = F(\pi)$ by Corollary 2 of (2.9). Let

$$f(X) = X^n + a_{n-1}X^{n-1} + \dots + a_0$$

be the irreducible polynomial of π over F. Then

$$n = e, \quad nv_L(\pi) = \min_{0\leqslant i\leqslant n-1} \big(nv_F(a_i) + i\big),$$

hence $v_F(a_i) > 0$, and $n = nv_F(a_0)$, $v_F(a_0) = 1$. □

Exercises.

1. a) Let π be a prime element in F, and let $\overline{F}^{\text{sep}}$ be of infinite degree over $\overline{F}$ (e.g. $\overline{F} = \mathbb{F}_p, F = \mathbb{Q}_p$). Let F_i be finite unramified extensions of F, $F_i \subset F_j$, $F_i \neq F_j$ for $i < j$. Put

$$\alpha_n = \sum_{i=1}^{n} \theta_i \pi^i,$$

where $\theta_i \in \mathcal{O}_{F_{i+1}}, \notin \mathcal{O}_{F_i}$. Show that the sequence $\{\alpha_n\}_{n\geqslant 0}$ is a Cauchy sequence in F^{ur}, but $\lim \alpha_n \notin F^{\text{ur}}$.

   b) Show that F^{sep} is not complete, but the completion C of F^{sep} is separably closed (use Exercise 5b section 2).

2. a) Let L_1, L_2 be finite extensions of F and let $\overline{L}_1/\overline{F}$, $\overline{L}_2/\overline{F}$ be separable. Show that $\overline{L}_1 \cap \overline{L}_2 = \overline{L_1 \cap L_2}$.

   b) Does $\overline{L}_1 \cap \overline{L}_2 = \overline{L_1 \cap L_2}$ hold without the assumption of the residue fields?

   c) Prove or refute: if L_1, L_2 are finite extensions of F and $\overline{L}_1$, $\overline{L}_2$ are separable extensions of $\overline{F}$, then $\overline{L}_1\overline{L}_2 = \overline{L_1 L_2}$.

3. Show that in general the compositum of two totally (totally tamely) ramified extensions is not a totally (totally tamely) ramified extension.

4. Let L be a finite extension of F.

   a) Show that if $\mathcal{O}_L = \mathcal{O}_F[\alpha]$ with $\alpha \in \mathcal{O}_L$, then $\overline{L} = \overline{F}(\overline{\alpha})$.

   b) Find an example: $L = F(\alpha)$ with $\alpha \in \mathcal{O}_L$ and $\overline{L} \neq \overline{F}(\overline{\alpha})$.

5. Let L be a finite separable extension of F and let $\overline{L}/\overline{F}$ be separable. Let $\overline{L} = \overline{F}(\theta)$, let $g(X) \in \overline{F}[X]$ be the monic irreducible polynomial of θ over $\overline{F}$ and let $f(X) \in \mathcal{O}_F[X]$ be the monic polynomial of the same degree such that $\overline{f}(X) = g(X)$. Let $\alpha \in \mathcal{O}_L$ be such that $\overline{\alpha} = \theta$. Show that $\mathcal{O}_L = \mathcal{O}_F[\alpha]$ if $f(\alpha)$ is a prime element in L, and $\mathcal{O}_L = \mathcal{O}_F[\alpha+\pi]$ otherwise, where π is a prime element in L.
6. Let L be a separable totally ramified extension of F, and π a prime element in L. Show that $f(X) = N_{L/F}(X - \pi) = \prod(X - \sigma_i\pi)$ is the Eisenstein polynomial of π over F.

4. Galois Extensions

We study Galois extensions of Henselian discrete valuation fields and introduce a ramification filtration on the Galois group.

In this section F is a Henselian discrete valuation field.

(4.1). LEMMA. *Let L be a finite Galois extension of F. Then $v \circ \sigma = v$ for the discrete valuation v on L and $\sigma \in \mathrm{Gal}(L/F)$. If π is a prime element in L, then $\sigma\pi$ is a prime element and $\sigma\mathcal{O}_L = \mathcal{O}_L$, $\sigma\mathcal{M}_L = \mathcal{M}_L$.*

Proof. It follows from Corollary 3 of (2.9). □

PROPOSITION. *Let L be a finite Galois extension of F and let L_0/F be the maximal unramified subextension in L/F. Then L_0/F and $\overline{L_0}/\overline{F}$ are Galois, and the map $\sigma \to \overline{\sigma}$ defined in Proposition (3.3) induces the surjective homomorphism $\mathrm{Gal}(L/F) \to \mathrm{Gal}(L_0/F) \to \mathrm{Gal}(\overline{L_0}/\overline{F})$. If, in addition, $\overline{L}/\overline{F}$ is separable, then $\overline{L} = \overline{L_0}$ and $\overline{L}/\overline{F}$ is Galois, and L/L_0 is totally ramified.*

The extension L^{ur}/F is Galois and the group $\mathrm{Gal}(L^{\mathrm{ur}}/L_0)$ is isomorphic with $\mathrm{Gal}(L^{\mathrm{ur}}/L) \times \mathrm{Gal}(L^{\mathrm{ur}}/F^{\mathrm{ur}})$, and

$$\mathrm{Gal}(L^{\mathrm{ur}}/F^{\mathrm{ur}}) \simeq \mathrm{Gal}(L/L_0), \quad \mathrm{Gal}(L^{\mathrm{ur}}/L) \simeq \mathrm{Gal}(F^{\mathrm{ur}}/L_0).$$

Proof. Recall that in (3.4) we got an agreement $F^{\mathrm{alg}} = L^{\mathrm{alg}}$. Let $\sigma \in \mathrm{Gal}(L/F)$. Corollary 3 of (2.9) implies that σL_0 is unramified over F, hence $L_0 = \sigma L_0$ and L_0/F is Galois. The surjectivity of the homomorphism $\mathrm{Gal}(L/F) \to \mathrm{Gal}(\overline{L_0}/\overline{F})$ follows from Proposition (3.3). Since L/F and F^{ur}/F are Galois extensions, we obtain that LF^{ur}/F is a Galois extension. Then $L^{\mathrm{ur}} = LF^{\mathrm{ur}}$ by Proposition (3.4). The remaining assertions are easily deduced by Galois theory. □

Thus, a Galois extension L/F induces the Galois extension $L^{\mathrm{ur}}/F^{\mathrm{ur}}$. The converse statement can be formulated as follows.

(4.2). PROPOSITION. *Let M be a finite extension of F^{ur} of degree n. Then there exist a finite unramified extension L_0 of F and an extension L/L_0 of degree n such* *that* $L \cap F^{\mathrm{ur}} = L_0, LF^{\mathrm{ur}} = M$. *If* M/F^{ur} *is separable (Galois) then one can find* L_0 *and* L, *such that* L/L_0 *is separable (Galois).*

Proof. Assume that L_0 is a finite unramified extension of F, L is a finite extension of L_0 of the same degree as M/F^{ur} and $M = LF^{\mathrm{ur}}$. Then for a finite unramified extension N_0 of L_0 and $N = N_0L$ we get $|M : F^{\mathrm{ur}}| \leqslant |N : N_0| \leqslant |L : L_0|$, hence $|N : N_0| = |L : L_0|$ and $|N : L| = |N_0 : L_0|$. This shows $L \cap F^{\mathrm{ur}} = L_0$ and L_0, L are such as desired. Moreover, N_0, N are also valid for the Proposition. Therefore, it suffices to consider a case of $M = F^{\mathrm{ur}}(\alpha)$.

Let $f(X) \in F^{\mathrm{ur}}[X]$ be the irreducible monic polynomial of α over F^{ur}. In fact, its coefficients belong to some finite subextension L_0/F in F^{ur}/F. Put $L = L_0(\alpha)$. Then $f(X)$ is irreducible over L_0, L is the finite extension of L_0 of the same degree as M/F^{ur} and $M = LF^{\mathrm{ur}}$. This proves the first assertion of the Proposition. If α is separable over F^{ur}, then it is separable over L_0. If M/F^{ur} is a Galois extension, then $M = F^{\mathrm{ur}}(\alpha)$ for a suitable α and $\sigma_i(\alpha)$ for $\sigma_i \in \mathrm{Gal}(M/F^{\mathrm{ur}})$ can be expressed as polynomials in α with coefficients in F^{ur}. All these coefficients belong to some finite extension L_0' of L_0 in F^{ur}. The pair L_0', $L' = L_0'(\alpha)$ is the desired one. □

COROLLARY. *If* $\overline{M} = \overline{F^{\mathrm{ur}}}$, *then* L/L_0 *and* M/F^{ur} *are totally ramified.*

Proof. It follows from Proposition (3.4). □

(4.3). Let L be a finite Galois extension of F, $G = \mathrm{Gal}(L/F)$. For an integer $i \geqslant -1$ put

$$G_i = \left\{\sigma \in G : \sigma\alpha - \alpha \in \mathcal{M}_L^{i+1} \text{ for all } \alpha \in \mathcal{O}_L\right\}.$$

Then $G_{-1} = G$ by Lemma (4.1) and G_{i+1} is a subset of G_i.

Let v_L be the discrete valuation of L. For a real number x define

$$G_x = \{\sigma \in G : v_L(\sigma\alpha - \alpha) \geqslant x + 1 \text{ for all } \alpha \in \mathcal{O}_L\}.$$

Certainly each of G_x is equal to G_i with the least integer $i \geqslant x$.

LEMMA. G_i *are normal subgroups of* G.

Proof. Let $\sigma \in G_i, \alpha \in \mathcal{O}_L$. Then $\sigma\alpha - \alpha \in \mathcal{M}_L^{i+1}$. Hence $\alpha - \sigma^{-1}(\alpha) \in \sigma^{-1}(\mathcal{M}_L^{i+1}) = \mathcal{M}_L^{i+1}$ by Lemma (4.1), i.e., $\sigma^{-1} \in G_i$. Let $\sigma, \tau \in G_i$. Then

$$\sigma\tau(\alpha) - \alpha = \sigma(\tau(\alpha) - \alpha) + \sigma(\alpha) - \alpha \in \mathcal{M}_L^{i+1},$$

i.e., $\sigma\tau \in G_i$. Furthermore, let $\sigma \in G_i, \tau \in G$. Then $\tau(\alpha) \in \mathcal{O}_L$ for $\alpha \in \mathcal{O}_L$ and $\sigma(\tau\alpha) - \tau\alpha \in \mathcal{M}_L^{i+1}$, $\tau^{-1}\sigma\tau(\alpha) - \alpha \in \mathcal{M}_L^{i+1}$, $\tau^{-1}\sigma\tau \in G_i$. □

The groups G_x are called *(lower) ramification groups* of $G = \mathrm{Gal}(L/F)$.

Proposition. *Let L be a finite Galois extension of F, and let $\overline{L}$ be a separable extension of $\overline{F}$. Then $G_0 = \mathrm{Gal}(L/L_0)$ and the ith ramification groups of G_0 and G coincide for $i \geqslant 0$. Moreover,*

$$G_i = \left\{\sigma \in G_0 : \sigma\pi - \pi \in \mathcal{M}_L^{i+1}\right\}$$

for a prime element π in L, and $G_i = \{1\}$ for sufficiently large i.

Proof. Note that $\sigma \in G_0$ if and only if $\overline{\sigma} \in \mathrm{Gal}(\overline{L}/\overline{F})$ is trivial. Then G_0 coincides with the kernel of the homomorphism $\mathrm{Gal}(L/F) \to \mathrm{Gal}(\overline{L}/\overline{F})$. Proposition (4.1) and Proposition (3.3) imply that this kernel is equal to $\mathrm{Gal}(L/L_0)$. Since G_i is a subgroup of G_0 for $i \geqslant 0$, we get the assertion on the ith ramification group of G_0. Finally, using Corollary 2 of (2.9) we obtain $\mathcal{O}_L = \mathcal{O}_{L_0}[\pi]$. Let

$$\alpha = \sum_{i=0}^{n} a_i\pi^i$$

be an expansion of $\alpha \in \mathcal{O}_L$ with coefficients in $\mathcal{O}_{L_0}$. As $\sigma a_i = a_i$ for $\sigma \in G_0$ it follows that

$$\sigma\alpha - \alpha = \sum_{i=0}^{n} a_i\left(\sigma(\pi^i) - \pi^i\right).$$

Now we deduce the description of G_i, since $\sigma(\pi^i) - \pi^i \in G_i$. If $i \geqslant \max\{v_L(\sigma\pi - \pi) : \sigma \in G\}$, then $G_i = \{1\}$. □

The group G_0 is called the *inertia group* of G, and the field L_0 is called the *inertia subfield* of L/F.

(4.4). Proposition. *Let L be a finite Galois extension of F, $\overline{L}$ a separable extension of $\overline{F}$, and π a prime element in L. Introduce the maps*

$$\psi_0 : G_0 \longrightarrow \overline{L}^*, \quad \psi_i : G_i \longrightarrow \overline{L} \quad (i > 0)$$

by the formulas $\psi_i(\sigma) = \lambda_i(\sigma\pi/\pi)$, where the maps

$$\lambda_0 : U_L \longrightarrow \overline{L}^*, \quad \lambda_i : 1 + \mathcal{M}_L^i \longrightarrow \overline{L}$$

were defined in Proposition (5.4) Ch. I. Then ψ_i is a homomorphism with the kernel G_{i+1} for $i \geqslant 0$.

Proof. The proof follows from the congruence

$$\frac{\sigma\tau(\pi)}{\pi} = \sigma\left(\frac{\tau\pi}{\pi}\right)\cdot\frac{\sigma\pi}{\pi} \equiv \frac{\tau\pi}{\pi}\cdot\frac{\sigma\pi}{\pi} \mod U_{i+1}$$

for $\sigma, \tau \in G_i$ and Proposition (5.4) Ch. I. The kernel of ψ_i consists of those automorphisms $\sigma \in G_i$, for which $\sigma\pi/\pi \in 1 + \mathcal{M}_L^{i+1}$, i.e., $\sigma\pi - \pi \in \mathcal{M}_L^{i+2}$. □

COROLLARY 1. *Let L be a finite Galois extension of F, and $\overline{L}$ a separable extension of $\overline{F}$. If* $\operatorname{char}(\overline{F}) = 0$, *then* $G_1 = \{1\}$ *and* G_0 *is cyclic. If* $\operatorname{char}(\overline{F}) = p > 0$, *then the group* G_0/G_1 *is cyclic of order relatively prime to* p, G_i/G_{i+1} *are abelian* p*-groups, and* G_1 *is the maximal* p*-subgroup of* G_0.

Proof. The previous Proposition permits us to transform the assertions of this Corollary into the following: a finite subgroup in $\overline{L}^*$ is cyclic (of order relatively prime to $\operatorname{char}(\overline{L})$ when $\operatorname{char}(\overline{L}) \neq 0$); there are no nontrivial finite subgroups in the additive group of $\overline{L}$ if $\operatorname{char}(\overline{L}) = 0$; if $\operatorname{char}(\overline{L}) = p > 0$ then a finite subgroup in $\overline{L}$ is a p-group. □

COROLLARY 2. *Let L be a finite Galois extension of F and $\overline{L}$ a separable extension of $\overline{F}$. Then the group G_1 coincides with* $\operatorname{Gal}(L/L_1)$, *where L_1/F is the maximal tamely ramified subextension in L/F.*

Proof. The extension L_1/L_0 is totally ramified by Proposition (4.1) and is the maximal subextension in L/L_0 of degree relatively prime with $\operatorname{char}(\overline{F})$. Now Corollary 1 implies $G_1 = \operatorname{Gal}(L/L_1)$. □

COROLLARY 3. *Let L be a finite Galois extension of F and $\overline{L}$ a separable extension of $\overline{F}$. Then G_0 is a solvable group. If, in addition, $\overline{L}/\overline{F}$ is a solvable extension, then L/F is solvable.*

Proof. It follows from Corollary 1. □

REMARK. G_0 is solvable in the case of an inseparable extension $\overline{L}/\overline{F}$; see Exercise 2.

(4.5). DEFINITION. Let L/F be a finite Galois extension with separable residue field extension; let $G = \operatorname{Gal}(L/F)$. Integers i such that $G_i \neq G_{i+1}$ are called *ramification numbers of L/F* or *lower ramification jumps of L/F*.

One of the first properties of ramification numbers if supplied by the following

PROPOSITION. *Let L/F be a finite Galois extension with separable residue field extension. Let $\sigma \in G_i \setminus G_{i+1}$ and $\tau \in G_j \setminus G_{j+1}$ with $i, j \geqslant 1$. Then $\sigma\tau\sigma^{-1}\tau^{-1} \in G_{i+j+1}$ and $i \equiv j \mod p$.*

Proof. Let π_L be a prime element of L. Then

$$\frac{\sigma\pi_L}{\pi_L} = 1 + \alpha\pi_L^i, \qquad \frac{\tau\pi_L}{\pi_L} = 1 + \beta\pi_L^j \qquad \text{with } \alpha, \beta \in \mathcal{O}_L^*.$$

Therefore

$$\begin{aligned}\sigma\tau\pi_L &= \sigma\pi_L + (\sigma\beta)(\sigma\pi_L)^{j+1} \\ &\equiv \pi_L + \alpha\pi_L^{i+1} + \beta\pi_L^{j+1} + (j+1)\alpha\beta\pi_L^{i+j+1} \mod \mathcal{M}_L^{i+j+2}.\end{aligned}$$

Hence $(\sigma\tau - \tau\sigma)\pi_L \equiv (j-i)\alpha\beta\pi_L^{i+j+1} \mod \mathcal{M}_L^{i+j+2}$. Substituting instead of π_L the other prime element $\sigma^{-1}\tau^{-1}\pi_L$ of L we deduce that

$$\frac{\sigma\tau\sigma^{-1}\tau^{-1}\pi_L}{\pi_L} \equiv 1 + (j-i)\alpha\beta\pi_L^{i+j} \mod \mathcal{M}_L^{i+j+1}.$$

Now if j is the maximal ramification number of L/F, then $G_{j+1} = \{1\}$. Therefore the last formula in the previous paragraph shows that every positive ramification number i of L/F is congruent to j modulo p. Therefore every two positive ramification number of L/F are congruent to each other modulo p. Finally, from the same formula we deduce that $\sigma\tau\sigma^{-1}\tau^{-1} \in G_{i+j+1}$. □

REMARK. For more properties of ramification groups see sections 3–5 Chapter III and sections 3 and 6 Chapter IV.

Exercises.

1. Let F be a complete discrete valuation field and let L/F be a finite totally ramified Galois extension. For integers $i, j \geqslant 0$ define the (i,j)-th ramification group $G_{i,j}$ of $G = \mathrm{Gal}(L/F)$ as

$$G_{i,j} = \{\sigma \in G : v_L(\sigma\alpha - \alpha) \geqslant i + j \text{ for all } \alpha \in \mathcal{M}_L^j\}.$$

   Show that
   a) $G_{i,j}$ consists of those automorphisms which act trivially on $\mathcal{M}_L^j/\mathcal{M}_L^{i+j}$.
   b) $G_i = G_{i+1,0}$.
   c) $G_{i+1,1} \leqslant G_i \leqslant G_{i,1}$.
   d) $G_i = G_{i,1}$ if $\overline{L}/\overline{F}$ is separable.
   e) $G_i = G_{i+1,1}$ if $|L:F| = |\overline{L}:\overline{F}|$.
   For more properties of this double filtration see [dSm1].
2. (*I.B. Zhukov*) Let L/F be a finite Galois extension, $G = \mathrm{Gal}(L/F)$. Let π be a prime element in L. Put

$$G_{(0)} = G_0, \quad G_{(i)} = \{\sigma \in G_{(0)} : \sigma\pi - \pi \in \mathcal{M}_L^{i+1}\}.$$

   a) Show that $G_{(i)}/G_{(i+1)}$ is abelian and that $\cap G_{(i)}$ is a subgroup of $\mathrm{Gal}(L/L_0(\pi))$.
   b) Show that $L = L_0(\pi)(\pi')$ for a suitable prime element π' in L, and that the group $\mathrm{Gal}(L/L_0(\pi))$ is solvable.
   Thus, G_0 is solvable by a) and b).
3. Find an example of a finite separable extension L/F such that $\overline{L}/\overline{F}$ is separable, and for every nontrivial finite extension M/L with M/F being a Galois extension, the extension $\overline{M}/\overline{F}$ is not separable.
4. (◇) Prove that for every finite extension of complete discrete valuation fieds L/F there is a finite extension K' of a maximal complete discrete valuation subfield K of F with perfect residue field such that $e(K'L|K'F) = 1$ following the steps below (this statement is called elimination of wild ramification, see [Ep], [KZ]).
   a) Prove the assertion for an inseparable extension L/F of degree p.
   b) Reduce the problem to the case of Galois extensions.

c) Reduce the problem using solvability of G_0 (see Exercise 2) to the case $e(L|F) = |L : F| = l$ with prime l.
d) Prove the assertion in the latter case.

5. Structure Theorems for Complete Fields

In this section we shall describe classical structural results on complete discrete valuation fields [HSch], [Te], [Mc], [Co].

Lemma (3.2) Ch. I shows that there are three cases: two equal-characteristic cases, when $\operatorname{char}(F) = \operatorname{char}(\overline{F}) = 0$ or $\operatorname{char}(F) = \operatorname{char}(\overline{F}) = p > 0$, and one unequal-characteristic case, when $\operatorname{char}(F) = 0, \operatorname{char}(\overline{F}) = p > 0$.

(5.1). LEMMA. *The ring of integers $\mathcal{O}_F$ contains a nontrivial field M if and only if* $\operatorname{char}(F) = \operatorname{char}(\overline{F})$.

Proof. Since $M \cap \mathcal{M}_F = (0)$, M is mapped isomorphically onto the field $\overline{M} \subset \overline{F}$, therefore $\operatorname{char}(F) = \operatorname{char}(\overline{F})$. Conversely, let A be the subring in $\mathcal{O}_F$ generated by 1. Then A is a field if $\operatorname{char}(F) = p$, and $A \cap \mathcal{M}_F = (0)$ if $\operatorname{char}(\overline{F}) = 0$. Hence, the quotient field of A is the desired one. □

A field $M \subset \mathcal{O}_F$, that is mapped isomorphically onto the residue field $\overline{F} = \overline{M}$ is called a *coefficient field* in $\mathcal{O}_F$. Such a field, if it exists, is a set of representatives of $\overline{F}$ in $\mathcal{O}_F$ (see (5.1) Ch. I). Proposition (5.2) Ch. I implies immediately that in this case F is isomorphic (algebraically and topologically) with the field $M((X))$: a prime element π in F corresponds to X. Note that this isomorphism depends on the choice of a coefficient field (which is sometimes unique, see Proposition (5.4)) and the choice of a prime element of F.

We shall show below that a coefficient field exists in an equal-characteristic case.

(5.2). The simplest case is that of $\operatorname{char}(F) = \operatorname{char}(\overline{F}) = 0$.

PROPOSITION. *Let* $\operatorname{char}(\overline{F}) = 0$. *Then there exists a coefficient field in $\mathcal{O}_F$. A coefficient field can be selected in infinitely many ways if and only if $\overline{F}$ is not algebraic over $\mathbb{Q}$.*

Proof. Let M be a maximal subfield in $\mathcal{O}_F$, in other words, M be not contained in any other larger subfield of $\mathcal{O}_F$. We assert that $\overline{M} = \overline{F}$, i.e., M is a coefficient field. Indeed, if $\theta \in \overline{F}$ is algebraic over $\overline{M}$, then θ is separable over $\overline{M}$ and we can apply the arguments of the proof of Proposition (3.4) to show that there exists an element $\alpha \in \mathcal{O}_F$ which is algebraic over M and such that $\overline{\alpha} = \theta$. Since $M(\alpha) = M$, by the maximality of M, we get $\alpha \in M, \theta \in \overline{M}$. Furthermore, let $\theta \in \overline{F}$ be transcendental over $\overline{M}$. Let $\alpha \in \mathcal{O}_F$ be such that $\overline{\alpha} = \theta$. Then α is not algebraic over M, because if $\sum_{i=0}^{n} a_i\alpha^i = 0$ with $a_i \in M$, then $\sum_{i=0}^{n} \overline{a}_i\theta^i = 0$. Hence, $\overline{a}_i = 0$ and $a_i = 0$ (M is mapped isomorphically onto $\overline{M}$). By the same reason $M[\alpha] \cap \mathcal{M} = (0)$. Hence, the quotient field $M(\alpha)$ is contained in $\mathcal{O}_F$ and $M \neq M(\alpha)$, contradiction. Thus, we have been convinced ourselves in the existence of a coefficient field.

If $\overline{F}$ is not algebraic over $\mathbb{Q}$, let $\alpha \in \mathcal{O}_F$ be an element transcendental over the prime subfield $\mathbb{Q}$ in $\mathcal{O}_F$. Then the maximal subfield in $\mathcal{O}_F$, which contains $\mathbb{Q}(\alpha + a\varepsilon)$ with $\varepsilon \in \mathcal{M}_F, a \in \mathbb{Q}$, is a coefficient field. If $\overline{F}$ is algebraic over $\mathbb{Q}$, then M is algebraic over $\mathbb{Q}$ and is uniquely determined by our previous constructions. □

(5.3). To treat the case $\operatorname{char}(\overline{F}) = p$ we consider the following notion: elements θ_i of $\overline{F}$ are called a *p-basis* of $\overline{F}$ if

$$\overline{F} = \overline{F}^p[\{\theta_i\}] \quad \text{and} \quad |\overline{F}^p[\theta_1, \dots, \theta_n] : \overline{F}^p| = p^n$$

for every distinct elements $\theta_1, \dots, \theta_n$. The empty set is a p-basis if and only if $\overline{F}$ is perfect. For an imperfect $\overline{F}$, a p-basis $\Theta = \{\theta_i\}$ exists by Zorn's Lemma, because every maximal set of elements θ_i satisfying the second condition possesses the first property. The definition of a p-basis implies that $\overline{F} = \overline{F}^{p^n}[\{\theta_i\}]$ for $n \geqslant 1$.

LEMMA. *Let F be a complete discrete valuation field with the residue field $\overline{F}$ of characteristic p, and $\Theta = \{\theta_i\}$ be a p-basis of $\overline{F}$. Let $\alpha_i \in \mathcal{O}_F$ be such that $\overline{\alpha}_i = \theta_i$. Then there exists an extension L/F with $e(L|F) = 1$, such that L is a complete discrete valuation field, $\overline{L} = \bigcup_{n\geqslant 0} \overline{F}^{p^{-n}}$ and α_i are the multiplicative representatives of θ_i in L (see section 7 Ch. I).*

Proof. Let I be an index-set for Θ. One can put $F_n = F_{n-1}(\{\alpha_{i,n}\})$ with $\alpha_{i,n}^p = \alpha_{i,n-1}$, $i \in I$, and $F_0 = F$, $\alpha_{i,0} = \alpha_i$. Then the completion of $L' = \bigcup_{n\geqslant 0} F_n$ is the desired field. Since $\alpha_i \in \bigcap_{n\geqslant 0} L^{p^n}$, we obtain that α_i is the multiplicative representative of θ_i. □

(5.4). Now we treat the case $\operatorname{char}(F) = \operatorname{char}(\overline{F}) = p$. If $\overline{F}$ is perfect, then Corollaries 1 and 2 of (7.3) Ch. I show that the set of the multiplicative representatives of $\overline{F}$ in $\mathcal{O}_F$ forms a coefficient field. Moreover, this is the unique coefficient field in $\mathcal{O}_F$ because if M is such a field and $\alpha \in M$, then, as M is perfect, $\alpha \in \bigcap_{n\geqslant 0} M^{p^n}$ is the multiplicative representative of $\overline{\alpha}$. (Note that in general there are infinitely many maximal fields as well as in the case of $\operatorname{char}(\overline{F}) = 0$, therefore in general a maximal field is not a coefficient field).

PROPOSITION. *Let $\operatorname{char}(F) = p$. If $\overline{F}$ is perfect. Then a coefficient field exists and is unique; it coincides with the set of multiplicative representatives of $\overline{F}$ in $\mathcal{O}_F$. If $\overline{F}$ is imperfect then there are infinitely many coefficient fields.*

Proof. If $\overline{F}$ is imperfect we apply the construction of the previous Lemma. Then $\overline{L}$ is perfect and there is the unique coefficient field N of $\overline{L}$ in $\mathcal{O}_L$. Let M be the subfield of N corresponding to $\overline{F}$. If $\gamma \in M$ then $\overline{\gamma} \in \overline{F}^{p^n}[\Theta]$ and there exists an element $\beta_n \in \mathcal{O}_F[\{\alpha_{i,n}]$, where $\alpha_{i,n}$ are as in the proof of Lemma (5.3), such that $\overline{\beta}_n = \overline{\gamma}^{p^{-n}}$. It follows that $\beta_n \equiv \gamma^{p^{-n}} \mod \mathcal{M}_L$, and by Lemma (7.2) Ch. I we deduce $\gamma \equiv \beta_n^{p^n} \mod \mathcal{M}_L^{n+1}$. Since $\beta_n^{p^n} \in \mathcal{O}_F^{p^n}[\{\alpha_i\}] \subset \mathcal{O}_F$, we obtain $\gamma = \lim \beta_n^{p^n} \in \mathcal{O}_F$. This proves the existence of a coefficient field of $\overline{F}$ in $\mathcal{O}_F$. If we apply this construction for another set of elements $\alpha_i' \in \mathcal{O}_F$ with $\overline{\alpha}_i' = \overline{\alpha}_i$, then we get a coefficient field M' containing α_i'. Since $\mathcal{M}_F \cap M = \mathcal{M}_F \cap M' = (0)$ we deduce $M \neq M'$. □

(5.5). We conclude with the case of unequal characteristic: $\operatorname{char}(F) = 0$, $\operatorname{char}(\overline{F}) = p$. For the discrete valuation v_F such that $v_F(F^*) = \mathbb{Z}$ recall that $e(F) = v_F(p)$ is called the absolute index of ramification of F, see (5.7) Ch. I. The preceding assertions show that in equal-characteristic case for an arbitrary field K there exists a complete discrete valuation field F with the residue field $\overline{F}$ isomorphic to K. Here is an analog:

PROPOSITION. *Let F be a complete discrete valuation field of characteristic 0 with residue field K of characteristic p. Let K_1 be any extension of K. Then there exists a complete discrete valuation field F_1 which is an extension of F, such that $e(F_1|F) = 1$ and $\overline{F}_1 = K_1$.*

Proof. It is suffices to consider two cases: $K_1 = K(a)$ is an algebraic extension over K and $K_1 = K(y)$ is a transcendental extension over K. If, in addition, in the first case K_1/K is separable, then let $g(X)$ be the monic irreducible polynomial of a over K, and let $f(X)$ be a monic polynomial over the ring of integers of K such that $\overline{f}(X) = g(X)$. By the Hensel Lemma (1.2) there exists a root α of $f(X)$ such that $\overline{\alpha} = a$. Then $F_1 = F(\alpha)$ is the desired extension of F. Next, if $a^p = b \in K$ and β is an element in the ring of integers of F such that $\overline{\beta} = b$, then $F_1 = F(\alpha)$ is the desired extension of F for $\alpha^p = \beta$. Finally, in the second case let w be the discrete valuation on $F(y)$ defined in Example 4 in (2.3) Ch. I. Then $F_1 = F(y)$ is the desired extension of F. □

COROLLARY. *There exists a complete discrete valuation field of characteristic 0 with any given residue field of characteristic p and the absolute index of ramification is equal to 1.*

Proof. One can set $F = \mathbb{Q}_p$ and apply the Proposition. □

(5.6). PROPOSITION. *Let L be a complete discrete valuation field of characteristic 0 with the residue field $\overline{L}$ of characteristic p. Let F be a complete discrete valuation field of characteristic 0 with p as a prime element. Suppose that there is an isomorphism*

$\overline{\omega}\colon \overline{F} \to \overline{L}$. Then there exists a field embedding $\omega\colon F \to L$, such that $v_L \circ \omega = e(L)v_F$ and the image of $\omega(\alpha) \in \mathcal{O}_L$ for $\alpha \in \mathcal{O}_F$ in the residue field $\overline{L}$ coincides with $\overline{\omega}(\overline{\alpha})$.

Proof. Assume first that $\overline{F}$ is perfect. By Corollary 1 of (7.3) Chapter I any element $\theta \in \overline{F}$ has the unique multiplicative representative $r_F(\theta)$ in F and $r_L(\overline{\omega}(\theta))$ in L. Put

$$\omega\left(\sum r_F(\theta_i)p^i\right) = \sum r_L(\overline{\omega}(\theta_i))p^i.$$

Proposition (5.2) Ch. I shows that the map ω is defined on F, Proposition (7.6) Ch. I shows that ω is a homomorphism of fields. Evidently $v_L \circ \omega = e(L)v_F$ and $\overline{\omega(\alpha)} = \overline{\omega}(\overline{\alpha})$ for $\alpha \in \mathcal{O}_F$.

Further, assume that $\overline{F}$ is imperfect. Let $\Theta = \{\theta_i\}_{i\in I}$ be a p-basis of $\overline{F}$. Let $\mathrm{A} = \{\alpha_i\}_{i\in I}$ be a set of elements $\alpha_i \in \mathcal{O}_F$ with $\overline{\alpha}_i = \theta_i$, and let $\mathrm{B} = \{\beta_i\}_{i\in I}$ be a set of elements $\beta_i \in \mathcal{O}_L$ with $\overline{\beta}_i = \theta_i$. For a map

$$\nu\colon I \longrightarrow \{0, 1, \dots, p^n - 1\}$$

such that $\nu(i) = 0$ for almost all $i \in I$, put

$$\Theta^\nu = \prod_{i\in I} \theta_i^{\nu(i)}.$$

The same meaning will be used for $\mathrm{A}^\nu, \mathrm{B}^\nu$. By Lemma (5.3) there exist complete discrete valuation fields F', L' for F, L, such that $e(F'|F) = e(L'|L) = 1$, and $\overline{F'}$ is perfect and isomorphic to $\overline{L'}$, and α_i (resp. β_i) are multiplicative representatives of θ_i in $\mathcal{O}_{F'}$ (resp. of $\overline{\omega}(\theta_i)$ in $\mathcal{O}_{L'}$). The previous arguments show the existence of a homomorphism $\omega'\colon F' \to L'$ with $v_{L'} \circ \omega' = e(L)v_{F'}$ and $\overline{\omega'(\alpha)} = \overline{\omega}(\overline{\alpha})$ for $\alpha \in \mathcal{O}_{F'}$. Moreover, ω' maps α_i in β_i, since they are the multiplicative representatives of θ_i and $\overline{\omega}(\theta_i)$. Let $\gamma \in \mathcal{O}_F$ and $\overline{\gamma} = \sum a_\nu^{p^n}\Theta^\nu$ with $a_\nu \in \overline{F}$. Let b_ν be an element of $\mathcal{O}_F$ with the property $\overline{b}_\nu = a_\nu$, and c_ν an element of $\mathcal{O}_L$ with the property $\overline{c}_\nu = \overline{\omega'(b_\nu)}$. Then $\gamma \equiv \sum b_\nu^{p^n}\mathrm{A}^\nu \mod p\mathcal{O}_F$, i.e.,

$$\gamma = \sum b_\nu^{p^n}\mathrm{A}^\nu + p\gamma_1$$

with $\gamma_1 \in \mathcal{O}_F$. We get $\omega'(\mathrm{A}^\nu) = \mathrm{B}^\nu$ and using Lemma (7.2) Ch. I, we have

$$\omega'(b_\nu^{p^n}) \equiv c_\nu^{p^n} \mod \mathcal{M}_{L'}^{n+1}.$$

Therefore,

$$\omega'(\gamma) \equiv \sum c_\nu^{p^n}\mathrm{B}^\nu + p\omega'(\gamma_1) \mod \mathcal{M}_{L'}^{n+1}.$$

Repeating this reasoning for γ_1, we conclude that $\omega'(\gamma) \equiv \delta_n \mod \mathcal{M}_{L'}^{n+1}$ for some $\delta_n \in \mathcal{O}_L$. Then $\omega'(\gamma) = \lim \delta_n$ and since $\mathcal{O}_L$ is complete, we deduce $\omega'(\gamma) \in \mathcal{O}_L$. Thus, ω' maps $\mathcal{O}_F$ in $\mathcal{O}_L$, and we finally put $\omega = \omega'|_F$ to obtain the desired homomorphism. □

COROLLARY 1. *Let F_1, F_2 be complete discrete valuation fields of characteristic 0 with p as a prime element. Let there be an isomorphism $\overline{\omega}$ of the residue field $\overline{F_1}$ to $\overline{F_2}$. Let $\overline{F_2}$ be of characteristic p. Then there exists a field embedding $\omega\colon F_1 \to F_2$ such that $\overline{\omega(\alpha)} = \overline{\omega}(\overline{\alpha})$ for $\alpha \in \mathcal{O}_{F_1}$.*

Proof. Apply the Proposition for $F = F_1, L = F_2$ and $F = F_2, L = F_1$. □

COROLLARY 2. *The image $\omega(F)$ is uniquely determined in the field L if and only if $\overline{F}$ is perfect or $e(L) = 1$.*

Proof. Let $\overline{F}$ be imperfect and $e(L) > 1$. Let $\omega(F)$ be uniquely determined in L. Then, in the proof of the Proposition we can replace β_i by $\beta_i + \pi_L$ and obtain that $\beta_i \in \omega(\mathcal{O}_F)$, $\beta_i + \pi_L \in \omega(\mathcal{O}_F)$. Hence, $\pi_L \in \omega(\mathcal{O}_F)$ which is impossible because $v_L \circ \omega = e(L) v_F$. □

REMARK. If $\overline{F}$ is perfect then we can identify $\omega(F)$ with the field of fractions of Witt vectors $W(\overline{F})$ (see (8.3) Ch. I and Exercise 6 below).

Exercises.

1. Let F be a complete discrete valuation field of characteristic p with a residue field $\overline{F}$. Let $\Theta = \{\theta_i\}$ be a p-basis of $\overline{F}$. Let $\mathrm{A} = \{\alpha_i\}$ be a set of elements in $\mathcal{O}_F$ such that $\overline{\alpha}_i = \theta_i$. Put $R_n = \mathcal{O}_F^{p^n}[\mathrm{A}]$ and let S_n be the completion of R_n in $\mathcal{O}_F$. Show that $\bigcap_{n\geqslant 0} S_n$ is a coefficient field.
2. Let F be as in Exercise 1 and let $\overline{F}$ be imperfect. Show that a maximal subfield in $\mathcal{O}_F$ contains the largest perfect subfield in $\mathcal{O}_F$, but is not necessarily a coefficient field. Show that a coefficient field contains the largest perfect subfield in $\mathcal{O}_F$ as well.
3. Let $K = \mathbb{F}_p(X)$, and let F be the completion of $K(Y)$ with respect to the discrete valuation corresponding to the irreducible polynomial $Y^p - X$. Show that $\overline{F} = K(X^{1/p})$, but K is not contained in any coefficient field of $\overline{F}$ in $\mathcal{O}_F$.
4. (◇) Let F be a complete discrete valuation field, and let L be a finite extension of F. Show that if $\overline{F}$ is perfect, then coefficient fields M_F of $\overline{F}$ in $\mathcal{O}_F$ and M_L of $\overline{L}$ in $\mathcal{O}_L$ can be chosen so that $M_F \subset M_L$. Show that if $\overline{F}$ is imperfect, this assertion does not hold in general.
5. Find another proof of Corollary (5.5) using Witt vectors.
6. (◇) Let F be a complete discrete valuation field with a prime element p and $\operatorname{char}(\overline{F}) = p$. Show that for a subfield $K \subset \overline{F}$ there exists a subfield F' in F which is a complete discrete valuation field with respect to the induced valuation and is such that $\overline{F'} = K$. Show that if K is perfect, then such a field is unique.

CHAPTER 3

The Norm Map

In this chapter we study the norm map acting on Henselian discrete valuation fields. Section 1 studies the behaviour of the norm map on the factor filtration introduced in section 5 Chapter I for cyclic extensions of prime degree. Section 2 demonstrates that almost all cyclic extensions of degree p can be described by explicit equations of Artin–Schreier type. Section 3 associates to the norm map a real function called the Hasse–Herbrand function; properties of this function and applications to ramification groups are studied in sections 3 and 4. The long section 5 presents a relatively recent theory of a class of infinite Galois extensions of local fields: arithmetically profinite extensions and their fields of norms. The latter establishes a relation between the fields of characteristic 0 and characteristic p.

We will work with complete discrete valuation fields leaving the Henselian case to Exercises.

1. Cyclic Extensions of Prime Degree

In this section we describe the norm map on the factor filtration of the multiplicative group in a cyclic extension of prime degree. The most difficult and interesting case is of totally ramified p-extensions which is treated in subsections (1.4) and (1.5). Using these results we will be able to simplify expositions of theories presented in several other sections of this book.

Let F be a complete discrete valuation field and L its Galois extension of prime degree n. Then there are four possible cases:

L/F is unramified;
L/F is tamely and totally ramified;
L/F is totally ramified of degree $p = \text{char}(\overline{F}) > 0$;
$\overline{L}/\overline{F}$ is inseparable of degree $p = \text{char}(\overline{F}) > 0$.

Since the fourth case is outside the subject of this book, we restrict our attention to the first three cases (still, see Exercise 2).

The following results are classical and essentially due to *H. Hasse*.

(1.1). LEMMA. *Let L/F be a separable extension of prime degree n, $\gamma \in \mathcal{M}_L$. Then*

$$N_{L/F}(1+\gamma) = 1 + N_{L/F}(\gamma) + \mathrm{Tr}_{L/F}(\gamma) + \mathrm{Tr}_{L/F}(\delta)$$

with some $\delta \in \mathcal{O}_L$ such that $v_L(\delta) \geqslant 2v_L(\gamma)$ ($N_{L/F}$ and $\mathrm{Tr}_{L/F}$ are the norm and the trace maps, respectively).

Proof. Recall that for distinct embeddings σ_i of L over F into the algebraic closure of F, $1 \leqslant i \leqslant n$, one has (see [La1, Ch. VIII])

$$N_{L/F}\alpha = \prod_{i=1}^{n} \sigma_i(\alpha), \quad \mathrm{Tr}_{L/F}\,\alpha = \sum_{i=1}^{n} \sigma_i(\alpha), \qquad \alpha \in L.$$

Hence

$$\begin{aligned} N_{L/F}(1+\gamma) &= \prod_{i=1}^{n}(1+\sigma_i(\gamma)) \\ &= 1 + \sum_{i=1}^{n}\sigma_i(\gamma) + \Big(\sum_{i=1}^{n}\sigma_i\Big)\Big(\sum_{1\leqslant j\leqslant n}\gamma\sigma_j(\gamma) + \cdots\Big) + \prod_{i=1}^{n}\sigma_i(\gamma). \end{aligned}$$

For $\delta = \sum_{1\leqslant j\leqslant n}\gamma\sigma_j(\gamma) + \cdots$ we get $v_L(\delta) \geqslant 2v_L(\gamma)$. □

Our nearest purpose is to describe the action of the norm map $N_{L/F}$ with respect to the filtration discussed in section 5 Ch. I.

(1.2). PROPOSITION. *Let L/F be an unramified extension of degree n. Then a prime element π_F in F is a prime element in L. Let $U_{i,L} = 1 + \pi_F^i\mathcal{O}_L$, $U_{i,F} = 1 + \pi_F^i\mathcal{O}_F$ and let $\lambda_{i,L}$, $\lambda_{i,F}$ ($i \geqslant 0$), be identical to those of Proposition* (5.4) *Ch. I, for $\pi = \pi_F$. Then the following diagrams are commutative:*

$$\begin{array}{ccc} L^* & \xrightarrow{v_L} & \mathbb{Z} \\ {\scriptstyle N_{L/F}}\downarrow & & \downarrow{\scriptstyle \times n} \\ F^* & \xrightarrow{v_F} & \mathbb{Z} \end{array} \qquad \begin{array}{ccc} U_L & \xrightarrow{\lambda_{0,L}} & \overline{L}^* \\ {\scriptstyle N_{L/F}}\downarrow & & \downarrow{\scriptstyle N_{\overline{L}/\overline{F}}} \\ U_F & \xrightarrow{\lambda_{0,F}} & \overline{F}^* \end{array} \qquad \begin{array}{ccc} U_{i,L} & \xrightarrow{\lambda_{i,L}} & \overline{L} \\ {\scriptstyle N_{L/F}}\downarrow & & \downarrow{\scriptstyle \mathrm{Tr}_{\overline{L}/\overline{F}}} \\ U_{i,F} & \xrightarrow{\lambda_{i,F}} & \overline{F} \end{array}$$

Proof. The first commutativity follows from (2.3) Ch. II. Proposition (3.3) Ch. II implies that $\overline{N_{L/F}(\alpha)} = N_{\overline{L}/\overline{F}}(\overline{\alpha})$ for $\alpha \in \mathcal{O}_L$, i.e., the second diagram is commutative. The preceding Lemma shows that

$$N_{L/F}(1+\theta\pi_F^i) = 1 + (\mathrm{Tr}_{L/F}\,\theta)\pi_F^i + (N_{L/F}\theta)\pi_F^{pi} + \mathrm{Tr}_{L/F}(\delta)$$

with $v_L(\delta) \geqslant 2i$ and, consequently, $v_F\,\mathrm{Tr}_{L/F}(\delta) \geqslant 2i$. Thus, we get

$$N_{L/F}(1+\theta\pi_F^i) \equiv 1 + (\mathrm{Tr}_{L/F}\,\theta)\pi_F^i \mod \pi_F^{i+1}$$

and the commutativity of the third diagram. □

COROLLARY. *In the case under consideration* $N_{L/F}U_{1,L} = U_{1,F}$.

(1.3). PROPOSITION. *Let* L/F *be a totally and tamely ramified cyclic extension of degree* n. *Then for some prime element* π_L *in* L, *the element* $\pi_F = \pi_L^n$ *is prime in* F *(Proposition* (3.5) *Ch. II) and* $\overline{F} = \overline{L}$. *Let* $U_{i,L} = 1 + \pi_L^i \mathcal{O}_L$, $U_{i,F} = 1 + \pi_F^i \mathcal{O}_F$, *and let* $\lambda_{i,L}$, $\lambda_{i,F}$ *be identical to those of Proposition* (5.4) *Ch. I, for* $\pi = \pi_L$ *and* $\pi = \pi_F$. *Then the following diagrams*

$$\begin{array}{ccc} L^* & \xrightarrow{v_L} & \mathbb{Z} \\ {\scriptstyle N_{L/F}}\downarrow & & \downarrow{\scriptstyle \mathrm{id}} \\ F^* & \xrightarrow{v_F} & \mathbb{Z} \end{array} \qquad \begin{array}{ccc} U_L & \xrightarrow{\lambda_{0,L}} & \overline{L}^* \\ {\scriptstyle N_{L/F}}\downarrow & & \downarrow\uparrow{\scriptstyle n} \\ U_F & \xrightarrow{\lambda_{0,F}} & \overline{F}^* \end{array}$$

$$\begin{array}{ccc} U_{ni,L} & \xrightarrow{\lambda_{ni,L}} & \overline{L} = \overline{F} \\ {\scriptstyle N_{L/F}}\downarrow & & \downarrow{\scriptstyle \times\overline{n}} \\ U_{i,F} & \xrightarrow{\lambda_{i,F}} & \overline{F} \end{array}$$

are commutative, where id *is the identity map,* $\uparrow n$ *takes an element to its* n*th power,* $\times\overline{n}$ *is the multiplication by* $\overline{n} \in \overline{F}$, $i \geqslant 1$. *Moreover,* $N_{L/F}U_{i,L} = N_{L/F}U_{i+1,L}$ *if* $n \nmid i$.

Proof. Since $\pi_L^n = \pi_F$ and L/F is Galois, then $\mathrm{Gal}(L/F)$ is cyclic of order n and $\sigma(\pi_L) = \zeta\pi_L$ for a generator σ of $\mathrm{Gal}(L/F)$, where ζ is a primitive nth root of unity, $\zeta \in F$. The first diagram is commutative in view of Theorem (2.5) Ch. II. Proposition (4.1) Ch. II shows that $\overline{\sigma(\alpha)} = \overline{\alpha}$ for $\sigma \in \mathrm{Gal}(L/F)$, $\alpha \in \mathcal{O}_L$, and we get the commutativity of the second diagram. If $j = ni$, then $1 + \theta\pi_L^j \in F$ for $\theta \in \mathcal{O}_F$, and

$$N_{L/F}(1 + \theta\pi_L^j) = (1 + \theta\pi_F^i)^n \equiv 1 + n\theta\pi_F^i \mod \pi_F^{i+1}$$

by Proposition (5.4) Ch. I. Applying Corollary (5.5) Ch. I, we deduce

$$U_{i,F} = U_{i,F}^n = N_{L/F}U_{ni,L}.$$

Finally, $X^n - 1 = \prod_{j=0}^{n-1}(X - \zeta^j)$, therefore for $n \nmid i$ and for $\theta \in \mathcal{O}_F$ one has

$$N_{L/F}(1 + \theta\pi_L^i) = \prod_{j=0}^{n-1}(1 + \zeta^j\theta\pi_L^i) = 1 - (-\theta)^n\pi_F^i.$$

Thus $N_{L/F}U_{i,L} = N_{L/F}U_{i+1,L}$. □

COROLLARY. *In the case under consideration* $N_{L/F}U_{1,L} = U_{1,F}$.

If $\overline{F}$ *is algebraically closed then* $N_{L/F}L^* = F^*$.

(1.4). Now we treat the most complicated case when L/F is a totally ramified Galois extension of degree $p = \operatorname{char}(\overline{F}) > 0$. Then Corollary 2 of (2.9) Ch. II shows that $\mathcal{O}_L = \mathcal{O}_F[\pi_L]$, $L = F(\pi_L)$ for a prime element π_L in L, and $\overline{L} = \overline{F}$. Let σ be a generator of $\operatorname{Gal}(L/F)$, then $\sigma(\pi_L)/\pi_L \in U_L$. One can write $\sigma(\pi_L)/\pi_L = \theta\varepsilon$ with $\theta \in U_F, \varepsilon \in 1 + \mathcal{M}_L$. Then

$$\sigma^2(\pi_L)/\pi_L = \sigma(\theta\varepsilon) \cdot \theta\varepsilon = \theta^2\varepsilon \cdot \sigma(\varepsilon),$$

and

$$1 = \sigma^p(\pi_L)/\pi_L = \theta^p\varepsilon \cdot \sigma(\varepsilon) \cdot \dots \cdot \sigma^{p-1}(\varepsilon).$$

This shows that $\theta^p \in 1 + \mathcal{M}_L$ and $\theta \in 1 + \mathcal{M}_F$, because raising to the pth power is an injective homomorphism of $\overline{F}$. Thus, we obtain $\sigma(\pi_L)/\pi_L \in 1 + \mathcal{M}_L$. Put

$$\frac{\sigma(\pi_L)}{\pi_L} = 1 + \eta\pi_L^s \qquad \text{with} \qquad \eta \in U_L, \ s = s(L|F) \geqslant 1. \tag{$*$}$$

Note that s does not depend on the choice of the prime element π_L and of the generator σ of $G = \operatorname{Gal}(L/F)$. Indeed, we have

$$\frac{\sigma^i(\pi_L)}{\pi_L} \equiv 1 + i\eta\pi_L^s \mod \pi_L^{s+1} \qquad \text{and} \qquad \frac{\sigma(\rho)}{\rho} \equiv 1 \mod \pi_L^{s+1}$$

for an element $\rho \in U_L$. We also deduce that

$$\frac{\sigma(\alpha)}{\alpha} \in U_{s,L}$$

for every element $\alpha \in L^*$. This means that $G = G_s$, $G_{s+1} = \{1\}$ (see (4.3) Ch. II).

LEMMA. *Let $f(X) = X^p + a_{p-1}X^{p-1} + \dots + a_0$ be the irreducible polynomial of π_L over F. Then*

$$\operatorname{Tr}_{L/F}\left(\frac{\pi_L^j}{f'(\pi_L)}\right) = \begin{cases} 0 & \text{if} \quad 0 \leqslant j \leqslant p-2, \\ 1 & \text{if} \quad j = p-1. \end{cases}$$

Proof. Since $\sigma^i(\pi_L)$ for $0 \leqslant i \leqslant p-1$ are all the roots of the polynomial $f(X)$, we get

$$\frac{1}{f(X)} = \sum_{i=0}^{p-1} \frac{1}{f'\big(\sigma^i(\pi_L)\big)\big(X - \sigma^i(\pi_L)\big)}.$$

Putting $Y = X^{-1}$ and performing the calculations in the field $F((Y))$, we consequently deduce

$$f(X) = Y^{-p}(1 + a_{p-1}Y + \dots + a_0Y^p),$$
$$\frac{1}{f(X)} = \frac{Y^p}{1 + a_{p-1}Y + \dots + a_0Y^p} \equiv Y^p \mod Y^{p+1},$$
$$\frac{1}{X - \sigma^i(\pi_L)} = \frac{Y}{1 - \sigma^i(\pi_L)Y} = \sum_{j\geqslant 0} \sigma^i(\pi_L^j)Y^{j+1}$$

(because $1/(1-Y) = \sum_{i\geqslant 0} Y^i$ in $F((Y))$). Hence

$$\sum_{j\geqslant 0}\sum_{i=0}^{p-1} \frac{\sigma^i(\pi_L^j)Y^{j+1}}{f'\big(\sigma^i(\pi_L)\big)} \equiv Y^p \mod Y^{p+1},$$

or

$$\operatorname{Tr}_{L/F}\left(\frac{\pi_L^j}{f'(\pi_L)}\right) = \sum_{i=0}^{p-1} \frac{\sigma^i(\pi_L^j)}{f'\big(\sigma^i(\pi_L)\big)} = \begin{cases} 0 & \text{if} \quad 0 \leqslant j \leqslant p-2, \\ 1 & \text{if} \quad j = p-1, \end{cases}$$

as desired. □

PROPOSITION. *Let $[a]$ denote the maximal integer $\leqslant a$. For an integer $i \geqslant 0$ put $j(i) = s + 1 + \big[(i-1-s)/p\big]$. Then*

$$\operatorname{Tr}_{L/F}(\pi_L^i \mathcal{O}_L) = \pi_F^{j(i)}\mathcal{O}_F.$$

Proof. One has $f'(\pi_L) = \prod_{i=1}^{p-1}\big(\pi_L - \sigma^i(\pi_L)\big)$ and $\sigma^i(\pi_L)/\pi_L \equiv 1 + i\eta\pi_L^s \mod \pi_L^{s+1}$. Then

$$f'(\pi_L) = (p-1)!(-\eta)^{p-1}\pi_L^{(p-1)(s+1)}\varepsilon$$

with some $\varepsilon \in 1 + \mathcal{M}_L^{(p-1)(s+1)+1}$. Since $\overline{F} = \overline{L}$, for a prime element π_F in F one has the representation $\pi_F = \pi_L^p\varepsilon'$ with $\varepsilon' \in U_L$. The previous Lemma implies

$$\operatorname{Tr}_{L/F}\left(\pi_L^{j+s+1}\varepsilon_{j+s+1}\right) = \begin{cases} 0 & \text{if} \quad 0 \leqslant j < p-1, \\ \pi_F^{s+1} & \text{if} \quad j = p-1 \end{cases}$$

for $\varepsilon_{j+s+1} = (\varepsilon')^{s+1}/\big((p-1)!(-\eta)^{p-1}\varepsilon\big)$. Taking into consideration the evident equality $\operatorname{Tr}_{L/F}(\pi_F^i\alpha) = \pi_F^i \operatorname{Tr}_{L/F}(\alpha)$ we can choose the units ε_{j+s+1}, for every integer j, such that $\operatorname{Tr}_{L/F}(\pi_L^{j+s+1}\varepsilon_{j+s+1}) = 0$ if $p \nmid (j+1)$ and $= \pi_F^{s+(j+1)/p}$ if $p | (j+1)$. Thus, since the $\mathcal{O}_F$-module $\pi_L^i\mathcal{O}_L$ is generated by $\pi_L^j\varepsilon_j$, $j \geqslant i$, we conclude that $\operatorname{Tr}_{L/F}(\pi_L^i\mathcal{O}_L) = \pi_F^{j(i)}\mathcal{O}_F$. □

(1.5). PROPOSITION. *Let L/F be a totally ramified Galois extension of degree $p = \operatorname{char}(\overline{F}) > 0$. Let π_L be a prime element in L. Then $\pi_F = N_{L/F}\pi_L$ is a prime element in F. Let $U_{i,L} = 1 + \pi_L^i \mathcal{O}_L, U_{i,F} = 1 + \pi_F^i \mathcal{O}_F$ and let $\lambda_{i,L}, \lambda_{i,F}$ be identical to those in Proposition (5.4) Ch. I, for $\pi = \pi_L$ and $\pi = \pi_F$. Then the following diagrams are commutative:*

$$\begin{array}{ccc} L^* & \xrightarrow{v_L} & \mathbb{Z} \\ {\scriptstyle N_{L/F}}\downarrow & & \downarrow{\scriptstyle \mathrm{id}} \\ F^* & \xrightarrow{v_F} & \mathbb{Z} \end{array} \qquad \begin{array}{ccc} U_L & \xrightarrow{\lambda_{0,L}} & \overline{L}^* \\ {\scriptstyle N_{L/F}}\downarrow & & \downarrow{\scriptstyle \uparrow p} \\ U_F & \xrightarrow{\lambda_{0,F}} & \overline{F}^* \end{array}$$

$$\begin{array}{ccc} U_{i,L} & \xrightarrow{\lambda_{i,L}} & \overline{L} = \overline{F} \\ {\scriptstyle N_{L/F}}\downarrow & & \downarrow{\scriptstyle \uparrow p} \\ U_{i,F} & \xrightarrow{\lambda_{i,F}} & \overline{F} \end{array} \qquad \text{if} \qquad 1 \leqslant i < s,$$

$$\begin{array}{ccc} U_{s,L} & \xrightarrow{\lambda_{s,L}} & \overline{L} = \overline{F} \\ {\scriptstyle N_{L/F}}\downarrow & & \downarrow{\scriptstyle \overline{\theta} \mapsto \overline{\theta}^p - \overline{\eta}^{p-1}\overline{\theta}} \\ U_{s,F} & \xrightarrow{\lambda_{s,F}} & \overline{F} \end{array}$$

$$\begin{array}{ccc} U_{s+pi,L} & \xrightarrow{\lambda_{s+pi,L}} & \overline{L} = \overline{F} \\ {\scriptstyle N_{L/F}}\downarrow & & \downarrow{\scriptstyle \times(-\overline{\eta}^{p-1})} \\ U_{s+i,F} & \xrightarrow{\lambda_{s+i,F}} & \overline{F} \end{array} \qquad \text{if} \qquad i > 0.$$

Moreover, $N_{L/F}(U_{s+i,L}) = N_{L/F}(U_{s+i+1,L})$ for $i > 0, p \nmid i$.

Proof. The commutativity of the first and the second diagrams can be verified similarly to the proof of Proposition (1.3). In order to look at the remaining diagrams, put $\varepsilon = 1 + \theta\pi_L^i$ with $\theta \in U_L$. Then, by Lemma (1.1), we get

$$N_{L/F}\varepsilon = 1 + N_{L/F}(\theta)\pi_F^i + \operatorname{Tr}_{L/F}(\theta\pi_L^i) + \operatorname{Tr}_{L/F}(\theta\delta)$$

with $v_L(\delta) \geqslant 2i$. The previous Proposition implies that

$$v_F\left(\operatorname{Tr}_{L/F}(\pi_L^i)\right) \geqslant s + 1 + \left[\frac{i-1-s}{p}\right], \; v_F\left(\operatorname{Tr}_{L/F}(\delta)\right) \geqslant s + 1 + \left[\frac{2i-1-s}{p}\right]$$

and for $i < s$

$$v_F\left(\operatorname{Tr}_{L/F}(\pi_L^i)\right) \geqslant i + 1, \quad v_F\left(\operatorname{Tr}_{L/F}(\delta)\right) \geqslant i + 1.$$

Therefore, the third diagram is commutative. Further, using $(*)$ of (1.4), one can write

$$1 = N_{L/F}\left(\frac{\sigma(\pi_L)}{\pi_L}\right) \equiv 1 + N_{L/F}(\eta)\pi_F^s + \mathrm{Tr}_{L/F}(\eta\pi_L^s) \mod \pi_F^{s+1}.$$

We deduce that

$$\mathrm{Tr}_{L/F}(\eta\pi_L^s) \equiv -N_{L/F}(\eta)\pi_F^s \mod \pi_F^{s+1}.$$

Since $N_{L/F}(\eta) \equiv \eta^p \mod \pi_L$ in view of $U_L \subset U_F U_{1,L}$, we conclude that

$$N_{L/F}(1+\theta\eta\pi_L^s) - 1 - \eta^p\pi_F^s(\theta^p - \theta) \in \pi_L^{ps+1}\theta\mathcal{O}_L$$

for $\theta \in \mathcal{O}_F$. This implies the commutativity of the fourth (putting $\theta \in \mathcal{O}_F$) and the fifth (when $\theta \in \pi_F^i\mathcal{O}_F$) diagrams. Finally, if $p \nmid i, \theta \in \mathcal{O}_F$, then

$$\frac{\sigma(1+\theta\pi_L^i)}{1+\theta\pi_L^i} \equiv 1 + i\theta\eta\pi_L^{i+s} \mod \pi_L^{i+s+1}.$$

This means that $N_{L/F}(1+i\theta\eta\pi_L^{i+s}) \in N_{L/F}U_{s+i+1,L}$ and $N_{L/F}(U_{s+i,L}) = N_{L/F}(U_{s+i+1,L})$. □

Corollary. $U_{s+1,F} = N_{L/F}U_{s+1,L}$.

If $\overline{F}$ is algebraically closed then $N_{L/F}L^* = F^*$.

Proof. It follows immediately from the last diagram of the Proposition, since the multiplication by $(-\overline{\eta})^{p-1}$ is an isomorphism of the additive group $\overline{F}$. □

Exercises.

1. a) Let F be a Henselian discrete valuation field, and L/F a cyclic extension of prime degree. Show that $U_{i,F} \subset N_{L/F}U_L$ for sufficiently large i.
   b) Show that all assertions of this section hold for a Henselian discrete valuation field.
2. Let L/F be a Galois extension of degree $p = \mathrm{char}(\overline{F})$, and let $\overline{L}/\overline{F}$ be an inseparable extension of degree p. Let $\theta \in U_L$ be such that $\overline{L} = \overline{F}(\overline{\theta})$. Let σ be a generator of $\mathrm{Gal}(L/F)$.
   a) Show that $v_L(\sigma(\theta) - \theta) > 0$. Put

   $$\frac{\sigma(\theta)}{\theta} = 1 + \eta\pi_F^s$$

   for a prime element π_F in F and some $\eta \in U_L, s \geqslant 1$.
   b) Show that $\mathrm{Tr}_{L/F}(\pi_F^i\mathcal{O}_L) = \pi_F^{j(i)}\mathcal{O}_F$ with $j(i) = (p-1)s + i$.
   c) Show that $N_{L/F}(1+\eta\pi_F^i) \equiv 1 + (N_{L/F}\eta)\pi_F^{pi} \mod \pi_F^{pi+1}$ if $i < s$.
   d) Show that $\mod \pi_F^{ps+1}$

   $$N_{L/F}(1+c\theta^i\eta\pi_F^s) \equiv \begin{cases} 1 + c^p\pi_F^{ps}N_{L/F}(\theta^i\eta), & 0 < i \leqslant p-1, \\ 1 + (c^p - c)\pi_F^{ps}N_{L/F}(\eta), & i = 0, \end{cases}$$

where $c \in \mathcal{O}_F$.

e) Show that $U_{ps+1,F} \subset N_{L/F}U_{s+1,L}$.

3. Let L/F be a Galois extension of degree $p = \operatorname{char}(\overline{F}) > 0$, that is not unramified. Show that

$$v_L\left(\frac{\gamma}{\operatorname{Tr}_{L/F}(\gamma)}\right) = \max_{\alpha \in \mathcal{O}_L} \{v_L(\alpha) : \operatorname{Tr}_{L/F}(\alpha) = 1\},$$

where $\gamma = \alpha^{-1}\sigma(\alpha) - 1$ for a generator σ of $\operatorname{Gal}(L/F)$ and an element $\alpha \in \mathcal{O}_L$, such that $p \nmid v_L(\alpha)$ when $e(L|F) = p$ and $\bar{\alpha} \notin \bar{F}$ when $e(L|F)$ is equal to 1.

2. Artin–Schreier Extensions

A theorem of *E. Artin* and *O. Schreier* asserts that every cyclic extension of degree p over a field K of characteristic p is generated by a root of the polynomial $X^p - X - \alpha$, $\alpha \in K$ (see Exercise 6 section 5 Ch. V or [La1, Ch. VIII]). In this section we show in Proposition (2.4) and (2.5), following *R. MacKenzie* and *G. Whaples* ([MW]), how to extend this result to complete discrete valuation fields of characteristic 0. An alternative proof of the main results of this section can be obtained by using formal groups, see for example [FVZ].

(2.1). First we treat the case of unramified extensions. The polynomial $X^p - X$ is denoted by $\wp(X)$ (see (6.3) Ch. I).

LEMMA. *Let L/F be an unramified Galois extension of degree $p = \operatorname{char}(\overline{F})$. Then $L = F(\lambda)$, where λ is a root of the polynomial $X^p - X - \alpha = 0$ for some $\alpha \in U_F$ with $\overline{\alpha} \notin \wp(\overline{F})$.*

Proof. Let $\overline{L} = \overline{F}(\theta)$, where θ is a root of the polynomial $X^p - X - \eta = 0$ for some $\eta \notin \wp(\overline{F})$. Then the polynomial $X^p - X - \alpha = 0$, with $\alpha \in U_F$, such that $\overline{\alpha} = \eta$, has a root λ in L, by Hensel Lemma (1.2) Ch. II. Thus, $L = F(\lambda)$. □

(2.2). Now we study the case of totally ramified extensions.

Let L/F be a totally ramified Galois extension of degree $p = \operatorname{char}(\overline{F})$. Let σ be a generator of $\operatorname{Gal}(L/F)$, π_L a prime element in L and $s = v_L(\pi_L^{-1}\sigma(\pi_L) - 1)$.

LEMMA. *For $\beta \in L$ there exists an element $b \in F$ such that $v_L(\sigma\beta - \beta) = v_L(\beta - b) + s$.*

Proof. Let $\beta = a_0 + a_1\pi_L + \cdots + a_{p-1}\pi_L^{p-1}$ with $a_i \in F$ (see Proposition (3.6) Ch. II). Then

$$\sigma(\beta) - \beta = a_1\pi_L\gamma + \cdots + a_{p-1}\pi_L^{p-1}\left((1+\gamma)^{p-1} - 1\right),$$

where $\gamma = \pi_L^{-1}\sigma(\pi_L) - 1$. Since $v_L(\gamma) = s > 0$, we get

$$(1+\gamma)^i - 1 \equiv i\gamma \mod \pi_L^{s+1} \quad \text{for } i \geqslant 0.$$

Hence, $v_L\left(a_i\pi_L^i\left((1+\gamma)^i - 1\right)\right)$ are distinct for $1 \leqslant i \leqslant p-1$. Put $b = a_0$. Then $v_L(\sigma(\beta) - \beta) = v_L((\beta - b)\gamma) = v_L(\beta - b) + s$, as desired. □

(2.3). PROPOSITION. *Let F be a complete discrete valuation field with residue field of characteristic $p > 0$. Let L be a totally ramified Galois extension of degree p. If* $\operatorname{char}(F) = p$ *then $p \nmid s$. If* $\operatorname{char}(F) = 0$, *then $s \leqslant pe/(p-1)$, where $e = e(F)$ is the absolute index of ramification of F. In this case, if $p|s$, then a primitive pth root of unity belongs to F, and $s = pe/(p-1)$, $L = F(\sqrt[p]{\alpha})$ with some $\alpha \in F^*, \alpha \notin U_F F^{*p}$.*

Proof. First assume that $\operatorname{char}(F) = p$ and $s = pi$. Then $(1 + \theta\pi_F^i)^p = 1 + \theta^p\pi_F^{pi}$ for $\theta \in U_F$. One can take $\pi_F = N_{L/F}\pi_L$ for a prime element π_L in L. Then it follows from (1.4) that $\pi_F \equiv \pi_L^p \mod \pi_L^{p+1}$. Since $N_{L/F}U_{pi+1,L} \subset U_{pi+1,F}$, we get the congruence $1 + \theta^p\pi_F^{pi} \equiv N_{L/F}(1 + \theta\pi_L^{pi}) \mod \pi_F^{pi+1}$, which contradicts the fourth diagram of Proposition (1.5). Hence, $p \nmid s$.

Assume now that $\operatorname{char}(F) = 0$ and $s > pe/(p-1)$. Let $\varepsilon = 1 + \theta\pi_F^s \in U_{s,F}$ with $\theta \in U_F$. Corollary 2 of (5.8) Ch. I shows that $\varepsilon = \varepsilon_1^p$ for some $\varepsilon_1 = 1 + \theta_1\pi_F^{s-e} \in U_F$ with $\theta_1 \in U_F$. Then $N_{L/F}U_{p(s-e),L} \not\subset U_{s+1,F}$, but $p(s-e) \geqslant s+1$, which is impossible because of Corollary (1.5). Hence, $s \leqslant pe/(p-1)$. By the same reasons as in the case of $\operatorname{char}(F) = p$, it is easy to verify that if $s = pi < pe/(p-1)$, then $1+\theta^p\pi_F^{pi} \equiv N_{L/F}(1+\theta\pi_L^{pi}) \mod \pi_F^{pi+1}$, which is impossible. Therefore, in this case we get $s = pe/(p-1)$. One can write $\sigma(\pi_L)\pi_L^{-1} \equiv 1+\theta\pi_F^{e/(p-1)} \mod \pi_L^{pe/(p-1)+1}$. Then, acting by $N_{L/F}$, we get $1 \equiv (1+\theta\pi_F^{e/(p-1)})^p \mod \pi_F^{pe/(p-1)+1}$. But $U_{pe/(p-1)+1,F} \subset U_{e/(p-1)+1,F}^p$ (see Corollary 2 of (5.8) Ch. I), that permits us to find an element $\zeta \equiv 1 + \theta\pi_F^{e/(p-1)} \mod \pi_F^{e/(p-1)+1}$, such that $\zeta^p = 1$; ζ is a primitive pth root of unity in F, hence $L = F(\sqrt[p]{\alpha})$ for some $\alpha \in F^*$, by the Kummer theory. Writing $\alpha = \pi_F^a\varepsilon_1$ with $\varepsilon_1 \in U_F$ and assuming $p|a$, we can replace α with ε_1. Since $\overline{L} = \overline{F}$ we obtain $\overline{\varepsilon}_1 \in \overline{F}^p$ (otherwise L/F would not be totally ramified) and $\varepsilon_1 \equiv \varepsilon_2^p \mod \pi_L$ for some $\varepsilon_2 \in U_F$. Replacing ε_1 with $\varepsilon_3 = \varepsilon_1\varepsilon_2^{-p}$, we get $\varepsilon_3 \in U_{1,F}$, $L = F(\sqrt[p]{\varepsilon_3})$. Note that

$$\frac{\sigma(1+\rho\pi_L^i)}{1+\rho\pi_L^i} \equiv 1 + \rho i\eta\pi_L^{i+pe/(p-1)} \mod \pi_L^{1+i+pe/(p-1)}$$

for $\rho \in U_F$. Hence $\varepsilon_3^{-1}\sigma(\varepsilon_3) \equiv 1 \mod \pi_L^{1+pe/(p-1)}$, but $\varepsilon_3^{-1}\sigma(\varepsilon_3)$ is a primitive pth root of unity. This contradiction proves that $\alpha \notin U_F F^{*p}$. □

(2.4). PROPOSITION. *Let F be a complete discrete valuation field with residue field of characteristic $p > 0$. Let L be a Galois totally ramified extension of degree p. Let*

$s \neq pe/(p-1)$ *if* char$(F) = 0$, $e = e(F)$. *Then* $L = F(\lambda)$, *where* λ *is a root of some polynomial* $X^p - X - \alpha$ *with* $\alpha \in F$, $v_F(\alpha) = -s$.

Proof. The previous Proposition shows that $p \nmid s$. First consider the case of char$(F) = p$. Then, by the Artin–Schreier theory, $L = F(\lambda)$, where λ is a root of a suitable polynomial $X^p - X - \alpha$ with $\alpha \in F$. Let σ be a generator of Gal(L/F). Then $(\sigma(\lambda) - \lambda)^p = \sigma\lambda - \lambda$. Since $\lambda \notin F$, we get $\sigma(\lambda) - \lambda = a$ with an integer a, $p \nmid a$. Then $\lambda^{-1}\sigma(\lambda) = 1 + a\lambda^{-1}$, and hence Proposition (1.5) implies $1 + a\lambda^{-1} \in U_{s,L}$. This shows $v_L(\lambda) \leqslant -s$ and $v_F(\alpha) \leqslant -s$. Put $t = v_F(\alpha)$. If $t = pt'$, then we can write $\lambda \equiv \pi_L^t \theta \mod \pi_L^{t+1}$ with $\theta \in U_F$ and a prime element π_L in L. Therefore, $\alpha \equiv \pi_L^{pt}\theta^p \equiv \pi_F^{pt'}\theta^p \mod \pi_L^{pt+1}$, where $\pi_F = N_{L/F}\pi_L \equiv \pi_L^p \mod \pi_L^{p+1}$ is a prime element in F. Replacing λ by $\lambda' = \lambda - \pi_F^{t'}\theta$ and α by $\alpha' = \alpha - \pi_F^{pt'}\theta^p + \pi_F^{t'}\theta$, we get $\lambda'^p - \lambda' = \alpha'$ and $L = F(\lambda'), v_F(\alpha') > v_F(\alpha)$. Proceeding in this way we can assume $p \nmid t$ because $v_F(\alpha') \leqslant -s$. Then it follows from (1.4) that $v_L(\lambda^{-1}\sigma(\lambda) - 1) = s$ and $v_F(\alpha) = -s$.

Now we consider the case of char$(F) = 0$.

First we will show that there is an element $\lambda_1 \in L$, such that $v_L(\lambda_1) = -s$ and $v_L(\sigma(\lambda_1) - \lambda_1 - 1) > 0$. Indeed, put $\beta = -\pi_L^{-s}\rho s^{-1}$ with $\rho \in U_F$. Then

$$\sigma(\beta) - \beta = -\pi_L^{-s}\rho s^{-1}\left((1 + \eta\pi_L^s)^{-s} - 1\right) \equiv \rho\eta \mod \pi_L.$$

Hence, if we choose $\overline{\rho} = \overline{\eta}^{-1}$, then $v_L(\sigma(\beta) - \beta - 1) > 0$. Put $\lambda_1 = \beta - b$.

Since $s < pe/(p-1) = e(L)/(p-1)$, we get

$$v_L(\sigma(\lambda_1^p) - \lambda_1^p - 1) > 0 \qquad \text{and} \qquad v_L(\sigma\wp(\lambda_1) - \wp(\lambda_1)) > 0.$$

Second we will construct a sequence $\{\lambda_n\}$ of elements in L satisfying the conditions

$$v_L(\lambda_n) = -s, \quad v_L(\lambda_{n+1} - \lambda_n) \geqslant v_L(\lambda_n - \lambda_{n-1}) + 1,$$
$$v_L(\sigma\wp(\lambda_{n+1}) - \wp(\lambda_{n+1})) \geqslant v_L\left(\sigma\wp(\lambda_n) - \wp(\lambda_n)\right) + 1.$$

Then for $\lambda = \lim \lambda_n$ we obtain $\sigma\wp(\lambda) = \wp(\lambda)$, or in other words $\lambda^p - \lambda = \alpha \in F$ and $v_F(\alpha) = -s$.

Put $\lambda_0 = 0$. Let $\delta_n = \sigma\wp(\lambda_n) - \wp(\lambda_n)$. Then $v_L(\delta_n) > 0$. If $\delta_n = 0$, then put $\lambda_m = \lambda_n$ for $m > n$. Otherwise, by Lemma (2.2), there exists an element $c_n \in F$ such that

$$v_L(\sigma\wp(\lambda_n) - \wp(\lambda_n)) = v_L(\wp(\lambda_n) - c_n) + s.$$

Put $\mu_n = \wp(\lambda_n) - c_n$, $\lambda_{n+1} = \lambda_n + \mu_n$. Then $\sigma\mu_n = \mu_n + \delta_n$, $v_L(\sigma(\lambda_{n+1}) - \lambda_{n+1} - 1) > 0$ and $v_L(\mu_n) > -s$, $v_L(\lambda_{n+1}) = -s$. So

$$v_L(\lambda_{n+1} - \lambda_n) = v_L(\mu_n) = -s + v_L(\sigma\wp(\lambda_n) - \wp(\lambda_n))$$
$$\geqslant -s + 1 + v_L(\sigma\wp\left(\lambda_{n-1}\right) - \wp\left(\lambda_{n-1}\right)) = v_L(\lambda_n - \lambda_{n-1}) + 1$$

for $n > 1$, and $v_L(\lambda_2 - \lambda_1) = -s + v_L(\sigma\wp(\lambda_1) - \wp(\lambda_1)) \geqslant v_L(\lambda_1 - \lambda_0) + 1$. Furthermore,

$$\sigma\wp(\mu_n) - \wp(\mu_n) = \wp(\mu_n + \delta_n) - \wp(\mu_n) = -\delta_n + \sum_{i=1}^{p} \binom{p}{i} \mu_n^{p-i}\delta_n^i.$$

We also get

$$v_L(\sigma\wp(\mu_n) - \wp(\mu_n) + \delta_n) \geqslant v_L(\delta_n) + 1.$$

Moreover,

$$\sigma\wp(\lambda_{n+1}) - \wp(\lambda_{n+1}) = \sigma\wp(\lambda_n) - \wp(\lambda_n) + \sigma\wp(\mu_n) - \wp(\mu_n) + \sum_{i=1}^{p-1} \binom{p}{i} \left(\sigma(\lambda_n^{p-i}\mu_n^i) - \lambda_n^{p-i}\mu_n^i\right)$$

and

$$\sigma(\lambda_n^{p-i}\mu_n^i) - \lambda_n^{p-i}\mu_n^i = \lambda_n^{p-i}\mu_n^i\left(\varepsilon_n^{p-i}(1 + \delta_n\mu_n^{-1})^i - 1\right),$$

where $\lambda_n^{-1}\sigma\lambda_n = \varepsilon_n \in U_{s,L}, v_L(\delta_n\mu_n^{-1}) = v_L(\delta_n) + s - v_L(\delta_n) = s$. Hence, for $1 \leqslant i \leqslant p-1$ we get

$$v_L\left(\sigma(\lambda_n^{p-i}\mu_n^i) - \lambda_n^{p-i}\mu_n^i\right) \geqslant -(p-1)s + v_L(\delta_n) \geqslant -pe + v_L(\delta_n) + 1.$$

As a result we obtain the following inequality

$$v_L\left(\sigma\wp(\lambda_{n+1}) - \wp(\lambda_{n+1})\right) \geqslant v_L(\delta_n) + 1,$$

which completes the proof. □

(2.5). The assertions converse to Propositions (2.1) and (2.4) can be formulated as follows.

PROPOSITION. *Let F be a complete discrete valuation field with a residue field of characteristic $p > 0$. Then every polynomial $X^p - X - \alpha$ with $\alpha \in F$, $v_F(\alpha) > -pe/(p-1)$ if* char$(F) = 0$ *and $e = e(F)$, either splits completely or has a root λ which generates a cyclic extension $L = F(\lambda)$ over F of degree p. In the last case $v_L(\sigma(\lambda) - \lambda - 1) > 0$ for some generator σ of* Gal(L/F). *If $\alpha \in U_F, \overline{\alpha} \notin \wp(\overline{F})$, then L/F is unramified; if $\alpha \in \mathcal{M}_F$, then $\lambda \in F$; if $\alpha \notin \mathcal{O}_F$ and $p \nmid v_F(\alpha)$, then L/F is totally ramified with $s = -v_F(\alpha)$.*

Proof. Let $\alpha \in \mathcal{M}_F$, $f(X) = X^p - X - \alpha$. Then $f(0) \in \mathcal{M}_F$, $f'(0) \notin \mathcal{M}_F$, and, by Hensel Lemma (1.2) Ch. II, for every integer a there is $\lambda \in \mathcal{M}_F$ with $f(\lambda) = 0$, $\lambda - a \in \mathcal{M}_F$. This means that $f(X)$ splits completely in F. If $\alpha \in U_F, \overline{\alpha} \notin \wp(\overline{F})$, then Proposition (3.2) Ch. II shows that $F(\lambda)/F$ is an unramified extension and Proposition (3.3) Ch. II shows that $F(\lambda)/F$ is Galois of degree p. The generator $\sigma \in$ Gal(L/F), for which $\bar{\sigma}\bar{\alpha} = \overline{\alpha} + 1$, is the required one.

If $\alpha \notin \mathcal{O}_F$, then let λ be a root of the polynomial $X^p - X - \alpha$ in F^{alg} and $L = F(\lambda)$. Put

$$g(Y) = (\lambda + Y)^p - (\lambda + Y) - \alpha = Y^p + \binom{p}{1}\lambda Y^{p-1} + \dots + \binom{p}{p-1}\lambda^{p-1}Y - Y.$$

If $\mathrm{char}(F) = p$, then L/F is evidently cyclic of degree p when $\alpha \notin \wp(F)$. If $\mathrm{char}(F) = 0$, then $v_L\big(\binom{p}{i}\lambda^i\big) > e(L|F)(e - ei/(p-1)) \geqslant 0$ for $i \leqslant p-1$ and $\overline{g}(Y) = Y^p - Y$ over $\overline{L}$. Hence by Hensel Lemma $g(Y)$ splits completely in L. Therefore, L/F is cyclic of degree p if $f(X)$ does not split over F. Let σ be a generator of $\mathrm{Gal}(L/F)$, such that $\sigma(\lambda) - \lambda$ is a root of $g(Y)$ and is congruent to 1 mod π_L. Then $v_L(\sigma(\lambda) - \lambda - 1) > 0$. If $p \nmid v_F(\alpha)$, then the equality $pv_L(\lambda) = v_L(\alpha)$ implies $e(L|F) = p$, and L/F is totally ramified. It follows from the definition of s in (1.4) that $s = v_L(\sigma(\lambda)\cdot\lambda^{-1} - 1)$, and consequently $s = v_L(\sigma(\lambda) - \lambda) - v_L(\lambda) = -v_L(\lambda) = -v_F(\alpha)$. □

COROLLARY. *Let λ be a root of the polynomial $X^p - X + \theta^p\alpha$ with $\theta \in U_F$, $v_F(\alpha) = -s > -pe/(p-1)$, $p \nmid s$. Let $L = F(\lambda)$. Then $\alpha \in N_{L/F}L^*$ and $1 + \theta^{-p}\wp(\mathcal{O}_F)\alpha^{-1} + \pi_F^{s+1}\mathcal{O}_F \subset N_{L/F}L^*$, where $\wp(\mathcal{O}_F) = \{\wp(\beta) : \beta \in \mathcal{O}_F\}$.*

Proof. The preceding Proposition shows that L/F is a totally ramified extension of degree p and that $v_L(\sigma(\pi_L)\pi_L^{-1} - 1) = s$ for a generator σ of $\mathrm{Gal}(L/F)$ and a prime element π_L in L. Put $f(X) = X^p - X + \theta^p\alpha$. Then we get $N_{L/F}(-\lambda) = f(0) = \theta^p\alpha$ and $\alpha = N_{L/F}(-\lambda\theta^{-1})$. For $\beta \in \mathcal{O}_F$ put

$$g(Y) = f(\beta - Y) = (\beta - Y)^p - (\beta - Y) + \theta^p\alpha.$$

Then

$$N_{L/F}(\beta - \lambda) = g(0) = \wp(\beta) + \theta^p\alpha.$$

Therefore, $1 + \wp(\beta)\theta^{-p}\alpha^{-1} \subset N_{L/F}L^*$. It remains to use Corollary (1.5). □

(2.6). REMARK. Another description of totally ramified extensions of degree p can be found in [Am]. For a treatment of Artin–Schreier extensions by using *Lubin–Tate formal groups* and a generalization to n-dimensional local fields see [FVZ].

Exercises.

1. Let L/F be a Galois extension of degree $p = \mathrm{char}(\overline{F})$, and let $\overline{L}/\overline{F}$ be an inseparable extension of degree p. Let θ, σ, s be as in Exercise 2 section 1. Let $\mathrm{char}(F) = 0$ and $e = e(F)$ the absolute index of ramification of F.
   a) Prove an analog of Lemma (2.2) (with θ instead of π_L).
   b) Show that $s \leqslant e/(p-1)$.
   c) Show that $s < e/(p-1)$ if and only if there exists an element $\lambda_1 \in L$, such that $v_L(\sigma(\lambda_1) - \lambda_1 - 1) > 0, v_L(\lambda_1) = -ps$.

d) Show that if $s < e/(p-1)$, then $L = F(\lambda)$, where the element λ is a root of the polynomial $X^p - X - \alpha$ with $\alpha \in F$, $v_F(\alpha) = -ps$, $v_L(\sigma(\lambda) - \lambda - 1) > 0$.

e) Maintaining the conditions in d) show that $\alpha = \beta_1\beta_2^p$ with $\beta_1 \in U_F$, $\overline{\beta}_1 \notin \overline{F}^p$, $v_F(\beta_2) = -s$.

f) Show that if $L = F(\lambda)$, where $\lambda^p - \lambda = \alpha$ and α is as in e), then L/F is Galois of degree p and $\overline{L}/\overline{F}$ is inseparable of degree p.

2. (*R. MacKenzie* and *G. Whaples* [MW])

a) Let $F = \mathbb{Q}$, and let L be the unique cyclic subextension of prime degree p in $F(\zeta_{p^2})/F$ (ζ_{p^2} is a primitive p^2th root of unity). Show that the equation $X^p - X - \alpha = 0$ for $\alpha \in F$ can have at most three real roots. However, for $p \geqslant 3$ any defining equation of L over F splits into real linear factors in $\mathbb{C}$. Hence, L/F is not generated by a root of any Artin–Schreier equation for $p > 3$.

b) Let $F = \mathbb{Q}$, and let L be the splitting field of the polynomial $X^p - X - 1$. Show that L/F is not a cyclic extension when $p \geqslant 3$.

3. Let $L = F(\gamma)$, $\gamma^p - \gamma = \alpha \in F$, be a cyclic extension of degree p over F. Assume that F itself is a cyclic extension of K with a generator σ. Describe what condition should satisfy $\sigma\alpha$ so that L/K is a Galois (abelian) extension?

4. (*V.A. Abrashkin* [Ab1]) Let F be a complete discrete valuation field of characteristic 0 with residue field $\overline{F}$ of characteristic p. Let $\mathbb{F}_{p^n} \subset \overline{F}$ for some integer $n \geqslant 1$. Let λ be a root of the polynomial $X^{p^n} - X - \alpha$ with $\alpha \in F$, $v_F(\alpha) > -p^n e(F)/(p^n - 1)$. Then the extension $F(\lambda)/F$ is said to be elementary.

a) Show that $F(\lambda)/F$ is Galois.

b) Show that if $p \nmid v_F(\alpha)$, $v_F(\alpha) < 0$, then $F(\lambda)/F$ is a totally ramified Galois extension of degree p^n, and if $G = \mathrm{Gal}(F(\lambda)/F)$, then $G = G_0 = \dots = G_s, G_{s+1} = \dots = \{1\}$ for the ramification groups of G, where $s = -v_F(\alpha)$.

c) Show that if $\alpha_1 - \alpha_2 \in \mathcal{M}_F$, then $F(\lambda_1) = F(\lambda_2)$.

d) Show that if $\alpha_3 - \alpha_1 - \alpha_2 \in \mathcal{M}_F$, then $F(\lambda_3)$ is contained in the compositum of $F(\lambda_1)$ and $F(\lambda_2)$.

e) Show that if $\overline{F}$ is algebraically closed, then $\mathcal{M}_F$ can be replaced with $\mathcal{O}_F$ in c), d).

For additional properties of elementary extensions see [Ab]. The theory of such extensions is used to show that there are no abelian schemes over $\mathbb{Z}$.

5. Generalize the results of this section to Henselian discrete valuation fields.

3. The Hasse–Herbrand Function

In this section we associate to a finite separable extension L/F a certain real function $h_{L/F}$ which partially describes the behaviour of the norm map from arithmetical point of view. In subsections (3.1), (3.2) we study the case of Galois extensions and in subsection (3.3) the case of separable extensions. In (3.4) we derive first applications. Then we relate the function $h_{L/F}$ which was originally introduced in a different way by *H. Hasse* and *J. Herbrand* to properties of ramification subgroups and prove in section (3.5) a theorem of *J. Herbrand* on the behaviour of ramification groups in extensions; further properties of ramification subgroups are studied in (3.6) and (3.7).

We maintain the hypothesis of the preceding sections concerning F, and assume in addition that all residue field extensions are separable.

(3.1). PROPOSITION. *Let the residue field $\overline{F}$ be infinite. Let L/F be a finite Galois extension, $N = N_{L/F}$. Then there exists a unique function*

$$h = h_{L/F}: \mathbb{N} \to \mathbb{N}$$

such that $h(0) = 0$ and

$$NU_{h(i),L} \subset U_{i,F}, \quad NU_{h(i),L} \not\subset U_{i+1,F}, \quad NU_{h(i)+1,L} \subset U_{i+1,F}.$$

Proof. The uniqueness of h follows immediately. Indeed, for $j > h(i)$ $NU_{j,L} \subset U_{i+1,F}$, hence if $\widetilde{h}$ is another function with the required properties, then $\widetilde{h}(i) \leqslant h(i), h(i) \leqslant \widetilde{h}(i)$, i.e., $h = \widetilde{h}$.

As for the existence of h, we first consider the case of an unramified extension L/F. Then Proposition (1.2) shows that in this case $h(i) = i$ (because $N_{\overline{L}/\overline{F}}(\overline{L}^*) \neq 1$ and $\mathrm{Tr}_{\overline{L}/\overline{F}}\overline{L} = \overline{F}$). The next case to consider is a totally ramified cyclic extension L/F of prime degree. In this case Proposition (1.3) and Proposition (1.5) describe the behavior of $N_{L/F}$. By means of the homomorphisms $\lambda_{i,L}$, the map $N_{L/F}$ is determined by some nonzero polynomials over $\overline{L}$. The image of $\overline{L}$ under the action of such a polynomial is not zero since $\overline{L}$ is infinite. Hence, we obtain

$$h(i) = |L : F|i,$$

if L/F is totally tamely ramified, and

$$h(i) = \begin{cases} i, & i \leqslant s, \\ s(1-p) + pi, & i \geqslant s, \end{cases}$$

if L/F is totally ramified of degree $p = \mathrm{char}(\overline{F}) > 0$.

Now we consider the general case. Note that if we have the functions $h_{L/M}$ and $h_{M/F}$ for the Galois extensions $L/M, M/F$, then for the extension L/F one can put $h_{L/F} = h_{L/M} \circ h_{M/F}$. Indeed,

$$N_{L/F}U_{h_{L/F}(i),L} \subset N_{M/F}U_{h_{M/F}(i),M} \subset U_{i,F}.$$

Furthermore, the behavior of $N_{L/F}$ is determined by some nonzero polynomials (the composition of the polynomials for $N_{L/M}$ and $N_{M/F}$, the existence of which can be assumed by induction). Hence

$$N_{L/F}U_{h_{L/F}(i),L} \not\subset U_{i+1,F}.$$

Since

$$N_{L/F}U_{h_{L/F}(i)+1,L} \subset N_{M/F}U_{h_{M/F}(i)+1,M} \subset U_{i+1,M},$$

we deduce that $h = h_{L/F}$ is the desired function.

In the general case we put $h_{L/F} = h_{L/L_0}$ for $L_0 = L \cap F^{\mathrm{ur}}$ and determine h_{L/L_0} by induction using Corollary 3 of (4.4) Ch. II, which shows that L/L_0 is solvable. □

(3.2). To treat the case of finite residue fields we need

LEMMA. *Let L/F be a finite separable totally ramified extension. Then for an element $\alpha \in L$ we get*

$$N_{L/F}(\alpha) = N_{\widehat{L^{\mathrm{ur}}}/\widehat{F^{\mathrm{ur}}}}(\alpha)$$

where $\widehat{F^{\mathrm{ur}}}$ is the completion of F^{ur}, $\widehat{L^{\mathrm{ur}}} = L\widehat{F^{\mathrm{ur}}}$.

Proof. Let $L = F(\pi_L)$ with a prime element π_L in L, and let $\alpha \in L$. Let

$$\alpha\pi_L^i = \sum_{j=0}^{n-1} c_{ij}\pi_L^j \quad \text{with } c_{ij} \in F, 0 \leqslant i \leqslant n-1, n = |L:F|.$$

Then $N_{L/F}(\alpha) = \det(c_{ij})$ (see [La1, Ch. VIII]). Since $L^{\mathrm{ur}} = F^{\mathrm{ur}}(\pi_L)$ and

$$|L^{\mathrm{ur}} : F^{\mathrm{ur}}| = e(L^{\mathrm{ur}}|F^{\mathrm{ur}}) = e(L^{\mathrm{ur}}|F) = e(L|F) = |L:F|,$$

we get

$$N_{L^{\mathrm{ur}}/F^{\mathrm{ur}}}(\alpha) = \det(c_{ij}) = N_{L/F}(\alpha).$$

Finally, let E/F^{ur} be a finite totally ramified Galois extension with $E \supset L^{\mathrm{ur}}$. Let $G = \mathrm{Gal}(E/F^{\mathrm{ur}}), H = \mathrm{Gal}(E/L^{\mathrm{ur}})$, and let G be the disjoint union of $\sigma_i H$ with $\sigma_i \in G, 1 \leqslant i \leqslant |L^{\mathrm{ur}} : F^{\mathrm{ur}}|$. Then

$$N_{L^{\mathrm{ur}}/F^{\mathrm{ur}}}(\alpha) = \prod \sigma_i(\alpha) = N_{\widehat{L^{\mathrm{ur}}}/\widehat{F^{\mathrm{ur}}}}(\alpha),$$

because G and H are isomorphic to $\mathrm{Gal}(\widehat{E}/\widehat{F^{\mathrm{ur}}})$ and $\mathrm{Gal}(\widehat{E}/\widehat{L^{\mathrm{ur}}})$ by (4) in Theorem (2.8) Ch. II. □

This Lemma shows that for a finite totally ramified Galois extension L/F the functions $h_{L/F}$ and $h_{\widehat{L^{\mathrm{ur}}}/\widehat{F^{\mathrm{ur}}}}$ coincide. Now, if L/F is a finite Galois extension, we put

$$h_{L/F} = h_{L/L_0} = h_{\widehat{L^{\mathrm{ur}}}/\widehat{F^{\mathrm{ur}}}}.$$

In particular, if $\overline{F}$ is finite we put $h_{L/F} = h_{\widehat{L^{\mathrm{ur}}}/\widehat{F^{\mathrm{ur}}}}$ (the residue field of $\widehat{F^{\mathrm{ur}}}$ is infinite as the separable closure of a finite field).

It is useful to extend this function to real numbers. For unramified extension, or tamely totally ramified extension of prime degree, or totally ramified extension of degree $p = \mathrm{char}(\overline{F}) > 0$ put

$$h_{L/F}(x) = x, \quad h_{L/F}(x) = |L:F|x, \quad h_{L/F}(x) = \begin{cases} x, & x \leqslant s, \\ s(1-p) + px, & x \geqslant s \end{cases}$$

for real $x \geqslant 0$ respectively. Using the solvability of L/L_0 (Corollary 3 of (4.4) Ch. II) and the equality $h_{L/F} = h_{L/M} \circ h_{M/F}$ define now $h_{L/F}(x)$ as the composite of the functions for a tower of cyclic subextensions in L/L_0.

PROPOSITION. *Thus defined function* $h_{L/F}\colon [0,+\infty) \to [0,+\infty)$ *is independent on the choice of a tower of subfields. The function* $h_{L/F}$ *is called the Hasse–Herbrand function of* L/F. *It is piecewise linear, continuous and increasing.*

Proof. It suffices to show that if M_1/M, M_2/M are cyclic extensions of prime degree, then

$$(*) \qquad h_{E/M_1} \circ h_{M_1/M} = h_{E/M_2} \circ h_{M_2/M}$$

where $E = M_1M_2$.

Note that each of $h_{M_1/M}(x)$, $h_{M_2/M}(x)$ has at most one point at which its derivate is not continuous. Therefore there at most two points at which the function of the left (resp. right) hand side of $(*)$ has discontinuous derivative. By looking at graphs of the functions it is obvious that at such points the derivative strictly increases and there is at most one such noninteger point for at most one of the composed functions of the left hand side and the right hand side of $(*)$. At this point (if it exists) the derivative jumps from p to p^2.

From the uniqueness in the preceding Proposition we deduce that the left and right hand sides of $(*)$ are equal at all nonnegative integers. Thus, elementary calculus shows that the left and right hand sides of $(*)$ are equal at all nonnegative real numbers. □

(3.3). Let the residue field of F be perfect. For a finite separable extension L/F put

$$h_{L/F} = h_{E/L}^{-1} \circ h_{E/F},$$

where E/F is a finite Galois extension with $E \supset L$. Then $h_{L/F}$ is well defined, since if E'/F is a Galois extension with $E' \supset L$ and $E'' = E'E$, then

$$h_{E''/L}^{-1} \circ h_{E''/F} = \left(h_{E''/E'} \circ h_{E'/L}\right)^{-1} \circ \left(h_{E''/E'} \circ h_{E'/F}\right) = h_{E'/L}^{-1} \circ h_{E'/F}$$

and, similarly, $h_{E''/L}^{-1} \circ h_{E''/F} = h_{E/L}^{-1} \circ h_{E/F}$. We can easily deduce from this that the equality

$$h_{L/F} = h_{L/M} \circ h_{M/F} \qquad (*)$$

holds for separable extensions.

PROPOSITION. *Let* L/F *be a finite separable extension, and let* $\overline{F}$ *be perfect. Then* $h_{L/F}(\mathbb{N}) \subset \mathbb{N}$ *and the left and right derivatives of* $h_{L/F}$ *at any point are positive integers.*

Proof. Let E/F be a finite Galois extension with $E \supset L$. Then from Lemma (3.2) we get

$$h_{L/F} = h_{E/L}^{-1} \circ h_{E/F} = h_{\widehat{E^{\mathrm{ur}}}/\widehat{L^{\mathrm{ur}}}}^{-1} \circ h_{\widehat{E^{\mathrm{ur}}}/\widehat{F^{\mathrm{ur}}}} = h_{\widehat{L^{\mathrm{ur}}}/\widehat{F^{\mathrm{ur}}}}.$$

Put $G = \mathrm{Gal}(\widehat{E^{\mathrm{ur}}}/\widehat{F^{\mathrm{ur}}})$, $H = \mathrm{Gal}(\widehat{E^{\mathrm{ur}}}/\widehat{L^{\mathrm{ur}}})$. Since G is a solvable group, there exists a chain of normal subgroups

$$G \triangleright G_{(1)} \triangleright \cdots \triangleright G_{(m)} = \{1\},$$

such that $G_{(i)}/G_{(i+1)}$ is a cyclic group of prime order. Then we obtain the chain of subgroups

$$G \geqslant G_{(1)}H \geqslant \ldots \geqslant G_{(m)}H = H,$$

for which $G_{(i+1)}H$ is of prime index or index 1 in $G_{(i)}H$. This shows the existence of a tower of fields

$$\widehat{F^{\mathrm{ur}}} - M_1 - \cdots - M_{n-1} - M_n = \widehat{L^{\mathrm{ur}}},$$

such that M_{i+1}/M_i is a separable extension of prime degree. Therefore, it suffices to prove the statements of the Proposition for such an extension.

If M_{i+1}/M_i is a totally tamely ramified extension of degree l, then $\pi = \pi_1^l$ is a prime element in M_i for some prime element π_1 in M_{i+1}. Since l is relatively prime with $\mathrm{char}(\overline{F})$, we obtain, using the Henselian property of M_i and the equality $\overline{M}_i = \overline{F}^{\mathrm{sep}}$, that a primitive l th root of unity belongs to M_i. This means that M_{i+1}/M_i is a Galois extension and

$$h_{M_{i+1}/M_i}(x) = lx.$$

If M_{i+1}/M_i is an extension of degree $p = \mathrm{char}(\overline{F}) > 0$, then let K/M_i be the smallest Galois extension, for which $K \supset M_{i+1}$. Let K_1 be the maximal tamely ramified extension of M_i in K; then $l = e(K_1|M_i) = e(K|M_{i+1})$ is relatively prime to p. Choose prime elements π and π_1 in M_{i+1} and K such that $\pi = \pi_1^l$. Let $f(X) \in M_i[X]$ be the monic irreducible polynomial of π over M_i. Then

$$f'(\pi) = \prod_{i=1}^{p-1} \left(\pi - \sigma^i(\pi)\right) = \prod_{i=1}^{p-1} \left(\pi_1^l - \sigma^i(\pi_1^l)\right),$$

where σ is a generator of $\mathrm{Gal}(K/M_i)$. Let s be defined for K/K_1 as in (1.4). Then $v_K\left(\pi_1^l - \sigma^i(\pi_1^l)\right) = l + s$ for $1 \leqslant i \leqslant p-1$, and $(p-1)(l+s) = v_K\left(f'(\pi)\right)$ is divisible by l. We deduce that $l|(p-1)s$ and

$$h_{M_{i+1}/M_i}(x) = \frac{1}{l} h_{K/K_1}(lx) = \begin{cases} x, & x \leqslant sl^{-1}, \\ s(1-p)l^{-1} + px, & x \geqslant sl^{-1}. \end{cases}$$

These considerations complete the proof. □

COROLLARY. *The function* $h_{L/F}$ *is piecewise linear, continuous and increasing.*

REMARK. $h_{L/F}$ possesses the properties of Proposition (3.1) in the general case of a separable extension L/F (see Exercise 1).

(3.4). The following assertion clarifies relation between the Hasse–Herbrand function and the norm map.

PROPOSITION. *Let* L/F *be a finite separable extension.*
Then for $\varepsilon \in \mathcal{O}_L$

$$h_{L/F}\Big(v_F\big(N_{L/F}(\varepsilon) - 1\big)\Big) \geqslant v_L(\varepsilon - 1),$$

and if $v_L(\alpha - \beta) > 0$ *for* $\alpha, \beta \in L$, *then*

$$h_{L/F}\Big(v_F\big(N_{L/F}(\alpha) - N_{L/F}(\beta)\big)\Big) \geqslant v_L(\alpha - \beta).$$

Proof. First we show that the second inequality is a consequence of the first one. Indeed, if $v_L(\beta) \geqslant v_L(\alpha - \beta)$, then $v_L(\alpha) \geqslant v_L(\alpha - \beta)$, and applying Theorem (2.5) Ch. II we get

$$v_F\big(N_{L/F}(\alpha) - N_{L/F}(\beta)\big) \geqslant v_L(\alpha - \beta).$$

Since $h_{L/F}(x) \geqslant x$, we obtain the second inequality. If $v_L(\beta) < v_L(\alpha - \beta)$, then $v_L(1 - \alpha\beta^{-1}) \geqslant v_L(\alpha - \beta) - v_L(\beta) > 0$, and putting $\varepsilon = \alpha\beta^{-1}$, we deduce

$$h_{L/F}\Big(v_F\big(N_{L/F}(\alpha) - N_{L/F}(\beta)\big)\Big) \geqslant v_L(\beta) + v_L(1 - \alpha\beta^{-1}) \geqslant v_L(\alpha - \beta).$$

We now verify the first inequality of the Proposition. By the proof of the previous Proposition, we may assume that L/F is totally ramified and $\overline{F}$ is algebraically closed. It is easy to show that if the first inequality holds for L/M and M/F, then it holds for L/F. The arguments from the proof of the previous Proposition imply now that it suffices to verify the first inequality for a separable extension L/F of prime degree. If L/F is tamely ramified, then L/F is Galois, and the inequality follows from Proposition (1.3). If $|L : F| = p = \operatorname{char}(\overline{F}) > 0$, then Proposition (1.5) implies the required inequality for the Galois case. In general, assume that E/F is the minimal Galois extension such that $E \supset L$, and let E_1 is the maximal tamely ramified subextension of F in E. Let $l = |E : L| = |E_1 : F|$. Then $N_{L/F}(U_{i,L}) = N_{E/F}(U_{li,E}) \subset N_{E_1/F}(U_{j,E_1})$ with $j \geqslant h^{-1}_{E/E_1}(li)$. Hence, $N_{L/F}(U_{i,L}) \subset U_{k,F}$ with $lk \geqslant h^{-1}_{E/E_1}(li)$, i.e., $k \geqslant h^{-1}_{L/F}(i)$, as desired. □

(3.5). We will relate the Hasse–Herbrand function to ramification groups which are defined in (4.3) Ch. II.

If H is a subgroup of the Galois group G, then $H_x = H \cap G_x$. As for the quotients, the description is provided by the following

THEOREM (HERBRAND). *Let L/F be a finite Galois extension and let M/F be a Galois subextension. Let x, y be nonnegative real numbers related by $y = h_{L/M}(x)$.*

Then the image of $\mathrm{Gal}(L/F)_y$ *in* $\mathrm{Gal}(M/F)$ *coincides with* $\mathrm{Gal}(M/F)_x$.

Proof. The cases $x \leqslant 1$ or $e(L|M) = 1$ are easy and left to the reader. Due to solvability of Galois groups of totally ramified extensions it is sufficient to prove the assertion in the case of a ramified cyclic extension L/M of prime degree l.

If $l \neq p$, then using Proposition (3.5) Ch. II choose a prime element π of L such that $\pi_M = \pi^l$ is a prime element of M. Then for every $\tau \in \mathrm{Gal}(L/F)_1$ we have $\pi_M^{-1}\tau\pi_M = (\pi^{-1}\tau\pi)^l$ and therefore

$$v_L(\pi^{-1}\tau\pi - 1) = v_L\big((\pi^{-1}\tau\pi)^l - 1\big) = l v_M(\pi_M^{-1}\tau\pi_M - 1).$$

Consider now the most interesting case $l = p$, $x \geqslant 1$. Let π_L be a prime element of L. Put $s = s(L|M)$, see (1.4).

The element $\pi_M = N_{L/M}\pi_L$ is a prime element of M. Let $\tau \in \mathrm{Gal}(L/F)_y$. We have $\pi_M^{-1}\tau\pi_M = N_{L/M}(\pi_L^{-1}\tau\pi_L)$.

From Proposition (3.4) we get

$$h_{L/M}(v_M(\pi_M^{-1}\tau\pi_M) - 1) = h_{L/M}(v_M(N_{L/M}(\pi_L^{-1}\tau\pi_L) - 1)) \geqslant y,$$

so $\tau|_M$ belongs to $\mathrm{Gal}(M/F)_x$.

Conversely, if $\tau|_M \in \mathrm{Gal}(M/F)_x$, then $i = v_M(\pi_M^{-1}\tau\pi_M - 1) \geqslant x$. If $i \leqslant s$ then applying (1.5) we deduce that $\tau \in \mathrm{Gal}(L/F)_i = \mathrm{Gal}(L/F)_y$. If $i > s$ then Proposition (4.5) Ch. II and (1.5) show that $j = v_L(\pi_L^{-1}\tau\pi_L - 1) = s + pr$ for some nonnegative integer r.

If $r > 0$ then Proposition (1.5) implies that $i = s + r$ and $\tau \in \mathrm{Gal}(L/F)_j = \mathrm{Gal}(L/F)_y$. If $r = 0$ then since $i > s$ from the same Proposition we deduce that

$$\frac{\tau\pi_L}{\pi_L} \equiv \frac{\sigma\pi_L}{\pi_L} \mod \mathcal{M}_L^{s+1}$$

for an appropriate generator σ of $\mathrm{Gal}(L/M)$. Then $\tau\sigma^{-1}$ belongs to $\mathrm{Gal}(L/F)_k$ for $k > s$. Due to the previous discussions (view k as $j > s$ above) $k = h_{L/M}(i)$ and τ belongs to $\mathrm{Gal}(L/F)_y\,\mathrm{Gal}(L/M)$, as required. □

COROLLARY. *Define the upper ramification filtration of* $G = \mathrm{Gal}(L/F)$ *as*

$$G(x) = \mathrm{Gal}(L/F)_{h_{L/F}(x)}.$$

Then for a normal subgroup H of G the previous theorem shows that

$$(G/H)(x) = G(x)H/H.$$

Definition. For an infinite Galois extension L/F define *upper ramification subgroups* of $G = \mathrm{Gal}(L/F)$ as

$$G(x) = \varprojlim \mathrm{Gal}(M/F)(x)$$

where M/F runs through all finite Galois subextensions of L/F. Real numbers x such that $G(x) \neq G(x+\delta)$ for every $\delta > 0$ are called *upper ramification jumps of* L/F.

(3.6). The following Proposition is a generalization of results of section 1.

Suppose that L/F is a finite totally ramified Galois extension and that $|L:F|$ is a power of $p = \mathrm{char}(\overline{F})$. Put $G = \mathrm{Gal}(L/F)$. For the chain of normal ramification groups

$$G = G_1 \geqslant G_2 \geqslant \ldots \geqslant G_n > G_{n+1} = \{1\}$$

let L_m be the fixed field of G_m; then we get the tower of fields

$$F = L_1 - L_2 - \cdots - L_n - L_{n+1} = L.$$

Proposition. *Let $1 \leqslant m \leqslant n$. Then $\mathrm{Gal}(L_{m+1}/L_m)$ coincides with the ramification group $\mathrm{Gal}(L_{m+1}/L_m)_m$, $\mathrm{Gal}(L_{m+1}/L_m)_{m+1} = \{1\}$, and $h_{L_{m+1}/L_m}(m) = m$.*

Moreover, if $i < m$, then $h_{L_{m+1}/L_m}(i) = i$ and the homomorphism

$$U_{i,L_{m+1}}/U_{i+1,L_{m+1}} \longrightarrow U_{i,L_m}/U_{i+1,L_m}$$

induced by N_{L_{m+1}/L_m} is injective;
if $i > m$, then the homomorphism

$$U_{h(i),L_{m+1}}/U_{h(i)+1,L_{m+1}} \longrightarrow U_{i,L_m}/U_{i+1,L_m}$$

induced by N_{L_{m+1}/L_m} for $h = h_{L_{m+1}/L_m}$ is bijective.

Furthermore, the homomorphism

$$U_{h(i),L}/U_{h(i)+1,L} \longrightarrow U_{i,F}/U_{i+1,F}$$

induced by $N_{L/F}$ for $h = h_{L/F}$, is bijective if $h(i) > n$.

Proof. Induction on m. Base of induction $m = n$. Since $\mathrm{Gal}(L/L_n)_x$ is equal to the group $\mathrm{Gal}(L/F)_x \cap \mathrm{Gal}(L/L_n)$, we deduce that $\mathrm{Gal}(L/L_n)_n = \mathrm{Gal}(L/L_n)$ and $\mathrm{Gal}(L/L_n)_{n+1} = \{1\}$, and $h_{L/L_n}(x) = x$ for $x \leqslant n$. All the other assertions for $m = n$ follow from Proposition (1.5).

Induction step $m+1 \to m$. The transitivity property of the Hasse–Herbrand function implies that $h_{L/L_{m+1}}(x) = x$ for $x \leqslant m+1$. Now from the previous Theorem

$$\mathrm{Gal}(L_{m+1}/L_m)_x = \mathrm{Gal}(L/L_m)_{h_{L/L_{m+1}}(x)}\,\mathrm{Gal}(L_{m+1}/L_m)/\,\mathrm{Gal}(L_{m+1}/L_m).$$

We deduce that $\mathrm{Gal}(L_{m+1}/L_m)_m = \mathrm{Gal}(L_{m+1}/L_m)$ and $\mathrm{Gal}(L_{m+1}/L_m)_{m+1} = \{1\}$. The rest follows from Proposition (1.5).

To deduce the last assertion note that $k = h_{L/F}(i) > n$ implies $j = h_{L_m/F}(i) > m$. □

COROLLARY. *The word "injective" in the Proposition can be replaced by "bijective" if $\overline{F}$ is perfect.*

(3.7). PROPOSITION. *Let L/F be a finite Galois extension, and let $G = \mathrm{Gal}(L/F)$, $h = h_{L/F}$. Let h'_l and h'_r be the left and right derivatives of h. Then $h'_l(x) = |G_0 : G_{h(x)}|$, and*

$$h'_r(x) = \begin{cases} |G_0 : G_{h(x)}| & \text{if } h(x) \text{ is not integer,} \\ |G_0 : G_{h(x)+1}| & \text{if } h(x) \text{ is integer.} \end{cases}$$

Therefore

$$h_{L/F}(x) = \int_0^x |G_0 : G_{h(t)}| dt.$$

Proof. Using the equality (∗) of (3.3), we may assume that L/F is a totally ramified extension the degree of which is a power of $p = \mathrm{char}(\overline{F}) > 0$. Then $G = G_0 = G_1$. We proceed by induction on the degree $|L : F|$. Let L_n be identical to that of (3.6); then $|L_n : F| < |L : F|$. Since $(G/G_n)_m = G_m/G_n$ for $m \leqslant n$ due to (3.6), we deduce the following series of claims.

If $h_{L_n/F}(x) \leqslant n$, then, by Proposition (3.6), $h_{L/F}(x) = h_{L_n/F}(x)$ and

$$h'_l(x) = |(G/G_n) : (G/G_n)_{h(x)}| = |G : G_{h(x)}|.$$

If $h_{L_n/F}(x) < n$ and $h_{L/F}(x) = h_{L_n/F}(x)$ is not integer, then $h'_r(x) = |G : G_{h(x)}|$. If $h_{L_n/F}(x)$ is an integer $< n$, then

$$h'_r(x) = |(G/G_n) : (G/G_n)_{h(x)+1}| = |G : G_{h(x)+1}|.$$

Since the derivative (right derivative) of $h_{L/L_n}(x)$ for $x > n$ (resp. $x \geqslant n$) is equal to $|G_n : (G_n)_{n+1}| = |G_n|$, we deduce that if $h_{L_n/F}(x) > n$, then

$$h'_l(x) = |G_n| \cdot |G : G_n| = |G| = |G : G_{h(x)}|.$$

So if $h_{L_n/F}(x) \geqslant n$, then $h'_r(x) = |G_n| \cdot |G : G_n| = |G|$. This completes the proof. □

REMARKS.

1. The function $h_{L/F}$ often appears under the notation $\psi_{L/F}$; in which case it is defined in quite a different way by using ramification groups, not the norm map. This function is inverse to the function $\varphi_{L/F} = \int_0^x \frac{dt}{|G_0:G_t|}$.

2. Information encoded in the Hasse–Herbrand function can be extended using some additional ramification invariants introduced by *V. Heiermann* [Hei]. These arise when

one investigates more closely Eisenstein polynomials corresponding to prime elements (see also Exercise 6 in section 4).

Exercises.

1. Show that the three properties of the Hasse–Herbrand function obtained in Proposition (3.1) hold for a finite separable extension L/F with a separable residue extension.
2. In terms of the proof of Proposition (3.2) show that $h_{E/M_1} \circ h_{M_1/M} = h_{M_1M_2/M_2} \circ h_{M_2/M}$ by calculating the functions in accordance with the steps below.
   a) Suppose that $|M_1 : M| = l$ is prime to p and $|M_2 : M| = p$. Choose a prime element π of E such that π^l is a prime element of M_2 and calculate all the functions.
   b) Suppose that M_1/M and M_2/M are totally ramified extensions of prime degree p and $M_1 \cap M_2 = M$. Using Proposition (4.5) Ch. II deduce that $s_1 = s(E|M_2)$ is congruent to $s_2 = s(E|M_1)$ modulo p. Show that if $s(M_2|M) > s(M_1|M)$, then $s(M_1|M) = s_1$ and $s(M_2|M) = s_1 + r$. Show that if $s = s(M_2|M) = s(M_1|M)$, then $s_1 = s_2 \leqslant s$.
3. (*Y. Kawada* [Kaw1]) Let L be an infinite Galois extension of a local field F.
   a) Let M_1/F, M_2/F be finite Galois subextensions of L/F. Show that the set of upper ramification jumps of M_1/F is a subset of upper ramification jumps of M_2/F. Denote by $B(L/F)$ the union of all upper ramification jumps of finite Galois subextensions of L/F.
   b) For a real x define $L(x) = \cup_M M(x)$ where M runs over all finite Galois extensions of F in L and $M(x)$ is the fixed field of $\operatorname{Gal}(M/F)(x)$ inside M. Show that if $x_1 < x_2$, then $L(x_1) \neq L(x_2)$ if and only if $[x_1, x_2) \cap B(L/F) \neq \emptyset$.
   c) Show that if x is the limit of a monotone increasing sequence x_n, then $L(x) = \cup L(x_n)$.
   d) Show that if x is the limit of a monotone decreasing sequence x_n and $x \notin B(L/F)$, then $L(x) = \cap L(x_n)$.
   e) Let x be the limit of a strictly monotone decreasing sequence x_n. Define $L[x] = \cup_M(\cap_n M(x_n))$ where M runs over all finite Galois extensions of F in L. Show that $L[x] = \cap_n L(x_n)$. Show that $L[x] = L(x)$ is and only if $x \notin B(L/F)$.
   f) A subfield E of L, $F \subset E$ is called a ramification subfield if for every finite Galois subextension M/F of L/F there is y such that $E \cap M = M(y)$. Show that every ramifications subfield of L over F coincides either with some $L(x)$ or with some $L[x]$.
   g) Deduce that the set of all upper ramification jumps of L/F is the union of $B(L/F)$ and the limits of strictly monotone decreasing sequences of elements of $B(L/F)$.

4. The Norm and Ramification Groups

We continue the study of ramification groups and the norm map. After recalling Satz 90 in (4.1) we further generalize results of section 1 as Theorem (4.2). In subsection (4.3) we study ramification numbers of abelian extensions.

In this section F is a complete discrete valuation field.

(4.1). The following assertion is of general interest.

PROPOSITION ("SATZ 90"). *Let L/F be a cyclic Galois extension, and let $N_{L/F}(\alpha) = 1$ for some $\alpha \in L$. Then there exists an element $\beta \in L$ such that $\alpha = \beta^{\sigma-1}$, where σ is a generator of* Gal(L/F).

Proof. Let $\beta(\gamma)$ denote

$$\gamma + \alpha^{-1}\sigma(\gamma) + \alpha^{-1}\sigma(\alpha^{-1})\sigma^2(\gamma) + \dots + \alpha^{-1}\sigma(\alpha^{-1})\cdot \ldots \cdot \sigma^{n-2}(\alpha^{-1})\sigma^{n-1}(\gamma)$$

for $\gamma \in L$, $n = |L : F|$. If $\beta(\gamma)$ were equal to 0 for all γ, then we would have a nontrivial solution $1, \alpha^{-1}, \alpha^{-1}\sigma(\alpha^{-1}), \dots$ for the $n \times n$ system of linear equations with the matrix $\left(\sigma^i(\gamma_j)\right)_{0\leqslant i,j\leqslant n-1}$, where $(\gamma_j)_{0\leqslant j\leqslant n-1}$ is a basis of L over F. This is impossible because L/F is separable (see [La1, sect. 5 Ch. VIII]). Hence $\beta(\gamma) \neq 0$ for some $\gamma \in L$. Then $\beta = \beta(\gamma)$ is the desired element. □

COROLLARY. *If L is a cyclic unramified extension of F and $N_{L/F}(\alpha) = 1$ for $\alpha \in L$, then $\alpha = \gamma^{\sigma-1}$ for some element $\gamma \in U_L$.*

Proof. In this case a prime element π in F is also a prime one in L. By the Proposition, $\alpha = \beta^{-1}\sigma(\beta)$ with $\beta = \pi^i\varepsilon$, $\varepsilon \in U_L$. Then $\alpha = \varepsilon^{-1}\sigma(\varepsilon)$. □

Recall that in section 4 Ch. II we employed the homomorphisms

$$\psi_i \colon G_i \to U_{i,L}/U_{i+1,L}$$

(we put $U_{0,L} = U_L$), where $G = \mathrm{Gal}(L/F)$, π_L is a prime element in L, $i \geqslant 0$. Obviously these homomorphisms do not depend on the choice of π_L if L/F is totally ramified. The induced homomorphisms $G_i/G_{i+1} \to U_{i,L}/U_{i+1,L}$ will be also denoted by ψ_i.

(4.2). THEOREM. *Let L/F be a finite totally ramified Galois extension with group G. Let $h = h_{L/F}$. Then for every integer $i \geqslant 0$ the sequence*

$$1 \to G_{h(i)}/G_{h(i)+1} \xrightarrow{\psi_{h(i)}} U_{h(i),L}/U_{h(i)+1,L} \xrightarrow{N_i} U_{i,F}/U_{i+1,F}$$

is exact (the right homomorphism N_i is induced by the norm map).

Proof. The injectivity of $\psi_{h(i)}$ follows from the definitions. It remains to show that if $N_{L/F}\alpha \in U_{i+1,F}$ for $\alpha \in U_{h(i),L}$, then

$$\alpha \equiv \frac{\sigma(\pi_L)}{\pi_L} \mod U_{h(i)+1,L}$$

for some $\sigma \in G_{h(i)}$.

If L/F is a tamely ramified extension of degree l, then the fourth commutative diagram of Proposition (1.3) shows that N_i is injective for $i \geqslant 1$, and the kernel of N_0

coincides with the group of l th roots of unity which is contained in F. Since $\pi_L = \sqrt[l]{\pi_F}$ is a prime element in L for some prime element π_F in F, we get $\ker(N_0) \subset \operatorname{im}(\psi_0)$, and in this case the sequence of the Theorem is commutative.

If L/F is a cyclic extension of degree $p = \operatorname{char}(\overline{F}) > 0$, then the fourth commutative diagram of Proposition (1.5) shows that $\ker(N_s) \subset \operatorname{im}(\psi_s)$ for $s = v_L(\pi_L^{-1}\sigma(\pi_L))$ and a generator σ of $\operatorname{Gal}(L/F)$. Other diagrams of Proposition (1.5) show that N_i is injective for $i \neq s$.

We proceed by induction on the degree $|L:F|$. Since we have already considered the tamely ramified case, we may assume that the maximal tamely ramified extension L_1 of F in L does not coincide with L. Since $|L:L_1|$ is a power of p, the homomorphism induced by N_{L/L_1}

$$U_{0,L}/U_{1,L} \longrightarrow U_{0,L_1}/U_{1,L_1}$$

is the raising to this power of p, and $\ker(N_0)$ is equal to the preimage under this homomorphism of the kernel of $U_{0,L_1}/U_{1,L_1} \longrightarrow U_{0,F}/U_{1,F}$. In other words $\ker(N_0)$ coincides with the group of all l th roots of unity for $l = |L_1:F|$ which is contained in F. Hence the kernel of N_0 is contained in the image of ψ_0, since ψ_0 is injective and $|G_0:G_1| = l$.

Now suppose $i \geqslant 1$. In this case we may assume $L_1 = F$ because the homomorphism N_i induced by $N_{L_1/F}$ is injective for $i \geqslant 1$. Let L_n be as in Proposition (3.6). Then one can express N_i as the composition

$$U_{h(i),L}/U_{h(i)+1,L} \xrightarrow{N'} U_{h_1(i),L_n}/U_{h_1(i)+1,L_n} \xrightarrow{N''} U_{i,F}/U_{i+1,F},$$

where N' and N'' are induced by N_{L/L_n} and $N_{L_n/F}$ respectively, and $h_1(i) = h_{L_n/F}(i)$. If $h_1(i) \geqslant n$, then by Proposition (3.6) $\operatorname{Gal}(L_n/F)_{h_1(i)} = \{1\}$, and we may assume that N'' is injective. Then by the induction assumption $\ker N_i = \ker N'$ coincides with the set of elements $\pi_L^{-1}\sigma(\pi_L) \bmod U_{h(i)+1,L}$, where σ runs over $\operatorname{Gal}(L/L_n)_n = G_n$. If $h_1(i) < n$ and $N_i(\alpha) \in U_{i+1,F}$ for some $\alpha \in U_{h(i),L}$, then $h(i) = h_1(i)$, and by the induction assumption,

$$N'(\alpha) \equiv \frac{\sigma(\pi_{L_n})}{\pi_{L_n}} \mod U_{h_1(i)+1,L_n}$$

for a prime element π_{L_n} in L_n and some $\sigma \in \operatorname{Gal}(L/F)$. We can take $\pi_{L_n} = N_{L/L_n}\pi_L$. Hence

$$N'(\alpha) \equiv N'\left(\frac{\sigma(\pi_L)}{\pi_L}\right) \mod U_{h_1(i)+1,L_n}.$$

The homomorphisms

$$U_{j,L}/U_{j+1,L} \longrightarrow U_{j,L_n}/U_{j+1,L_n}$$

induced by N_{L/L_n}, are injective for $j < n$ by Proposition (3.6). Therefore, the element $\pi_L^{-1}\sigma(\pi_L)$ belongs to $U_{h(i),L}$ and so $\sigma \in G_{h(i)}$,

$$\alpha \equiv \frac{\sigma(\pi_L)}{\pi_L} \mod U_{h(i)+1,L}.$$

□

(4.3). Now we study ramification numbers of abelian extensions. We shall see that these satisfy much stronger congruences than that of Proposition (4.5) Ch. II.

THEOREM (HASSE–ARF). *Let L/F be a finite abelian extension, and let the residue extension $\overline{L}/\overline{F}$ be separable. Let $G = \mathrm{Gal}(L/F)$. Then $G_j \neq G_{j+1}$ for an integer $j \geqslant 0$ implies $j = h_{L/F}(j')$ for an integer $j' \geqslant 0$. In other words, upper ramification jumps of abelian extensions are integers.*

Proof. We may assume that $j > 0$ and that L/F is totally ramified. Let E/F be the maximal p-subextension in L/F, and $m = |L : E|$. Let π_L be a suitable prime element in L such that $\pi_L^m \in E$. For $\sigma \in G_j$, $\sigma \notin G_{j+1}$ we get $\pi_L^{-m}\sigma\pi_L^m = 1+m\theta\pi_L^j$ for some $\theta \in U_L$; therefore $j = mj_1$, and $\sigma|_E \in \mathrm{Gal}(E/F)_{j_1}$, $\sigma \notin \mathrm{Gal}(E/F)_{j_1+1}$. If we verify that $j_1 = h_{E/F}(j')$ for some integer j', then $j = h_{L/F}(j')$. Thus, we may also assume $G = G_1$.

If L/F is cyclic of degree $p = \mathrm{char}(\overline{F})$, then the required assertion follows from Proposition (1.5). In the general case we proceed by induction on the degree of L/F. In terms of Proposition (3.6) it suffices to show that $n \in h_{L_n/F}(\mathbb{N})$ where $G_n \neq \{1\} = G_{n+1}$. Let $\sigma \in G_n, \sigma \neq 1$. Assume that there is a cyclic subgroup H of order p such that $\sigma \notin H$. Then denote the fixed field of H by M. For a prime element π_L in L the element $\pi_M = N_{L/M}(\pi_L)$ is prime in M, and $M = F(\pi_M)$ by Corollary 2 of (2.9) Ch. II. Then $\varepsilon = N_{L/M}(\pi_L^{-1}\sigma(\pi_L)) = N_{L/M}(\pi_L^{-1})\sigma(N_{L/M}(\pi_L)) \neq 1$, since $\sigma(\pi_M) \neq \pi_M$. Put $n' = v_M(\varepsilon - 1)$; then $\sigma|_M \in (G/H)_{n'}$, $\sigma|_M \notin (G/H)_{n'+1}$. By the induction hypothesis, $n' = h_{M/F}(n'')$ for some $n'' \in \mathbb{N}$. Proposition (1.5) implies $n \leqslant h_{L/M}(n')$, and we obtain $n \leqslant h_{L/F}(n'')$. If $n < h_{L/F}(n'')$, then, by Proposition (3.7) the left derivative of $h_{L/F}$ at n'' is equal to $|L : F|$, and the left derivative of $h_{L/M}$ at n' is equal to $|L : M|$. Therefore, the left derivative of $h_{M/F}$ at n'', which is equal to $|(G/H) : (G/H)_{n'}|$ by Proposition (3.7), coincides with $|M : F|$. This contradiction shows that $n = h_{L/F}(n'')$.

It remains to consider the case when there are no cyclic subgroups H of order p, such that $\sigma \notin H$. This means that G is itself cyclic. Let τ be a generator of G. The choice of n and Theorem (4.2) imply that $\sigma = \tau^{ip^{m-1}}$, where $p \nmid i, p^m = |G|$. We can assume $m \geqslant 2$ because the case of $m = 1$ has been considered above. Let $n_1 = v_L(\pi_L^{-1}\tau^{p^{m-2}}(\pi_L) - 1)$. Since $|G : G_n| = p^{m-1}$, Proposition (3.7) shows now that it suffices to prove that $p^{m-1}|(n - n_1)$. This is, in fact, a part of the third statement of the following Proposition.

Proposition. *Let L/F be a totally ramified cyclic extension of degree p^m. Let π_L be a prime element in L. For $\sigma \in \mathrm{Gal}(L/F)$ and integer k put*

$$c_k = c_k(\sigma) = v_L\left(\frac{\sigma^k(\pi_L)}{\pi_L} - 1\right).$$

Then

(1) *c_k depends only on $v_p(k)$, where v_p is the p-adic valuation (see section 1 Ch. I);*
(2) *there exists an element $\alpha_k \in L^*$ such that*

$$v_L(\alpha_k) = k, \qquad v_L\left(\frac{\sigma(\alpha_k)}{\alpha_k} - 1\right) = c_k;$$

(3) *if $v_p(k_1 - k_2) \geqslant a$, then $v_p(c_{k_1} - c_{k_2}) \geqslant a + 1$.*

Proof. (After *Sh. Sen* [Sen1])

(1) Note that c_k does not depend on the choice of a prime element in L by the same reasons as s in (1.4). Let $k = ip^j$ with $p \nmid i$, $j \geqslant 0$. Then $\sigma^k - 1 = (\rho - 1)\mu$ for $\rho = \sigma^{p^j}$, $\mu = \rho^{i-1} + \rho^{i-2} + \dots + 1$. As c_k does not depend on the choice of a prime element in L and $v_L(\mu(\pi_L)) = 1$, then $c_k = c_{p^j}$.

(2) Put $\alpha_k = \prod_{i=0}^{k-1} \sigma^i(\pi_L)$ for $k \geqslant 0$ and $\alpha_k = \alpha_{-k}^{-1}$ for $k < 0$.

(3) Assume, by induction, that if $v_p(k_1 - k_2) \geqslant a$ for $a \leqslant n-2$, then $v_p(c_{k_1}(\sigma) - c_{k_2}(\sigma)) \geqslant a + 1$ for $\sigma \in \mathrm{Gal}(L/F)$.

First we show that all the integers $c_{p^{n-1}}$, $k + c_k$ for $v_p(k) \leqslant n-1$ are distinct. Indeed, let $k_1 + c_{k_1} = k_2 + c_{k_2}$, $v_p(k_1) \neq v_p(k_2)$. Then $v_p(k_1 - k_2) = v_p(c_{k_2} - c_{k_1}) \geqslant v_p(k_1 - k_2) + 1$, and thus $k_1 = k_2$. We also obtain that $v_p(c_{p^{n-1}} - c_k) \geqslant v_p(p^{n-1} - k) + 1 > v_p(k)$ and $c_{p^{n-1}} \neq c_k + k$.

Assume that $v_p(c_{p^{n-1}}(\tau) - c_{p^n}(\tau)) < n$ for a generator τ of $\mathrm{Gal}(L/F)$. Our purpose is to show that this leads to a contradiction. Then, obviously, $v_p(c_{k_1}(\sigma) - c_{k_2}(\sigma)) \geqslant a + 1$ for $v_p(k_1 - k_2) \geqslant a, a \leqslant n - 1$.

Put $d = c_{p^{n-1}}(\tau) - c_{p^n}(\tau)$. Since $v_p(d) = v_p(c_{p^{n-2}}(\tau^p) - c_{p^{n-1}}(\tau^p)) \geqslant n - 1$, we get $v_p(d) = n - 1$. By (2), there exists an element $\alpha \in L$ such that $v_L(\alpha) = d$,

$$v_L(\tau^p(\alpha) - \alpha) = d + c_d(\tau^p) = d + c_{p^n}(\tau) = c_{p^{n-1}}(\tau).$$

Put $\beta = (\tau^{p-1} + \tau^{p-2} + \dots + 1)\alpha$. Since $v_L(\tau^p(\alpha) - \alpha) = c_{p^{n-1}}(\tau) > 0$, we get $v_L(\tau(\alpha) - \alpha) > v_L(\alpha)$ and $v_L(\beta) > d$. We also obtain $v_L(\tau(\beta) - \beta) = v_L(\tau^p(\alpha) - \alpha) = c_{p^{n-1}}(\tau)$. Recalling that $\mathcal{O}_L = \mathcal{O}_F[\pi_L]$, we deduce that β can be expanded as

$$\beta = \sum_{k \geqslant v_L(\beta)} \beta_k,$$

with $\beta_k \in L$ possessing the same properties with respect to τ as α_k of (2). Then

$$\tau(\beta) - \beta = \sum_{\substack{k \geqslant v_L(\beta) \\ v_p(k) < n}} (\tau(\beta_k) - \beta_k) + \sum_{\substack{k \geqslant v_L(\beta) \\ v_p(k) \geqslant n}} (\tau(\beta_k) - \beta_k).$$

The elements of the first sum in the right-hand expression do not cancel among themselves because $v_L(\tau(\beta_k)-\beta_k)=k+c_k(\tau)$ are all distinct and none of them coincides with $c_{p^{n-1}}(\tau)=v_L(\tau(\beta)-\beta)$. Therefore,

$$c_{p^{n-1}}(\tau)=v_L(\sum_{\substack{k\geqslant v_L(\beta)\\ v_p(k)\geqslant n}}(\tau(\beta_k)-\beta_k)).$$

In this sum

$$v_L(\tau(\beta_k)-\beta_k)=k+c_k(\tau)\geqslant v_L(\beta)+c_{p^n}(\tau)>d+c_{p^n}(\tau)=c_{p^{n-1}}(\tau),$$

a contradiction.

□

REMARKS.

1. This Theorem can be naturally proved using local class field theory (see (3.5) Ch. IV and (4.7) Ch. V). In addition, one can show that a finite Galois totally ramified extension L/F is abelian if and only if for every finite abelian totally ramified extension M/F the extension LM/F has integer upper ramification jumps [Fe8]. For several other proofs of the Hasse–Arf Theorem see [Se3], [N2].

2. The arguments of the previous Proposition are valid for the more general situation of so called wildly ramified automorphisms, see Remark 3 in (5.7) and [Sen1].

3. In the study of properties of ramification subgroup of finite Galois extensions of local fields one can use a theorem of *F. Laubie* [Lau1] which claims that for every finite Galois totally ramified extension of a local field there exists a Galois totally ramified extension of a local field with finite residue field such that the Galois groups are isomorphic and the ramification groups of the extensions are mapped to each other under this isomorphism.

Exercises.

1. Prove Proposition (4.1) for a complete discrete valuation field and a cyclic extension L/F of prime degree using explicit calculations in section 1.
2. Show that if L/F is a finite totally ramified Galois extension, then
$$\sum_{i\geqslant 0}|\ker N_i|\leqslant |G|.$$
3. In terms of (4.3) show that
$$c_k=\max\left\{v_L\left(\frac{\sigma(\alpha)}{\alpha}-1\right):v_L(\alpha)=k\right\},$$
$$k=\max\left\{v_L(\alpha):v_L(\sigma(\alpha)-\alpha)=k+c_k\right\}.$$
4. (◇) (*Sh. Sen*) Let L/F be a cyclic totally ramified extension of degree p^n, $p=\operatorname{char}(\overline{F})>0$. Let σ be a generator of $\operatorname{Gal}(L/F)$, and let π be a prime element in F. Let c_k be identical to those of Proposition (4.3). Let $\mathrm{A}=\{\alpha\in\mathcal{O}_L:\operatorname{Tr}_{L/F}(\alpha)=0\}$, $\mathrm{B}=(1-\sigma)\mathcal{O}_L$.

a) Show that A/B is isomorphic $\bigoplus_{k=1}^{k=p^n-1} \mathcal{O}_F/\pi^{g_k}\mathcal{O}_F$, where $g_k = [p^{-n}k + p^{-n}c_k]$.

b) Show that $p\mathrm{A} \subset \mathrm{B}$.

This assertion can be generalized to the case of arbitrary Galois extensions. It implies *J. Tate*'s Theorem on "invariants": let $\mathrm{char}(F) = 0$, $\mathrm{char}(\overline{F}) = p$, and let $L = \widehat{F^{\mathrm{sep}}}$ be the completion of F^{sep}. The Galois group $G_F = \mathrm{Gal}(F^{\mathrm{sep}}/F)$ operates on L by continuity. Then $L^{G_F} = F$ ([T2], [Sen1], [Ax]).

5. ($\diamond$) (*B.F. Wyman* [Wy]) Let L/F be a cyclic totally ramified extension of complete discrete valuation fields, $|L:F| = p^n$. Let $\mathrm{char}(F) = 0$, $\mathrm{char}(\overline{F}) = p$, and let $\overline{F}$ be perfect.

a) Show that L/F has n ramification numbers $x_1 < x_2 < \dots < x_n$.

b) Show that if x_i are divisible by p, then $x_i = x_1 + (i-1)e$ for $1 \leqslant i \leqslant n$, where $e = e(F)$.

c) For the rest of this Exercise assume that a primitive pth root of unity ζ belongs to F. Let $N_{L/F}(\alpha) = \zeta$ and $v_L(\alpha - 1) = i$. Show that if $x_1 < e/(p-1)$, then $x_1 \leqslant i \leqslant h_{L/F}(e/(p-1))$ and if $x_1 \geqslant e/(p-1)$, then $i = e/(p-1)$.

d) Assume that M/F is cyclic of degree p^{n-1} and $L = M(\sqrt[p]{\alpha})$ with $\alpha \in M^*$. Let $\alpha^{-1}\sigma(\alpha) = \beta^p$ for a generator σ of $\mathrm{Gal}(L/F)$. Show that $N_{M/F}(\beta)$ is a primitive pth root of unity.

e) Show that if $x_1 \geqslant e/(p-1)$, then $x_i = x_1 + (i-1)e$ for $1 \leqslant i \leqslant n$.

f) Let $n \geqslant 2$. Show that if $x_{n-1} \geqslant p^{n-2}e/(p-1)$, then $x_n = x_{n-1} + p^{n-1}e$, and if $x_{n-1} \leqslant p^{n-2}e/(p-1)$, then

$$(1 + p(p-1))x_{n-1} \leqslant x_n \leqslant p^n e/(p-1) - (p-1)x_{n-1}.$$

6. ($\diamond$) Let L/F be a Galois totally ramified p-extension. Let π_L be a prime element of L and put $\pi_F = N_{L/F}\pi_L$. Investigating the Eisenstein polynomial of π_L over F show that

a) For every $i > 0$ there exists $j = j(i)$ and $g_i \in \mathcal{O}_F[X]$ such that $\overline{g}_i \neq 0$ and

$$N_{L/F}(1 - \alpha\pi_L^i) = 1 + g_i(\alpha)\pi_F^j \quad \text{for every } \alpha \in \mathcal{O}_F\ .$$

Show that $j(h_{L/F}(k)) = k$ for every integer $k > 0$.

b) Show that the sequence

$$1 \to \mathrm{Gal}(L/F)_i/\,\mathrm{Gal}(L/F)_{i+1} \to U_{i,L}/U_{i+1,L} \to U_{j,F}/U_{j+1,F}$$

is exact where the left arrow is induced by the norm map and is described by the polynomial $\overline{g}_i$.

c) Put $a_i = \deg \overline{g}_i$. Show that $\prod_i a_i = |L:F|$.

d) Let $i_1 < \dots < i_m$ be the indices of all $a_{i_1}, \dots a_{i_m}$ which are > 1. Show that $j(i_1) < \dots < j(i_m)$.

e) Assume in addition that $\mathrm{char}(F) = p$. Put $b_k = \log_p(a_{i_{k+1}}) + \dots + \log_p(a_{i_m})$ for $0 \leqslant k \leqslant m-1$ and $b_m = 0$. Prove that for all $\alpha \in \mathcal{O}_F$

$$N_{L/F}(1 - \alpha\pi_L) = 1 + \alpha^{p^n}\pi_F + f_1(\alpha)\pi_F^{j_1} + \dots + f_m(\alpha)\pi_F^{j_m}$$

with $f_k(X) = h_k(X^{p^{b_k}})$ where $h_k(X) \in \mathcal{O}_F[X]$ is such that

$$\overline{h}_k(X) = \sum_{l=1}^{b_{k-1}-b_k} \sum_{1 \leqslant r \leqslant d_{k,l}} c_{r,l} X^{p^{l-1}r}$$

where all $d_{k,l}$ are prime to p and $c_{d_{k,l},l} \neq 0$. The n numbers $d_{k,l}$, $1 \leqslant k \leqslant m$, $1 \leqslant l \leqslant b_{k_1} - b_k$ correspond to Heiermann's ramification numbers [Hei].

7. Let L/F be a finite separable extension, and let $\overline{F}$ be perfect. Let M/L be a finite extension such that M/F is Galois. For an embedding $\sigma\colon L \to M$ over F put

$$S_{L/F}(\sigma) = \min_{\alpha \in \mathcal{O}_L} \frac{v_M(\alpha - \sigma\alpha)}{v_M(\pi_L)} \in \mathbb{Q} \cup \{+\infty\},$$

where π_L is a prime element in L. Let L_0 be the inertia subfield in L/F.

a) Show that $S_{L/F}$ does not depend on the choice of M.

b) Show that if $\sigma|_{L_0} \neq \mathrm{id}$, then $S_{L/F}(\sigma) = 0$.

c) Show that if $\sigma|_{L_0} = \mathrm{id}$, then $S_{L/F}(\sigma) = \frac{v_M(\pi_L - \sigma\pi_L)}{v_M(\pi_L)} \geqslant 1$.

d) Let $f(X)$ be the Eisenstein polynomial of π_L over L_0. Show that

$$v_L(f'(\pi_L)) = \sum S_{L/F}(\sigma),$$

where σ runs over all distinct nontrivial embeddings of L into M over F.

e) Let N/F be a subextension of L/F. Show that for an embedding $\sigma\colon N \to M$ over F

$$S_{N/F}(\sigma) = \frac{\sum S_{L/F}(\tau)}{e(L|N)},$$

where τ runs over all the embeddings of L into M, the restriction of which on N coincides with σ.

5. The Field of Norms

Whereas arithmetically profinite extensions (for example almost totally ramified $\mathbb{Z}_p$-extensions) were in use for a long time, the notion of a field of norms ("corps des normes") was introduced by *J.-M. Fontaine* and *J.-P. Wintenberger* [FW], [Win3]. Below we follow [Win3].

In this section F is a local field with perfect residue field of characteristic $p > 0$. In subsection (5.1) we introduce arithmetically profinite extensions. In subsection (5.2) we introduce a useful invariant of an arithmetically profinite extension which indicates the point from which "ramification starts". In subsection (5.3) we look at the projective limit of multiplicative groups with respect to norm maps. To introduce addition on that limit (with zero added) we study the norm map of the sum of two elements in (5.4). The main theorem on the field of norms $N(L|F)$ is proved in (5.5). Sections (5.6) and (5.7) aim to prove that separable extensions of the field of norms $N(L|F)$ (which is a local field of characteristic p) are in one-to-one correspondence with separable extensions of L; the latter correspondence is compatible with ramification filtrations.

(5.1). DEFINITION. Let L be a separable extension of F with finite residue field extension $\overline{L}/\overline{F}$. We can view L as the union of an increasing directed family of subfields L_i, which are finite extensions of F, $i \geqslant 0$. The extension L/F is said to be *arithmetically profinite* if the composite $\dots \circ h_{L_i/L_{i-1}} \circ \dots \circ h_{L_0/F}(a)$ is a real number for every real $a > 0$.

In other words, taking into consideration Proposition (3.3), L/F is arithmetically profinite if and only if it has finite residue field extension and for every real $a > 0$ there exists an integer j, such that the derivative (left or right) of h_{L_i/L_j} for $x < h_{L_j/F}(a)$, $i > j$, is equal to 1. Define the *Hasse–Herbrand function* of L/F as

$$h_{L/F} = \dots \circ h_{L_i/L_{i-1}} \circ \dots \circ h_{L_0/F}.$$

PROPOSITION. *The function $h_{L/F}$ is well defined. It is a piecewise linear, continuous and increasing function. If E/L is a finite separable extension, then E/F is arithmetically profinite. If M/F is a subextension of L/F, then M/F is arithmetically profinite. If, in addition, M/F is finite, then*

$$h_{L/F} = h_{L/M} \circ h_{M/F}.$$

Proof. Let L'_i be another increasing directed family of subfields in L such that $L = \cup L'_i$. Let a be a real number > 0. There exist integers j and k such that

$$h_{L_i/L_j}(x) = x \qquad \text{for } x < h_{L_j/F}(a), i > j$$

and

$$h_{L'_i/L'_k}(x) = x \qquad \text{for } x < h_{L'_k/F}(a), i > k.$$

Since there exists an integer $m \geqslant j$ such that $L_jL'_k \subset L_m$, we obtain by (3.3) that

$$h_{L_jL'_k/L_j}(x) = x \qquad \text{for } x < h_{L_j/F}(a).$$

Then

$$h_{L_j/F}(x) = h_{L_jL'_k/F}(x) \qquad \text{for } x < a$$

and similarly,

$$h_{L'_k/F}(x) = h_{L_jL'_k/F}(x) \qquad \text{for } x < a.$$

Therefore,

$$h_{L_i/F}(x) = h_{L'_i/F}(x) \qquad \text{for } x < a \text{ and sufficiently large } i,$$

and the function $h_{L/F}$ is well defined.

Let $E = L(\beta)$, and let $P = L(\alpha)$ be a finite Galois extension of L with $P \supset E$. Using the same arguments as in the proof of Proposition (4.2) Ch. II, one can show that

$L_i(\alpha) \cap L = L_i$ and $L_i(\alpha)/L_i$ is a Galois extension of the same degree as P/L for a sufficiently large i. Then $\mathrm{Gal}(L_i(\alpha)/L_i)$ and $\mathrm{Gal}(L_i(\alpha)/L_i(\beta))$ are isomorphic with $\mathrm{Gal}(P/L)$ and $\mathrm{Gal}(P/E)$ for $i > m$, respectively.

Put $E_i = L_i$ for $i \leqslant m$ and $E_i = L_i(\beta)$ for $i > m$. Then $E = \cup E_i$. If the left derivative of $h_{L_i/F}(x)$ is bounded by d for $x < a$ and $c = |E : L|$, then the left derivative of $h_{E_i/F}(x)$ is bounded by cd for $x < a$, $i > m$. This means that E/F is arithmetically profinite.

If M/F is a finite subextension of L/F, then we can take $L_0 = M$. Therefore L/M is arithmetically profinite and

$$h_{L/F} = h_{L/M} \circ h_{M/F}.$$

If M/F is a separable subextension of L/F, then there exists an increasing directed family of subfields $M_i, i \geqslant 0$, which are finite extensions of F and such that $M = \cup M_i$. If $L = \cup L_i$, then also $L = \cup L_i M_i$, and the left derivative of $h_{L_i M_i/F}(x)$ for $x < a$ is bounded. Hence, the left derivative of $h_{M_i/F}(x)$ for $x < a$ is bounded, i.e., M/F is arithmetically profinite. □

REMARKS.

1. Translating to the language of ramification groups by using the two previous sections, we deduce that a Galois extension L/F with finite residue field extension is arithmetically profinite extension if and only if its upper ramification jumps form a discrete unbounded set and for every upper ramification jump x the index of $\mathrm{Gal}(L/F)(x+\delta)$ in $\mathrm{Gal}(L/F)(x)$ is finite. Alternatively, a Galois extension L/F is arithmetically profinite if and only if for every x the upper ramification group $\mathrm{Gal}(L/F)(x)$ is open (i.e. of finite index) in $\mathrm{Gal}(L/F)$. More generally, a separable extension L/F is arithmetically profinite if and only if for every x the group $\mathrm{Gal}(F^{\mathrm{sep}}/F)(x)\,\mathrm{Gal}(F^{\mathrm{sep}}/L)$ is open in $\mathrm{Gal}(F^{\mathrm{sep}}/F)$.

Since the Hasse–Herbrand function relates upper and lower ramification filtrations, we can define lower ramification groups of an infinite Galois arithmetically profinite extension L/F as $\mathrm{Gal}(L/F)_x = \mathrm{Gal}(L/F)(h_{L/F}^{-1}(x))$.

2. By Corollary of (6.2) Ch. IV every abelian extension of a local field with finite residue field and finite residue field extension is arithmetically profinite.

An important property of a totally ramified $\mathbb{Z}_p$-extension L/F in characteristic zero is that its upper ramification jumps form an arithmetic progression with difference $e = e(F)$ for sufficiently large jumps, see Exercises 1 and 2 below.

3. *E. Maus* and *Sh. Sen*'s theorem on ramification filtration of p-adic Lie extensions L/F in characteristic zero with finite residue field extension proves a conjecture of *J.-P. Serre* that the p-adic Lie filtration is equivalent to the upper ramification filtration of the Galois group of such extensions (see [Mau4], [Sen2], and for a leisure exposition [dSF]). This theorem implies that every such extension is an arithmetically profinite ex-

tension. In positive characteristic the analogous result was proved by *J.-P. Wintenberger* [Win1].

There are arithmetically profinite extensions in characteristic zero which are very far from being related to p-adic Lie extensions [Fe12], see Remark 3 in (5.7).

4. An extension L/F of local fields is called *deeply ramified* if the set of its upper ramification jumps is unbounded. This class of these extensions was studied by *J. Coates* and *R. Greenberg* [CG] from the point of view of a generalization of *J. Tate*'s results [T2] (which hold for $\mathbb{Z}_p$-extensions) and applications to Kummer theory for abelian varieties. For a discussion of links between arithmetically profinite and deeply ramified extensions see [Fe11].

(5.2). Let L/F be arithmetically profinite. Put

$$q(L|F) = \sup\{x \geqslant 0 : h_{L/F}(x) = x\}.$$

LEMMA.

(1) *if M/F is a subextension in L/F, then $q(L|F) \leqslant q(M|F)$.*

(2) *if M/F is a finite subextension in L/F, then $q(L|M) \geqslant q(L|F)$.*

(3) *if $L = \cup L_i$ as in (5.1), then $q(L_j|L_i) \to +\infty$ as $j \geqslant i$, $i, j \to +\infty$.*

(4) *$q(L|F) = +\infty$ if and only if L/F is unramified; $q(L|F) = 0$ if and only if L/F is totally and tamely ramified, and $q(L|F) \leqslant pv_F(p)/(p-1)$ if L/F is totally ramified.*

Proof. (1) Let $L = \cup L_i$, $M = \cup M_i$ and $L_i' = L_i M_i$. As $h_{L_i'/F}(x) \leqslant h_{L/F}(x)$ by (3.3), we get $h_{L_i'/F}(x) = x$ for $x \leqslant q(L|F)$ and $h_{M_i/F}(x) = x$ for $x \leqslant q(L|F)$. Therefore, $q(L|F) \leqslant q(M|F)$. (2) The previous Proposition shows that

$$h_{L/M}(x) = x \qquad \text{for } x \leqslant h_{M/F}(q(L|F)).$$

This means that $q(L|M) \geqslant h_{M/F}(q(L|F))$. But by Proposition (3.3), $h_{M/F}(x) \geqslant x$, hence $q(L|M) \geqslant q(L|F)$. (3) It follows from the definition. (4) The first two assertions follow from Proposition (3.3). Proceeding as in the proof of Proposition (3.3) and using (1), it suffices to verify the last assertion for a separable totally ramified extension of degree p. Now the computations in the proof of Proposition (3.3) and Proposition (2.3) lead to the required inequality. □

(5.3). Let L be an infinite arithmetically profinite extension of F, and let L_i, $i \geqslant 0$, be an increasing directed family of subfields, which are finite extensions of F, $L = \cup L_i$. Let

$$N(L|F)^* = \varprojlim L_i^*$$

be the projective limit of the multiplicative groups with respect to the norm homomorphisms $N_{L_i/L_j}, i \geqslant j$. Put $N(L|F) = N(L|F)^* \cup \{0\}$.

LEMMA. *The group $N(L|F)^*$ does not depend on the choice of L_i.*

Proof. Let L'_i be another increasing directed family of finite extensions of F and $L = \cup L'_i$. For every i there exists an index j, such that $L'_i \subset L_j$ and $N_{L_j/F} = N_{L_j/L'_i} \circ N_{L'_i/F}$. This immediately implies the desired assertion. □

Therefore

$$N(L|F)^* = \varprojlim{}_{M \in S_{L/F}} M^*,$$

where $S_{L/F}$ is the partially ordered family of all finite subextensions in L/F and the projective limit is taken with respect to the norm maps. If $\mathrm{A} = (\alpha_M) \in N(L|F)$ with $\alpha_M \in M$, then $N_{M_1/M_2}\alpha_{M_1} = \alpha_{M_2}$ for $M_2 \subset M_1$.

We will show that $N(L|F)$ is in fact a field (the *field of norms*). Moreover, one can define a natural discrete valuation on $N(L|F)$, which makes $N(L|F)$ a complete field with residue field $\overline{L}$.

(5.4). The following statement plays a central role.

PROPOSITION. *Let M'/M be totally ramified of degree a power of p. Then*

$$v_M\left(N_{M'/M}(\alpha+\beta) - N_{M'/M}(\alpha) - N_{M'/M}(\beta)\right) \geqslant \frac{(p-1)q(M'|M)}{p}$$

for $\alpha, \beta \in \mathcal{O}_{M'}$. For $\alpha \in \mathcal{O}_M$ there exists an element $\beta \in \mathcal{O}_{M'}$ such that

$$v_M\left(N_{M'/M}(\beta) - \alpha\right) \geqslant \frac{(p-1)q(M'|M)}{p}.$$

Proof. Assume first that M'/M is a cyclic extension of degree p. Then we get $q(M'|M) = s(M'|M)$ (see (1.4) and (3.1)) and, by Proposition (1.4),

$$\mathrm{Tr}_{M'/M}(\mathcal{O}_{M'}) = \pi_M^r \mathcal{O}_M$$

with $r = s + 1 + [(-1-s)/p] \geqslant (p-1)s(M'|M)/p$. Then Lemma (1.1) shows that

$$v_M\left(N_{M'/M}(1+\gamma) - 1 - N_{M'/M}(\gamma)\right) \geqslant \frac{(p-1)q(M'|M)}{p}$$

for $\gamma \in \mathcal{O}_{M'}$. Substituting $\gamma = \alpha\beta^{-1}$ if $v_{M'}(\alpha) \geqslant v_{M'}(\beta)$ and $\beta \neq 0$, we obtain the desired inequality.

In the general case we proceed by induction on the degree of M'/M. Let E/M be a finite Galois extension with $E \supset M'$, and let E_1 be the maximal tamely ramified extension of M in E. Then E_1 and M' are linearly disjoint over M, and

$$\begin{aligned} &N_{M'/M}(\alpha+\beta) - N_{M'/M}(\alpha) - N_{M'/M}(\beta) \\ &\qquad = N_{E_1M'/E_1}(\alpha+\beta) - N_{E_1M'/E_1}(\alpha) - N_{E_1M'/E_1}(\beta). \end{aligned}$$

The group $G = \mathrm{Gal}(E/E_1)$ is a p-group, and hence for $H = \mathrm{Gal}(E/E_1M')$ there exists a chain of subgroups

$$G' = G_{(0)} \geqslant G_{(1)} \geqslant \ldots \geqslant G_{(m)} = H,$$

such that $G_{(i+1)}$ is a normal subgroup of index p in $G_{(i)}$. For the fields we obtain the tower $E_{(0)} - E_{(1)} - \cdots - E_{(m)} = E_1 M'$, in which $E_{(i+1)}$ is a cyclic extension of degree p over $E_{(i)}$. Let E_2 be some $E_{(i)}$ for $1 \leqslant i < m$. By the induction assumption,

$$N_{E_1 M'/E_2}(\alpha + \beta) = N_{E_1 M'/E_2}(\alpha) + N_{E_1 M'/E_2}(\beta) + \delta$$

with $v_{E_2}(\delta) \geqslant (p-1)q(E_1 M'|E_2)/p$. We deduce also that

$$N_{E_1 M'/E_1}(\alpha + \beta) = N_{E_1 M'/E_1}(\alpha) + N_{E_1 M'/E_1}(\beta) + N_{E_2/E_1}(\delta) + \delta'$$

with $v_{E_1}(\delta') \geqslant (p-1)q(E_2|E_1)/p$. Then

$$v_{E_1}\left(N_{E_2/E_1}(\delta)\right) \geqslant \frac{(p-1)q(E_1 M'|E_2)}{p} \geqslant \frac{(p-1)q(E_1 M'|E_1)}{p}$$

and

$$v_{E_1}(\delta') \geqslant \frac{(p-1)q(E_1 M'|E_1)}{p}$$

by Lemma (5.2). These two inequalities imply that

$$v_M\left(N_{M'/M}(\alpha + \beta) - N_{M'/M}(\alpha) - N_{M'/M}(\beta)\right) \geqslant \frac{(p-1)q(M'|M)}{p},$$

as required.

To prove the second inequality of the Proposition, we choose a prime element π' in M' and put $\pi = N_{M'/M}\pi'$. Then π is a prime element in M. Let $n = |M' : M|$ (a power of p). Writing the element α of M as

$$\alpha = \sum_{i \geqslant a} \theta_i \pi^i$$

with multiplicative representatives θ_i, put

$$\beta = \sum_{i \geqslant a} \theta_i^{1/n} \pi'^i \in M.$$

Then $N_{M'/M}\left(\theta_i^{1/n}\pi'\right) = \theta_i \pi$ and, by the first inequality of the Proposition,

$$v_M(N_{M'/M}(\beta) - \alpha) \geqslant \frac{(p-1)q(M'|M)}{p},$$

as required. □

(5.5). Let L/F be an arithmetically profinite extension. Let L_0 be the maximal unramified extension of F in L, and let L_1 be the maximal tamely ramified extension of F in L. Then L_0/F is finite by the definition, and L_1/F is finite because of the relation $h_{L_1/L_0}(x) = |L_1 : L_0|x$. So one can choose L_i for $i \geqslant 2$ as finite extensions of L_1 in L with $L_i \subset L_{i+1}$ and $L = \cup L_i$.

For an element $\mathrm{A} \in N(L|F)$ put

$$v(\mathrm{A}) = v_{L_0}(\alpha_{L_0}).$$

Then $v(\mathrm{A}) = v_{L_i}(\alpha_{L_i})$ for $i \geqslant 0$.

Let a be an element of the residue field $\overline{L} = \overline{L}_0$, and $\theta = r(a)$ the multiplicative representative of a in L_0 (see section 7 Ch. I). Put $\theta_{L_i} = \theta^{1/n_i}$, where $n_i = |L_i : L_1|$ for $i \geqslant 1$ and $\theta_{L_0} = N_{L_1/L_0}\theta$. Then $\Theta = (\theta_{L_i})$ is an element of $N(L|F)$. Denote the map $a \mapsto \Theta$ by R.

THEOREM. *Let L/F be an infinite arithmetically profinite extension. Let $\mathrm{A} = (\alpha_M)$ and $\mathrm{B} = (\beta_M)$ be elements of $N(L|F)$, $M \in S_{L/F}$. Then the sequence $N_{M'/M}(\alpha_{M'} + \beta_{M'})$ is convergent in M when $M \subset M' \subset L, |M' : M| \to +\infty$. Let γ_M be the limit of this sequence. Then $\Gamma = (\gamma_M)$ is an element of $N(L|F)$. Put $\Gamma = \mathrm{A} + \mathrm{B}$.*

Then $N(L|F)$ is a field with respect to the multiplication and addition defined above. The map v is a discrete valuation of $N(L|F)$ and $N(L|F)$ is a complete field of characteristic p. The map R is an isomorphism of $\overline{L}$ onto a subfield in $N(L|F)$ which maps isomorphically onto the residue field of $N(L|F)$.

Proof. Let L_i be as above of (5.5) in the context of Lemma (5.3).

Let k be an integer such that $(p-1)q(L_j|L_i)/p \geqslant a$ for $j \geqslant i \geqslant k$, where a is a positive integer (see Lemma (5.2)). Let $\mathrm{A} = (\alpha_{L_i}), \mathrm{B} = (\beta_{L_i})$ be elements of $N(L|F)$ and $\alpha_{L_0}, \beta_{L_0} \in \mathcal{O}_{L_0}$. Then Proposition (5.4) shows that

$$N_{L_i/L_k}(\alpha_{L_i} + \beta_{L_i}) \equiv \alpha_{L_k} + \beta_{L_k} \mod \mathcal{M}_{L_k}^a. \qquad (*)$$

Let $a_k \geqslant 0$ be a sequence of integers such that

$$a_k \leqslant a_{k+1}, \quad a_k \leqslant (p-1)q(L|L_k)/p, \quad \lim a_k = +\infty$$

(the existence of the sequence follows from Lemma (5.2)). Let an index $k \geqslant 1$ be in addition such that $a_k > 1$. Suppose that β_{L_k} is a prime element in L_k. Proposition (5.4) and Lemma (5.2) show that one can construct a sequence $\beta_{L_i} \in L_i$, $i \geqslant k$, such that

$$v_{L_i}(N_{L_{i+1}/L_i}\beta_{L_{i+1}} - \beta_{L_i}) \geqslant a_i.$$

Then β_{L_i} is prime in L_i, and applying $(*)$, we get

$$v_{L_i}(N_{L_j/L_i}\beta_{L_j} - \beta_{L_i}) \geqslant a_i \qquad \text{for } j \geqslant i \geqslant k.$$

Now Proposition (3.4) and Proposition (5.1) imply that

$$v_{L_s}(N_{L_j/L_s}\beta_{L_j} - N_{L_i/L_s}\beta_{L_i}) \geqslant h_{L_i/L_s}^{-1}(a_i) \geqslant h_{L/L_s}^{-1}(a_i)$$

for $j \geqslant i \geqslant s \geqslant k$. Since $h_{L/L_s}^{-1}(a_i) \to +\infty$ as $i \to +\infty$, we obtain that there exists $\gamma_{L_s} = \lim_{i \to +\infty} N_{L_i/L_s}\beta_{L_i}$ and γ_{L_s} is prime in L_s. Putting $\gamma_{L_j} = N_{L_k/L_j}\gamma_{L_k}$ for $j < k$, we get the element $\Gamma = (\gamma_{L_i}) \in N(L|F)$ with $v(\Gamma) = 1$.

Furthermore, by Proposition (3.4) and $(*)$ we obtain:

$$v_{L_j}\left(N_{L_i/L_j}(\alpha_{L_i}+\beta_{L_i})-N_{L_k/L_j}(\alpha_{L_k}+\beta_{L_k})\right)\geqslant h^{-1}_{L_k/L_j}(a)\geqslant h^{-1}_{L/L_j}(a).$$

This means that the sequence $N_{L_i/L_j}(\alpha_{L_i}+\beta_{L_i})$ is convergent. In the general case let $c=v_{L_0}(\alpha_{L_0}), d=v_{L_0}(\beta_{L_0})$. Taking prime elements π_{L_i} in L_i such that $\Pi=(\pi_{L_i})\in N(L|F)$ with $v(\Pi)=1$ and replacing $\mathrm{A}=(\alpha_{L_i})$ by $\mathrm{A}'=(\alpha_{L_i}\pi_{L_i}^{-g})$ and $\mathrm{B}=(\beta_{L_i})$ by $\mathrm{B}'=(\beta_{L_i}\pi_{L_i}^{-g})$, where $g=\min(c,d)$, we deduce that $N_{L_i/L_j}(\alpha_{L_i}+\beta_{L_i})$ is convergent. Put $\gamma_{L_j}=\lim_{i\to+\infty}N_{L_i/L_j}(\alpha_{L_i}+\beta_{L_i})$. Obviously, $(\gamma_{L_i})=\Gamma\in N(L|F)$ and $N(L|F)$ is a field. As

$$v(\Gamma)=v_{L_k}(\gamma_{L_k})=\lim_{i\to+\infty}v_{L_k}(N_{L_i/L_k}(\alpha_{L_i}+\ be_{L_i})),$$

we get $v(\Gamma)\geqslant\min(v(\mathrm{A}),v(\mathrm{B}),a)$. Choosing $a\geqslant\max(v(\mathrm{A}),v(\mathrm{B}))$, we obtain $v(\Gamma)\geqslant\min(v(\mathrm{A}),v(\mathrm{B}))$. Since $1=(1_{L_i})$, for $p=(\alpha_{L_i})$ we get that

$$\alpha_{L_i}=\lim_{i\to+\infty}N_{L_i/L_j}(p)=\lim_{i\to+\infty}p^{|L_i:L_j|}=0.$$

Therefore, $N(L|F)$ is a discrete valuation field of characteristic p.

To verify the completeness of $N(L|F)$ with respect to v, take a Cauchy sequence $\mathrm{A}^{(n)}=(\alpha^{(n)}_{L_i})\in N(L|F)$. We may assume $v(\mathrm{A}^{(n)})\geqslant 0$. For any i there exists an integer n_i such that $v(\mathrm{A}^{(n)}-\mathrm{A}^{(m)})\geqslant a_i$ for $n,m\geqslant n_i$ (a_i as above). One may assume that $(n_i)_i$ is an increasing sequence. Applying $(*)$, we get

$$v_{L_i}(\alpha^{(n)}_{L_i}-\alpha^{(m)}_{L_i})\geqslant a_i\qquad\text{for } n,m\geqslant n_i.$$

Let α_{L_i} be an element in L_i such that

$$v_{L_i}(\alpha_{L_i}-\alpha^{(n_i)}_{L_i})\geqslant a_i.$$

Then, by $(*)$,

$$v_{L_i}(N_{L_j/L_i}\alpha_{L_j}-\alpha_{L_i})\geqslant a_i.$$

Proposition (3.4) and Proposition (5.1) imply now that

$$v_{L_s}(N_{L_i/L_s}\alpha_{L_i}-N_{L_j/L_s}\alpha_{L_j})\geqslant h^{-1}_{L/L_s}(a_j)\to+\infty$$

when $i\geqslant j\to+\infty$. Putting $\alpha'_{L_s}=\lim_{i\to+\infty}N_{L_i/L_s}\alpha_{L_i}$, we obtain an element $\mathrm{A}'=(\alpha'_{L_i})\in N(L|F)$ with $\mathrm{A}'=\lim\mathrm{A}^{(n)}$. Therefore, $N(L|F)$ is complete with respect to the discrete valuation v.

Finally, R is multiplicative. If $R(a)=\Theta$, $R(b)=\Lambda$, $R(a+b)=\Omega$, then it follows immediately from (7.3) Ch. I, that $\omega_{L_i}\equiv\theta_{L_i}+\lambda_{L_i}\mod p$. By Lemma (5.2) and the definition of a_i we get $v_{L_i}(p)\geqslant a_i$. Then by $(*)$ and Proposition (3.4) we obtain

$$v_{L_i}(\omega_{L_i}-N_{L_j/L_i}(\theta_{L_j}+\lambda_{L_j}))\to+\infty$$

as $j \to +\infty$. This means that $\Omega = \Theta + \Lambda$ and R is an isomorphism of $\overline{L}$ onto a subfield in $N(L|F)$. The latter subfield is mapped onto the residue field of $N(L|F)$, hence it is isomorphic to the residue field $\overline{N(L|F)}$. □

COROLLARY. *Let* $\mathrm{A} = (\alpha_{L_i}), \mathrm{B} = (\beta_{L_i})$ *belong to the ring of integers of* $N(L|F)$. *Then* $\gamma_{L_i} \equiv \alpha_{L_i} + \beta_{L_i} \mod \mathcal{M}_{L_i}^{a_i}$, *where* a_i *are those defined in the proof of the Theorem. Moreover, for any* $\alpha \in \mathcal{O}_{L_j}$ *there exists an element* $\mathrm{A} = (\alpha_{L_i})$ *in the ring of integers of* $N(L|F)$ *such that* $\alpha \equiv \alpha_{L_j} \mod \mathcal{M}_{L_j}^{a_j}$.

Proof. The first assertion follows from $(*)$ and the second from Proposition (5.4). □

(5.6). An immediate consequence of the definitions is that if M/F is a finite subextension of an arithmetically profinite extension L/F, then $N(L|F) = N(L|M)$. On the other hand, if E/L is a finite separable extension, then, as shown in Proposition (5.1), E/F is an arithmetically profinite extension. Let M be a finite extension of F such that $ML = E$. Then $N_{L_jM/L_iM}(\alpha) = N_{L_j/L_i}(\alpha)$ for $\alpha \in L_j$, $j \geqslant i \geqslant m$, and sufficiently large m, we deduce that $N(L|F)$ can be identified with a subfield of $N(E|F)$: $\mathrm{A} = (\alpha_{L_i}) \mapsto \mathrm{A}' \in N(E|F)$ with $\mathrm{A}' = (\alpha'_{L_iM})$, $\alpha'_{L_iM} = \alpha_{L_i}$ for $i \geqslant m$, $\alpha'_{L_iM} = N_{L_mM/L_iM}(\alpha_{L_m})$ for $i < m$. In fact the discrete valuation topology of $N(L|F)$ coincides with the induced topology from $N(E|F)$, and $N(E|F)/N(L|F)$ is an extension of complete discrete valuation fields. For an arbitrary separable extension E/L denote by $N(E,L|F)$ the injective limit of $N(E'|F)$ for finite separable subextensions E'/L in E/L. Obviously, $N(E,L|F) = N(E|F)$ if E/L is finite.

Let L/F be infinite arithmetically profinite, and let L'/L be a finite separable extension. Let τ be an automorphism in $G_F = \mathrm{Gal}(F^{\mathrm{sep}}/F)$ with $\tau(L) \subset L'$. There exists a tower of increasing subfields L'_i in L' such that L'_i/F is finite, $\tau(L)L'_i = L'$, $L' = \cup L'_i$, and $N_{L'_j/L'_i}(\tau\alpha) = \tau N_{\tau^{-1}L'_j/\tau^{-1}L'_i}(\alpha)$ for $j \geqslant i, \alpha \in \tau^{-1}L'_j$; see the proof of Proposition (5.1). Let $\mathrm{T}\colon N(L|F) \to N(L'|F)$ denote the homomorphism of fields, which is defined for $\mathrm{A} = (\alpha_{L_i}) \in N(L|F)$ as $\mathrm{T(A)} = \mathrm{A}' = (\alpha'_{L'_i})$ with $\alpha'_{L'_i} = \tau(\alpha_{\tau^{-1}L'_i})$. Then $\mathrm{A}' \in N(L'|F)$. This notion is naturally generalized for $N(E,L|F)$ and $N(E',L|F)$ with $\tau(E) \subset E'$.

PROPOSITION. *Let* E_1 *and* E_2 *be separable extensions of* L. *Then the set of all automorphisms* $\tau \in G_L$ *with* $\tau(E_1) \subset E_2$ *is identified (by* $\tau \to \mathrm{T}$*) with the set of all automorphisms* $\mathrm{T} \in G_{N(L|F)}$ *with* $\mathrm{T}(N(E_1,L|F)) \subset N(E_2,L|F)$. *In particular, if* E/L *is a Galois extension, then* $\mathrm{Gal}(E/L)$ *is isomorphic to* $\mathrm{Gal}(N(E,L|F)/N(L|F))$.

Proof. First we verify the second assertion for a finite Galois extension E/L. Let $\tau \in \mathrm{Gal}(E/F)$ and T act trivially on $N(E|F)$. Then $\overline{\mathrm{T}}$ acts trivially on the residue field of $N(E|F)$, which coincides with $\overline{E}$, and hence τ belongs to the inertia subgroup $\mathrm{Gal}(E/F)_0$. Let $E = L(\beta)$ and L_i form a standard tower of fields for L over F, as in (5.5). Then one can show that there exists an index m, such that $L_i(\beta)/L_i$ is Galois and Gal$(L_i(\beta)/L_i)$ is isomorphic to Gal(E/L) for $i > m$. Let $\Pi = (\pi_{L_i(\beta)})_{i>m}$ be a prime element of $N(E|F)$. Then $\mathrm{T}(\Pi) = \Pi$ and $\tau\pi_{L_i(\beta)} = \pi_{L_i(\beta)}$ for $i > m$. We obtain now that $\tau = 1$ because τ acts trivially on the residue field $\overline{L_i(\beta)} = \overline{E}$.

We conclude that Gal(E/L) can be identified with a subgroup of

$$\mathrm{Gal}(N(E|F)/N(L|F)).$$

Since the field of the fixed elements under the action of the image of Gal(E/L) is contained in $N(L|F)$, these two groups are isomorphic.

From this we easily deduce the second assertion of the Proposition for an arbitrary Galois extension E/L.

Finally, if E/L is a Galois extension such that $E_1, E_2 \subset E$, denote the Galois groups of E/E_1 and E/E_2 by H_1 and H_2. These two groups H_1 and H_2 can be identified with Gal$(N(E,L|F)/N(E_1,L|F))$, and Gal$(N(E,L|F)/N(E_2,L|F))$ respectively. Since the set of $\tau \in G_L$ with $\tau(E_1) \subset E_2$ coincides with $\{\tau \in G_L : \tau H_1 \tau^{-1} \supset H_2\}$, the proof is completed. □

(5.7). The preceding Proposition shows that the group Gal(F^{sep}/L) can be considered as a quotient group of Gal$(N(L|F)^{\mathrm{sep}}/N(L|F))$. We will show in what follows that the former group coincides with the latter.

THEOREM. *Let Q be a separable extension of $N(L|F)$. Then there exists a separable extension E/L and an $N(L|F)$-isomorphism of $N(E,L|F)$ onto Q.*

Thus, the absolute Galois group of L is naturally isomorphic to the absolute Galois group of $N(L|F)$.

Proof. One can assume that $Q/N(L|F)$ is a finite Galois extension. Using the description of Galois extensions of (4.4) Ch. II we must consider the following three cases: $Q/N(L|F)$ is unramified, cyclic tamely totally ramified, and cyclic totally ramified of degree $p = \mathrm{char}(\overline{F})$.

Let $\mathcal{O}_Q = \mathcal{O}_{N(L|F)}[\Gamma]$. Let $f(X)$ be the monic irreducible polynomial of Γ over $N(L|F)$. It suffices to find a separable extension E'/L such that $f(X)$ has a root in $N(E',L|F)$. Let L_i and a_i be identical to those in the proof of Theorem (5.5). By Lemma (3.1) Ch. II, we can write

$$f(X) = X^n + \mathrm{A}^{(n-1)}X^{n-1} + \dots + \mathrm{A}^{(0)}$$

with $\mathrm{A}^{(m)} = (\alpha_{L_i}^{(m)}) \in \mathcal{O}_{N(L|F)}$, $n = |Q : N(L|F)|$. Denote by $f_i(X) \in \mathcal{O}_{L_i}[X]$ the polynomial $X^n + \alpha_{L_i}^{(n-1)}X^{n-1} + \dots + \alpha_{L_i}^{(0)}$. Let α_i be a root of $f_i(X)$ and $M_i = L_i(\alpha_i), E_i = L(\alpha_i)$.

The following assertion will be useful in our considerations.

LEMMA. *Let $\Delta = \prod_{m<l}(\Gamma_m - \Gamma_l)^2$ be the discriminant of $f(X)$ (Γ_m for $1 \leqslant m \leqslant n$ are all distinct roots of $f(X)$. Then $\Delta = (-1)^{\frac{n(n-1)}{2}} \prod_{m=1}^n \sigma_m f'(\Gamma)$ where $\sigma_1, \dots, \sigma_n$*

are elements of $\mathrm{Gal}(Q/N(L|F))$, $\Delta \in N(L|F)$). *Let* $d_i \in L_i$ *be the discriminants of* $f_i(X)$. *Then there exists an index* i_1 *such that* $v_{L_i}(d_i) = v(\Delta)$ *for* $i \geqslant i_1$.

Proof. Let $\Delta = (\delta_{L_i})$, and let i_1 be such that $a_i > v(\Delta)$ for $i \geqslant i_1$. Then $v(\Delta) = v_{L_i}(\delta_{L_i})$, and Corollary (5.5) shows that $v_{L_i}(\delta_{L_i} - d_i) \geqslant a_i$. Hence, $v_{L_i}(d_i) = v_{L_i}(\delta_{L_i}) = v(\Delta)$ for $i \geqslant i_1$. □

This Lemma implies that M_i/L_i is separable for $i \geqslant i_1$. Now we shall verify that in the three cases under consideration, there exists an index i_2, such that M_i/L_i and L/L_i are linearly disjoint and $q(E_i|M_i) \geqslant q(L|L_i)$ for $i \geqslant i_2$.

If $Q/N(L|F)$ is unramified, then the residue polynomial $\overline{f}_i \in \overline{L}[X]$ is irreducible of degree n and M_i/L_i is an unramified extension of the same degree. Hence, M_i/L_i and L/L_i are linearly disjoint and $h_{E_i/M_i}(x) = h_{L/L_i}(x)$, so $q(E_i|M_i) = q(L|L_i)$.

If $Q/N(L|F)$ is totally and tamely ramified, then one can take $f(X) = X^n - \Pi$, where Π is a prime element in $N(L|F)$ (see (3.5) Ch. II). Hence, M_i/L_i is tamely and totally ramified of degree n for $i \geqslant 1$. We deduce that $L \cap M_i = L_i$ and $h_{E_i/M_i}(nx) = nh_{L/L_i}(x)$, and hence $q(E_i|M_i) \geqslant nq(L|L_i)$ for $i \geqslant 1$.

If $Q/N(L|F)$ is totally ramified of degree $n = p = \mathrm{char}(\overline{F})$, then one may assume that $f(X)$ is an Eisenstein polynomial (see (3.6) Ch. II). Then $f_i(X)$ is a separable Eisenstein polynomial in $L_i[X]$, and α_i is prime in M_i. Let N_i be the minimal finite extension of M_i such that N_i/L_i is Galois, and M_i' the maximal tamely unramified extension of L_i in N_i. Then $|N_i : L_i| \leqslant p!$. One has $N_i = M_i'(\alpha_i)$ and $s_i = s(N_i|M_i') = v_{N_i}(\sigma\alpha_i - \alpha_i) - v_{N_i}(\alpha_i)$ for a generator σ of $\mathrm{Gal}(N_i/M_i')$ (see (1.4) and the proof of Proposition (3.3)). Note that

$$v_{N_i}(\sigma\alpha_i - \alpha_i) = \frac{1}{p(p-1)} v_{N_i}(d_i) \leqslant \frac{p!}{p(p-1)} v_{L_i}(d_i) = (p-2)!v(\Delta)$$

for $i \geqslant i_1$. Furthermore, in the same way as in the proof of Proposition (3.3), we get $h_{M_i/L_i}(x) = l^{-1}h_{N_i/M_i'}(lx)$, where $l = e(M_i'|L_i)$. Consequently,

$$q(M_i|L_i) = s_i l^{-1} < (p-2)!v(\Delta).$$

Since $h_{L_j(\alpha_i)/M_i} \circ h_{M_i/L_i} = h_{L_j(\alpha_i)/L_j} \circ h_{L_j/L_i}$ for $j \geqslant i$, we deduce that $q(E_i|M_i) = h_{M_i/L_i}(q(L|L_i)) \geqslant q(L|L_i)$.

Now we construct the desired field E'. Let $v\colon N(L|F)^{\mathrm{sep}*} \to \mathbb{Q}$ be the extension of the discrete valuation $v\colon N(L|F)^* \to \mathbb{Z}$ (see Corollary 1 of (2.9) Ch. II). According to Corollary (5.5) there is an element $\mathrm{B}^{(j)} = (\beta^{(j)}_{L_i(\alpha_j)})_{i\geqslant j} \in N(E_j|F)$ such that $v_{M_j}(\alpha_j - \beta^{(j)}_{M_j}) \geqslant b_j$, where b_j is the maximal integer $\leqslant (p-1)q(E_j|M_j)/p$. Note that $b_j \geqslant a_j$. We claim that $v(f(\mathrm{B}^{(j)})) \to +\infty$ as $j \to +\infty$.

Indeed, E_j/M_j is totally ramified. Therefore, if $f(\mathrm{B}^{(j)}) = (\rho_{L_i(\alpha_j)})_{i\geqslant j}$ then $v(f(\mathrm{B}^{(j)})) \geqslant v_{M_j}(\rho_{M_j})/n$.

By using Corollary (5.5) we deduce

$$v_{M_j}(\rho_{M_j} - f_j(\beta_{M_j}^{(j)})) \geqslant (p-1)q(E_j|M_j)/p \geqslant a_j.$$

This means that

$$v(f(\mathrm{B}^{(j)})) \geqslant \frac{a_j}{n} \qquad \text{for } j \geqslant i_2.$$

Since $a_j \to +\infty$ when $j \to +\infty$, we conclude that $v(f(\mathrm{B}^{(j)})) \to +\infty$.

By the same arguments we obtain that for $f'(\mathrm{B}^{(j)}) = (\mu_{L_i(\alpha_j)})_{i \geqslant j}$

$$v(f'(\mathrm{B}^{(j)})) \leqslant v_{M_j}(\mu_{M_j}), \quad v_{M_j}(\mu_{M_j} - f_j'(\alpha_j)) \geqslant a_j, \quad v_{M_j}(f_j'(\alpha_j)) \leqslant nv(\Delta)$$

for $j \geqslant i_2$. This implies that for a sufficiently large j

$$v(f'(\mathrm{B}^{(j)})) \leqslant nv(\Delta) < \frac{1}{2}v(f(\mathrm{B}^{(j)})).$$

Corollary 3 of (1.3) Ch. II shows the existence of a root of $f(X)$ in $N(E_j|F)$. Putting $E' = E_j$ we complete the proof of the Theorem. □

DEFINITION. *The functor of fields of norms* associates to every arithmetically profinite extension L over F its field of norms $N(L|F)$, to every separable extension E of L the field $N(E, L|F)$ and to every element of G_F the corresponding element of the group of automorphisms of the field $N(L|F)^{\mathrm{sep}}$ (so that elements of $G_L \leqslant G_F$ are mapped isomorphically to elements of $G_{N(L|F)}$).

REMARKS.

1. The isomorphism between the absolute Galois groups is compatible with their upper ramification filtrations (see Exercises 4 and 5).

2. Fields of norms are related to various rings introduced by *J.-M. Fontaine* in his study of Galois representations over local fields, some of which are briefly introduced in Exercises 6 and 8. For more details see [A] and [Colm].

3. A local field F with finite residue field $\mathbb{F}_q$ has infinitely many *wild automorphisms*, i.e., continuous homomorphisms $\sigma\colon F \to F$ such that $\pi_F^{-1}\sigma(\pi_F) \in U_1$, if and only if F is of positive characteristic. The group R of wild automorphisms of F has a natural filtration $R_i = \{\sigma \in R : \pi_F^{-1}\sigma\pi_F \in U_i\}$ and R is isomorphic to $\varprojlim R/R_i$. Therefore *the wild group* R is a pro-p-group. It has finitely many generators. One can check that every nontrivial closed normal subgroup of an open subgroup of R is open; so R is a so-called hereditarily just infinite pro-p-group. Those are of importance for the theory of infinite pro-p-groups [dSSS].

Every Galois totally ramified and arithmetically profinite p-extension of a local field with residue field $\mathbb{F}_q$ is mapped under the functor of fields of norms to a closed subgroup of R. Using this functor and realizability of pro-p-groups as Galois groups of arithmetically profinite extensions in positive characteristic one can easily show that every finitely generated pro-p-group is isomorphic to a closed subgroup of R ([Fe12], for the first, different proof see [Cam]).

Define a closed subgroup T of R

$$T = \{\sigma \in R: \pi_F^{-1}\sigma\pi_F = f(\pi_F) \quad \text{with } f(X) \in \mathbb{F}_q[[X^{p^r}]]\ \}.$$

For $p > 2$ the group T is hereditarily just infinite. It can be proved that T does not have infinite subquotients isomorphic to p-adic Lie groups. The group T for $r > 1$ can be realized as the Galois group of an arithmetically profinite extension of a finite extension of $\mathbb{Q}_p$ [Fe12].

4. One can ask what is the image with respect to the functor of fields of norms of p-adic Lie extensions in R? *J.–P. Wintenberger* proved [Win2, 4] that every closed subgroup of R isomorphic to $\mathbb{Z}_p$ is the image of an appropriate $\mathbb{Z}_p$-extension either in characteristic 0 or characteristic p. For the study of the image of p-adic Lie extensions see *F. Laubie*'s works [Lau2–4].

5. General ramification theory of infinite extensions is far from being complete, despite many deep investigations including [Mau1–5]; see references in Bibliography.

Exercises.

1.
   a) Let L_n be a cyclic totally ramified extension of F of degree p^n, $p = \operatorname{char}(\overline{F})$ and $L_n \subset L_{n+1}$. Let $L = \cup L_n$. Show that $i(L_{n+1}|L_n) \geqslant i(L_n|L_{n-1}) + 1$. [Hint: show that for a prime $\pi \in L_{n+1}$ and a generator σ of $\operatorname{Gal}(L_{n+1}/L_{n-1})$, $v_{L_{n+1}}(\pi^{-1}\sigma^p(\pi)-1) \geqslant 1+v_{L_{n+1}}(\pi^{-1}\sigma(\pi)-1)$.] Deduce that L/F is arithmetically profinite.
   b) Let $\pi_0 = \pi$ be a prime element of F and let $\pi_i^p = \pi_{i-1}$ for $i \geqslant 1$. Show that the extension $L = F(\{\pi_i\})$ is an arithmetically profinite extension of F. This extension L/F plays an important role in *V. Abrashkin*'s approach to explicit formulas for the Hilbert pairing, see Remark 2 (3.5) Ch. VIII and [Ab5–6].
2. ($\diamond$) Let L/F be as in Exercise 1 and $\operatorname{char}(F) = 0$, $\overline{F}$ perfect. Using Exercise 5 of section 4 show that there exists an index j depending only on F (not on L), such that the upper ramification jumps $x_1 < x_2 < \dots$ of L/F satisfy relations $x_i = x_j + (i-j)e(F)$ for $i \geqslant j$. This assertion was employed by *J. Tate* in [T2].
3. Let L_i and a_i be such as in (5.5). Show that the norm map N_{L_j/L_i} for $j \geqslant i$ induces the surjective ring homomorphism

$$\mathcal{O}_{L_j}/\mathcal{M}_{L_j}^{a_j} \longrightarrow \mathcal{O}_{L_i}/\mathcal{M}_{L_i}^{a_i}.$$

Put $\mathcal{O}_F(L) = \varprojlim \mathcal{O}_{L_i}/\mathcal{M}_{L_i}^{a_i}$. For $\mathrm{A} = (\alpha_{L_i} \mod \mathcal{M}_{L_i}^{a_i}) \neq 0$ one can find an index $i \geqslant 1$ such that $\alpha_{L_i} \notin \mathcal{M}_{L_i}^{a_i}$. Then we put $w(\mathrm{A}) = v_{L_i}(\alpha_{L_i})$. For $a \in \overline{L}$ let $\theta \in L_i$, $i \geqslant 1$, be its multiplicative representative and $\theta_{L_i} = \theta^{1/n_i}$, where $n_i = |L_i : L_1|$. Put

$$R'(a) = (\theta_{L_i} \mod \mathcal{M}_{L_i}^{a_i})_{i \geqslant 1}.$$

Show that $\mathcal{O}_F(L)$ is a ring of characteristic p. The extension of the map w on the quotient field $N_F(L)$ of $\mathcal{O}_F(L)$ is a discrete valuation, and $N_F(L)$ is complete with respect to it. The map R' is an isomorphism of $\overline{L}$ onto a subfield of $N_F(L)$, which is isomorphic to the residue fiel d of $N_F(L)$. Show that the map

$$\mathcal{O}_{N(L|F)} \to \mathcal{O}_F(L) \quad (\alpha_{L_i}) \mapsto (\alpha_{L_i} \mod \mathcal{M}_{L_i}^{a_i})$$

is an isomorphism, preserving the discrete valuation topology.

4. ($\diamond$) [Win3] Let L/F be infinite arithmetically profinite and let $\tau: L \to L$ be an F-automorphism.
   a) Show that there exists an increasing tower of finite extensions L_i/F with $\tau(L_i) \subset L_i$ and $L = \cup L_i$. Show that for $\mathrm{T}: N(L|F) \to N(L|F)$ there exists an index i_0 such that for $i \geqslant i_0$

$$v_{L_i}\left(\frac{\tau\pi_{L_i}}{\pi_{L_i}} - 1\right) = v\left(\frac{\mathrm{T}\Pi}{\Pi} - 1\right)$$

   for a prime element $\Pi \in N(L|F)$ and a prime element π_{L_i} in L_i.
   b) Deduce that if L/F is Galois, then the image of $\mathrm{Gal}(L/F)$ under the homomorphism $\tau \to \mathrm{T}$ is a subgroup in the group of continuous with respect to the discrete valuation v automorphisms $\mathrm{Aut}\, N(L|F)$ of $N(L|F)$.
   c) Show that the image of the upper ramification group $\mathrm{Gal}(L/F)(x)$ in $\mathrm{Aut}\, N(L|F)$ is equal to the intersection of the image of $\mathrm{Gal}(L/F)$ and the subgroup

$$\{\mathrm{T} \in \mathrm{Aut}\, N(L|F) : \Pi^{-1}\mathrm{T}\Pi \in h_{L/F}(x)\}.$$

5. ($\diamond$) [Win3] Let L/F be an infinite arithmetically profinite extension, and let E/L be a finite separable extension.
   a) Show that for a tower of fields L_i such as in (5.5), there exists a tower of finite extensions E_i of F such that $E_i \subset E_{i+1}$, $E = \cup E_i$, $L_i \subset E_i$, and an index i_0 such that

$$h_{N(E|F)/N(L|F)} = h_{E_i/L_i} \qquad \text{for } i \geqslant i_0.$$

   b) Show that if E/L is a separable extension (not necessarily finite), then E/F is an arithmetically profinite extension if and only if $N(E, L|F)/N(L|F)$ is arithmetically profinite. Show that in this case the field $N(E|F)$ can be identified with $N(N(E, L|F)|N(L|F))$ and

$$h_{E/F} = h_{N(E,L|F)/N(L|F)} \circ h_{L/F}.$$

   c) Assume in addition that E/F and E/L are Galois extensions. Show that

$$\mathrm{Gal}(N(E, L|F)/N(L|F))(h_{L/F}(x)) = \mathrm{Gal}(E/F)(x) \cap \mathrm{Gal}(N(E, L|F)/N(L|F))$$

   where we identified $\mathrm{Gal}(N(E, L|F)/N(L|F))$ with $\mathrm{Gal}(E/L)$.

6. ($\diamond$) [Win3] Let F be a complete field with respect to some nontrivial valuation $v: F^* \to \mathbb{Q}$ (in particular, if $v(F^*) = \mathbb{Z}$, then v is discrete). Let the perfect residue field $\overline{F}$ be of characteristic $p > 0$. Put $F^{(n)} = F$, and let $R^*(F) = \varprojlim F^{(n)*}$ with respect to the homomorphism of the raising to the pth power $F^{(n+1)} \xrightarrow{\uparrow p} F^{(n)}$. Put $R(F) = R^*(F) \cup \{0\}$.

a) Show that if $\mathrm{A} = (\alpha^{(n)}), \mathrm{B} = (\beta^{(n)}) \in R(F)$, then the sequence $(\alpha^{(n+m)}+\beta^{(n+m)})^{p^m}$ converges as $m \to +\infty$. Put $\gamma^{(n)} = \lim_{m\to+\infty}(\alpha^{(n+m)} + \beta^{(n+m)})^{p^m}$ and define $\mathrm{A} + \mathrm{B} = \Gamma = (\gamma^{(n)})$; put $\delta^{(n)} = \alpha^{(n)}\beta^{(n)}$ and define $\mathrm{A} \cdot \mathrm{B} = \Delta = (\delta^{(n)})$. Show that $R(F)$ is a perfect field of characteristic p.

b) For $\mathrm{A} = (\alpha^{(n)})$ put $v(\mathrm{A}) = v(\alpha^{(0)})$. Show that v possesses the properties of a valuation. Let $\theta \in F$ be the multiplicative representative of $a \in \overline{F}$ and $\Theta = (\theta^{(n)})$ with $\theta^{(n)} = \theta^{1/p^n}$. Show that $R: a \to \Theta$ is an isomorphism of $\overline{F}$ onto a subfield in $R(F)$ which is isomorphic to the residue field of $R(F)$.

c) Show that if $v: F^* \to \mathbb{Z}$ is discrete, then $R(F)$ can be identified with $\overline{F}$.

d) Show that if F is of characteristic p, then the homomorphism $\mathrm{A} = (\alpha^{(n)}) \mapsto \alpha^{(0)}$ is an isomorphism of $R(F)$ with the maximal perfect subfield in F.

7. ($\diamond$) [Win3] Let L be an infinite arithmetically profinite extension of a local field F with residue field of characteristic p. Assume that the Hasse–Herbrand function $h_{L/F}$ grows relatively fast, i.e., there exists a positive c such that $h_{L/F}(x_0)/h'_{L/F}(x_0) > c$ for all x_0 where the derivative is defined. Let C be the completion of the separable closure of F.

a) For $(\alpha_E) \in N(L/F)$ show that there exists $\beta^{(n)} = \lim_E \alpha_E^{|E:L_1|/p^n} \in C$ where L_1/F is the maximal tamely ramified subextension of L/F and E runs over all finite extensions of L_1 in L. Show that $(\beta^{(n)})$ belongs to $R(C)$.

b) Show that the homomorphism $N(L|F) \longrightarrow R(C)$ is a continuous (with respect to the discrete valuation topology on $N(L|F)$ and the topology associated with the valuation v defined in the previous exercise) field homomorphism.

c) Let E be a separable extension of L. Let S be the completion of the (p-)radical closure of $N(E, L|F)$, i.e., the completion (with respect to the extension of the valuation) of the subfield of $N(E, L|F)^{\mathrm{alg}}$ generated by $\sqrt[p^n]{\alpha}$ for all n and $\alpha \in N(E, L|F)$. Show that there is a field isomorphism from S to $R(\widehat{E})$ where $\widehat{E}$ is the completion of E. Deduce that if F is of positive characteristic, then $\widehat{E}$ is a perfect field.

8. ($\diamond$) [Win3] Let K be a discrete valuation field of characteristic 0 with residue field of characteristic p, and let C be the completion of the separable closure of K. Define the map

$$g: W(\mathcal{O}_{R(C)}) \to \mathcal{O}_C$$

by the formula $g(\mathrm{A}_0, \mathrm{A}_1, \dots) = \sum_{n\geqslant 0} p^n \alpha_n^{(n)}$, where $\mathrm{A}_m = (\alpha_m^{(n)}) \in \mathcal{O}_{R(C)}$.

a) Show that g is a surjective homomorphism. Show that its kernel is a principal ideal in $W(\mathcal{O}_{R(C)})$, generated by some element $(\mathrm{A}_0, \mathrm{A}_1, \dots)$ for which, in particular, $v(\alpha_0^{(0)}) = v(p)$.

b) Let $W_K(R) = W(\mathcal{O}_{R(C)}) \otimes_{W(\overline{K})} K$. Then g can be uniquely extended to a surjective homomorphism of K-algebras $g: W_K(R) \to C$. Show that the kernel I of this homomorphism is a principal ideal. Let B^+ be the completion of $W_K(R)$ with respect to I-adic topology and let B be its quotient field. Show that B does not depend on the choice of K and is a complete discrete valuation field with residue field C.

The ring B plays a role in the theory of p-adic representations and p-adic periods [A].

CHAPTER 4

Local Class Field Theory I

In this chapter we develop the theory of abelian extensions of a local field with finite residue field. The main theorem establishes a correspondence between abelian extensions of such a local field F and subgroups in its multiplicative group F^*; moreover we construct the so called local reciprocity homomorphism from F^* to the maximal abelian quotient of the absolute Galois group of F which has the property that for every finite Galois extension L/F it induces an isomorphism between $F^*/N_{L/F}L^*$ and the maximal abelian quotient of $\mathrm{Gal}(L/F)$. In our approach we use simultaneously two explicit constructions of the reciprocity maps and its inverse, one suggested by *M. Hazewinkel* (we use it only for totally ramified extensions) and another suggested by *J. Neukirch*. The origin of the former approach is a Theorem of *B. Dwork* [Dw, p.185] with a proof by *J. Tate*, see Exercise 4 in section 3. In our exposition it will be an interplay between the two constructions which provides an easy proof of all main results of local class field theory. Our approach can also be extended to other generalized local class field theories, like those described in section 8 of this Chapter and in Chapter V.

Section 1 lists properties of the local fields as a corollary of results of the previous chapters; it also provides an important for the subsequent sections information on some properties of the maximal unramified extension of the field and its completion. Section 2 presents the Neukirch map which is at first defined as a map from the set of Frobenius automorphisms in the Galois group of the maximal unramified extension of L over F to the factor group $F^*/N_{L/F}L^*$. To show that this map factorizes through the Galois group and that it is a homomorphism is not entirely easy. We choose a route which involves the second reciprocity map by Hazewinkel which is defined in section 3 as a homomorphism from $F^*/N_{L/F}L^*$ to the maximal abelian quotient of the Galois group of L/F in the case where the latter is a totally ramified extension. We show in section 3 that the two maps are inverse to each other and then prove that for a finite Galois extension L/F the Neukirch map induces an isomorphism

$$\Upsilon_{L/F}^{\mathrm{ab}}\colon \mathrm{Gal}(L/F)^{\mathrm{ab}} \longrightarrow F^*/N_{L/F}L^*.$$

In section 4 we extend the reciprocity maps from finite extensions to infinite Galois extensions and derive first properties of the norm groups. Section 5 presents two important pairings of the multiplicative group of a local field with finite residue field: the *Hilbert symbol* and *Artin–Schreier pairing*; the latter is defined in positive characteristic.

We apply them to the proof of the Existence Theorem in section 6. There we clarify properties of the correspondence between abelian extensions and their norm groups. In section 7 we review other approaches to local class field theory. Finally, in section 8 we introduce as a generalization of the reciprocity maps in the previous sections a non-abelian reciprocity map and review results on absolute Galois groups.

For the case of Henselian discrete valuation fields with finite residue field see Exercises.

1. Useful Results on Local Fields

This section focuses on local fields with finite residue field in (1.1)–(1.5). Many of results are just partial cases of more general assertions of the previous chapters.

Keeping in mind applications to reciprocity maps we describe several properties of the maximal unramified extension of the field under consideration and its completion in more the general context of a Henselian or complete discrete valuation field with algebraically closed residue field in subsections (1.6)–(1.9).

(1.1). Let F be a local field with finite residue field $\overline{F} = \mathbb{F}_q$, $q = p^f$ elements. The number f is called the *absolute residue degree* of F. Since $\operatorname{char}(\mathbb{F}_q) = p$, Lemma (3.2) Ch. I shows that F is of characteristic 0 or of characteristic p.

In the first case $v(p) > 0$ for the discrete valuation v in F, hence the restriction of v on $\mathbb{Q}$ is equivalent to the p-adic valuation by Ostrowski's Theorem of (1.1) Ch. I. Then we can view the field $\mathbb{Q}_p$ of p-adic numbers as a subfield of F (another way to show this is to use the quotient field of the Witt ring of a finite field and Proposition (5.6) Ch. II). Let $e = v(p) = e(F)$ be the absolute ramification index of F as defined in (5.7) Ch. I. Then by Proposition (2.4) Ch. II we obtain that F is a finite extension of $\mathbb{Q}_p$ of degree $n = ef$. In (4.6) Ch. I such a field was called a *local number field*.

In the second case Propositions (5.4) Ch. II and (5.1) Ch. II show that F is isomorphic (with respect to the field structure and the discrete valuation topology) to the field of formal power series $\mathbb{F}_p((X))$ with prime element X. In (4.6) Ch. I such a field was called a *local functional field*.

LEMMA. *F is a locally compact topological space with respect to the discrete valuation topology. The ring of integers $\mathcal{O}$ and the maximal ideal $\mathcal{M}$ are compact. The multiplicative group F^* is locally compact, and the group of units U is compact.*

Proof. Assume that $\mathcal{O}$ is not compact. Let $(V_i)_{i\in I}$ be a covering by open subsets in $\mathcal{O}$, i.e., $\mathcal{O} = \cup V_i$, such that $\mathcal{O}$ isn't covered by a finite union of V_i. Let π be a prime element of $\mathcal{O}$. Since $\mathcal{O}/\pi\mathcal{O}$ is finite, there exists an element $\theta_0 \in \mathcal{O}$ such that the set $\theta_0 + \pi\mathcal{O}$ is not contained in the union of a finite number of V_i. Similarly, there exist elements $\theta_1, \dots, \theta_n \in \mathcal{O}$ such that

$$\theta_0 + \theta_1\pi + \cdots + \theta_n\pi^n + \pi^{n+1}\mathcal{O}$$

is not contained in the union of a finite number of V_i. However, the element

$$\alpha = \lim_{n\to+\infty} \sum_{m=0}^{n} \theta_m \pi^m$$

belongs to some V_i, a contradiction. Hence, $\mathcal{O}$ is compact and U, as the union of $\theta + \pi\mathcal{O}$ with $\overline{\theta} \neq 0$, is compact. □

(1.2). LEMMA. *The Galois group of every finite extension of F is solvable.*

Proof. Follows from Corollary 3 of (4.4) Ch. II . □

PROPOSITION. *For every $n \geqslant 1$ there exists a unique unramified extension L of F of degree n: $L = F(\mu_{q^n-1})$. The extension L/F is cyclic and the maximal unramified extension F^{ur} of F is a Galois extension.* $\mathrm{Gal}(F^{\mathrm{ur}}/F)$ *is isomorphic to $\widehat{\mathbb{Z}}$ and topologically generated by an automorphism φ_F, such that*

$$\varphi_F(\alpha) \equiv \alpha^q \mod \mathcal{M}_{F^{\mathrm{ur}}} \qquad \textit{for } \alpha \in \mathcal{O}_{F^{\mathrm{ur}}}.$$

The automorphism φ_F is called the Frobenius automorphism of F.

Proof. First we note that, by Corollary 1 of (7.3) Ch. I, F contains the group μ_{q-1} of $(q-1)$th roots of unity which coincides with the set of nonzero multiplicative representatives of $\overline{F}$ in $\mathcal{O}$. Moreover, Proposition (5.4) and section 7 of Ch. I imply that the unit group U_F is isomorphic to $\mu_{q-1} \times U_{1,F}$.

The field $\mathbb{F}_q$ has the unique extension $\mathbb{F}_{q^n}$ of degree n, which is cyclic over $\mathbb{F}_q$. Propositions (3.2) and (3.3) Ch. II show that there is a unique unramified extension L of degree n over F and hence $L = F(\mu_{q^n-1})$.

Now let E be an unramified extension of F and $\alpha \in E$. Then $F(\alpha)/F$ is of finite degree. Therefore, F^{ur} is contained in the union of all finite unramified extensions of F. We have

$$\mathrm{Gal}(F^{\mathrm{ur}}/F) \simeq \varprojlim \mathrm{Gal}(\mathbb{F}_{q^n}/\mathbb{F}_q) \simeq \widehat{\mathbb{Z}}.$$

It is well known that $\mathrm{Gal}(\mathbb{F}_q^{\mathrm{sep}}/\mathbb{F}_q)$ is topologically generated by the automorphism σ such that $\sigma(a) = a^q$ for $a \in \mathbb{F}_q^{\mathrm{sep}}$. Hence, $\mathrm{Gal}(F^{\mathrm{ur}}/F)$ is topologically generated by the Frobenius automorphism φ_F. □

REMARK. If $\theta \in \mu_{q^n-1}$, then

$$\varphi_F(\theta) \equiv \theta^q \mod \mathcal{M}_L$$

and $\varphi_F(\theta) \in \mu_{q^n-1}$. The uniqueness of the multiplicative representative for $\overline{\theta}^q \in \overline{F}$ implies now that $\varphi_F(\theta) = \theta^q$.

(1.3). EXAMPLE. Let ζ_{p^m} be a primitive p^m th root of unity. Put $\mathbb{Q}_p^{(m)} = \mathbb{Q}_p(\zeta_{p^m})$. Then

$$v_{\mathbb{Q}_p^{(m)}}(\zeta_{p^m}) = 0$$

and ζ_{p^m} belongs to the ring of integers of $\mathbb{Q}_p^{(m)}$. Let

$$f_m(X) = \frac{X^{p^m} - 1}{X^{p^{m-1}} - 1} = X^{(p-1)p^{m-1}} + X^{(p-2)p^{m-1}} + \dots + 1.$$

Then ζ_{p^m} is a root of $f_m(X)$, and hence $|\mathbb{Q}_p^{(m)} : \mathbb{Q}_p| \leqslant (p-1)p^{m-1}$. The elements $\zeta_{p^m}^i$, $0 < i < p^m, p \nmid i$, are roots of $f_m(X)$. Hence

$$f_m(X) = \prod_{\substack{p \nmid i \\ 0<i<p^m}} (X - \zeta_{p^m}^i) \quad \text{and} \quad p = f_m(1) = \prod_{\substack{p \nmid i \\ 0<i<p^m}} (1 - \zeta_{p^m}^i).$$

However,

$$(1 - \zeta_{p^m}^i)(1 - \zeta_{p^m})^{-1} = 1 + \zeta_{p^m} + \dots + \zeta_{p^m}^{i-1}$$

belongs to the ring of integers of $\mathbb{Q}_p^{(m)}$. For the same reason, $(1 - \zeta_{p^m})(1 - \zeta_{p^m}^i)^{-1}$ belongs to the ring of integers of $\mathbb{Q}_p^{(m)}$. Thus, $(1 - \zeta_{p^m}^i)(1 - \zeta_{p^m})^{-1}$ is a unit and $p = (1 - \zeta_{p^m})^{p^{m-1}(p-1)}\varepsilon$ for some unit ε. Therefore, $e(\mathbb{Q}_p^{(m)}|\mathbb{Q}_p) \geqslant (p-1)p^{m-1}$, and $\mathbb{Q}_p^{(m)}$ is a cyclic totally ramified extension with the prime element $1 - \zeta_{p^m}$, and of degree $(p-1)p^{m-1}$ over $\mathbb{Q}_p$. In particular,

$$\mathcal{O}_{\mathbb{Q}_p^{(m)}} = \mathcal{O}_{\mathbb{Q}_p}[1 - \zeta_{p^m}] = \mathcal{O}_{\mathbb{Q}_p}[\zeta_{p^m}].$$

(1.4). In order to describe the group $U_1 = U_{1,F}$ of principal units we can apply assertions of sections 5, 6 Ch. I.

If $\operatorname{char}(F) = p$, then Proposition (6.2) Ch. I shows that every element $\alpha \in U_1$ can be uniquely expressed as the convergent product

$$\alpha = \prod_{\substack{p \nmid i \\ i>0}} \prod_{j \in J} (1 + \theta_j \pi_i)^{a_{ij}},$$

where the index-set J numerates f elements in $\mathcal{O}_F$, such that their residues form a basis of $\mathbb{F}_q$ over $\mathbb{F}_p$, and the elements θ_j belong to this set of f elements; π_i are elements of $\mathcal{O}_F$ with $v(\pi_i) = i$, and $a_{ij} \in \mathbb{Z}_p$. Thus, U_1 has the infinite topological basis $1 + \theta_j \pi_i$.

Now let $\operatorname{char}(F) = 0$. (6.4) and (6.5) Ch. I show that every element $\alpha \in U_1$ can be expressed as a convergent product

$$\alpha = \prod_{i \in I} \prod_{j \in J} (1 + \theta_j \pi_i)^{a_{ij}} \omega_*^a$$

where $I = \{1 \leqslant i < pe/(p-1),\ p \nmid i\}$, $e = e(F)$; the index-set J numerates f elements in $\mathcal{O}_F$, such that their residues form a basis of $\mathbb{F}_q$ over $\mathbb{F}_p$, and the elements θ_j belong to this set of f elements; π_i are elements of $\mathcal{O}_F$ with $v(\pi_i) = i$, and $a_{ij} \in \mathbb{Z}_p$.

If a primitive pth root of unity does not belong to F, then $\omega_* = 1, a = 0$ and the above expression for α is unique; U_1 is a free $\mathbb{Z}_p$-module of rank $n = ef = |F : \mathbb{Q}_p|$.

If a primitive pth root of unity belongs to F, then $\omega_* = 1 + \theta_* \pi_{pe/(p-1)}$ is a principal unit such that $\omega_* \notin F^{*p}$, and $a \in \mathbb{Z}_p$. In this case the above expression for α is not unique. Subsections (5.7) and (5.8) Ch. I imply that U_1 is isomorphic to the product of n copies of $\mathbb{Z}_p$ and the p-torsion group μ_{p^r}, where $r \geqslant 1$ is the maximal integer such that $\mu_{p^r} \subset F$.

LEMMA. *If* char$(F) = 0$, *then* F^{*n} *is an open subgroup of finite index in* F^* *for* $n \geqslant 1$. *If* char$(F) = p$, *then* F^{*n} *is an open subgroup of finite index in* F^* *for* $p \nmid n$. *If* char$(F) = p$ *and* $p|n$, *then* F^{*n} *is not open and is not of finite index in* F^*.

Proof. It follows from Proposition (5.9) Ch. I and the previous considerations. □

(1.5). Now we have a look at the norm group $N_{L/F}(L^*)$ for a finite extension L of F. Recall that the norm map

$$N_{\mathbb{F}_{q'}/\mathbb{F}_q} : \mathbb{F}_{q'}^* \longrightarrow \mathbb{F}_q^*$$

is surjective when $\mathbb{F}_{q'} \supset \mathbb{F}_q$. Then the second and third diagrams of Proposition (1.2) Ch. III show that $N_{L/F} U_L = U_F$ in the case of an unramified extension L/F. Further, the first diagram there implies that

$$N_{L/F} L^* = \langle \pi^n \rangle \times U_F,$$

where π is a prime element in F, $n = |L : F|$. This means, in particular, that $F^*/N_{L/F}L^*$ is a cyclic group of order n in the case under consideration. Conversely, every subgroup of finite index in F^* that contains U_F coincides with the norm group $N_{L/F}L^*$ for a suitable unramified extension L/F.

The next case is a totally and tamely ramified Galois extension L/F of degree n. Proposition (1.3) Ch. III and its Corollary show that

$$N_{L/F} U_{1,L} = U_{1,F}, \qquad \pi \in N_{L/F} L^*,$$

for a suitable prime element π in F (e.g. such that $L = F(\sqrt[n]{-\pi})$, and $\theta \in N_{L/F}L^*$ for $\theta \in U_F$ if and only if $\overline{\theta} \in \mathbb{F}_q^{*n}$. Since L/F is Galois, we get $\mu_n \subset F^*$ and $n|(q-1)$. Hence, the subgroup $\mathbb{F}_q^{*n}$ is of index n in $\mathbb{F}_q^*$, and the quotient group $\mathbb{F}_q^*/\mathbb{F}_q^{*n}$ is cyclic. We conclude that

$$N_{L/F} L^* = \langle \pi \rangle \times \langle \theta \rangle \times U_{1,F}$$

with an element $\theta \in U_F$, such that its residue $\overline{\theta}$ generates $\mathbb{F}_q^*/\mathbb{F}_q^{*n}$. In particular, $F^*/N_{L/F}L^*$ is cyclic of order n. Conversely, every subgroup of index n relatively prime to char($\overline{F}$) coincides with the norm group $N_{L/F}L^*$ for a suitable cyclic extension L/F.

The last case to be considered is the case of a totally ramified Galois extension L/F of degree p. Preserving the notations of (1.4) Ch. III, we apply Proposition (1.5) Ch. III. The right vertical homomorphism of the fourth diagram

$$\overline{\theta} \to \overline{\theta}^p - \overline{\eta}^{p-1}\overline{\theta}$$

has a kernel of order p; therefore its cokernel is also of order p. Let $\theta^* \in U_F$ be such that $\overline{\theta}^*$ does not belong to the image of this homomorphism. Since $\overline{F}$ is perfect, we deduce, using the third and fourth diagrams, that $1 + \theta^*\pi_F^s \notin N_{L/F}U_{1,L}$. The other diagrams imply that $F^*/N_{L/F}L^*$ is a cyclic group of order p and generated by

$$1 + \theta^*\pi_F^s \mod N_{L/F}L^*.$$

If char(F) = 0, then, by Proposition (2.3) Ch. III, $s \leqslant pe/(p-1)$, where $e = e(F)$. That Proposition also shows that if $p|s$, then $s = pe/(p-1)$ and a primitive pth root of unity ζ_p belongs to F, and $L = F(\sqrt[p]{\pi})$ for a suitable prime element π in F. In this case $F^*/N_{L/F}L^*$ is generated by ω_* mod $N_{L/F}L^*$.

Conversely, note that every subgroup of index p in the additive group $\mathbb{F}_q$ can be written as $\overline{\eta}\wp\left(\mathbb{F}_q\right)$ for some $\overline{\eta} \in \mathbb{F}_q$. Let N be an open subgroup of index p in F^* such that some prime element $\pi_F \in N$ and $\omega_* \in N$ (if char(F) = 0). Then, in terms of the cited Corollary (2.5) Ch. III, if s is the maximal integer relatively prime to p such that $U_{s,F} \not\subset N$ and $U_{s+1,F} \subset N$, then $1+\eta\wp(\mathcal{O}_F)\pi^s+\pi^{s+1}\mathcal{O}_F \subset N$ for some element $\eta \in \mathcal{O}_F$. By that Corollary we obtain that $1+\eta\wp(\mathcal{O}_F)\pi^s+\pi^{s+1}\mathcal{O}_F \subset N_{L/F}L^*$, where $L = F(\lambda)$ and λ is a root of the polynomial $X^p - X + \theta^p\alpha$, with $\alpha = \theta^{-p}\eta^{-1}\pi^{-s}$ for a suitable $\theta \in U_F$. Since $s = s(L|F)$ (see (1.4) Ch. III), we get

$$U_{i,F} \subset U_{i+1,F}N_{L/F}U_L \quad \text{for } i < s$$

by Proposition (1.5) Ch. III. In terms of the homomorphism λ_i of section 5 Ch. I we obtain that

$$\lambda_i\left((N \cap U_{i,F})U_{i+1,F}/U_{i+1,F}\right) = \lambda_i\left((N_{L/F}L^* \cap U_{i,F})U_{i+1,F}/U_{i+1,F}\right)$$

for $i \geqslant 0$. If $\omega_* \notin N$ and char(F) = 0, then one can put $L = F(\sqrt[p]{\pi})$. Then we obtain the same relations for N and $N_{L/F}L^*$ as just above. Later we shall show that, moreover, for every open subgroup N of finite index in F^*, $N = N_{L/F}L^*$ for a suitable abelian extension L/F.

(1.6). Now we prove several properties of the maximal unramified extension F^{ur} of F and its completion. The field F^{ur} is a Henselian discrete valuation field with algebraically closed residue field and its completion is a local field with algebraically closed residue field $\mathbb{F}_q^{\text{sep}}$.

If fact, the case of complete fields will be enough in the main text, and we have included the Henselian case for the sake of completeness, especially because the two cases can be handled almost similarly. The field F^{ur} as an algebraic extension of F is perhaps easier to deal with than its completion which is a transcendental extension of F. All results of sections 2–6 except Corollary (3.2) can be proved without using the completion of F^{ur}, see Exercise 6 section 3.

We consider, keeping in mind applications in Ch. V, the more general situation of a Henselian or complete discrete valuation fields with algebraically closed residue field. We denote any of these fields by $\mathcal{F}$.

Let R be the set of multiplicative representatives of the residue field of $\mathcal{F}$ if its characteristic is p or a coefficient field, see section 5 Ch. II, if that characteristic is zero.

In the case where the residue field is $\mathbb{F}_q^{\mathrm{sep}}$, R is the union of all sets $\mu_{q^n-1}, n \geqslant 1$ (which coincides with the set of all roots of unity of order relatively prime to p) and of 0. Then R is the set of the multiplicative representatives of $\mathbb{F}_q^{\mathrm{sep}}$ in $\mathcal{F}$.

Let $\mathcal{L}$ be a finite separable extension of $\mathcal{F}$. Since the residue field of $\mathcal{F}$ is algebraically closed, $\mathcal{L}/\mathcal{F}$ is totally ramified.

LEMMA. *The norm maps*

$$N_{\mathcal{L}/\mathcal{F}}\colon \mathcal{L}^* \to \mathcal{F}^*, \qquad N_{\mathcal{L}/\mathcal{F}}\colon U_{\mathcal{L}} \to U_{\mathcal{F}}$$

are surjective.

Proof. Since the Galois group of $\mathcal{L}/\mathcal{F}$ is solvable by Corollary 3 (4.4) Ch. II, it suffices to consider the case of a Galois extension of prime degree l. Certainly, the norm of a prime element of $\mathcal{L}$ is a prime element of $\mathcal{F}$. Now if $\mathcal{F}$ is complete, then from results of section 1 Ch. III we deduce the surjectivity of the norm maps. If $\mathcal{F} = F^{\mathrm{ur}}$ then from its Henselian properties in Corollary 4 (2.9) Ch. II and (1.3) Ch. II we deduce that all sufficiently small units are l-th powers. Therefore we again deduce the surjectivity of the norm map from section 1 Ch. III. □

REMARK. If the extension $\mathcal{L}/\mathcal{F}$ is totally ramified of degree a power of p and the residue field of $\mathcal{F}$ is not algebraically closed but just a perfect field without nontrivial separable p-extensions, then similarly to the proof of the Lemma we deduce that the norm $N_{\mathcal{L}/\mathcal{F}}$ is still surjective.

(1.7). DEFINITION. For a finite Galois extension $\mathcal{L}/\mathcal{F}$ denote by $U(\mathcal{L}/\mathcal{F})$ the subgroup of $U_{1,\mathcal{L}}$ generated by $u^{\sigma-1}$ where u runs through all elements of $U_{1,\mathcal{L}}$ and σ runs through all elements of $\mathrm{Gal}(\mathcal{L}/\mathcal{F})$.

Every unit in $U_{\mathcal{L}}$ can be factorized as $\theta\varepsilon$ with $\theta \in R^*$, $\varepsilon \in U_{1,\mathcal{L}}$ Since $\theta^{\sigma-1} = 1$ we deduce that $U(\mathcal{L}/\mathcal{F})$ coincides with the subgroup of $U_{\mathcal{L}}$ generated by $u^{\sigma-1}$, $u \in U_{\mathcal{L}}$, $\sigma \in \mathrm{Gal}(\mathcal{L}/\mathcal{F})$.

PROPOSITION. *Let $\mathcal{L}$ be a finite Galois extension of $\mathcal{F}$. For a prime element π of $\mathcal{L}$ define*

$$\ell\colon \mathrm{Gal}(\mathcal{L}/\mathcal{F}) \to U_{\mathcal{L}}/U(\mathcal{L}/\mathcal{F}), \qquad \ell(\sigma) = \pi^{\sigma-1} \mod U(\mathcal{L}/\mathcal{F}).$$

The map ℓ is a homomorphism which does not depend on the choice of π. It induces a monomorphism $\ell\colon \mathrm{Gal}(\mathcal{L}/\mathcal{F})^{\mathrm{ab}} \to U_{\mathcal{L}}/U(\mathcal{L}/\mathcal{F})$ where for a group G the notation G^{ab} stands for the maximal abelian quotient of G.

The sequence

$$1 \to \mathrm{Gal}(\mathcal{L}/\mathcal{F})^{\mathrm{ab}} \xrightarrow{\ \ \ell\ \ } U_{\mathcal{L}}/U(\mathcal{L}/\mathcal{F}) \xrightarrow{N_{\mathcal{L}/\mathcal{F}}} U_{\mathcal{F}} \to 1$$

is exact.

Proof. Since $\pi^{\tau-1}$ belongs to $U_{\mathcal{L}}$, we deduce that $(\pi^{\tau-1})^{\sigma-1} \in U(\mathcal{L}/\mathcal{F})$ and

$$\pi^{\sigma\tau-1} \equiv \pi^{\tau-1}\pi^{\sigma-1} \mod U(\mathcal{L}/\mathcal{F}).$$

Thus, the map ℓ is a homomorphism. It does not depend on the choice of π, since $(\pi\varepsilon)^{\sigma-1} \equiv \pi^{\sigma-1} \mod U(\mathcal{L}/\mathcal{F})$.

Surjectivity of the norm map has already been proved.

Suppose first that $\mathrm{Gal}(\mathcal{L}/\mathcal{F})$ is cyclic with generator σ. Proposition (4.1) Ch. III shows that the kernel of $N_{\mathcal{L}/\mathcal{F}}$ coincides with $\mathcal{L}^{*\sigma-1}$. Since ℓ is a homomorphism, we have $\pi^{\sigma^m-1} \equiv (\pi^{\sigma-1})^m \mod U(\mathcal{L}/\mathcal{F})$. So we deduce that $\mathcal{L}^{*\sigma-1}$ is equal to the product of $U(\mathcal{L}/\mathcal{F})$ and the image of ℓ. This shows the exactness in the middle term.

Note that $u^{\sigma^m-1} = (u^{1+\sigma+\dots+\sigma^{m-1}})^{\sigma-1}$, so $U(\mathcal{L}/\mathcal{F}) = U_{\mathcal{L}}^{\sigma-1}$. If $\pi^{\sigma^m-1} \in U(\mathcal{L}/\mathcal{F})$, then $(\pi^{\sigma-1})^m = u^{\sigma-1}$ for some $u \in U_{\mathcal{L}}$. Hence $\pi^m u^{-1}$ belongs to $\mathcal{F}$ and therefore $|L:F|$ divides m and $\sigma^m = 1$. This shows the injectivity of ℓ.

Now in the general case we use the solvability of $\mathrm{Gal}(\mathcal{L}/\mathcal{F})$ and argue by induction. Let $\mathcal{M}/\mathcal{F}$ be a Galois cyclic subextension of $\mathcal{L}/\mathcal{F}$ such that $\mathcal{L} \neq \mathcal{M} \neq \mathcal{F}$. Put $\pi_M = N_{\mathcal{L}/\mathcal{M}}\pi$. Since $N_{\mathcal{L}/\mathcal{M}}\colon U_{\mathcal{L}} \to U_{\mathcal{M}}$ is surjective, we deduce that $N_{\mathcal{L}/\mathcal{M}}U(\mathcal{L}/\mathcal{F}) = U(\mathcal{M}/\mathcal{F})$.

Let $N_{\mathcal{L}/\mathcal{F}}u = 1$ for $u \in U_{\mathcal{L}}$. Then by the induction hypothesis there is $\tau \in \mathrm{Gal}(\mathcal{L}/\mathcal{F})$ such that $N_{\mathcal{L}/\mathcal{M}}u = \pi_M^{\tau-1}\eta$ with $\eta \in U(\mathcal{M}/\mathcal{F})$. Write $\eta = N_{\mathcal{L}/\mathcal{M}}\xi$ with $\xi \in U(\mathcal{L}/\mathcal{F})$. Then $u^{-1}\pi^{\tau-1}\xi$ belongs to the kernel of $N_{\mathcal{L}/\mathcal{M}}$ and therefore by the induction hypothesis can be written as $\pi^{\sigma-1}\rho$ with $\sigma \in \mathrm{Gal}(\mathcal{L}/\mathcal{M})$, $\rho \in U(\mathcal{L}/\mathcal{M})$. Altogether, $u \equiv \pi^{\sigma\tau-1} \mod U(\mathcal{L}/\mathcal{F})$ which shows the exactness in the middle term.

To show the injectivity of ℓ assume that $\pi^{\sigma-1} \in U(\mathcal{L}/\mathcal{F})$. Then $\pi_M^{\sigma-1} \in U(\mathcal{M}/\mathcal{F})$ and by the previous considerations of the cyclic case σ acts trivially on $\mathcal{M}$. So σ belongs to $\mathrm{Gal}(\mathcal{L}/\mathcal{M})$. Now the maximal abelian extension of $\mathcal{F}$ in $\mathcal{L}$ is the compositum of all cyclic extensions $\mathcal{M}$ of $\mathcal{F}$ in $\mathcal{L}$. Since σ acts trivially on each $\mathcal{M}$, we conclude that ℓ is injective. □

REMARK. If the extension $\mathcal{L}/\mathcal{F}$ is totally ramified of degree a power of p and the residue field of $\mathcal{F}$ is not algebraically closed but just a perfect field without nontrivial separable p-extensions, then the Proposition still holds.

(1.8). For every n every element $\alpha \in \mathcal{F}$ can be uniquely expanded as

$$\alpha = \sum_{a \leqslant i \leqslant n-1} \theta_i \pi^i \mod \pi^n, \qquad \theta_i \in R,$$

where π is a prime element in F. If $\mathcal{F}$ is complete, then the same holds with $n = \infty$.

Suppose from now on that $\mathcal{F}$ is the maximal unramified extension F^{ur}, or its completion, of a local field F with perfect residue field, such that the Galois group $\mathrm{Gal}(F^{\mathrm{ur}}/F)$ is isomorphic to $\widehat{\mathbb{Z}}$. Fix a generator φ of $\mathrm{Gal}(F^{\mathrm{ur}}/F)$.

For example, if the residue field of $\mathcal{F}$ is $\mathbb{F}_q^{\mathrm{sep}}$, then F is just a local number field. In the situation of the previous paragraph we can take the Frobenius automorphism φ_F as the generator φ.

Since $\varphi\colon F^{\mathrm{ur}} \to F^{\mathrm{ur}}$ is continuous, it has exactly one extension $\varphi\colon \mathcal{F} \to \mathcal{F}$.

If the residue field of $\mathcal{F}$ is $\mathbb{F}_q^{\mathrm{sep}}$ then the continuous extension φ of the Frobenius automorphism φ_F acts as $\sum_{i \geqslant a} \theta_i \pi^i \mapsto \sum_{i \geqslant a} \theta_i^q \pi^i$, since Remark in (1.2) shows that $\varphi(\theta_i) = \varphi_F(\theta_i) = \theta_i^q$ for $\theta_i \in R$.

We shall study the action of ${}^{\varphi-1}$ on the multiplicative group.

Denote by $T_{\mathcal{F}}$ the group of roots of unity in $\mathcal{F}$ of order not divisible by the characteristic of the residue field of $\mathcal{F}$. If the residue field of $\mathcal{F}$ is $\mathbb{F}_q^{\mathrm{sep}}$ then $T_{\mathcal{F}} = R \setminus \{0\}$.

PROPOSITION.

(1) *The kernel of the homomorphism*

$$\mathcal{F}^* \to \mathcal{F}^*, \qquad \alpha \mapsto \alpha^{\varphi-1}$$

is equal to F *and the image is contained in* $U_{\mathcal{F}}$.

(2) $U_{0,\mathcal{F}}^{\varphi-1} \supset T_{\mathcal{F}}$.

(3) *For every* $n, m \geqslant 1$ *the sequence*

$$1 \to U_{n,F} U_{n+m,\mathcal{F}}/U_{n+m,\mathcal{F}} \to U_{n,\mathcal{F}}/U_{n+m,\mathcal{F}} \xrightarrow{\varphi-1} U_{n,\mathcal{F}}/U_{n+m,\mathcal{F}} \to 1$$

is exact.

(4) *If* $\mathcal{F}$ *is complete, then* $U_{n,\mathcal{F}}^{\varphi-1} = U_{n,\mathcal{F}}$ *for every* $n \geqslant 1$.

(5) *If* $\mathcal{F}$ *is complete, then* $\mathcal{F}^{*\varphi-1}$ *contains* $T_{\mathcal{F}} U_{1,\mathcal{F}}$.

(6) *If the residue field of* $\mathcal{F}$ *is* $\mathbb{F}_q^{\mathrm{sep}}$ *then the sequence*

$$1 \to U_F U_{n+1,\mathcal{F}}/U_{n+1,\mathcal{F}} \to U_{\mathcal{F}}/U_{n+1,\mathcal{F}} \xrightarrow{\varphi-1} U_{\mathcal{F}}/U_{n+1,\mathcal{F}} \to 1$$

is exact, and $\mathcal{F}^{*\varphi-1} = U_{\mathcal{F}}$.

Proof. If $\mathcal{F} = F^{\mathrm{ur}}$ then every element of it belongs to a finite extension of F, and the kernel of ${}^{\varphi-1}$ is F. If $\mathcal{F}$ is complete then for $\alpha = \sum_{i \geqslant a} \theta_i \pi^i \in \mathcal{F}$ with $\theta_i \in R$ the condition $\varphi(\alpha) = \alpha$ implies that $\overline{\varphi}(\overline{\theta}_i) = \overline{\theta}_i$ for $i \geqslant a$. Hence, $\overline{\theta}_i$ belongs to the residue field of F and $\alpha \in F$. Similarly one shows the exactness of the sequence in the central term $U_{n,\mathcal{F}}/U_{n+m,\mathcal{F}}$.

Since every prime element π of F belongs to the kernel of ${}^{\varphi-1}$, we deduce that the image of the homomorphism is contained in $U_{\mathcal{F}}$.

Let $\varepsilon \in T_{\mathcal{F}} U_{1,\mathcal{F}}$. We shall show the existence of a sequence $\beta_n \in U_{\mathcal{F}}$ such that $\varepsilon \equiv \beta_n^{\varphi-1} \mod U_{n+1,\mathcal{F}}$ and $\beta_{n+1}\beta_n^{-1} \in U_{n+1,\mathcal{F}}$.

Let $\varepsilon = \theta\varepsilon_0$ with $\theta \in T_{\mathcal{F}}$, $\varepsilon_0 \in U_{1,\mathcal{F}}$. The element θ is an lth root of unity and belongs to some finite extension K of F. Let K' be the extension of degree l over K. Then $N_{K'/K}\theta = 1$, and Proposition (4.1) Ch. III shows that $\theta = \eta^{\sigma-1}$ for some element $\eta \in K'^*$ and automorphism σ of F^{ur} over K. Then $\sigma = \varphi^m$ for a positive integer m and we conclude that $\theta = \rho^{\varphi-1}$ where $\rho = \prod_{i=0}^{m-1} \varphi^i(\eta)$. Put $\beta_0 = \rho$.

Now assume that the elements $\beta_0, \beta_1, \dots, \beta_n \in U_{\mathcal{F}}$ have already been constructed. Define the element $\theta_{n+1} \in R$ from the congruence

$$\varepsilon^{-1}\beta_n^{\varphi-1} \equiv 1 + \theta_{n+1}\pi^{n+1} \mod \pi^{n+2}.$$

We claim that there is an element $\eta_{n+1} \in R$ such that

$$\varphi(\eta_{n+1}) - \eta_{n+1} + \theta_{n+1} \equiv 0 \mod \pi.$$

Indeed, consider the element $\overline{\theta}_{n+1}$ as an element of some finite extension $\overline{K}$ over $\overline{F}$. Let $\overline{K}'$ be the extension of degree p over $\overline{K}$. Now $\mathrm{Tr}_{\overline{K}'/\overline{K}}\overline{\theta}_{n+1} = 0$. Since $\overline{K}'/\overline{K}$ is separable, one can find an element $\overline{\xi}$ in $\overline{K}'$ with $\mathrm{Tr}_{\overline{K}'/\overline{K}}\overline{\xi} = 1$. Then, setting $\overline{\delta} = -\overline{\theta}_{n+1}\sum_{i=1}^{p-1} i\sigma^i(\overline{\xi})$, where σ is a generator of $\mathrm{Gal}(\overline{K}'/\overline{K})$, we conclude that $\sigma(\overline{\delta}) - \overline{\delta} = \overline{\theta}_{n+1}$. If $\sigma = \overline{\varphi}^m$ with positive integer m then put $\overline{\xi}' = \sum_{i=0}^{m-1}\overline{\varphi}^i(\overline{\delta})$. Then $\overline{\varphi}(\overline{\xi}') - \overline{\xi}' = \overline{\theta}_{n+1}$. So the required element η_{n+1} can be taken as any element of R whose residue is equal to $\overline{\xi}'$.

Now put $\beta_{n+1} = \beta_n(1 + \eta_{n+1}\pi^{n+1})$. Then $\varepsilon^{-1}\beta_{n+1}^{\varphi-1} \in U_{n+2,\mathcal{F}}$ and $\beta_{n+1}\beta_n^{-1} \in U_{n+1,\mathcal{F}}$.

If $\mathcal{F}$ is complete, then there exists $\beta = \lim \beta_n \in U_{\mathcal{F}}$, and $\beta^{\varphi-1} = \varepsilon$. When $\varepsilon \in U_{n,\mathcal{F}}$ the element β can be chosen in $U_{n,\mathcal{F}}$ as well. □

REMARKS.

1. If the residue field of $\mathcal{F}$ is $\mathbb{F}_q^{\mathrm{sep}}$ then the existence of β_0 and η_{n+1} also follows from Remark in (1.2), because the polynomials $X^{q-1} - \overline{\theta}$, $X^q - X + \overline{\theta}_{n+1}$ are completely split in $\mathbb{F}_q^{\mathrm{sep}}$.

2. If the residue field of $\mathcal{F}$ is the maximal abelian unramified p-extension of a local field F with perfect residue field such that the Galois group $\mathrm{Gal}(F^{\mathrm{ur}}/F)$ is isomorphic to $\mathbb{Z}_p$ and φ is its generator, then assertions (1), (3), (4) of the Proposition still hold. This follows from the proof of the Proposition in which for principal units we used unramified extensions of degree p.

(1.9). Let L/F be a finite Galois totally ramified extension. By (4.1) Ch. II the extension $L^{\mathrm{ur}}/F^{\mathrm{ur}}$ is Galois with the group isomorphic to that of L/F. We may assume that the completion of F^{ur} is a subfield of the completion of L^{ur}.

The extension $\mathcal{L}/\mathcal{F}$ is totally ramified of the same degree as L/F. Using for example (2.6)–(2.7) Ch. II we deduce that the extension $\mathcal{L}/\mathcal{F}$ is Galois with the group isomorphic to that of L/F.

PROPOSITION. *Let $\gamma \in \mathcal{L}^*$ be such that $\gamma^{\varphi-1} \in U(\mathcal{L}/\mathcal{F})$. Then $N_{\mathcal{L}/\mathcal{F}}\gamma$ belongs to the group $N_{L/F}L^*$.*

Proof. We have $\gamma^{\varphi-1} = \prod \varepsilon_j^{\tau_j-1}$ for some $\varepsilon_j \in U_{1,\mathcal{L}}$ and $\tau_j \in \mathrm{Gal}(\mathcal{L}/\mathcal{F})$. By Proposition (1.8) (applied to $\mathcal{L}$) for every positive integer r we have $\varepsilon_j = \eta_j^{\varphi-1} \bmod U_{r,\mathcal{L}}$ for some $\eta_j \in U_{\mathcal{L}}$. So the element $(\gamma^{-1}\prod \eta_j^{\tau_j-1})^{\varphi-1}$ belongs to $U_{r,\mathcal{L}}$. By the same Proposition (applied to $\mathcal{L}$) $\gamma^{-1}\prod \eta_j^{\tau_j-1} = a\delta$ with $a \in L^*$ and $\delta \in U_{r,\mathcal{L}}$. Then $N_{\mathcal{L}/\mathcal{F}}\gamma = N_{L/F}a\, N_{\mathcal{L}/\mathcal{F}}\delta$. From the description of the norm map in section 3 Ch. III we know that as soon as r tends to infinity, the element $N_{\mathcal{L}/\mathcal{F}}\delta$ of U_F tends to 1 and therefore belongs to the norm group $N_{L/F}L^*$ for sufficiently large r. Thus, $N_{\mathcal{L}/\mathcal{F}}\gamma$ belongs to $N_{L/F}L^*$. □

REMARKS.

1. Due to Remark 2 in (1.8) if the residue field of $\mathcal{F}$ is the maximal abelian unramified p-extension of a local field F with perfect residue field such that the Galois group $\mathrm{Gal}(F^{\mathrm{ur}}/F)$ is isomorphic to $\mathbb{Z}_p$ and φ is its generator, then the Proposition still holds.

2. Since $N_{\mathcal{L}/\mathcal{F}}\gamma = N_{L/F}b$ for some $b \in L^*$, we deduce that $\gamma = b\lambda$ for some $\lambda \in \ker N_{\mathcal{L}/\mathcal{F}}$. From Proposition (1.7) $\lambda = \pi^{\sigma-1}u$ with $u \in U(\mathcal{L}/\mathcal{F})$ and so $\gamma^{\varphi-1} = u^{\varphi-1} \in U(\mathcal{L}/\mathcal{F})^{\varphi-1}$. Thus, $\mathcal{L}^{*\varphi-1} \cap U(\mathcal{L}/\mathcal{F}) = U(\mathcal{L}/\mathcal{F})^{\varphi-1}$. It will be this property and its extension that we use in section 4 Ch. V for p-class field theory of local fields with perfect residue field.

Exercises.

1. Show that a discrete valuation field F is locally compact if and only if it is complete and its residue field is finite.
2. Let F be a finite extension of $\mathbb{Q}_p$. Show, using Exercise 5b) section 2 Ch. II, that there exists a finite extension E over $\mathbb{Q}$ such that $F = E\mathbb{Q}_p$, $|F : \mathbb{Q}_p| = |E : \mathbb{Q}|$, and E is dense in F. This means that local number fields are completions of algebraic number fields (finite extensions of $\mathbb{Q}$).
3. a) Compute the index of F^{*n} in F^*.
   b) Show that if $F \subset L, F \neq L$, then the index of F^* in L^* is infinite.
4. a) Show that $\mathbb{Q}_p^{(1)} = \mathbb{Q}_p(\sqrt[p-1]{-p})$.
   b) Find a local number field F for $n \geqslant 0$ such that $\mu_{p^n} \subset F, \mu_{p^{n+1}} \not\subset F$, and the extension $F(\mu_{p^{n+1}})/F$ is unramified.

5. Let F be a local number field, and let L/F be a Galois totally ramified extension of degree n. Let M be the unramified extension of F of degree n. Show using (1.5) that $F^* \subset N_{LM/M}(LM)^*$.
6. Prove the *local Kronecker–Weber Theorem*: every finite abelian extension L of $\mathbb{Q}_p$ is contained in a field $\mathbb{Q}_p(\zeta_n)$ for a suitable primitive nth root of unity, following the steps below. Denote by $\mathbb{Q}_p^{\text{cycl}}$ the extension generated by roots of unity over $\mathbb{Q}_p$. For a prime l let E_l be the maximal l-subextension in $\mathbb{Q}_p^{\text{cycl}}/\mathbb{Q}_p$, i.e., the compositum of all finite extensions of degree a power of l of $\mathbb{Q}_p$ in $\mathbb{Q}_p^{\text{cycl}}$.
   a) Show that E_p is the compositum of linearly disjoint over $\mathbb{Q}_p$ extensions K_p and M_p where $K_p/\mathbb{Q}_p$ is totally ramified with Galois group isomorphic to $\mathbb{Z}_p$ (if $p > 2$) or $\mathbb{Z}_2 \times \mathbb{Z}/2\mathbb{Z}$ (if $p = 2$) and $M_p/\mathbb{Q}_p$ is unramified with Galois group isomorphic to $\mathbb{Z}_p$.
   b) Show that every abelian tamely totally ramified extension L of $\mathbb{Q}_p$ is contained in $\mathbb{Q}_p(\sqrt[p-1]{pa})$ where a is a $(p-1)$st root of unity. Deduce that $L \subset \mathbb{Q}_p^{\text{cycl}}$.
   c) Show that every abelian totally ramified extension of $\mathbb{Q}_p$ of degree p if $p > 2$ and degree 4 if $p = 2$ is contained in $\mathbb{Q}_p^{\text{cycl}}$.
   d) Denote by $\mathbb{Q}_p^{\text{pab}}$ the maximal abelian p-extension of $\mathbb{Q}_p$. Let $\sigma \in \text{Gal}(\mathbb{Q}_p^{\text{pab}}/\mathbb{Q}_p)$ be a lifting of a generator of $\text{Gal}(E_p/M_p)$ if $p > 2$ and of $\text{Gal}(E_p/M_p(\sqrt{-1}))$ if $p = 2$. Let $\varphi \in \text{Gal}(\mathbb{Q}_p^{\text{pab}}/\mathbb{Q}_p)$ be a lifting of a generator of $\text{Gal}(E_p/K_p)$. Let R be the fixed field of σ and φ in $\mathbb{Q}_p^{\text{pab}}$. Deduce from c) that $R = \mathbb{Q}_p$ if $p > 2$ and $R = \mathbb{Q}_p(\sqrt{-1})$ if $p = 2$ and therefore $\mathbb{Q}_p^{\text{pab}} \subset \mathbb{Q}_p^{\text{cycl}}$.

   For another elementary proof see for example [Ro].
7. Let $\mu_{p^n} \subset F$, where F is a local number field. An element ω of F is said to be p^n-*primary* if the extension $F(\sqrt[p^n]{\omega})/F$ is unramified of degree p^n.
   a) Show, using Kummer theory ([La1, Ch. VIII]), that the set of p^n-primary elements forms in F^*/F^{*p^n} a cyclic group of order p^n.
   b) Show that if ω is p^n-primary, then it is p^m-primary for $m \leqslant n$.
   c) Show that a p-primary element ω can be written as $\omega = \omega_*^i \varepsilon^p$, where $\varepsilon \in U_{1,F}$ and ω_* is as in (1.4).
8. Let L be a finite Galois extension of a local number field F. Show that if L/F is tamely ramified, then the ring of integers $\mathcal{O}_L$ is a free $\mathcal{O}_F[G]$-module of rank 1, where $G = \text{Gal}(L/F)$. The converse assertion was proved by *E. Noether*.
9. ($\diamond$) Let F be a local number field, $n = |F : \mathbb{Q}_p|$. Let L/F be a finite Galois extension, $G = \text{Gal}(L/F)$. A field L is said to possess a normal basis over F, if the group $U_{1,L}$ of principal units decomposes, as a multiplicative $\mathbb{Z}_p[G]$-module, into the direct product of a finite group and a free $\mathbb{Z}_p[G]$-module of rank n.
   a) (*M. Krasner*) Show that if G is of order relatively prime to p, then L possesses a normal basis over F.
   b) (*M. Krasner, D. Gilbarg*) Let $\mu_p \cap F^* = \{1\}$. Show that L possesses a normal basis over F if and only if L/F is tamely ramified.

   For further information on the group of principal units as a $\mathbb{Z}_p[G]$-module see [Bor1–2], [BSk].
10. ($\diamond$) (*C. Chevalley, K. Yamamoto*) Let F be a local number field.

a) Let L/F be a totally ramified cyclic extension of prime degree. Let $\sigma: L \to L$ be a field automorphism. Show that $\varepsilon^{\sigma-1} \in N_{L/F}U_{1,L}$ for $\varepsilon \in U_{1,F}$.

b) Let L/F be a cyclic extension, and let σ be a generator of $\mathrm{Gal}(L/F)$. Let M/F be a subextension in L/F. Show that $\alpha^{\sigma-1} \in N_{L/M}L^*$ for $\alpha \in M^*$ and $M^* \subset F^* N_{L/M}L^*$.

c) From now on let L/F be a cyclic extension of degree n. Prove that the quotient group $F^*/N_{L/F}L^*$ is of order $\geqslant n$.

d) Show that the group $F^*/N_{L/F}L^*$ is of order $\leqslant n$, and deduce that $F^*/N_{L/F}L^*$ is of order $|L : F|$.

11. Let $\mathcal{F}$ be the maximal unramified extension of F. Show that $U_{1,\mathcal{F}} \neq U_{1,\mathcal{F}}^{\varphi-1}$.
12. Let F be a local field with finite residue field. Prove that there is a nontrivial character $\psi: F \to \mathbb{C}^*$ and that every character of F is of the form $x \mapsto \psi(ax)$ for a uniquely defined $a \in F$. This means that the additive group of F is selfdual. It is one of the first observations which lead to the theory of *J. Tate* and *K. Iwasawa* on harmonic analysis interpretation of zeta function, see [T1], [W].
13. Check which assertions of this section hold for a Henselian discrete valuation field with finite residue field.

2. The Neukirch Map

In this section F is a local field with finite residue field. Following J. Neukirch [N3–4] we introduce and study for a finite Galois extension L/F a map

$$\widetilde{\Upsilon}_{L/F}: \mathrm{Frob}(L/F) \longrightarrow F^*/N_{L/F}L^*$$

where the set $\mathrm{Frob}(L/F)$ consists of the Frobenius automorphisms φ_Σ where Σ runs through all finite extensions Σ of F in L^{ur} with $\mathrm{Gal}(L^{\mathrm{ur}}/\Sigma) \simeq \widehat{\mathbb{Z}}$.

(2.1). Let L be a finite Galois extension of F. According to Proposition (3.4) Ch. II $L^{\mathrm{ur}} = LF^{\mathrm{ur}}$.

DEFINITION. Put

$$\mathrm{Frob}(L/F) = \{\widetilde{\sigma} \in \mathrm{Gal}(L^{\mathrm{ur}}/F) : \widetilde{\sigma}|_{F^{\mathrm{ur}}} \text{ is a positive integer power of } \varphi_F\}$$

(recall that $\mathrm{Gal}(F^{\mathrm{ur}}/F)$ consists of $\widehat{\mathbb{Z}}$-powers of φ_F).

PROPOSITION. *The set* $\mathrm{Frob}(L/F)$ *is closed with respect to multiplication; it is not closed with respect to inversion and* $1 \notin \mathrm{Frob}(L/F)$.

The fixed field Σ *of* $\widetilde{\sigma} \in \mathrm{Frob}(L/F)$ *is of finite degree over* F, $\Sigma^{\mathrm{ur}} = L^{\mathrm{ur}}$, *and* $\widetilde{\sigma}$ *is the Frobenius automorphism of* Σ.

Thus, the set $\mathrm{Frob}(L/F)$ *consists of the Frobenius automorphisms* φ_Σ *of finite extensions* Σ *of* F *in* L^{ur} *with* $\mathrm{Gal}(L^{\mathrm{ur}}/\Sigma) \simeq \widehat{\mathbb{Z}}$.

The map $\mathrm{Frob}(L/F) \longrightarrow \mathrm{Gal}(L/F)$, $\widetilde{\sigma} \mapsto \widetilde{\sigma}|_L$ *is surjective.*

Proof. The first assertion is obvious.

Since $F \subset \Sigma \subset L^{\mathrm{ur}}$ we deduce that $F^{\mathrm{ur}} \subset \Sigma^{\mathrm{ur}} \subset L^{\mathrm{ur}}$. The Galois group of L^{ur}/Σ is topologically generated by $\widetilde{\sigma}$ and isomorphic to $\widehat{\mathbb{Z}}$, therefore it does not have nontrivial closed subgroups of finite order. So the group $\mathrm{Gal}(L^{\mathrm{ur}}/\Sigma^{\mathrm{ur}})$ being a subgroup of the finite group $\mathrm{Gal}(L^{\mathrm{ur}}/F^{\mathrm{ur}})$ should be trivial. So $L^{\mathrm{ur}} = \Sigma^{\mathrm{ur}}$.

Put $\Sigma_0 = \Sigma \cap F^{\mathrm{ur}}$. This field is the fixed field of $\widetilde{\sigma}|_{F^{\mathrm{ur}}} = \varphi_F^m$, therefore $|\Sigma_0 : F| = m$ is finite. From Corollary (3.4) Ch. II we deduce that

$$|\Sigma : \Sigma_0| = |\Sigma^{\mathrm{ur}} : F^{\mathrm{ur}}| = |L^{\mathrm{ur}} : F^{\mathrm{ur}}| = |L : L_0|$$

is finite. Thus, Σ/F is a finite extension.

Now $\widetilde{\sigma}$ is a power of φ_Σ and $\varphi_\Sigma|_{\Sigma_0} = \varphi_F^{|\Sigma_0:F|}|_{\Sigma_0} = \varphi_F^m|_{\Sigma_0} = \widetilde{\sigma}|_{\Sigma_0}$. Therefore, $\widetilde{\sigma} = \varphi_\Sigma$. Certainly, the Frobenius automorphism φ_Σ of a finite extension Σ of F in L^{ur} with $\mathrm{Gal}(L^{\mathrm{ur}}/\Sigma) \simeq \widehat{\mathbb{Z}}$ belongs to $\mathrm{Frob}(L/F)$.

Denote by $\widetilde{\varphi}$ an extension in $\mathrm{Gal}(L^{\mathrm{ur}}/F)$ of φ_F. Let $\sigma \in \mathrm{Gal}(L/F)$, then $\sigma|_{L_0}$ is equal to φ_F^n for some positive integer n. Hence $\sigma^{-1}\widetilde{\varphi}^n|_L$ acts trivially on L_0, and so $\tau = \sigma\widetilde{\varphi}^{-n}|_L$ belongs to $\mathrm{Gal}(L/L_0)$. Let $\widetilde{\tau} \in \mathrm{Gal}(L^{\mathrm{ur}}/F^{\mathrm{ur}})$ be such that $\widetilde{\tau}|_L = \tau$ (see Proposition (4.1) Ch. II). Then for $\widetilde{\sigma} = \widetilde{\tau}\widetilde{\varphi}^n$ we deduce that $\widetilde{\sigma}|_{F^{\mathrm{ur}}} = \varphi_F^n$ and $\widetilde{\sigma}|_L = \tau\widetilde{\varphi}^n|_L = \sigma$. Then the element $\widetilde{\sigma} \in \mathrm{Frob}(L/F)$ is mapped to $\sigma \in \mathrm{Gal}(L/F)$. □

(2.2). DEFINITION. Let L/F be a finite Galois extension. Define

$$\widetilde{\Upsilon}_{L/F}\colon \mathrm{Frob}(L/F) \longrightarrow F^*/N_{L/F}L^*, \quad \widetilde{\sigma} \mapsto N_{\Sigma/F}\pi_\Sigma \mod N_{L/F}L^*,$$

where Σ is the fixed field of $\widetilde{\sigma} \in \mathrm{Frob}(L/F)$ and π_Σ is any prime element of Σ.

LEMMA. *The map $\widetilde{\Upsilon}_{L/F}$ is well defined. If $\widetilde{\sigma}|_L = \mathrm{id}_L$ then $\widetilde{\Upsilon}_{L/F}(\widetilde{\sigma}) = 1$.*

Proof. Let π_1, π_2 be prime elements in Σ. Then $\pi_1 = \pi_2\varepsilon$ for a unit $\varepsilon \in U_\Sigma$. Let E be the compositum of Σ and L. Since $\Sigma \subset E \subset \Sigma^{\mathrm{ur}}$, the extension E/Σ is unramified. From (1.5) we know that $\varepsilon = N_{E/\Sigma}\eta$ for some $\eta \in U_E$. Hence

$$N_{\Sigma/F}\pi_1 = N_{\Sigma/F}(\pi_2\varepsilon) = N_{\Sigma/F}\pi_2 \cdot N_{\Sigma/F}(N_{E/\Sigma}\eta) = N_{\Sigma/F}\pi_2 \cdot N_{L/F}(N_{E/L}\eta).$$

We obtain that $N_{\Sigma/F}\pi_1 \equiv N_{\Sigma/F}\pi_2 \mod N_{L/F}L^*$.

If $\widetilde{\sigma}|_L = \mathrm{id}_L$ then $L \subset \Sigma$ and therefore $N_{\Sigma/F}\pi_\Sigma \in N_{L/F}L^*$. □

(2.3). The definition of the Neukirch map is very natural from the point of view of the well known principle that a prime element in an unramified extension should correspond to the Frobenius automorphism (see Theorem (2.4) below) and the functorial property of the reciprocity map (see (2.5) and (3.4)) which forces the reciprocity map $\Upsilon_{L/F}$ to be defined as it is.

Already at this stage and even without using results of subsections (1.6)–(1.9) one can prove (see Exercises 1 and 2) that the map $\widetilde{\Upsilon}_{L/F}\colon \mathrm{Frob}(L/F) \longrightarrow F^*/N_{L/F}L^*$

induces the Neukirch homomorphism

$$\Upsilon_{L/F}\colon \mathrm{Gal}(L/F) \longrightarrow F^*/N_{L/F}L^*.$$

In other words, $\widetilde{\Upsilon}_{L/F}(\widetilde{\sigma})$ does not depend on the choice of $\widetilde{\sigma} \in \mathrm{Frob}(L/F)$ which extends $\sigma \in \mathrm{Gal}(L/F)$, and moreover, $\widetilde{\Upsilon}_{L/F}(\widetilde{\sigma}_1)\widetilde{\Upsilon}_{L/F}(\widetilde{\sigma}_2) = \widetilde{\Upsilon}_{L/F}(\widetilde{\sigma_1\sigma_2})$. This is how the theory proceeds in the first edition of this book and how it goes in Neukirch's [N4–5] (where it is also extended to global fields). However, that proof does not seem to induce a lucid understanding of what is going on.

We will choose a different route, which is a little longer but more clarifying in the case of local or Henselian fields.

The plan is the following: first we easily show the existence of $\Upsilon_{L/F}$ for unramified extensions and even prove that it is an isomorphism. Then we deduce some functorial properties of $\widetilde{\Upsilon}_{L/F}$. To treat the case of totally ramified extensions in the next section, we introduce, using results of (1.6)–(1.7), the Hazewinkel homomorphism $\Psi_{L/F}$ which acts in the opposite direction to $\Upsilon_{L/F}$. Calculating composites of the latter with $\Psi_{L/F}$ we shall deduce the existence of $\Upsilon_{L/F}$ which is expressed by the commutative diagram

$$\begin{array}{ccc}
\mathrm{Frob}(L/F) & \xrightarrow{\widetilde{\Upsilon}_{L/F}} & F^*/N_{L/F}L^* \\
\downarrow & & \downarrow{\scriptstyle \mathrm{id}} \\
\mathrm{Gal}(L/F) & \xrightarrow{\Upsilon_{L/F}} & F^*/N_{L/F}L^*.
\end{array}$$

Then using $\Psi_{L/F}$ we prove that $\Upsilon_{L/F}$ is a homomorphism and that its abelian part

$$\Upsilon^{\mathrm{ab}}_{L/F}\colon \mathrm{Gal}(L/F)^{\mathrm{ab}} \to F^*/N_{L/F}L^*$$

is an isomorphism.

Then we treat the general case of abelian extensions and then Galois extensions reducing it to the two cases described above and using functorial properties of $\widetilde{\Upsilon}_{L/F}$. This route not only establishes the existence of $\Upsilon_{L/F}$, but also implies its isomorphism properties.

It is exactly this route (its totally ramified part) which can be used for construction of p-class field theory of local fields with arbitrary perfect residue field of positive characteristic and other generalizations in section 4 Ch. V.

(2.4). THEOREM. *Let L be an unramified extension of F of finite degree.*

Then $\widetilde{\Upsilon}_{L/F}(\widetilde{\sigma})$ does not depend on the choice of $\widetilde{\sigma}$ for $\sigma \in \mathrm{Gal}(L/F)$. It induces an isomorphism $\Upsilon_{L/F}\colon \mathrm{Gal}(L/F) \longrightarrow F^*/N_{L/F}L^*$ and

$$\Upsilon_{L/F}(\varphi_F|_L) \equiv \pi_F \mod N_{L/F}L^*$$

for a prime element π_F in F.

Proof. Since L/F is unramified, σ is equal to φ_F^n sor some $n \geqslant 1$. Let $m = |L : F|$. Then $\widetilde{\sigma}$ must be in the form φ_F^d with $d = n + lm > 0$ for some integer l. The fixed field Σ of $\widetilde{\sigma}$ is the unramified extension of F of degree d. We can take π_F as a prime element of Σ. Then

$$\widetilde{\Upsilon}_{L/F}(\widetilde{\sigma}) = N_{\Sigma/F}\pi_F = \pi_F^d \equiv \pi_F^n \mod N_{L/F}L^*,$$

since $\pi_F^m = N_{L/F}\pi_F$. Thus, $\widetilde{\Upsilon}_{L/F}(\widetilde{\sigma})$ does not depend on the choice of $\widetilde{\sigma}$.

It is now clear that $\Upsilon_{L/F}$ is a homomorphism and it sends φ_F to $\pi_F \mod N_{L/F}L^*$. Results of (1.5) show that $\pi_F \mod N_{L/F}L^*$ generates the group $F^*/N_{L/F}L^*$ which is cyclic of order $|L : F|$. Hence, $\Upsilon_{L/F}$ is an isomorphism. □

(2.5). Now we describe several functorial properties of $\widetilde{\Upsilon}_{L/F}$.

LEMMA. *Let M/F be a finite separable extension and let L/M be a finite Galois extension, $\sigma \in \mathrm{Gal}(F^{\mathrm{sep}}/F)$. Then the diagram of maps*

$$\begin{array}{ccc}
\mathrm{Frob}(L/M) & \xrightarrow{\widetilde{\Upsilon}_{L/M}} & M^*/N_{L/M}L^* \\
\sigma^* \downarrow & & \downarrow \sigma \\
\mathrm{Frob}(\sigma L/\sigma M) & \xrightarrow{\widetilde{\Upsilon}_{\sigma L/\sigma M}} & (\sigma M)^*/N_{\sigma L/\sigma M}(\sigma L)^*
\end{array}$$

is commutative; here $\sigma^*(\widetilde{\tau}) = \sigma\widetilde{\tau}\sigma^{-1}|_{\sigma L^{\mathrm{ur}}}$ for $\widetilde{\tau} \in \mathrm{Frob}(L/M)$.

Proof. If Σ is the fixed field of $\widetilde{\tau}$, then $\sigma\Sigma$ is the fixed field of $\sigma\widetilde{\tau}\sigma^{-1}$. For a prime element π in Σ, the element $\sigma\pi$ is prime in $\sigma\Sigma$ by Corollary 3 of (2.9) Ch. II. Since $N_{\sigma\Sigma/\sigma M}(\sigma\pi) = \sigma N_{\Sigma/M}\pi$, the proof is completed. □

PROPOSITION. *Let M/F and E/L be finite separable extensions, and let L/F and E/M be finite Galois extensions. Then the diagram of maps*

$$\begin{array}{ccc}
\mathrm{Frob}(E/M) & \xrightarrow{\widetilde{\Upsilon}_{E/M}} & M^*/N_{E/M}E^* \\
\downarrow & & \downarrow N_{M/F}^* \\
\mathrm{Frob}(L/F) & \xrightarrow{\widetilde{\Upsilon}_{L/F}} & F^*/N_{L/F}L^*
\end{array}$$

is commutative. Here the left vertical homomorphism is the restriction $\sigma|_{L^{\mathrm{ur}}}$ of $\sigma \in \mathrm{Frob}(E/M)$ and the right vertical homomorphism is induced by the norm map $N_{M/F}$.

The left vertical map is surjective if $M = F$.

Proof. Indeed, if $\widetilde{\sigma} \in \mathrm{Frob}(E/M)$ then for $\widetilde{\tau} = \widetilde{\sigma}|_{L^{\mathrm{ur}}} \in \mathrm{Gal}(L^{\mathrm{ur}}/F)$ we deduce that $\widetilde{\tau}|_{F^{\mathrm{ur}}} = \widetilde{\sigma}|_{F^{\mathrm{ur}}}$ is a positive power of φ_F, i.e., $\widetilde{\tau} \in \mathrm{Frob}(L/F)$. Let Σ be the fixed field of $\widetilde{\sigma}$. Then $\mathrm{T} = \Sigma \cap L^{\mathrm{ur}}$ is the fixed field of $\widetilde{\tau}$. The extension Σ/T is totally ramified, since $L^{\mathrm{ur}} = \mathrm{T}^{\mathrm{ur}}$ and so $\mathrm{T} = \Sigma \cap \mathrm{T}^{\mathrm{ur}}$. Hence for a prime element π_Σ in Σ the element $\pi_\mathrm{T} = N_{\Sigma/\mathrm{T}}\pi_\Sigma$ is prime in T and we get $N_{\mathrm{T}/F}\pi_\mathrm{T} = N_{\Sigma/F}\pi_\Sigma = N_{M/F}(N_{\Sigma/M}\pi_\Sigma)$.

If $M = F$, then the left vertical map is surjective, since every extension of $\widetilde{\sigma} \in \mathrm{Frob}(L/F)$ to $\mathrm{Gal}(E^{\mathrm{ur}}/F)$ belongs to $\mathrm{Frob}(E/F)$. □

COROLLARY. *Let M/F be a Galois subextension in a finite Galois extension L/F. Then the diagram of maps*

$$\begin{array}{ccccccc}
\mathrm{Frob}(L/M) & \longrightarrow & \mathrm{Frob}(L/F) & \longrightarrow & \mathrm{Frob}(M/F) & & \\
\Big\downarrow \widetilde{\Upsilon}_{L/M} & & \Big\downarrow \widetilde{\Upsilon}_{L/F} & & \Big\downarrow \widetilde{\Upsilon}_{M/F} & & \\
M^*/N_{L/M}L^* & \xrightarrow{N^*_{M/F}} & F^*/N_{L/F}L^* & \longrightarrow & F^*/N_{M/F}M^* & \longrightarrow & 1
\end{array}$$

is commutative; here the central homomorphism of the lower exact sequence is induced by the identity map of F^*.

Proof. An easy consequence of the preceding Proposition. □

Exercises.

1. Let $\widetilde{\sigma}_1, \widetilde{\sigma}_2 \in \mathrm{Frob}(L/F)$ and $\widetilde{\sigma}_3 = \widetilde{\sigma}_2\widetilde{\sigma}_1 \in \mathrm{Frob}(L/F)$. Let $\Sigma_1, \Sigma_2, \Sigma_3$ be the fixed fields of $\widetilde{\sigma}_1, \widetilde{\sigma}_2, \widetilde{\sigma}_3$. Let π_1, π_2, π_3 be prime elements in $\Sigma_1, \Sigma_2, \Sigma_3$. Show that

$$N_{\Sigma_3/F}\pi_3 \equiv N_{\Sigma_1/F}\pi_1 N_{\Sigma_2/F}\pi_2 \mod N_{L/F}L^*$$

following the steps below (*J. Neukirch* [N3]).

a) Let $\widetilde{\varphi} \in \mathrm{Frob}(L/F)$ be an extension of the Frobenius automorphism φ_F. Let Σ be the fixed field of $\widetilde{\varphi}$. Let L_1/F be the minimal Galois extension such that $\Sigma, \Sigma_1, \Sigma_2, \Sigma_3, L$ are contained in L_1 and $L_1 \subset L^{\mathrm{ur}}$. Then $L_1^{\mathrm{ur}} = L^{\mathrm{ur}}$ and the automorphisms $\widetilde{\sigma}_1, \widetilde{\sigma}_2, \widetilde{\sigma}_3$ can be considered as elements of $\mathrm{Frob}(L_1/F)$. Show that it suffices to prove that $N_{\Sigma_3/F}\pi_3 \equiv N_{\Sigma_1/F}\pi_1 N_{\Sigma_2/F}\pi_2 \mod N_{L_1/F}L_1^*$. Therefore, we may assume without loss of generality that L contains the fields Σ, Σ_1, Σ_2, Σ_3.

b) Let $\widetilde{\sigma}_i|_{F^{\mathrm{ur}}} = \varphi_F^{n_i}$, then $n_3 = n_1 + n_2$. Put $\widetilde{\sigma}_4 = \widetilde{\varphi}^{n_2}\widetilde{\sigma}_1\widetilde{\varphi}^{-n_2}$. Show that the fixed field Σ_4 of $\widetilde{\sigma}_4$ coincides with $\widetilde{\varphi}^{n_2}\Sigma_1$ and is contained in L. So it suffices to show that $N_{\Sigma_3/F}\pi_3 \equiv N_{\Sigma_2/F}\pi_2 N_{\Sigma_4/F}\pi_4 \mod N_{L/F}L^*$.

c) Let $\widetilde{\sigma}_i = \tau_i^{-1}\widetilde{\varphi}^{n_i}$ for $\tau_i \in \mathrm{Gal}(L^{\mathrm{ur}}/F^{\mathrm{ur}})$; then $\tau_3 = \tau_4\tau_2$ and $\tau_i(\pi_i) = \widetilde{\varphi}^{n_i}(\pi_i)$. Put

$$\widehat{\pi}_i = \prod_{j=0}^{n_i-1} \widetilde{\varphi}^j(\pi_i).$$

Show that

$$\varepsilon = \widehat{\pi}_3\widehat{\pi}_2^{-1}\widehat{\pi}_4^{-1} \in U_L, \quad \varepsilon^{\widetilde{\varphi}-1} = \varepsilon_2^{\tau_2-1}\varepsilon_4^{\tau_4-1}$$

where $\varepsilon_2 = \pi_3\pi_2^{-1} \in U_L$, $\varepsilon_4 = \pi_4^{-1}\tau_2(\pi_3) \in U_L$.

d) Let $L_0 = L \cap F^{\mathrm{ur}}$ and let M_1/L_0 be the unramified extension of degree $n = |L : F|$. Put $M = M_1 L$. Using (1.5) show that there are elements $\eta, \eta_2, \eta_4 \in U_M$ such that

$$N_{M/L}(\eta) = \varepsilon, \quad N_{M/L}(\eta_2) = \varepsilon_2, \quad N_{M/L}(\eta_4) = \varepsilon_4.$$

Deduce that

$$\varepsilon^{\widetilde{\varphi}-1} = N_{M/L}(\eta_2^{\tau_2-1}\eta_4^{\tau_4-1}).$$

e) Show that there is an element $\beta \in U_M$ such that

$$\eta^{\widetilde{\varphi}-1}\eta_2^{1-\tau_2}\eta_4^{1-\tau_4} = \beta^{\varphi_L-1}.$$

and so

$$(N_{M/M_1}\eta)^{\widetilde{\varphi}-1} = (N_{M/M_1}\beta)^{\varphi_L-1}.$$

f) Let $f = |L_0 : F|$, then $\varphi_L = \widetilde{\varphi}^f$. Show that $(N_{M/M_1}\eta)^{\widetilde{\varphi}-1} = \gamma^{\widetilde{\varphi}-1}$ where $\gamma = N_{M/M_1}\left(\prod_{j=0}^{f-1}\widetilde{\varphi}^j(\beta)\right) \in M_1^*$. Deduce that $\alpha = \gamma^{-1}N_{M/M_1}(\eta)$ belongs to F^* and

$$N_{L/L_0}(\varepsilon) = N_{M_1/L_0}(\gamma)\cdot\alpha^n, \quad N_{M_1/L_0}(\gamma) = N_{M/F}(\beta).$$

Conclude that

$$N_{\Sigma_3/F}\pi_3 N_{\Sigma_2/F}\pi_2^{-1} N_{\Sigma_4/F}\pi_4^{-1} = N_{L/L_0}(\varepsilon) = N_{L/F}(\alpha\cdot N_{M/L}(\beta)).$$

2. Deduce from Exercise 1 that the map $\widetilde{\Upsilon}_{L/F}$ induces the Neukirch homomorphism

$$\Upsilon_{L/F}\colon \mathrm{Gal}(L/F) \longrightarrow F^*/N_{L/F}L^*, \quad \sigma \mapsto \widetilde{\Upsilon}_{L/F}(\widetilde{\sigma})$$

where $\widetilde{\sigma} \in \mathrm{Frob}(L/F)$ is any extension of the element $\sigma \in \mathrm{Gal}(L/F)$.

3. Show that the assertions of this section hold for a Henselian discrete valuation field with finite residue field (see Exercise 12 in section 1).

3. The Hazewinkel Homomorphism

In this section we keep the notations of section 2. For a finite Galois totally ramified extension L/F using results of (1.6)–(1.7) we define in (3.1) the Hazewinkel homomorphism

$$\Psi_{L/F}\colon F^*/N_{L/F}L^* \longrightarrow \mathrm{Gal}(L/F)^{\mathrm{ab}}.$$

Simultaneous study of it and $\widetilde{\Upsilon}_{L/F}$ will lead to the proof that $\Psi_{L/F}$ is an isomorphism in (3.2). Using this result, Theorem (2.4) and functorial properties in (2.5) we shall show in (3.3) that $\Upsilon^{\mathrm{ab}}_{L/F}$ is an isomorphism for an arbitrary finite Galois extension L/F. In (3.4) we list some functorial properties of the reciprocity homomorphisms and as the first application of the obtained results reprove in (3.5) the Hasse–Arf Theorem of (4.3) Ch. III in the case of finite residue field. Finally, in (3.6) we discuss another functorial properties of $\Upsilon_{L/F}$ related to the transfer map in group theory.

(3.1). Let L be a finite Galois totally ramified extension of F. As in (1.6) we denote by $\mathcal{F}$ the maximal unramified extension of F or its completion. The Galois group of the extension $\mathcal{L}/\mathcal{F}$ is isomorphic to $\mathrm{Gal}(L/F)$.

DEFINITION. Let φ be the continuous extension on $\mathcal{L}$ of the Frobenius automorphism φ_L. Let π be a prime element of $\mathcal{L}$. Let E be the maximal abelian extension of F in L. For $\alpha \in F^*$ by Lemma in (1.6) there is $\beta \in \mathcal{L}^*$ such that $\alpha = N_{\mathcal{L}/\mathcal{F}}\beta$. Then $N_{\mathcal{L}/\mathcal{F}}\beta^{\varphi-1} = \alpha^{\varphi-1} = 1$ and by Proposition (1.7)

$$\beta^{\varphi-1} \equiv \pi^{1-\sigma} \mod U(\mathcal{L}/\mathcal{F})$$

for some $\sigma \in \mathrm{Gal}(\mathcal{L}/\mathcal{F})$ which is uniquely determined as an element of $\mathrm{Gal}(\mathcal{E}/\mathcal{F})$ where $\mathcal{E} = E\mathcal{F}$. Define the *Hazewinkel (reciprocity) homomorphism*

$$\Psi_{L/F}\colon F^*/N_{L/F}L^* \longrightarrow \mathrm{Gal}(L/F)^{\mathrm{ab}}, \quad \alpha \mapsto \sigma|_E.$$

LEMMA. *The map $\Psi_{L/F}$ is well defined and is a homomorphism.*

Proof. First, independence on the choice of π follows from Proposition (1.7). So we can assume that $\pi \in L$.

If $\alpha = N_{\mathcal{L}/\mathcal{F}}\gamma$ then $\gamma\beta^{-1}$ belongs to the kernel of $N_{\mathcal{L}/\mathcal{F}}$. Therefore by Proposition (1.7) $\gamma\beta^{-1} = \pi^{\tau-1}\xi$ with $\xi \in U(\mathcal{L}/\mathcal{F})$. Then $\gamma^{\varphi-1} = \beta^{\varphi-1}\xi^{\varphi-1} \equiv \beta^{\varphi-1} \mod U(\mathcal{L}/\mathcal{F})$ which proves correctness of the definition.

If $N_{\mathcal{L}/\mathcal{F}}(\beta_1) = \alpha_1$ and $N_{\mathcal{L}/\mathcal{F}}(\beta_2) = \alpha_2$, then we can choose $\beta_1\beta_2$ for $\alpha_1\alpha_2$ and then from Proposition (1.7) we deduce that $\Psi_{L/F}$ is a homomorphism. □

REMARKS.

1. Since L/F is totally ramified, the norm of a prime element of L is a prime element of F. So $F^*/N_{L/F}L^* = U_F/N_{L/F}U_L$. Moreover, if L/F is a totally ramified p-extension (i.e. its degree is a power of p), then $F^*/N_{L/F}L^* = U_{1,F}/N_{L/F}U_{1,L}$, since all multiplicative representatives are pth powers.

2. The Hazewinkel homomorphism can be defined for every finite Galois extension [Haz1–2], but it has the simplest form for totally ramified extensions.

(3.2). Now we prove that $\Psi_{L/F}$ is inverse to $\Upsilon^{\mathrm{ab}}_{L/F}$.

THEOREM. *Let L/F be a finite Galois totally ramified extension. Let E/F be the maximal abelian subextension of L/F. Then*

(1) *For every $\widetilde{\sigma} \in \mathrm{Frob}(L/F)$*

$$\Psi_{L/F}\big(\widetilde{\Upsilon}_{L/F}(\widetilde{\sigma})\big) = \widetilde{\sigma}|_E.$$

(2) *Let $\alpha \in F^*$ and let $\widetilde{\sigma} \in \mathrm{Frob}(L/F)$ be such that $\widetilde{\sigma}|_E = \Psi_{L/F}(\alpha)$. Then*

$$\widetilde{\Upsilon}_{L/F}(\widetilde{\sigma}) \equiv \alpha \mod N_{L/F}L^*.$$

Therefore, $\Psi_{L/F}$ is an isomorphism, $\widetilde{\Upsilon}_{L/F}(\widetilde{\sigma})$ does not depend on the choice of $\widetilde{\sigma}$ for $\sigma \in \mathrm{Gal}(L/F)$ and induces the Neukirch homomorphism

$$\Upsilon_{L/F} \colon \mathrm{Gal}(L/F) \longrightarrow F^*/N_{L/F}L^*.$$

The latter induces an isomorphism $\Upsilon^{\mathrm{ab}}_{L/F}$, between $\mathrm{Gal}(L/F)^{\mathrm{ab}} = \mathrm{Gal}(E/F)$ and $F^*/N_{L/F}L^*$, which is inverse to $\Psi_{L/F}$.

Proof. *To show* (1) note at first that since $\mathrm{Gal}(L^{\mathrm{ur}}/F)$ is isomorphic to $\mathrm{Gal}(L^{\mathrm{ur}}/L) \times \mathrm{Gal}(L^{\mathrm{ur}}/F^{\mathrm{ur}})$ the element $\widetilde{\sigma}$ is equal to $\sigma\varphi^m$ for some positive integer m and $\sigma \in \mathrm{Gal}(L^{\mathrm{ur}}/F^{\mathrm{ur}})$, where φ is the same as in (3.1). Let π_Σ be a prime element of the fixed field Σ of $\widetilde{\sigma}$. Since π_Σ is a prime element of $\Sigma^{\mathrm{ur}} = L^{\mathrm{ur}}$ we have $\pi_\Sigma = \pi\varepsilon$ for some $\varepsilon \in U_{L^{\mathrm{ur}}}$, where π is a prime element of L. Therefore $\pi^{1-\sigma} = \varepsilon^{\sigma\varphi^m - 1}$.

Let $\Sigma_0 = \Sigma \cap F^{\mathrm{ur}}$, then $|\Sigma_0 : F| = m$. Then $N_{\Sigma/F} = N_{\Sigma_0/F} \circ N_{\Sigma/\Sigma_0}$ and N_{Σ/Σ_0} acts as $N_{\Sigma^{\mathrm{ur}}/\Sigma_0^{\mathrm{ur}}} = N_{L^{\mathrm{ur}}/F^{\mathrm{ur}}} = N_{\mathcal{L}/\mathcal{F}}$, $N_{\Sigma_0/F}$ acts as $1 + \varphi + \dots + \varphi^{m-1}$. We have

$$N_{\Sigma/F}\pi_\Sigma = N_{L^{\mathrm{ur}}/F^{\mathrm{ur}}}\varepsilon_1 \, N_{L^{\mathrm{ur}}/F^{\mathrm{ur}}}\pi^m, \qquad \text{where } \varepsilon_1 = \varepsilon^{1+\varphi+\dots+\varphi^{m-1}}.$$

So $\alpha = N_{\Sigma/F}\pi_\Sigma \equiv N_{L^{\mathrm{ur}}/F^{\mathrm{ur}}}\varepsilon_1 \mod N_{L/F}L^*$ and $\Psi_{L/F}(\alpha)$ can be calculated by looking at $\varepsilon_1^{\varphi-1}$. We deduce

$$\varepsilon_1^{\varphi-1} = \varepsilon^{\varphi^m-1} \equiv \varepsilon^{\sigma\varphi^m-1} = \pi^{1-\sigma} = \pi^{1-\widetilde{\sigma}} \mod U(\mathcal{L}/\mathcal{F}).$$

This proves (1).

To show (2) let $\alpha = N_{\mathcal{L}/\mathcal{F}}\beta$ and $\beta^{\varphi-1} \equiv \pi^{1-\sigma} \mod U(\mathcal{L}/\mathcal{F})$ with $\sigma \in \mathrm{Gal}(L/F)$. Then again $\widetilde{\sigma} = \sigma\varphi^m$ and similarly to the previous

$$\widetilde{\Upsilon}_{L/F}(\widetilde{\sigma}) = N_{\Sigma/F}\pi_\Sigma \equiv N_{L^{\mathrm{ur}}/F^{\mathrm{ur}}}\,\varepsilon_1 \mod N_{L/F}L^*$$

and

$$\varepsilon_1^{\varphi-1} \equiv \pi^{1-\sigma} \equiv \beta^{\varphi-1} \mod U(\mathcal{L}/\mathcal{F}).$$

From Proposition (1.9) applied to $\gamma = \varepsilon_1\beta^{-1}$ we deduce that $N_{\mathcal{L}/\mathcal{F}}\gamma$ belongs to $N_{L/F}L^*$ and therefore $N_{\mathcal{L}/\mathcal{F}}\varepsilon_1 \equiv N_{\mathcal{L}/\mathcal{F}}\beta = \alpha \mod N_{L/F}L^*$ which proves (2).

Now from (1) we deduce the surjectivity of $\Psi_{L/F}$. From (2) and Lemma in (2.2) by taking $\widetilde{\sigma} = \varphi$, so that $\widetilde{|}_E = \mathrm{id}_E = \Psi_{L/F}(\alpha)$, we deduce that $\alpha \in N_{L/F}L^*$, i.e. $\Psi_{L/F}$ is injective. Hence $\Psi_{L/F}$ is an isomorphism. Now from (1) we conclude that $\widetilde{\Upsilon}_{L/F}$ does not depend on the choice of a lifting $\widetilde{\sigma}$ of $\sigma \in \mathrm{Gal}(L/F)$ and therefore determines the map $\Upsilon_{L/F}$.

Since we can take $\widetilde{\sigma_1\sigma_2} = \widetilde{\sigma}_1\widetilde{\sigma}_2$, from (1) we deduce that $\Upsilon_{L/F}$ is a homomorphism.

Proposition (2.1) and (2) show that this homomorphism is surjective. From (1) we deduce that its kernel is contained in $\mathrm{Gal}(L/E)$. The latter coincides with the kernel, since the image of $\Upsilon_{L/F}$ is abelian. □

Using the complete version of $\mathcal{F}$ we can give a very simple formula for the Neukirch and Hazewinkel maps in the case of totally ramified extensions.

COROLLARY. *Let $\mathcal{F}$ be the completion of the maximal unramified extension of F, and let $\mathcal{L} = L\mathcal{F}$.*

For $\sigma \in \mathrm{Gal}(L/F)$ there exists $\eta \in \mathcal{L}^*$ such that

$$\eta^{\varphi-1} = \pi^{1-\sigma}.$$

Then $\varepsilon = N_{\mathcal{L}/\mathcal{F}}\eta$ belongs to F^* and

$$\Upsilon_{L/F}(\sigma) = N_{\mathcal{L}/\mathcal{F}}\eta.$$

Conversely, for every $\varepsilon \in F^*$ there exists $\eta \in \mathcal{L}^*$ such that

$$\varepsilon \equiv N_{\mathcal{L}/\mathcal{F}}\eta \mod N_{L/F}L^*, \qquad \eta^{\varphi-1} = \pi^{1-\sigma} \quad \textit{for some } \sigma \in \mathrm{Gal}(\mathcal{L}/\mathcal{F}).$$

Then $\Psi_{L/F}(\varepsilon) = \sigma|_E$.

Proof. To prove the first assertion, we note that the homomorphism λ_0 in Proposition (4.4) Ch. II sends σ to a root of unity of order dividing the degree of the extension, so $\pi^{-1}\sigma\pi$ belongs to $T_{\mathcal{L}}U_{1,\mathcal{L}}$ and therefore, due to Proposition (1.8), η does exist. Its norm $\varepsilon = N_{\mathcal{L}/\mathcal{F}}\eta$ satisfies $\varepsilon^{\varphi-1} = N_{\mathcal{L}/\mathcal{F}}(\pi^{1-\sigma}) = 1$ so by Proposition (1.8) ε belongs to F^*.

We can assume that η is a unit, since $\pi^{\varphi-1} = 1$. Denote by the same notation σ the element of $\mathrm{Gal}(\mathcal{L}/\mathcal{F})$ which corresponds to σ. Let Σ be the fixed field of $\widetilde{\sigma} = \sigma\varphi$. Applying Proposition (1.8) to the continuous extension to $\mathcal{L}$ of the Frobenius automorphism $\widetilde{\sigma}$ we deduce that there is $\rho \in U_{\mathcal{L}}$ such that $\rho^{\sigma\varphi-1} = \pi^{1-\sigma}$. Now

$$\pi^{1-\sigma\varphi} = \pi^{1-\sigma} = \rho^{\sigma\varphi-1},$$

so $\pi\rho$ belongs to the fixed field of $\widetilde{\sigma}$ in $\mathcal{L}$ which by Proposition (1.8) equals to the fixed field Σ of $\widetilde{\sigma}$ in L^{ur}. The element $\pi_\Sigma = \pi\rho$ is a prime element of Σ. Note that $(\rho\eta^{-1})^{\varphi-1} = \rho^{\varphi-1}\pi^{\sigma-1} = (\rho^{1-\sigma})^{\varphi} \in U(\mathcal{L}/\mathcal{F})$; hence from Proposition (1.9) we deduce that $N_{\mathcal{L}/\mathcal{F}}\rho \equiv N_{\mathcal{L}/\mathcal{F}}\eta \mod N_{L/F}L^*$. Finally,

$$N_{\Sigma/F}\pi_\Sigma \equiv N_{\mathcal{L}/\mathcal{F}}\rho \equiv N_{\mathcal{L}/\mathcal{F}}\eta \mod N_{L/F}L^*.$$

To prove the second assertion use the first assertion and the congruence supplied by the Theorem: $\varepsilon \equiv \Upsilon_{L/F}(\sigma) \mod N_{L/F}L^*$ where $\sigma \in \mathrm{Gal}(L/F)$ is such that $\sigma|_E = \Psi_{L/F}(\varepsilon)$. □

REMARKS.

1. In the proof of Theorem (3.2) we did not use all the information on norm subgroups described in (1.5). We used the following two properties: the group of units U_F is contained in the image of the norm map of every unramified extension; for every finite Galois totally ramified extension L/F there is a finite unramified extension E/F such that U_E is contained in the image of the norm map $N_{LE/E}$.

2. The Theorem demonstrates that for a finite Galois totally ramified extension L/F in the definition of the Neukirch map one can fix the choice of Σ as the field invariant under the action of $\sigma\varphi$.

(3.3). The following Lemma will be useful in the proof of the main theorem.

LEMMA. *Let L/F be a finite abelian extension. Then there is a finite unramified extension M/L such that M is an abelian extension of F, M is the compositum of an unramified extension M_0 of F and an abelian totally ramified extension K of F. For every such M we have $N_{M/F}M^* = N_{K/F}K^* \cap N_{M_0/F}M_0^*$.*

Proof. Since L/F is abelian, the extension LF^{ur} is an abelian extension of F. Let $\widetilde{\varphi} \in \mathrm{Gal}(LF^{\mathrm{ur}}/F)$ be an extension of φ_F. Let K be the fixed field of $\widetilde{\varphi}$. Then $K \cap F^{\mathrm{ur}} = F$, so K is an abelian totally ramified extension of F. The compositum M of K and L is an unramified extension of L, since $K^{\mathrm{ur}} = L^{\mathrm{ur}}$. The field M is an abelian extension of F and $\mathrm{Gal}(M/F) \simeq \mathrm{Gal}(M/K) \times \mathrm{Gal}(M/M_0)$.

Now the left hand side of the formula of the Lemma is contained in the right hand side $\mathcal{N}$. We have $\mathcal{N} \cap U_F \subset N_{K/F}U_K \subset N_{M/F}U_M$, since $U_K \subset N_{M/K}U_M$ by (1.5). If π_M is a prime element of M, then $N_{M/F}\pi_M \in \mathcal{N}$. By (2.5) Ch. II $v_F(N_{M/F}\pi_M)\mathbb{Z} = v_F(N_{M_0/F}M_0^*)$. So every $\alpha \in \mathcal{N}$ can be written as $\alpha = N_{M/F}\pi_M^m\varepsilon$ with $\varepsilon \in \mathcal{N} \cap U_F$ and some m. Thence $\mathcal{N}$ is contained in $N_{M/F}M^*$ and we have $\mathcal{N} = N_{M/F}M^*$. □

Now we state and prove the first main theorem of local class field theory.

THEOREM. *Let L/F be a finite Galois extension. Let E/F be the maximal abelian subextension of L/F.*

Then $\Psi_{L/F}$ is an isomorphism, $\widetilde{\Upsilon}_{L/F}(\widetilde{\sigma})$ does not depend on the choice of $\widetilde{\sigma}$ for $\sigma \in \mathrm{Gal}(L/F)$ and induces the Neukirch (reciprocity) homomorphism

$$\Upsilon_{L/F}\colon \mathrm{Gal}(L/F) \longrightarrow F^*/N_{L/F}L^*.$$

The latter induces an isomorphism $\Upsilon_{L/F}^{\mathrm{ab}}$ between $\mathrm{Gal}(L/F)^{\mathrm{ab}} = \mathrm{Gal}(E/F)$ and $F^*/N_{L/F}L^*$ (which is inverse to $\Psi_{L/F}$ for totally ramified extensions).

Proof. First, we consider the case of an abelian extension L/F such that L is the compositum of the maximal unramified extension L_0 of F in L and an abelian totally ramified extension K of F. Then by the previous Lemma $N_{L/F}L^* = N_{K/F}K^* \cap N_{L_0/F}L_0^*$. From Proposition (2.5) applied to surjective maps

$$\mathrm{Frob}(L/F) \to \mathrm{Frob}(L_0/F) \quad \text{and} \quad \mathrm{Frob}(L/F) \to \mathrm{Frob}(K/F),$$

and from Theorem (2.4) and Theorem (3.2) we deduce that $\widetilde{\Upsilon}_{L/F}$ does not depend on the choice of $\widetilde{\sigma}$ modulo $N_{K/F}K^* \cap N_{L_0/F}L_0^*$, therefore, modulo $N_{L/F}L^*$. So we get the map $\Upsilon_{L/F}$.

Now from Proposition (2.5) and Theorem (2.4), Theorem (3.2) we deduce that $\Upsilon_{L/F}$ is a homomorphism modulo $N_{K/F}K^* \cap N_{L_0/F}L_0^*$, so it is a homomorphism modulo $N_{L/F}L^*$. It is injective, since if $\Upsilon_{L/F}(\sigma) \in N_{L/F}L^*$, then σ acts trivially on L_0 and K, and so on L. Its surjectivity follows from the commutative diagram of Corollary in (2.5).

Second, we consider the case of an arbitrary finite abelian extension L/F. By the previous Lemma and the preceding arguments there is an unramified extension M/L such that the map $\widetilde{\Upsilon}_{M/F}$ induces the isomorphism $\Upsilon_{M/F}$. The map $\mathrm{Frob}(M/F) \to \mathrm{Frob}(L/F)$ is surjective and we deduce using Proposition (2.5) that $\widetilde{\Upsilon}_{L/F}$ induces the well defined map $\Upsilon_{L/F}$, which is a surjective homomorphism. If $\sigma \in \mathrm{Gal}(M/F)$ is such that $\Upsilon_{L/F}(\sigma) = 1$, then from the commutative diagram of Corollary in (2.5) and surjectivity of Υ for every finite abelian extension we deduce that $\Upsilon_{M/F}(\sigma) = \Upsilon_{M/F}(\tau)$ for some $\tau \in \mathrm{Gal}(M/L)$. The injectivity of $\Upsilon_{M/F}$ now implies that $\sigma = \tau$ acts trivially on L.

Finally, we consider the general case of a finite Galois extension where we argue by induction on the degree of L/F. We can assume that L/F is not an abelian extension.

Every $\sigma \in \mathrm{Gal}(L/F)$ belongs to the cyclic subgroup of $\mathrm{Gal}(L/F)$ generated by it, and by what has already been proved and by Proposition in (2.5) $\widetilde{\Upsilon}_{L/F}(\widetilde{\sigma})$ does not depend on the choice of $\widetilde{\sigma}$ and therefore determines the map $\Upsilon_{L/F}$.

Since $\mathrm{Gal}(L/F)$ is solvable by Lemma (1.2), we conclude similarly to the second case above using the induction hypothesis that $\Upsilon_{L/F}$ is surjective. In the next several paragraphs we shall show that $\Upsilon_{L/F}(\mathrm{Gal}(L/E)) = 1$. Due to surjectivity of Υ this implies that the map $N^*_{E/F}$ in the diagram of Corollary (2.5) (where we put $M = E$) is zero. Since $\Upsilon_{E/F}$ is an isomorphism we see from the diagram of the Corollary that $\Upsilon_{L/F}$ is a surjective homomorphism with kernel $\mathrm{Gal}(L/E)$.

So it remains to prove that $\Upsilon_{L/F}$ maps every element of the derived group $\mathrm{Gal}(L/E)$ to 1. Since $\mathrm{Gal}(L/F)$ is solvable, we have $E \neq F$. Proposition (2.5) shows that $\Upsilon_{L/F}(\rho) = N^*_{E/F}(\Upsilon_{L/E}(\rho))$ for every $\rho \in \mathrm{Gal}(L/E)$. Since by the induction assumption $\Upsilon_{L/E}$ is a homomorphism, it suffices to show that

$$\Upsilon_{L/F}(\tau\sigma\tau^{-1}\sigma^{-1}) = N^*_{E/F}(\Upsilon_{L/E}(\tau\sigma\tau^{-1}\sigma^{-1})) = 1$$

for every $\sigma, \tau \in \mathrm{Gal}(L/F)$. To achieve that we use Lemma (2.5) and the induction hypothesis.

Suppose that the subgroup $\mathrm{Gal}(L/K)$ of $G = \mathrm{Gal}(L/F)$ generated by $\mathrm{Gal}(L/E)$ and τ is not equal to G. Then from the induction hypothesis and Lemma (2.5)

$$\Upsilon_{L/K}(\tau\sigma\tau^{-1}\sigma^{-1}) = \Upsilon_{L/K}(\tau)\Upsilon_{L/K}(\sigma\tau^{-1}\sigma^{-1}) = \Upsilon_{L/K}(\tau)^{1-\sigma},$$

and so

$$\Upsilon_{L/F}(\tau\sigma\tau^{-1}\sigma^{-1}) = N^*_{K/F}\left(\Upsilon_{L/K}(\tau)^{1-\sigma}\right) = 1.$$

In the remaining case the image of τ generates $\mathrm{Gal}(E/F)$. Hence $\sigma = \tau^m\rho$ for some $\rho \in \mathrm{Gal}(L/E)$ and integer m. We deduce $\tau\sigma\tau^{-1}\sigma^{-1} = \tau^m(\tau\rho\tau^{-1}\rho^{-1})\tau^{-m}$ and similarly to the preceding

$$\Upsilon_{L/F}(\tau^m(\tau\rho\tau^{-1}\rho^{-1})\tau^{-m}) = \Upsilon_{L/F}(\tau\rho\tau^{-1}\rho^{-1}) = N^*_{E/F}\big(\Upsilon_{L/E}(\rho)^{\tau-1}\big) = 1.$$

□

COROLLARY.

(1) *Let L/F be a finite Galois extension and let E/F be the maximal abelian subextension in L/F. Then $N_{L/F}L^* = N_{E/F}E^*$.*

(2) *Let L/F be a finite abelian extension, and M/F a subextension in L/F. Then $\alpha \in N_{L/M}L^*$ if and only if $N_{M/F}\alpha \in N_{L/F}L^*$.*

Proof. The first assertion follows immediately from the Theorem. The second assertion follows the diagram of Corollary in (2.5) (with Frob being replaced with Gal) in which the homomorphism $N^*_{M/F}$ is injective due to the Theorem. □

(3.4). We now list functorial properties of the homomorphism $\Upsilon_{L/F}$. Immediately from the previous Theorem and (2.5) we deduce the following

PROPOSITION.

(1) *Let M/F be a finite separable extension and let L/M be a finite Galois extension, $\sigma \in \mathrm{Gal}(F^{\mathrm{sep}}/F)$. Then the diagram*

$$\begin{array}{ccc}
\mathrm{Gal}(L/M) & \xrightarrow{\Upsilon_{L/M}} & M^*/N_{L/M}L^* \\
\Big\downarrow{\scriptstyle\sigma^*} & & \Big\downarrow{\scriptstyle\sigma} \\
\mathrm{Gal}(\sigma L/\sigma M) & \xrightarrow{\Upsilon_{\sigma L/\sigma M}} & (\sigma M)^*/N_{\sigma L/\sigma M}(\sigma L)^*
\end{array}$$

is commutative.

(2) *Let $M/F, E/L$ be finite separable extensions, and let L/F and E/M be finite Galois extensions. Then the diagram*

$$\begin{array}{ccc}
\mathrm{Gal}(E/M) & \xrightarrow{\Upsilon_{E/M}} & M^*/N_{E/M}E^* \\
\Big\downarrow & & \Big\downarrow{\scriptstyle N^*_{M/F}} \\
\mathrm{Gal}(L/F) & \xrightarrow{\Upsilon_{L/F}} & F^*/N_{L/F}L^*
\end{array}$$

is commutative.

(3.5). As the first application of Theorem (3.3) and functorial properties in (3.4) we describe ramification group of finite abelian extensions and reprove the Hasse–Arf Theorem of (4.3) Ch. III for local fields with finite residue field.

THEOREM. *Let L/F be a finite abelian extension, $G = \mathrm{Gal}(L/F)$. Denote by h the Hasse–Herbrand function $h_{L/F}$. Put $U_{-1,F} = F^*$, $U_{0,F} = U_F$, and $h(-1) = -1$. Then for every integer $n \geqslant -1$ the reciprocity map $\Psi_{L/F}$ maps the quotient group $U_{n,F}N_{L/F}L^*/N_{L/F}L^*$ isomorphically onto the ramification group $G(n) = G_{h(n)}$ and $U_{n,F}N_{L/F}L^*/U_{n+1,F}N_{L/F}L^*$ isomorphically onto $G_{h(n)}/G_{h(n)+1}$. Therefore*

$$G_{h(n)+1} = G_{h(n+1)},$$

i.e., upper ramification jumps of L/F are integers.

Proof. Let L_0 be the maximal unramified extension of F in L. We know from section 3 Ch. III that $h_{L/F} = h_{L/L_0}$, and from section 1 that the norm $N_{L_0/F}$ maps U_{n,L_0} onto $U_{n,F}$ for $n \geqslant 0$. Using the second commutative diagram of (3.4) (for $E = L, M = F, L = L_0$) we can therefore assume that L/F is totally ramified and $n \geqslant 0$.

We use the notations of Corollary (3.2), so $\mathcal{F}$ and $\mathcal{L}$ are complete fields. Let $\sigma \in G_{h(n)}$, then $\pi^{1-\sigma}$ belongs to $U_{h(n),L}$. Let $\eta \in \mathcal{L}^*$ be such that $\eta^{\varphi-1} = \pi^{1-\sigma}$. Proposition (1.8) shows that η can be chosen in $U_{h(n),\mathcal{L}}$. Now from Corollary (3.2) and section 3 Ch. III we deduce that $\Upsilon_{L/F}(\sigma) = \varepsilon = N_{\mathcal{L}/\mathcal{F}}(\eta)$ belongs to $U_{n,F}N_{L/F}L^*$. So $\Upsilon(G_{h(n)}) \subset U_{n,F}N_{L/F}L^*$. Similarly, we establish that $\Upsilon(G_{h(n)+1}) \subset U_{n+1,F}N_{L/F}L^*$.

Conversely, let ε belong to $U_{n,F}N_{L/F}L^*$. For the abelian extension L/F we will prove below a stronger assertion than that in Corollary (3.2):

> there exists $\eta \in U_{\mathcal{L}}$ such that $\varepsilon \equiv N_{\mathcal{L}/\mathcal{F}}(\eta) \mod N_{L/F}L^*$ and $\eta^{\varphi-1} = \pi^{1-\sigma}$ for some $\sigma \in \mathrm{Gal}(\mathcal{L}/\mathcal{F})$. For every such η we have $\eta^{\varphi-1} \in U_{h(n),\mathcal{L}}$.

From this assertion we deduce that $\Psi(U_{n,F}N_{L/F}L^*) \subset G_{h(n)}$. We conclude that $\Psi(U_{n,F}N_{L/F}L^*) = G_{h(n)}$ and $\Upsilon_{L/F}(G_{h(n)+1}) = \Upsilon_{L/F}(G_{h(n+1)})$, so $G_{h(n)+1} = G_{h(n+1)}$.

It remains to prove the assertion by induction on the degree of L/F. If $n = 0$, the assertion is obvious, so we assume that $n > 0$. If $\mathrm{Gal}(L/F)$ is of prime order with generator τ, then from Corollary (3.2) we know that there is $\eta \in U_{\mathcal{L}}$ such that $\varepsilon \equiv N_{\mathcal{L}/\mathcal{F}}(\eta) \mod N_{L/F}L^*$ and $\eta^{\varphi-1} = \pi^{1-\tau^m}$ for some integer m. So $j = v_{\mathcal{L}}(\eta^{\varphi-1} - 1) = v_{\mathcal{L}}(\pi^{1-\tau^m} - 1)$. If $\tau^m = 1$ then the assertion is obvious, so assume that $\tau^m \neq 1$. From section 1 Ch. III we know that $U_{j+1,F} \subset N_{L/F}L^*$. If $N_{\mathcal{L}/\mathcal{F}}(\eta)$ belongs to $U_{j+1,F}$, then $\Psi_{L/F}(N_{\mathcal{L}/\mathcal{F}}(\eta))$ is 1, not τ^m, a contradiction. Therefore, $v_F(N_{\mathcal{L}/\mathcal{F}}(\eta) - 1) = j = h_{L/F}(j)$.

For the induction step let M/F be a subextension of L/F such that $\mathrm{Gal}(L/M)$ is of prime degree l with generator τ. By Corollary (3.2) there is $\eta \in U_{\mathcal{L}}$ such that $\varepsilon \equiv N_{\mathcal{L}/\mathcal{F}}(\eta) \mod N_{L/F}L^*$ and $\eta^{\varphi-1} = \pi^{1-\sigma}$. By the induction hypothesis $N_{\mathcal{L}/\mathcal{M}}\eta^{\varphi-1}$ belongs to $U_{h_{M/F}(n),\mathcal{M}}$. By Proposition (1.8) the latter group is ${}^{\varphi-1}$-divisible, and therefore from the same Proposition we deduce that $N_{\mathcal{L}/\mathcal{M}}\eta^{\varphi-1} = \rho u$ with $\rho \in U_{h_{M/F}(n),\mathcal{M}}$ and $u \in U_M$. According to results of section 1 Ch. III, the definition of the Hasse–Herbrand function and Lemma (1.6) there is $\xi \in U_{h_{L/F}(n),\mathcal{L}}$ such that $N_{\mathcal{L}/\mathcal{M}}(\xi) = \rho$. Then $\xi^{\varphi-1} = \pi^{1-\sigma}\alpha$ for some α in the kernel of $N_{\mathcal{L}/\mathcal{M}}$.

Let τ be a generator of Gal(L/M). Using Proposition (1.7) we deduce that $\alpha \equiv \pi^{1-\tau^m} \mod U(\mathcal{L}/\mathcal{M})$ and so $\xi^{\varphi-1} = \pi^{1-\sigma\tau^m}\gamma^{1-\tau}$ for an appropriate $\gamma \in U_{\mathcal{L}}$ and some integer m. By Proposition (1.8) there is $\delta \in U_{\mathcal{L}}$ such that $\delta^{\varphi-1} = \gamma^{1-\tau}$. Then $\eta^{\varphi-1} = \pi^{\sigma\tau^m-1}$ where $\eta = \xi\delta^{-1}$. All we need to show is that $\gamma^{\tau-1}$ belongs to $U_{h(n),\mathcal{L}}$. If it does not, then $j = v_{\mathcal{L}}(\gamma^{\tau-1}-1) = v_L(\pi^{1-\sigma\tau^m}-1) > 0$. Let $s = s(L|M)$ as defined in (1.4) Ch. III. Since γ is a unit, from (1.4) Ch. III we deduce that $j-s$ is prime to p. On the other hand, Proposition (4.5) Ch. II implies that j is congruent to s modulo p, a contradiction. □

REMARK. For a similar result in the case of perfect residue field of positive characteristic see (4.7) Ch. V.

(3.6). Another functorial property involves the transfer map from group theory. Recall the notion of *transfer* (*Verlagerung*). Let G be a group and let G' be its commutator subgroup (derived group). Denote the quotient group G/G' by G^{ab}; it is abelian. Let H be a subgroup of finite index in G. Let

$$G = \cup_i H\rho_i, \qquad \rho_i \in G,\ 1 \leqslant i \leqslant |G:H|$$

be the decomposition of G into the disjoint union of sets $H\rho_i$.

Define the transfer

$$\mathrm{Ver}\colon G^{\mathrm{ab}} \to H^{\mathrm{ab}}, \quad \sigma \mod G' \mapsto \prod_i \rho_i\sigma\rho_{\sigma(i)}^{-1} \mod H',$$

where $\sigma(i)$ is determined by the condition $\rho_i\sigma \in H\rho_{\sigma(i)}$. So $\sigma(1), \dots, \sigma(|G:H|)$ is a permutation of $1, \dots, |G:H|$.

We shall verify that Ver is well defined. Let $\rho_i' = \kappa_i\rho_i$ with $\kappa_i \in H$. Then

$$\prod \rho_i'\sigma{\rho'}_{\sigma(i)}^{-1} = \prod \kappa_i\left(\rho_i\sigma\rho_{\sigma(i)}^{-1}\right)\kappa_{\sigma(i)}^{-1} \equiv \prod \rho_i\sigma\rho_{\sigma(i)}^{-1}\cdot\prod\kappa_i\cdot\prod\kappa_{\sigma(i)}^{-1} \mod H',$$

because H/H' is abelian. Hence

$$\prod \rho_i'\sigma{\rho'}_{\sigma(i)}^{-1} \equiv \prod \rho_i\sigma\rho_{\sigma(i)}^{-1} \mod H'.$$

Now we shall verify that Ver is a homomorphism. Let $\sigma, \tau \in G$; then

$$\rho_i\sigma\tau\rho_{\sigma\tau(i)}^{-1} \equiv \rho_i\sigma\rho_{\sigma(i)}^{-1}\rho_{\sigma(i)}\tau\rho_{\sigma\tau(i)}^{-1} \mod H'$$

and, as $\rho_i\sigma\rho_{\sigma(i)}^{-1} \in H$, $\rho_i\sigma\tau\rho_{\sigma\tau(i)}^{-1} \in H$, we get $\rho_{\sigma(i)}\tau\rho_{\sigma\tau(i)}^{-1} \in H$, i.e., $\sigma\tau(i) = \tau(\sigma(i))$. Hence

$$\prod \rho_i\sigma\tau\rho_{\sigma\tau(i)}^{-1} \equiv \prod \rho_i\sigma\rho_{\sigma(i)}^{-1}\cdot\prod\rho_i\tau\rho_{\tau(i)}^{-1} \mod H'.$$

Let σ be an element of G. For an element $\tau_1 \in G$ let $g_1 = g(\sigma, \tau_1)$ be the maximal integer such that all the sets $H\tau_1\sigma, H\tau_1\sigma^2, \dots, H\tau_1\sigma^{g_1}$ are distinct. Then, take an element $\tau_2 \in G$ such that all $H\tau_2\sigma, H\tau_1\sigma, \dots, H\tau_1\sigma^{g_1}$ are distinct and find $g_2 = g(\sigma, \tau_1, \tau_2)$ such that all the sets

$$H\tau_2\sigma, \dots, H\tau_2\sigma^{g_2}, H\tau_1\sigma, \dots, H\tau_1\sigma^{g_1}$$

are distinct. Repeating this construction, we finally obtain that G is the disjoint union of the sets $H\tau_n\sigma^{m_n}$, where $1 \leqslant n \leqslant k, 1 \leqslant m_n \leqslant g_n = g(\sigma, \tau_1, \tau_2, \dots, \tau_n)$. The number g_i can also be determined as the minimal positive integer, for which the element

$$\sigma[\tau_i] = \tau_i\sigma^{g_i}\tau_i^{-1}$$

belongs to H. The definition of Ver shows that in this case

$$\mathrm{Ver}(\sigma \mod G') \equiv \prod_n \sigma[\tau_n] \mod H'.$$

Since the image of $\Upsilon_{L/F}$ is abelian, one can define the homomorphism

$$\Upsilon_{L/F}\colon \mathrm{Gal}(L/F)^{\mathrm{ab}} \longrightarrow F^*/N_{L/F}L^*.$$

PROPOSITION. *Let L/F be a finite Galois extension and let M/F be a subextension in L/F. Then the diagram*

$$\begin{array}{ccc} \mathrm{Gal}(L/F)^{\mathrm{ab}} & \xrightarrow{\Upsilon_{L/F}} & F^*/N_{L/F}L^* \\ \big\downarrow \mathrm{Ver} & & \big\downarrow \\ \mathrm{Gal}(L/M)^{\mathrm{ab}} & \xrightarrow{\Upsilon_{L/M}} & M^*/N_{L/M}L^* \end{array}$$

is commutative; here the right vertical homomorphism is induced by the embedding $F \hookrightarrow M$.

Proof. Denote $\widetilde{G} = \mathrm{Gal}(L^{\mathrm{ur}}/F), \widetilde{H} = \mathrm{Gal}(L^{\mathrm{ur}}/M)$. Let $\sigma \in \mathrm{Gal}(L/F)$, and let $\widetilde{\sigma} \in \mathrm{Frob}(L/F)$ be its extension. Let $\widetilde{G}$ be the disjoint union of $\widetilde{H}\widetilde{\tau}_n\widetilde{\sigma}^{m_n}$ for $1 \leqslant n \leqslant k, 1 \leqslant m_n \leqslant g_n$, as above. Let $G = \mathrm{Gal}(L/F)$ and $H = \mathrm{Gal}(L/M)$; then G is the disjoint union of $H\tau_n\sigma^{m_n}$ for $\tau_n = \widetilde{\tau}_n|_L \in \mathrm{Gal}(L/F)$. This means that

$$\mathrm{Ver}(\sigma \mod G') \equiv \prod_n \sigma[\tau_n] \mod H'.$$

Let $\widetilde{\mathrm{A}}$ be the subgroup in $\widetilde{G}$ generated topologically by $\widetilde{\sigma}$ and

$$\widetilde{H}_n = \widetilde{H} \cap \widetilde{\tau}_n\widetilde{\mathrm{A}}\widetilde{\tau}_n^{-1}.$$

Then $\widetilde{H}_n$ is a subgroup in $\widetilde{H}$, which coincides with the subgroup in $\widetilde{H}$ topologically generated by $\widetilde{\sigma}[\widetilde{\tau}_n]$. Note that $\widetilde{\tau}_n\widetilde{\mathrm{A}}$ is the disjoint union of $\widetilde{H}_n\widetilde{\tau}_n\widetilde{\sigma}^{m_n}$ for $1 \leqslant m_n \leqslant g_n$.

Let $\widetilde{H}$ be the disjoint union of $\widetilde{\nu}_{n,l}\widetilde{H}_n$ for $\widetilde{\nu}_{n,l} \in \widetilde{H}, 1 \leqslant l \leqslant |\widetilde{H} : \widetilde{H}_n|$. Then

$$\widetilde{G} = \cup \cup \widetilde{\nu}_{n,l}\widetilde{H}_n\widetilde{\tau}_n\sigma^{m_n} = \cup \widetilde{\nu}_{n,l}\widetilde{\tau}_n\widetilde{\mathrm{A}}.$$

If Σ is the fixed field of $\widetilde{\sigma}$, then it is the fixed field of $\widetilde{\mathrm{A}}$, and we obtain that

$$N_{\Sigma/F}(\alpha) = \prod_{n,l} \widetilde{\nu}_{n,l}\widetilde{\tau}_n(\alpha) \qquad \text{for } \alpha \in \Sigma.$$

Let Σ_n be the fixed field of $\widetilde{\sigma}[\widetilde{\tau}_n] = \widetilde{\tau}_n\widetilde{\sigma}^{g_n}\widetilde{\tau}_n^{-1}$. Then $(\widetilde{\tau}_n\Sigma)^{\mathrm{ur}} = \widetilde{\tau}_n\Sigma^{\mathrm{ur}} = \widetilde{\tau}_n L^{\mathrm{ur}} = L^{\mathrm{ur}}$, $\widetilde{\tau}_n\Sigma \subset \Sigma_n$, and $\Sigma_n/\widetilde{\tau}_n\Sigma$ is the unramified extension of degree g_n. Hence, for a prime element π in Σ, the element $\widetilde{\tau}_n(\pi)$ is prime in Σ_n. Moreover, one can show as before that

$$N_{\Sigma_n/M}(\alpha) = \prod_{l} \widetilde{\nu}_{n,l}(\alpha) \qquad \text{for } \alpha \in \Sigma_n.$$

We deduce that

$$N_{\Sigma/F}(\pi) = \prod_{n,l} \widetilde{\nu}_{n,l}\widetilde{\tau}_n(\pi) = \prod_{n} N_{\Sigma_n/M}\big(\widetilde{\tau}_n(\pi)\big).$$

Since $\widetilde{\sigma}[\widetilde{\tau}_n] \in \mathrm{Frob}(L/M)$ extends the element $\sigma[\tau_n] \in \mathrm{Gal}(L/M)$, we conclude that

$$\Upsilon_{L/F}(\sigma) = \prod_{n} \Upsilon_{L/M}(\sigma[\tau_n]) = \Upsilon_{L/M}\Big(\prod_{n} \sigma[\tau_n]\Big)$$

and $\Upsilon_{L/F}(\sigma) = \Upsilon_{L/M}\big(\mathrm{Ver}\,(\sigma \bmod \mathrm{Gal}(L/F)')\big)$. □

Exercises.

1. Let F be of characteristic p, and let L/F be a purely inseparable extension of degree p^k. Let $\tau = \tau(L|F)$ be the automorphism of F^{alg} such that $\tau(\alpha) = \alpha^{1/p^k}$ for $\alpha \in F^{\mathrm{alg}}$. Show that $L^{p^k} = F$, L/F is totally ramified, $N_{L/F}L^* = F^*$, $\tau(F) = L$, and $v_F \circ \tau^{-1} = v_L$.
2. Show that the first assertion of Proposition in (3.4) holds if the condition $\sigma \in \mathrm{Gal}(F^{\mathrm{sep}}/F)$ is replaced by $\sigma \in \mathrm{Aut}(F^{\mathrm{alg}})$ and M/F is a finite extension. Show that the second assertion of the same Proposition holds if the condition "$M/F, E/L$ are finite separable extensions" is replaced by "$M/F, E/L$ are finite extensions".
3. a) Show that Ver does not depend on the choice of (right / left) cosets of H in G.
   b) Show that for a subgroup H_1 of finite index in G and for an intermediate subgroup H_2 the map Ver: $G^{\mathrm{ab}} \to H_1^{\mathrm{ab}}$ coincides with with the composition $G^{\mathrm{ab}} \to H_2^{\mathrm{ab}} \to H_1^{\mathrm{ab}}$.
   c) Show that if $G = H \times H_1$ and H_1 is of order n, then Ver: $G^{\mathrm{ab}} \to H^{\mathrm{ab}}$ maps an element $\sigma \in G$ to $\mathrm{pr}(\sigma)^n$, where pr: $G \to H$ is the natural projection.
   d) ($\diamond$) Let G be finitely generated, $H = G'$ and $|G : H| < \infty$. Show that the homomorphism Ver: $G' \to H'$ is the zero homomorphism.
4. (*B. Dwork* [Dw]) Let L/F be a finite Galois totally ramified extension and E be the maximal abelian extension of F in L. Let $\alpha \in F^*$ and $\alpha = N_{L^{\mathrm{ur}}/F^{\mathrm{ur}}}\beta$ for some

$\beta \in L^{\mathrm{ur}}$. Let $\beta^{\varphi-1} = \prod_{i=1}^m \gamma_i^{\widetilde{\sigma}_i - 1}$ with $\gamma_i \in L^{\mathrm{ur}*}$ and $\widetilde{\sigma}_i \in \mathrm{Gal}(L^{\mathrm{ur}}/F^{\mathrm{ur}})$. Show that

$$\Psi_{L/F}(\alpha)|_E = \widetilde{\sigma}^{-1}|_E$$

where $\widetilde{\sigma} = \widetilde{\sigma}_1^{v(\gamma_1)} \dots \widetilde{\sigma}_m^{v(\gamma_m)} \in \mathrm{Gal}(L^{\mathrm{ur}}/F^{\mathrm{ur}})$ and v is the discrete valuation of L^{ur}. Deduce that, in particular, if $\beta^{\varphi-1} = \pi^{\widetilde{\sigma}-1}$ for a prime element π of L^{ur}, then $\Psi_{L/F}(\alpha)|_E = \widetilde{\sigma}^{-1}|_E$.

In fact, from this theorem known already in the fifties one can deduce the construction of the Hazewinkel and Neukirch reciprocity maps for totally ramified extensions.

5. Let L/F be a finite abelian extension. Show that $U_{n,F} \cap N_{L/F}L^* = N_{L/F}U_{h_{L/F}(n),L}$ for every $n \geqslant 0$.
6. a) Show that if $\mathcal{F} = F^{\mathrm{ur}}$ then Corollary (3.2) holds if the equality $\eta^{\varphi-1} = \pi^{1-\sigma}$ is replaced with the congruence $\eta^{\varphi-1} \equiv \pi^{1-\sigma} \mod U_{r,\mathcal{L}}$ where r is any positive integer.
   b) Find a proof of Theorem (3.5) which uses only F^{ur} and its finite extensions.
7. Let L be a finite separable extension of F. Let M be the maximal abelian subextension of F in L. Show that $N_{L/F}L^* = N_{M/F}M^*$.
8. Show that the results of this section hold for a Henselian discrete valuation field with finite residue field.
9. Let L/F be a finite Galois extension with group G. Show, following the steps below, that $[G_i, G_j] \leqslant G_{i+pj}$ if $1 \leqslant j \leqslant i$.
   a) Reduce the problem to the following assertion: Let E/M be a finite Galois totally ramified p-extension and let K/M be its subextension of degree p. Let $j = s(K|M)$ as in sect. 1 Ch. III and let i be the minimal integer such that $\mathrm{Gal}(E/K)_i \neq \mathrm{Gal}(E/K)_{i+1}$. Suppose that E/K is abelian and $\mathrm{Gal}(E/K)_{i+pj} = \{1\}$. Then E/M is abelian.
   b) Using Proposition (3.6) Ch. III deduce that $U_{i+j,K} \subset N_{E/K}U_E$.
   c) By using b) and the formula $\tau\sigma\tau^{-1}\sigma^{-1} = \Upsilon_{E/K}(\tau)^{1-\sigma}$ prove the assertion of a).

   For more information on $k = k(i,j)$ such that $[G_i, G_j] \leqslant G_k$ see [Mau5].

4. The Reciprocity Map

In this section we define and describe properties of the reciprocity map

$$\Psi_F : F^* \longrightarrow \mathrm{Gal}(F^{\mathrm{ab}}/F)$$

using the Neukirch map $\Upsilon_{L/F}$ studied in the previous sections. We keep the conventions of the two preceding sections.

(4.1). The homomorphism inverse to $\Upsilon_{L/F}$ induces the surjective homomorphism

$$(\,\cdot\,, L/F) : F^* \longrightarrow \mathrm{Gal}(L/F)^{\mathrm{ab}}.$$

It coincides with $\Psi_{L/F}$ for totally ramified extensions.

Denote the maximal abelian extension of F in F^{sep} by F^{ab}.

PROPOSITION. *Let H be a subgroup in* $\mathrm{Gal}(L/F)^{\mathrm{ab}}$, *and let M be the fixed field of H in $L \cap F^{\mathrm{ab}}$. Then $(\cdot, L/F)^{-1}(H) = N_{M/F}M^*$.*

Let L_1, L_2 be abelian extensions of finite degree over F, and let $L_3 = L_1L_2$, $L_4 = L_1 \cap L_2$. Then

$$N_{L_3/F}L_3^* = N_{L_1/F}L_1^* \cap N_{L_2/F}L_2^*, \quad N_{L_4/F}L_4^* = N_{L_1/F}L_1^* \, N_{L_2/F}L_2^*.$$

The field L_1 is a subfield of the field L_2 if and only if $N_{L_2/F}L_2^* \subset N_{L_1/F}L_1^*$; in particular, $L_1 = L_2$ if and only if $N_{L_1/F}L_1^* = N_{L_2/F}L_2^*$.

If a subgroup N in F^* contains a norm subgroup $N_{L/F}L^*$ for some finite Galois extension L/F, then N itself is a norm subgroup.

Proof. The first assertion follows immediately from (3.3) and (3.4). Put $H_i = \mathrm{Gal}(L_3/L_i)$, $i = 1, 2$. Then

$$\begin{aligned} N_{L_3/F}L_3^* &= (\cdot, L_3/F)^{-1}(1) = (\cdot, L_3/F)^{-1}(H_1 \cap H_2) \\ &= (\cdot, L_3/F)^{-1}(H_1) \cap (\cdot, L_3/F)^{-1}(H_2) = N_{L_1/F}L_1^* \cap N_{L_2/F}L_2^*, \\ N_{L_4/F}L_4^* &= (\cdot, L_3/F)^{-1}(H_1H_2) = (\cdot, L_3/F)^{-1}(H_1)(\cdot, L_3/F)^{-1}(H_2) \\ &= N_{L_1/F}L_1^* N_{L_2/F}L_2^*. \end{aligned}$$

If $L_1 \subset L_2$, then $N_{L_2/F}L_2^* \subset N_{L_1/F}L_1^*$. Conversely, if $N_{L_2/F}L_2^* \subset N_{L_1/F}L_1^*$, then $N_{L_1L_2/F}(L_1L_2)^*$ coincides with $N_{L_2/F}L_2^*$, and Theorem (3.3) shows that the extension L_1L_2/F is of the same degree as L_2/F, hence $L_1 \subset L_2$.

Finally, if $N \supset N_{L/F}L^*$, then $N = N_{M/F}M^*$, where M is the fixed field of $(N, L/F)$. □

Passing to the projective limit, we get

$$\Psi_F\colon F^* \longrightarrow \varprojlim F^*/N_{L/F}L^* \longrightarrow \varprojlim \mathrm{Gal}(L/F)^{\mathrm{ab}} = \mathrm{Gal}(F^{\mathrm{ab}}/F)$$

where L runs through all finite Galois (or all finite abelian) extensions of F. The homomorphism Ψ_F is called *the reciprocity map.*

(4.2). THEOREM. *The reciprocity map is well defined.*

Its image is dense in $\mathrm{Gal}(F^{\mathrm{ab}}/F)$, *and its kernel coincides with the intersection of all norm subgroups $N_{L/F}L^*$ in F^* for finite Galois (or finite abelian) extensions L/F.*

If L/F is a finite Galois extension and $\alpha \in F^*$, then the automorphism $\Psi_F(\alpha)$ acts trivially on $L \cap F^{\mathrm{ab}}$ if and only if $\alpha \in N_{L/F}L^*$.

The restriction of $\Psi_F(\alpha)$ on F^{ur} coincides with $\varphi_F^{v_F(\alpha)}$ for $\alpha \in F^*$.

Let L be a finite separable extension of F, and let σ be an automorphism of $\mathrm{Gal}(F^{\mathrm{sep}}/F)$. *Then the diagrams*

$$\begin{array}{ccc} L^* & \xrightarrow{\Psi_L} & \mathrm{Gal}(L^{\mathrm{ab}}/L) \\ \downarrow{\scriptstyle\sigma} & & \downarrow{\scriptstyle\sigma^*} \\ (\sigma L)^* & \xrightarrow{\Psi_{\sigma L}} & \mathrm{Gal}\left((\sigma L)^{\mathrm{ab}}/\sigma L\right) \end{array}$$

$$\begin{array}{ccc} L^* & \xrightarrow{\Psi_L} & \mathrm{Gal}(L^{\mathrm{ab}}/L) \\ \downarrow{\scriptstyle N_{L/F}} & & \downarrow \\ F^* & \xrightarrow{\Psi_F} & \mathrm{Gal}\left(F^{\mathrm{ab}}/F\right) \end{array}$$

$$\begin{array}{ccc} F^* & \xrightarrow{\Psi_F} & \mathrm{Gal}(F^{\mathrm{ab}}/F) \\ \downarrow & & \downarrow{\scriptstyle \mathrm{Ver}} \\ L^* & \xrightarrow{\Psi_L} & \mathrm{Gal}\left(L^{\mathrm{ab}}/L\right) \end{array}$$

are commutative, where $\sigma^*(\tau) = \sigma\tau\sigma^{-1}$, the right vertical homomorphism of the second diagram is the restriction and

$$\mathrm{Ver}\colon \mathrm{Gal}(F^{\mathrm{sep}}/F)^{\mathrm{ab}} \longrightarrow \mathrm{Gal}(F^{\mathrm{sep}}/L)^{\mathrm{ab}} = \mathrm{Gal}(L^{\mathrm{ab}}/L).$$

Proof. Let $L_1/F, L_2/F$ be finite extensions and $L_1 \subset L_2$. Then Proposition (3.4) shows that the restriction of the automorphism

$$(\alpha, L_2/F) \in \mathrm{Gal}(L_2/F)^{\mathrm{ab}}$$

on the field $L_1 \cap F^{\mathrm{ab}}$ coincides with $(\alpha, L_1/F)$ for an element $\alpha \in F^*$. This means that Ψ_F is well defined.

The condition $\alpha \in N_{L/F}L^*$ is equivalent $(\alpha, L/F) = 1$ and the last relation means that $\Psi_F(\alpha)$ acts trivially on $L \cap F^{\mathrm{ab}}$.

Hence, the kernel of Ψ_F is equal to $\bigcap N_{L/F}L^*$, where L runs through all finite Galois extensions of F. Since $\Psi_F(F^*)|_L = \mathrm{Gal}(L/F)$ for a finite abelian extension L/F, we deduce that $\Psi(F^*)$ is dense in $\mathrm{Gal}(F^{\mathrm{ab}}/F)$.

Theorem (2.4) shows that $\Psi_F(\pi_F)|_{F^{\mathrm{ur}}} = \varphi_F$ for a prime element π_F in F. Hence, $\Psi_F(\alpha)|_{F^{\mathrm{ur}}} = \varphi_F^{v_F(\alpha)}$ and $\Psi_F(U_F)|_{F^{\mathrm{ur}}} = 1$.

The commutativity of the diagrams follow from Propositions (3.4) and (3.5). □

REMARK. See Exercise 4 for the case of Henselian discrete valuation fields.

Exercises.

1. Let L be a finite extension of F, let M be the maximal separable subextension of F in L, $p^k = |L : M|$. Using Exercises 1 and 2 of section 3 show that
$$\Psi_{\sigma L}(\sigma\alpha) = \sigma^*\Psi_L(\alpha) \quad \text{for } \alpha \in L^*, \sigma \in \operatorname{Aut}(F^{\text{alg}});$$
$$\Psi_F(N_{L/F}\alpha) = \Psi_L(\alpha)|_{F^{\text{ab}}} \quad \text{for } \alpha \in L^*;$$
$$\Psi_L(\alpha) = \tau \operatorname{Ver}(\Psi_F(\alpha^{p^k}))\tau^{-1} \quad \text{for } \alpha \in F^*,$$
where $\tau = \tau(L|F)$ was defined in Exercise 1 section 3.
2. a) Let ζ_1 be a primitive $(p^n - 1)$th root of unity, let ζ_2 be a primitive p^mth root of unity, and $L_1 = \mathbb{Q}_p(\zeta_1), L_2 = \mathbb{Q}_p(\zeta_2)$. Show that
$$N_{L_1/\mathbb{Q}_p} L_1^* = \langle p^n \rangle \times U_{\mathbb{Q}_p}, \qquad N_{L_2/\mathbb{Q}_p} L_2^* = \langle p \rangle \times U_{m,\mathbb{Q}_p}.$$
   b) Let L be contained in $\mathbb{Q}_p^{(k)}$ for some k (see (1.3)). Show that an element $\alpha \in L$ belongs to the intersection of all norm groups $N_{\mathbb{Q}_p^{(i)}/L}\mathbb{Q}_p^{(i)*}$ for $i \geqslant k$ if and only if $N_{L/\mathbb{Q}_p}\alpha = p^a$ for an integer a.
3. Let M/F be a cyclic extension with generator σ and L/M a finite abelian extension.
   a) Show that L/F is Galois if and only if $\sigma N_{L/M}L^* = N_{L/M}L^*$.
   b) Show that L/F is abelian if and only if $\{\alpha^{\sigma-1} : \alpha \in M^*\} \subset N_{L/M}L^*$.
4. Show that the assertions of this section hold for a Henselian discrete valuation field with finite residue field.

5. Pairings of the Multiplicative Group

In this section we define the Hilbert symbol associated to the local reciprocity map and study its properties in (5.1)–(5.3). Explicit formulas for the p^nth Hilbert symbol will be derived in Ch. VII.

In (5.4)–(5.7) we study the Artin–Schreier pairing which is important for the p-part of the theory in characteristic p.

These pairing will appear to be quite useful in the proof of the Existence Theorem in the next section.

We continue assuming that F is a local field with finite residue field $\overline{F}$.

(5.1). Let the group μ_n of all nth roots of unity in the separable closure F^{sep} be contained in F and let $p \nmid n$ if $\operatorname{char}(F) = p$.

The *norm residue symbol* or *Hilbert symbol* or *Hilbert pairing* $(\cdot,\cdot)_n \colon F^* \times F^* \to \mu_n$ is defined by the formula
$$(\alpha, \beta)_n = \gamma^{-1}\Psi_F(\alpha)(\gamma), \quad \text{where } \gamma^n = \beta, \gamma \in F^{\text{sep}}.$$
If $\gamma' \in F^{\text{sep}}$ is another element with $\gamma'^n = \beta$, then $\gamma^{-1}\gamma' \in \mu_n$ and
$$\gamma'^{-1}\Psi_F(\alpha)(\gamma') = \gamma^{-1}\Psi_F(\alpha)(\gamma).$$

This means that the Hilbert symbol is well defined.

PROPOSITION. *The norm residue symbol possesses the following properties:*

(1) $(\cdot,\cdot)_n$ *is bilinear;*
(2) $(1-\alpha,\alpha)_n = 1$ *for* $\alpha \in F^*, \alpha \neq 1$ *(Steinberg property);*
(3) $(-\alpha,\alpha)_n = 1$ *for* $\alpha \in F^*$ *;*
(4) $(\alpha,\beta)_n = (\beta,\alpha)_n^{-1}$ *;*
(5) $(\alpha,\beta)_n = 1$ *if and only if* $\alpha \in N_{F(\sqrt[n]{\beta})/F} F(\sqrt[n]{\beta})^*$ *and if and only if* $\beta \in N_{F(\sqrt[n]{\alpha})/F} F(\sqrt[n]{\alpha})^*$ *;*
(6) $(\alpha,\beta)_n = 1$ *for all* $\beta \in F^*$ *if and only if* $\alpha \in F^{*n}$, $(\alpha,\beta)_n = 1$ *for all* $\alpha \in F^*$ *if and only if* $\beta \in F^{*n}$ *;*
(7) $(\alpha,\beta)_{nm}^m = (\alpha,\beta)_n$ *for* $m \geqslant 1, \mu_{nm} \subset F^*$ *;*
(8) $(\alpha,\beta)_{n,L} = (N_{L/F}\alpha,\beta)_{n,F}$ *for* $\alpha \in L^*, \beta \in F^*$*, where* $(\cdot,\cdot)_{n,L}$ *is the Hilbert symbol in* L*,* $(\cdot,\cdot)_{n,F}$ *is the Hilbert symbol in* F*, and* L *is a finite separable extension of* F *;*
(9) $(\sigma\alpha,\sigma\beta)_{n,\sigma L} = \sigma(\alpha,\beta)_{n,L}$*, where* L *is a finite separable extension of* F*,* $\sigma \in \mathrm{Gal}(F^{\mathrm{sep}}/F)$*, and* $\mu_n \subset L^*$ *but not necessarily* $\mu_n \subset F^*$.

Proof. (1): For $\gamma \in F^{\mathrm{sep}}, \gamma^n = \beta$ we get

$$\begin{aligned}\gamma^{-1}\Psi_F(\alpha_1\alpha_2)(\gamma) &= \Psi_F(\alpha_1)\big(\gamma^{-1}\Psi_F(\alpha_2)(\gamma)\big)\cdot\big(\gamma^{-1}\Psi_F(\alpha_1)(\gamma)\big)\\ &= \big(\gamma^{-1}\Psi_F(\alpha_2)(\gamma)\big)\big(\gamma^{-1}\Psi_F(\alpha_1)(\gamma)\big),\end{aligned}$$

since $\Psi_F(\alpha_1)$ acts trivially on $(\alpha_2,\beta)_n \in \mu_n$. We also obtain

$$\begin{aligned}(\alpha,\beta_1\beta_2)_n &= \big(\gamma_1^{-1}\gamma_2^{-1}\Psi_F(\alpha)(\gamma_1\gamma_2)\big) = \big(\gamma_1^{-1}\Psi_F(\alpha)(\gamma_1)\big)\big(\gamma_2^{-1}\Psi_F(\alpha)(\gamma_2)\big)\\ &= (\alpha,\beta_1)_n(\alpha,\beta_2)_n.\end{aligned}$$

for $\gamma_1,\gamma_2 \in F^{\mathrm{sep}}$, $\gamma_1^n = \beta_1, \gamma_2^n = \beta_2$.
(5),(2),(3),(4): $(\alpha,\beta)_n = 1$ if and only if $\Psi_F(\alpha)$ acts trivially on $F(\sqrt[n]{\beta})$ and if and only if (by Theorem (4.2)) $\alpha \in N_{F(\sqrt[n]{\beta})/F} F(\sqrt[n]{\beta}))^*$.

Let $m|n$ be the maximal integer for which $\alpha \in F^{*m}$. Then $F(\sqrt[n]{\alpha})/F$ is of degree nm^{-1}. Let $\alpha = \alpha_1^m$ with $\alpha_1 \in F^*$ and let ζ_n be a primitive nth root of unity. Then for $\delta \in F^{\mathrm{sep}}, \delta^n = \alpha$, we get

$$\begin{aligned}1-\alpha = \prod_{i=1}^m (1-\zeta_n^i\delta) &= \prod_{i=1}^n \prod_{j=1}^{nm^{-1}} \left(1-\zeta_n^i\zeta_{nm^{-1}}^j\delta\right)\\ &= N_{F(\sqrt[n]{\alpha})/F}\left(\prod_{i=1}^m (1-\zeta_n^i\delta)\right) \in N_{F(\sqrt[n]{\alpha})/F} F(\sqrt[n]{\alpha})^*.\end{aligned}$$

Hence, $(1-\alpha,\alpha)_n = 1$. Further, $-\alpha = (1-\alpha)(1-\alpha^{-1})^{-1}$ for $\alpha \neq 0, \alpha \neq 1$. This means that $(-\alpha,\alpha)_n = (1-\alpha,\alpha)_n(1-\alpha^{-1},\alpha^{-1})_n^{-1} = 1$. Moreover,

$$1 = (-\alpha\beta,\alpha\beta)_n = (-\alpha,\alpha)_n(\alpha,\beta)_n(\beta,\alpha)_n(-\beta,\beta)_n = (\alpha,\beta)_n(\, be,\alpha)_n,$$

i.e., $(\alpha,\beta)_n = (\beta,\alpha)_n^{-1}$.

Finally, if $(\alpha,\beta)_n = 1$, then $(\beta,\alpha)_n = 1$, which is equivalent to

$$\beta \in N_{F(\sqrt[n]{\alpha})/F} F(\sqrt[n]{\alpha})^*.$$

(6): Let $\beta \in F^{*n}$; then $(\alpha,\beta)_n = 1$ for all $\alpha \in F^*$. Let $\beta \notin F^{*n}$, then $L = F(\sqrt[n]{\beta}) \neq F$, and L/F is a nontrivial abelian extension. By Theorem (4.2) the subgroup $N_{L/F}L^*$ does not coincide with F^*. If we take an element $\alpha \in F^*$ such that $\alpha \notin N_{L/F}L^*$ then, by property (5), we get $(\alpha,\beta)_n \neq 1$.
(7): For $\gamma \in F^{\mathrm{sep}}, \gamma^{nm} = \beta$, one has

$$(\alpha,\beta)_{nm}^m = \left(\gamma^{-1}\Psi_F(\alpha)(\gamma)\right)^m = \left(\gamma^{-m}\Psi_F(\alpha)(\gamma^m)\right) = (\alpha,\beta)_n,$$

because $(\gamma^m)^n = \beta$.
(8): Theorem (4.2) shows that

$$(\alpha,\beta)_{n,L} = \gamma^{-1}\Psi_L(\alpha)(\gamma) = \gamma^{-1}\Psi_F\left(N_{L/F}(\alpha)\right)(\gamma) = \left(N_{L/F}\alpha,\beta\right)_{n,F},$$

where $\gamma \in F^{\mathrm{sep}}, \gamma^n = \beta$.
(9): Theorem (4.2) shows that for $\gamma \in F^{\mathrm{sep}}, \gamma^n = \beta$,

$$(\sigma\alpha,\sigma\beta)_{n,\sigma L} = \sigma\left(\gamma^{-1}\Psi_L(\alpha)(\gamma)\right) = \sigma(\alpha,\beta)_{n,L}.$$

□

COROLLARY. *The Hilbert symbol induces the nondegenerate pairing*

$$(\cdot,\cdot)_n\colon F^*/F^{*n} \times F^*/F^{*n} \longrightarrow \mu_n.$$

(5.2). Kummer theory (see [La1, Ch. VIII]) asserts that abelian extensions L/F of exponent n ($\mu_n \subset F^*, p \nmid n$ if $\mathrm{char}(F) = p$) are in one-to-one correspondence with subgroups $\mathrm{B}_L \subset F^*$, such that $\mathrm{B}_L \supset F^{*n}$, $L = F(\sqrt[n]{\mathrm{B}_L}) = F(\gamma_i : \gamma_i^n \in \mathrm{B}_L)$ and the group B_L/F^{*n} is dual to $\mathrm{Gal}(L/F)$.

THEOREM. *Let $\mu_n \subset F^*, p \nmid n$, if* $\mathrm{char}(F) = p$. *Let* A *be a subgroup in F^* such that $F^{*n} \subset$* A. *Denote its orthogonal complement with respect to the Hilbert symbol $(\cdot,\cdot)_n$ by* $\mathrm{B} = \mathrm{A}^\perp$, *i.e.,*

$$\mathrm{B} = \{\beta \in F^* : (\alpha,\beta)_n = 1 \quad \textit{for all } \alpha \in \mathrm{A}\}.$$

Then $\mathrm{A} = N_{L/F}L^*$, *where $L = F(\sqrt[n]{\mathrm{B}})$ and $A = B^\perp$.*

Proof. We first recall that F^{*n} is of finite index in F^* by Lemma (1.4).

Let B be a subgroup in F^* with $F^{*n} \subset \mathrm{B}$ and $|\mathrm{B} : F^{*n}| = m$. Let $\mathrm{A} = \mathrm{B}^\perp$. Then $\Psi_F(\alpha)$, for $\alpha \in \mathrm{A}$, acts trivially on $F(\sqrt[n]{\beta})$ for $\beta \in \mathrm{B}$. This means that $\Psi_F(\alpha)$ acts trivially on $L = F(\sqrt[n]{\mathrm{B}})$ and, by Theorem (4.2), $\alpha \in N_{L/F}L^*$. Hence

$$\mathrm{A} \subset N_{L/F}L^*.$$

Conversely, if $\alpha \in N_{L/F}L^*$, then $\Psi_F(\alpha)$ acts trivially on $F(\sqrt[n]{\beta}) \subset L$ and

$$\alpha \in N_{F(\sqrt[n]{\beta})/F}F(\sqrt[n]{\beta})^*$$

for every $\beta \in \mathrm{B}$. Property (5) of (5.1) shows that $(\alpha,\beta)_n = 1$ and hence $N_{L/F}L^* \subset \mathrm{A}$. Thus, $\mathrm{A} = N_{L/F}L^*$.

Furthermore, to complete the proof it suffices to verify that a subgroup A in F^* with $F^{*n} \subset \mathrm{A}$ coincides with $(\mathrm{A}^\perp)^\perp$. Restricting the Hilbert symbol on $\mathrm{A} \times F^*$ we obtain that it induces the nondegenerate pairing $\mathrm{A}/F^{*n} \times F^*/\mathrm{A}^\perp \to \mu_n$. The theory of finite abelian groups (see [La1, Ch. I]) implies that the order of A/F^{*n} coincides with the order of $F^*/\mathrm{A}^\perp$. Similarly, one can verify that the order of $\mathrm{A}^\perp/F^{*n}$ is the same as that of $(\mathrm{A}^\perp)^\perp/F^{*n}$. As $\mathrm{A} \subset (\mathrm{A}^\perp)^\perp$, we deduce that $\mathrm{A} = (\mathrm{A}^\perp)^\perp$. □

(5.3). The problem to find explicit formulas for the norm residue symbol originates from Hilbert. In the case under consideration the challenge is to find a formula for the Hilbert symbol $(\alpha,\beta)_n$ in terms of the elements α,β of the field F. This problem is very complicated when $p|n$ and it will be discussed in Ch. VII. Nevertheless, there is a simple answer when $p \nmid n$.

THEOREM. *Let n be relatively prime with p and $\mu_n \subset F^*$. Then*

$$(\alpha,\beta)_n = c(\alpha,\beta)^{(q-1)/n},$$

where q is the cardinality of the residue field $\overline{F}$ and

$$c\colon F^* \times F^* \longrightarrow \mu_{q-1}$$

is the tame symbol defined by the formula

$$c(\alpha,\beta) = \mathrm{pr}\left(\beta^{v_F(\alpha)}\alpha^{-v_F(\beta)}(-1)^{v_F(\alpha)v_F(\beta)}\right),$$

with the projection $\mathrm{pr}\colon U_F \to \mu_{q-1}$ *induced by the decomposition* $U_F \simeq \mu_{q-1} \times U_{1,F}$ *from* (1.2) (*i.e.,* $\mathrm{pr}(u)$ *is the multiplicative representative of* $\overline{u} \in \overline{F}$).

Proof. Note that the elements of the group μ_n, for $p \nmid n$, are isomorphically mapped onto the subgroup in the multiplicative group $\mathbb{F}_q^*$. Hence, $n|(q-1)$. Note also that the prime elements generate F^*. Indeed, if $\alpha = \pi^a\varepsilon$ with $\varepsilon \in U_F$, then $\alpha = \pi_1\pi^{a-1}$ for the prime element $\pi_1 = \pi\varepsilon$, when $a \neq 1$, and $\alpha = \pi_2$ for the prime element $\pi_2 = \pi\varepsilon$, when $a = 1$. Using properties (1) and (7) of the Hilbert symbol it suffices to verify that $c(\pi,\beta) = (\pi,\beta)_{q-1}$ for $\beta \in F^*$.

Let $\beta = (-\pi)^a\theta\varepsilon$ with $a = v_F(\beta), \theta \in \mu_{q-1}, \varepsilon \in U_{1,F}$. Then, as $c(\pi, -\pi) = 1, c(\pi, \varepsilon) = 1$, we obtain $c(\pi, \beta) = c(\pi, \theta) = \theta$. Property (3) of the Hilbert symbol shows that $(\pi, -\pi)_{q-1} = 1$. Corollary (5.5) Ch. I implies that the group $U_{1,F}$ is $(q-1)$-divisible. Hence, $(\pi, \varepsilon)_{q-1} = 1$. Finally, since the extension $F(\sqrt[n]{\theta})/F$ is unramified, Remark in (1.2) shows that for $\eta \in F^{\rm sep}, \eta^{q-1} = \theta$,

$$(\pi, \theta)_{q-1} = \eta^{-1}\Psi_F(\pi)(\eta) = \eta^{-1}\varphi_F(\eta) = \eta^{q-1} = \theta.$$

We conclude that $(\pi, \beta)_{q-1} = \theta = c(\pi, \beta)$. □

(5.4). Abelian extensions of exponent p of a field F of characteristic p are described by the Artin–Schreier theory (see [La1, Ch. VIII]). Recall that the polynomial $X^p - X$ is denoted by $\wp(X)$ (see (6.3) Ch. I). This polynomial is additive, i.e.,

$$\wp(\alpha + \beta) = \wp(\alpha) + \wp(\beta)$$

for $\alpha, \beta \in F$. Abelian extensions L/F of exponent p are in one-to-one correspondence with subgroups $\mathrm{B} \subset F$ such that $\wp(F) \subset \mathrm{B}$. The quotient group $\mathrm{B}/\wp(F)$ is dual to $\mathrm{Gal}(L/F)$, where

$$L = F\big(\wp^{-1}(\mathrm{B})\big) = F\big(\gamma : \wp(\gamma) \in \mathrm{B}\big).$$

Since the kernel of the homomorphism $\wp\colon \mathbb{F}_q \to \mathbb{F}_q$ is of order p, the quotient group $\mathbb{F}_q/\wp\left(\mathbb{F}_q\right)$ is of order p. Note that the index of $\wp(F)$ in F is infinite. Indeed, we shall show that for a prime element π in F, the sets $\pi^{-i} + \wp(F)$ with $p \nmid i,\ i \geqslant 1$, are distinct cosets of $\wp(F)$ in F. If we had $\pi^{-i} + \wp(F) = \pi^{-j} + \wp(F)$ for $1 \leqslant i < j$, $p \nmid i,\ p \nmid j$, then we would have $\pi^{-j} - \pi^{-i} \in \wp(F)$. However, as $v_F\left(\wp\left(\pi^{-i}\right)\right) = -pi$ for $i \geqslant 0$, we obtain that

$$v_F\left(\wp(\alpha)\right) = pv_F(\alpha)$$

if $v_F(\alpha) \leqslant 0$. Hence, the relation $\pi^{-j} - \pi^{-i} \in \wp(F)$ is impossible.

For a complete discrete valuation field F of characteristic p with a finite residue field we define the map

$$(\cdot, \cdot]\colon F^* \times F \longrightarrow \mathbb{F}_p$$

by the formula

$$(\alpha, \beta] = \Psi_F(\alpha)(\gamma) - \gamma,$$

where γ is a root of the polynomial $X^p - X - \beta$. All the roots of this polynomial are $\gamma + c$ where c runs through $\mathbb{F}_p$, therefore we deduce that the pairing $(\cdot, \cdot]$ is well defined.

PROPOSITION. *The map $(\cdot, \cdot]$ has the following properties:*

(1) $(\alpha_1\alpha_2, \beta] = (\alpha_1, \beta] + (\alpha_2, \beta]$, $(\alpha, \beta_1 + \beta_2] = (\alpha, \beta_1] + (\alpha, \beta_2]$;

(2) $(-\alpha, \alpha] = 0$ *for* $\alpha \in F^*$;

(3) $(\alpha, \beta] = 0$ *if and only if* $\alpha \in N_{F(\gamma)/F}F(\gamma)^*$, *where* $\gamma^p - \gamma = \beta$;
(4) $(\alpha, \beta] = 0$ *for all* $\alpha \in F^*$ *if and only if* $\beta \in \wp(F)$;
(5) $(\alpha, \beta] = 0$ *for all* $\beta \in F$ *if and only if* $\alpha \in F^{*p}$;
(6) $(\pi, \beta] = \mathrm{Tr}_{\mathbb{F}_q/\mathbb{F}_p}\theta_0$, *where* π *is a prime element in* F *and* $\beta = \sum_{i \geqslant a} \theta_i \pi^i$ *with* $\theta_i \in \mathbb{F}_q$.

Proof.
(1): One has

$$\begin{aligned}\Psi_F(\alpha_1\alpha_2)(\gamma) - \gamma &= \Psi_F(\alpha_1)\big(\Psi_F(\alpha_2)(\gamma) - \gamma\big) + \Psi_F(\alpha_1)(\gamma) - \gamma\\ &= \Psi_F(\alpha_1)(\gamma) - \gamma + \Psi(\alpha_2)(\gamma) - \gamma,\end{aligned}$$

since $\Psi_F(\alpha_2)(\gamma) - \gamma \in F$. One also has

$$\Psi_F(\alpha)(\gamma_1 + \gamma_2) - (\gamma_1 + \gamma_2) = \Psi_F(\alpha)(\gamma_1) - \gamma_1 + \Psi_F(\alpha)(\gamma_2) - \gamma_2.$$

(3): $(\alpha, \beta] = 0$ if and only if $\Psi_F(\alpha)$ acts trivially on $F(\gamma)$, where $\gamma^p - \gamma = \beta$. Theorem (4.2) shows that this is equivalent to $\alpha \in N_{F(\gamma)/F}F(\gamma)^*$.
(2): If $\alpha \in \wp(F)$, then $(-\alpha, \alpha] = 0$ by property (3). If a root γ of the polynomial $X^p - X - \alpha$ does not belong to F, then $-\alpha = N_{F(\gamma)/F}(-\gamma)$ and property 3) shows that $(-\alpha, \alpha] = 0$.
(4): If $\beta \notin \wp(F)$, then $L = F(\gamma) \neq F$ for a root γ of the polynomial $X^p - X - \beta$; L/F is an abelian extension of degree p, and (1.5) shows that $N_{L/F}L^* \neq F^*$. For an element $\alpha \in F^*$, such that $\alpha \notin N_{L/F}L^*$, we deduce by Theorem (4.2) that $\Psi_F(\alpha)$ acts nontrivially on L, i.e., $\Psi_F(\alpha)(\gamma) \neq \gamma$ and $(\alpha, \beta] \neq 0$.
(5): Let A denote the set of those $\alpha \in F^*$, for which $(\alpha, \beta] = 0$ for all $\beta \in F$. Note that for $\alpha, \beta \in F^*$ properties (1) and (2) imply

$$(-\beta, \alpha\beta] = (-\alpha\beta, \alpha\beta] - (\alpha, \alpha\beta] = -(\alpha, \alpha\beta].$$

Hence, the condition $\alpha \in \mathrm{A}$ is equivalent to $(\alpha, \alpha\beta] = 0$ for all $\beta \in F^*$ and to $(-\beta, \alpha\beta] = 0$ for all $\beta \in F^*$. Then, if $\alpha_1, \alpha_2 \in \mathrm{A}$ we get $(-\beta, (\alpha_1 + \alpha_2)\beta] = (-\beta, \alpha_1\beta] + (-\beta, \alpha_2\beta] = 0$, and $(-\beta, -\alpha_1\beta] = -(-\beta, \alpha_1\beta] = 0$. This means that $\alpha_1 + \alpha_2, -\alpha_1 \in \mathrm{A}$. Obviously, $\alpha_1\alpha_2 \in \mathrm{A}, \alpha_1^{-1} \in \mathrm{A}$. Therefore, the set $\mathrm{A} \cup \{0\}$ is a subfield in F. Further, $F^p \subset \mathrm{A} \cup \{0\}$ by property (1), and we obtain $F^p \subset \mathrm{A} \cup \{0\} \subset F$.

One can identify the field F with $\mathbb{F}_q((\pi))$. Then the field F^p is identified with the field $\mathbb{F}_q((\pi^p))$ and we obtain that the extension $\mathbb{F}_q((\pi))/\mathbb{F}_q((\pi^p))$ is of degree p. Hence, $\mathrm{A} \cup \{0\} = F^p$ or $\mathrm{A} \cup \{0\} = F$. As we saw in (5.4) $\wp(F) \neq F$, and so property (4) shows that $(\alpha, \beta] \neq 0$ for some $\beta \in F, \alpha \in F^*$. Thus, $\mathrm{A} \cup \{0\} \neq F$, i.e., $\mathrm{A} = F^{*p}$.
(6): If $\theta \in \mathbb{F}_q$ and $\gamma \in F^{\mathrm{sep}}$, $\gamma^p - \gamma = \theta$, then $F(\gamma) = F$ or $F(\gamma)/F$ is the unramified extension of degree p. Remark in (1.2) and Theorem (4.2) imply

$$(\pi, \theta] = \varphi_F(\gamma) - \gamma = \gamma^q - \gamma = \theta^{q/p} + \theta^{q/p^2} + \cdots + \theta = \mathrm{Tr}_{\mathbb{F}_q/\mathbb{F}_p}\theta.$$

Let a be a positive integer and $\theta \in \mathbb{F}_q^*$. Then

$$a(\pi, \theta\pi^a] = (\pi^a, \theta\pi^a] = (\theta\pi^a, \theta\pi^a] = (-1, \theta\pi^a] = 0,$$

since the group $\mathbb{F}_q^*$ is p-divisible and $-1 \in \mathbb{F}_q^p$. Hence $(\pi, \theta\pi^a] = 0$ for $p \nmid a$. Finally, let $a = p^s b$, where $s > 0$ and $p \nmid b, b > 0$. Then

$$\theta\pi^a = (\theta_1\pi^{p^{s-1}b})^p - \theta_1\pi^{p^{s-1}b} + \theta_1\pi^{p^{s-1}b} \in \theta_1\pi^{p^{s-1}b} + \wp(F),$$

where $\theta_1^p = \theta$. Continuing in this way we deduce that $\theta\pi^a = \theta_s\pi^b + \wp(\lambda)$, where $\theta_s^{p^s} = \theta$ and $\lambda \in F$. Then $(\pi, \theta\pi^a] = (\pi, \theta_s\pi^b] = 0$. We obtain property (6) and complete the proof. □

COROLLARY. *The pairing* $(\cdot,\cdot]$ *determines the nondegenerate pairing*

$$F^*/F^{*p} \times F/\wp(F) \longrightarrow \mathbb{F}_p$$

(5.5). We introduce a map d_π which in fact coincides with $(\cdot,\cdot]$.

Let π be a prime element of a complete residue field F of characteristic p with the residue field $\mathbb{F}_q$. Then an element $\alpha \in F$ can be uniquely expanded as

$$\alpha = \sum_{i \geqslant a} \theta_i\pi^i, \quad \theta_i \in \mathbb{F}_q.$$

Put

$$\frac{d\alpha}{d\pi} = \sum_{i \geqslant a} i\theta_i\pi^{i-1}, \quad \operatorname{res}_\pi \alpha = \theta_{-1}.$$

Define the *Artin–Schreier pairing*

$$d_\pi \colon F^* \times F \to \mathbb{F}_p, \qquad d_\pi(\alpha, \beta) = \operatorname{Tr}_{\mathbb{F}_q/\mathbb{F}_p} \operatorname{res}_\pi(\beta\alpha^{-1}\frac{d\alpha}{d\pi}).$$

PROPOSITION. *The map* d_π *possesses the following properties:*

(1) *linearity*

$$d_\pi(\alpha_1\alpha_2, \beta) = d_\pi(\alpha_1, \beta) + d_\pi(\alpha_2, \beta),$$
$$d_\pi(\alpha, \beta_1 + \beta_2) = d_\pi(\alpha, \beta_1) + d_\pi(\alpha, \beta_2);$$

(2) *if* π_1 *is a prime element in* F, *then*

$$d_\pi(\pi_1, \beta) = d_{\pi_1}(\pi_1, \beta) = \operatorname{Tr}_{\mathbb{F}_q/\mathbb{F}_p} \theta_0,$$

where $\beta = \sum_{i \geqslant a} \theta_i\pi_1^i, \theta_i \in \mathbb{F}_q$.

Proof. (1): We have

$$\frac{d(\alpha_1\alpha_2)}{d\pi}\frac{1}{\alpha_1\alpha_2} = \frac{d\alpha_1}{d\pi}\frac{1}{\alpha_1} + \frac{d\alpha_2}{d\pi}\frac{1}{\alpha_2},$$

since $\frac{d\alpha}{d\pi}$ can be treated as a formal derivative $\left.\frac{d\alpha(X)}{dX}\right|_{X=\pi}$ for the series $\alpha(X) = \sum a_i X^i$. Hence, we get $d_\pi(\alpha_1\alpha_2, \beta) = d_\pi(\alpha_1, \beta) + d_\pi(\alpha_2, \beta)$.

The other formula follows immediately.

(2): Let $C = \mathbb{Z}[X_1, X_2, \dots]$, where $X_1, X_2, \dots$ are independent indeterminates. Let X be an indeterminate over C. Put

$$\alpha(X) = X_1 X + X_2 X^2 + X_3 X^3 + \dots \in C[[X]].$$

For an element $\sum_{j\geqslant a} \kappa_j X^j \in C[[X]], \kappa_i \in C$, we put

$$\frac{d(\sum_{j\geqslant a} \kappa_j X^j)}{dX} = \sum_{j\geqslant a} j\kappa_j X^{j-1}, \quad \operatorname{res}_X \sum_{j\geqslant a} \kappa_j X^j = \kappa_{-1}.$$

Note that

$$\operatorname{res}_X \frac{d\left(\sum_{j\geqslant a} \kappa_j X^j\right)}{dX} = 0.$$

Hence, for $i \neq 0$ we get

$$\operatorname{res}_X \left(\alpha(X)^{i-1} \frac{d\alpha(X)}{dX}\right) = \operatorname{res}_X \left(\frac{1}{i} \frac{d\left(\alpha(X)^i\right)}{dX}\right) = 0.$$

One can define a ring-homomorphism $C[[X]] \to F$ as follows: $X_i \in C \to \eta_i \in \mathbb{F}_q, X \to \pi$. The series $\alpha(X)$ is mapped to $\alpha(\pi) = \eta_1\pi + \eta_2\pi^2 + \dots \in F$, and we conclude that

$$\operatorname{res}_\pi \left(\alpha(\pi)^{i-1} \frac{d\alpha(\pi)}{d\pi}\right) = 0 \qquad \text{if} \quad i \neq 0.$$

Now let $\beta = \sum_{i\geqslant a} \theta_i \pi_1^i, \theta_i \in \mathbb{F}_q$. The definition of d_{π_1} shows that

$$d_{\pi_1}(\pi_1, \beta) = \operatorname{Tr}_{\mathbb{F}_q/\mathbb{F}_p} \theta_0.$$

Writing $\pi_1 = \eta_1\pi + \eta_2\pi^2 + \dots = \alpha(\pi)$ with $\eta_i \in \mathbb{F}_q$, we get

$$d_\pi(\pi_1, \theta_i\pi_1^i) = \operatorname{res}_\pi \left(\theta_i \pi_1^{i-1} \frac{d\pi_1}{d\pi}\right) = \operatorname{res}_\pi \left(\theta_i \alpha(\pi)^{i-1} \frac{d\alpha(\pi)}{d\pi}\right) = 0,$$

if $i \neq 0$, and

$$d_\pi(\pi_1, \theta_0) = \operatorname{res}_\pi \left(\theta_0 \alpha(\pi)^{-1} \frac{d\alpha(\pi)}{d\pi}\right) = \operatorname{res}_\pi(\theta_0\pi^{-1} + \delta) = \operatorname{Tr}_{\mathbb{F}_q/\mathbb{F}_p} \theta_0$$

where $\delta \in \mathcal{O}_F$. Thus $d_{\pi_1}(\pi_1, \beta) = d_\pi(\pi_1, \beta) = \operatorname{Tr}_{\mathbb{F}_q/\mathbb{F}_p} \theta_0$, as desired. □

(5.6). THEOREM. *Let F be a complete discrete valuation field of characteristic p with the residue field $\mathbb{F}_q$. Then the pairing $(\cdot, \cdot]$ defined in* (5.4) *coincides with d_π*

defined in (5.5). *In particular, the pairing* d_π *does not depend on the choice of the prime element* π.

Proof. As the prime elements generate F^*, it suffices to show, using property (1) of $(\cdot,\cdot]$ and property (1) of d_π, that for a prime element π_1 in F the following equality holds:

$$(\pi_1,\beta] = d_\pi(\pi_1,\beta), \qquad \beta \in F.$$

Let $\beta = \sum_{i\geqslant a} \theta_i \pi_1^i$. Then property (6) of $(\cdot,\cdot]$ and property (2) of d_π imply that

$$(\pi_1,\beta] = d_\pi(\pi_1,\beta) = \operatorname{Tr}_{\mathbb{F}_q/\mathbb{F}_p}\theta_0,$$

as desired. □

COROLLARY. *Let* i *be a positive integer, and* $\mathrm{B} = \mathbb{F}_q\pi^{-i} + \cdots + \mathbb{F}_q\pi^{-1} + \mathbb{F}_q + \wp(F)$. *Then* B *is an additive subgroup of* F *and the set*

$$\mathrm{A} = \mathrm{B}^{\perp} = \{\alpha \in F^* : (\alpha,\beta] = 0 \quad \textit{for all } \beta \in \mathrm{B}\}$$

coincides with $U_{i+1}F^{*p}$.

Proof. For $\theta,\eta \in \mathbb{F}_q$ one has

$$d_\pi(1+\eta\pi^j, \theta\pi^{-i}) = 0 \qquad \text{if } j > i > 0.$$

Hence, $U_{i+1}F^{*p} \subset \mathrm{A}$. If we fix $\eta \in \mathbb{F}_q$, then there exists an element $\theta \in \mathbb{F}_q$ such that $\operatorname{Tr}_{\mathbb{F}_q/\mathbb{F}_p}(\theta\eta) \neq 0$. Therefore,

$$d_\pi(1+\eta\pi^j, \theta\pi^{-j}) = \operatorname{Tr}_{\mathbb{F}_q/\mathbb{F}_p}(j\theta\eta) \neq 0 \qquad \text{for } p \nmid j.$$

We also get $d_\pi(\pi,\theta) = \operatorname{Tr}_{\mathbb{F}_q/\mathbb{F}_p}\theta$. Thus, for an element $\alpha \in F^*$ such that $\alpha \notin U_{i+1}F^{*p}$, there exists an element $\beta \in \mathrm{B}$ with $(\alpha,\beta] = d_\pi(\alpha,\beta) \neq 0$. This means that $U_{i+1}F^{*p} = \mathrm{A}$, as required. □

REMARK. *J. Tate* in [T8] gave an interpretation of the residues of differentials in terms of traces of linear operators acting on infinite dimensional vector spaces (like $K((X))$ over K). A generalization of this idea to the tame symbol by using the theory of infinite wedge representations is contained in [AdCK].

(5.7). Using Artin–Schreier extensions we saw in (1.5) the connection between an open subgroup of prime index in F^* and the norm subgroup $N_{L/F}L^*$ for a cyclic extension L/F of the same degree. Now we can refine this connection in a different way, applying the pairings of F^* defined above. We shall show, for instance, that every open subgroup N of prime index l in F^* contains the norm subgroup $N_{L/F}L^*$ for some abelian extension L/F if $\mu_l \subset F^*$. If $\operatorname{char}(F) = 0$ or l is relatively prime with p, then Theorem (5.2) shows that $N = N_{L/F}L^*$ for $L = F(\sqrt[l]{N^{\perp}})$, where $N^{\perp}$

is the orthogonal complement of N with respect to the Hilbert symbol $(\cdot,\cdot)_l$. If $l = p = \mathrm{char}(F)$ and $U_F \subset N$, then

$$N = \langle \pi^p \rangle \times U_F$$

for a prime element π in F. Taking an element $\theta \in \mathbb{F}_q$ with $\mathrm{Tr}_{\mathbb{F}_q/\mathbb{F}_p} \theta \neq 0$, we obtain

$$(\pi,\theta] \neq 0, \quad (\pi^p,\theta] = 0, \quad (\varepsilon,\theta] = 0$$

for $\varepsilon \in U_F$. Therefore, N coincides with the orthogonal complement of $\theta + \wp(F)$ with respect to the pairing $(\cdot,\cdot]$, and Proposition (5.4) shows that $N = N_{F(\gamma)/F}F(\gamma)^*$ for $\gamma \in F^{\mathrm{sep}}$ with $\gamma^p - \gamma = \theta$. If $U_F \not\subset N$, then one can find a positive integer i such that $U_{i+1} \subset N$ and $U_i \not\subset N$ (since N is open). If B is as in Corollary (5.6), then $\mathrm{B}^\perp = U_{i+1}F^{*p} \subset N$. Proposition (5.4) implies that $\mathrm{B}^\perp \supset N_{L/F}L^*$ for $L = F(\wp^{-1}(\mathrm{B}))$ and hence $N_{L/F}L^* \subset N$.

One can show that $N = N_{L/F}L^*$ for $L = F(\wp^{-1}(N^\perp))$, where $N^\perp$ is the orthogonal complement of N with respect to the pairing $(\cdot,\cdot]$ (see Exercise 5).

Exercises.

1. a) Let n be odd. Show that $(\beta,\beta)_n = (-1,\beta)_n = 1$ for $\beta \in F^*$.
   b) Show that $(\theta,\beta)_{p^m} = 1$ for $\theta \in \mu_{q-1}, \beta \in F^*$.
2. Let p be an odd prime, and let ζ_p be a primitive pth root of unity.
   a) Show that $X^p - Y^p = \prod_{i=0}^{p-1}\left(\zeta_p^i X - \zeta_p^{-i}Y\right)$ and $\prod_{i=1}^{p-1}\left(\zeta_p^i - \zeta_p^{-i}\right) = p$.
   b) Put $c(\zeta_p) = \prod_{i=1}^{\frac{p-1}{2}}\left(\zeta_p^i - \zeta_p^{-i}\right)$. Show that $c(\zeta_p)^2 = (-1)^{\frac{p-1}{2}}p$.
   c) For a natural b put

$$\left(\frac{b}{p}\right) = \begin{cases} 0 & \text{if } p|b, \\ 1 & \text{if } p\nmid b, b \equiv a^2 \mod p \text{ for} \\ -1, & \text{otherwise.} \end{cases}$$

   Show that

$$\left(\frac{b}{p}\right) = \frac{c(\zeta_p^b)}{c(\zeta_p)}.$$

   d) Let q be an odd prime, $q \neq p$, and let ζ_q be a primitive qth root of unity. Show that

$$\left(\frac{q}{p}\right) = \prod_{i=1}^{\frac{p-1}{2}}\prod_{j=1}^{\frac{q-1}{2}}\left(\zeta_p^i\zeta_q^j - \zeta_p^{-i}\zeta_q^{-j}\right).$$

   e) Prove the *quadratic reciprocity law*: if p, q are odd primes, $p \neq q$, then

$$\left(\frac{p}{q}\right)\left(\frac{q}{p}\right) = (-1)^{\frac{p-1}{2}\frac{q-1}{2}}, \quad \left(\frac{2}{p}\right) = (-1)^{\frac{p^2-1}{8}}.$$

   (there exist about 200 proofs of the quadratic reciprocity law, and the first of them are due to Gauss. See [IR, Ch. V] for more details.)

3. Let $(\cdot,\cdot)_{(p)}$ be the Hilbert symbol $(\cdot,\cdot)_{2,\mathbb{Q}_p}\colon \mathbb{Q}_p^* \times \mathbb{Q}_p^* \to \mu_2$. Show that if ε, η are units in $\mathbb{Z}_p$, then for $p > 2$

$$(\varepsilon,\eta)_{(p)} = 1, \quad (p,\varepsilon)_{(p)} = \left(\frac{\varepsilon_0}{p}\right), \quad (p,p)_{(p)} = (-1)^{\frac{p-1}{2}}$$

and

$$(\varepsilon,\eta)_{(2)} = (-1)^{\frac{\varepsilon-1}{2}\frac{\eta-1}{2}}, \quad (\varepsilon,2)_{(2)} = (-1)^{\frac{\varepsilon^2-1}{8}}, \quad (2,2)_{(2)} = 1,$$

where ε_0 is an integer such that $\varepsilon \equiv \varepsilon_0 \mod p$, $(-1)^a = (-1)^{a_0}$ for $a = a_0 + 2a_1 + 2^2a_2 + \dots \in \mathbb{Z}_2$ with integers $a_0, a_1, a_2, \dots$.

4. For $\alpha, \beta \in \mathbb{Q}^*$ put $(\alpha,\beta)_{(\infty)} = 1$, if $\alpha > 0, \beta > 0$, and $= -1$ otherwise. Show that $\prod_{p\in P'}(\alpha,\beta)_{(p)} = 1$ for $\alpha,\beta \in \mathbb{Q}^*$, where the set P' consists of all positive primes and the symbol ∞. Show that the last equality is equivalent to the quadratic reciprocity law.

5. Let $\operatorname{char}(F) = p$. Show that if A is an open subgroup of finite index in F^* such that $F^{*p} \subset \mathrm{A}$, and B is its orthogonal complement with respect to the pairing $(\cdot,\cdot]$, then $\mathrm{A} = N_{L/F}L^*$ for $L = F\big(\wp^{-1}(\mathrm{B})\big)$. Conversely: if B is a subgroup in F such that $\wp(F) \subset \mathrm{B}$ and $\mathrm{B}/\wp(F)$ is finite, then the orthogonal complement $\mathrm{A} = \mathrm{B}^\perp$ with respect to $(\cdot,\cdot]$ coincides with $N_{L/F}L^*$, where $L = F\big(\wp^{-1}(\mathrm{B})\big)$, and the index of A in F^* is equal to the order of $\mathrm{B}/\wp(F)$.

6. ($\diamond$) Let F be a field of characteristic p. Recall that the *Witt* theory establishes a one-to-one correspondence between subgroups B in $W_n(F)$ with $\wp W_n(F) \subset \mathrm{B}$ and abelian extensions L/F of exponent p^n $\mathrm{B} \leftrightarrow L = F\big(\wp^{-1}(\mathrm{B})\big)$, where the map $\wp$ was defined in Exercise 7 in section 8 Ch. I, and $F\big(\wp^{-1}(\mathrm{B})\big)$ is the compositum of the fields $F(\gamma_0,\gamma_1,\dots,\gamma_{n-1})$ such that $\wp\big((\gamma_0,\gamma_1,\dots,\gamma_{n-1})\big) \in \mathrm{B}$ (see [La1, Ch. VIII, Exercises 21–25]). This corresponds to *Witt pairing*

$$\operatorname{Gal}(F_n/F) \times W_n(F)/\wp(W_n(F)) \to W_n(\mathbb{F}_p)/\wp W_n(\mathbb{F}_p),$$

where F_n is the compositum of all extensions L/F as above, and there is an isomorphism

$$\operatorname{Hom}(G_F, \mathbb{Z}/p^n\mathbb{Z}) \simeq W_n(F)/\wp(W_n(F)).$$

For a complete discrete valuation field F of characteristic p with residue field $\mathbb{F}_q$ define a map

$$(\cdot,\cdot]_n\colon F^* \times W_n(F) \longrightarrow W_n(\mathbb{F}_p) \simeq \mathbb{Z}/p^n\mathbb{Z}$$

by the formula

$$(\alpha, x]_n = \Psi_F(\alpha)(z) - z,$$

where $z \in W_n(F^{\mathrm{sep}})$, and $\wp(z) = x$. In particular, $(\cdot,\cdot] = (\cdot,\cdot]_1$. Show that the map $(\cdot,\cdot]_n$ determines a nondegenerate pairing

$$F^*/F^{*p^n} \times W_n(F)/\wp W_n(F) \longrightarrow W_n(\mathbb{F}_p) \simeq \mathbb{Z}/p^n\mathbb{Z}.$$

Show that if A is an open subgroup of finite index in F^* such that $F^{*p^n} \subset \mathrm{A}$, then $\mathrm{A} = N_{L/F}L^*$ for $L = F\big(\wp^{-1}(\mathrm{A}^\perp)\big)$, where $\mathrm{A}^\perp$ is the orthogonal complement of A with respect to $(\cdot,\cdot]_n$. Conversely, if B is a subgroup in $W_n(F)$, such that $\wp W_n(F) \subset \mathrm{B}$

and $\mathrm{B}/\wp W_n(F)$ is finite, then $\mathrm{B}^\perp = N_{L/F}L^*$ for $L = F\big(\wp^{-1}(\mathrm{B})\big)$. Passing to the injective limit, we obtain the nondegenerate pairing

$$(\cdot,\cdot]_\infty\colon F^* \times W \longrightarrow \varinjlim \mathbb{Z}/p^n\mathbb{Z} \simeq \mathbb{Q}_p/\mathbb{Z}_p,$$

where $W = \varinjlim W_n(F)/\wp W_n(F)$ with respect to the homomorphisms

$$W_n(F)/\wp W_n(F) \to W_{n+1}(F)/\wp W_{n+1}(F),$$
$$(\alpha_0, \dots) + \wp W_n(F) \mapsto (0, \alpha_0, \dots) + \wp W_{n+1}(F).$$

Note that the group W is dual to the Galois group of the maximal abelian p-extension of F over F.

7. ($\diamond$) Let π be a prime element in F, $\mathrm{char}(F) = p$, and $|\overline{F} : \mathbb{F}_p| = f$. Let

$$d_{\pi,n}\colon F^* \times W_n(F) \to W_n(\mathbb{F}_p)$$

be the map defined by the formula $d_{\pi,n}(\alpha, x) = (1 + \mathbf{F} + \dots + \mathbf{F}^{f-1})y$, where the map $\mathbf{F}$ was defined in section 8 Ch. I, $y \in W_n(\mathbb{F}_q)$, and its ghost component $y^{(m)} = \mathrm{res}_\pi\left(\alpha^{-1}\dfrac{d\alpha}{d\pi}x^{(m)}\right)$, where $x^{(m)}$ is the ghost component of x (more precisely one needs to pass from F to a ring of characteristic zero from which there is a surjective homomorphism to F (e.g. $\mathbb{Z}_p((t))$) and operate with the ghost components at that level, returning afterwards to Witt vectors over F). Show that $d_{\pi,n} = (\cdot,\cdot]_n$.

8. ($\diamond$) (*Y. Kawada, I. Satake* [KwS]) Let F be of characteristic p with the residue field $\mathbb{F}_q$. Let π be a prime element in F, θ a generator of μ_{q-1}. Put $F_1 = F(\sqrt[q-1]{\pi}, \sqrt[q-1]{\theta}) = F(\sqrt[q-1]{F^*})$. Then Kummer theory and the tame symbol determine the homomorphism

$$\Psi_1\colon F^* \longrightarrow G_1 = \mathrm{Gal}(F_1/F).$$

Let F_2 be the maximal abelian p-extension of F. The Witt theory and the pairings $d_{\pi,n}$ determine the homomorphism $\Psi_2\colon F^* \to G_2 = \mathrm{Gal}(F_2/F)$ (the group G_2 is dual to W defined in Exercise 6). Introduce $\Psi_3\colon F^* \to G_3 = \mathrm{Gal}(F^{\mathrm{ur}}/F)$ by the formula $\Psi_3(\alpha) = \varphi_F^{v_F(\alpha)}$ for $\alpha \in F^*$.
Prove that Ψ_i are compatible with each other and therefore induce a homomorphism $\Psi\colon F^* \to \mathrm{Gal}(F^{\mathrm{ab}}/F)$ which coincides with the reciprocity map Ψ_F. This way one can construct class field theory for the fields of positive characteristic.

9. Show that the assertions of this section hold also for a Henselian discrete valuation field of characteristic 0 with finite residue field. What can be said about the case of positive characteristic?

6. The Existence Theorem

In this section we exhibit an additional feature of the reciprocity map which is expressed by the existence theorem. We show in (6.2) that the set of all open subgroups of finite index in F^* and the set of all norm subgroups $N_{L/F}L^*$ for finite Galois extensions L/F coincide. Then we discuss additional properties of the reciprocity map Ψ_F

in (6.3) and (6.4). A relation to the first continuous Galois cohomology group with coefficients in the completion of the separable closure of the field in characteristic zero is discussed in (6.5). Finally in (6.6) we describe two first generalizations of class field theory.

We continue to assume that F is a complete discrete valuation field with finite residue field.

(6.1). PROPOSITION. *Let L be a finite separable extension of F. Then the norm map $N_{L/F}\colon L^* \to F^*$ is continuous and $N_{L/F}L^*$ is an open subgroup of finite index in F^*.*

Proof. Let E/F be a finite Galois extension with $L \subset E$. Then, by Theorem (4.2), $N_{E/F}E^*$ is of finite index in F^*. The Galois group of the extension E/F is solvable by (1.2). Therefore, in order to show that $N_{L/F}L^*$ is open, it suffices to verify that the norm map for a cyclic extension of prime degree transforms open subgroups to open subgroups. This follows from the description of the behavior of the norm map in Propositions (1.2), (1.3), (1.5) Ch. III. Similarly, the same description of the norm map implies that the pre-image $N_{M/F}^{-1}$ of an open subgroup is an open subgroup for a cyclic extension M/F. Therefore, the pre-image $N_{E/F}^{-1}$ of an open subgroup N in F^* is an open subgroup in E^*. Since $N_{L/F}^{-1}(N) \supset N_{E/L}\bigl(N_{E/F}^{-1}(N)\bigr)$, we obtain that $N_{L/F}^{-1}(N)$ is open in L^* and $N_{L/F}$ is continuous. □

COROLLARY. *The Hilbert symbol $(\cdot,\cdot)_n$ is a continuous map of $F^* \times F^*$ to μ_n.*

Proof. It follows from property (5) of the Hilbert symbol and the Proposition. □

(6.2). THEOREM ("EXISTENCE THEOREM"). *There is a one-to-one correspondence between open subgroups of finite index in F^* and the norm subgroups of finite abelian extensions: $N \leftrightarrow N_{L/F}L^*$. This correspondence is an order reversing bijection between the lattice of open subgroups of finite index in F^* (with respect to the intersection $N_1 \cap N_2$ and the product N_1N_2) and the lattice of finite abelian extensions of F (with respect to the intersection $L_1 \cap L_2$ and the compositum L_1L_2).*

Proof. We verify that an open subgroup N of finite index in F^* coincides with the norm subgroup $N_{L/F}L^*$ of some finite abelian extension L/F. It suffices to verify that N contains the norm subgroup $N_{L/F}L^*$ of some finite separable extension L/F. Indeed, in this case N contains $N_{E/F}E^*$, where E/F is a finite Galois extension, $E \supset L$. Then by Proposition (4.1) we deduce that $N = N_{M/F}M^*$, where M is the fixed field of $(N, E/F)$ and M/F is abelian.

Assume $\operatorname{char}(F) \nmid n$, where n is the index of N in F^*. If $\mu_n \subset F^*$, then Theorem (5.2) shows that $F^{*n} = N_{L/F}L^*$ for some finite abelian extension L/F, since F^{*n} is of finite index in F^*. Then $N_{L/F}L^* \subset N$. If μ_n is not contained in

F^*, then put $F_1 = F(\mu_n)$. By the same arguments, $F_1^{*n} = N_{L/F_1}L^*$ for some finite abelian extension L/F_1. Then $N_{L/F}L^* \subset F^{*n} \subset N$.

Assume now that char$(F) = p$. We will show by induction on $m \geqslant 1$ that any open subgroup N of index p^m in F^* contains a norm subgroup. The arguments of (5.7) show that this is true for $m = 1$. Let $m > 1$, and let N_1 be an open subgroup of index p^{m-1} in F^* such that $N \subset N_1$. By the induction assumption, $N_1 \supset N_{L_1/F}L_1^*$. The subgroup $N \cap N_{L_1/F}L_1^*$ is of index 1 or p in $N_{L_1/F}L_1^*$. In the first case $N \supset N_{L_1/F}L_1^*$, and in the second case let L/L_1 be a finite separable extension with $N_{L_1/F}^{-1}\left(N \cap N_{L_1/F}L_1^*\right) \supset N_{L/L_1}L^*$; then $N \supset N_{L/F}L^*$.

For an open subgroup N of index np^m in F^* with $p\nmid n$ we now take open subgroups N_1 and N_2 of indices n and p^m, respectively, in F^* such that $N \subset N_1, N_2$. Then $N = N_1 \cap N_2 \supset N_{L_1/F}L_1^* \cap N_{L_2/F}L_2^* \supset N_{L_1L_2/F}(L_1L_2)^*$ and we have proved the desired assertion for N.

Finally, Proposition (4.1) implies all remaining assertions. □

COROLLARY.

(1) *The reciprocity map Ψ_F is injective and continuous.*
(2) *For $n \geqslant 0$ it maps $U_{n,F}$ isomorphically onto $G(n)$, where $G = \mathrm{Gal}(F^{\mathrm{ab}}/F)$.*
(3) *Every abelian extension with finite residue field extension is arithmetically profinite.*
(4) *Every abelian extension has integer upper ramification jumps.*

Proof.
(1) By Theorem (4.2) the preimage $\Psi_F^{-1}(G)$ of an open subgroup G of the group $\mathrm{Gal}(F^{\mathrm{ab}}/F)$ coincides with $N_{L/F}L^*$, where L is the fixed field of G. Hence, $\Psi_F^{-1}(G)$ is open and Ψ_F is continuous. Since the intersection of all norm subgroups coincides with the intersection of all open subgroups of finite index in F^*, we conclude that Ψ_F is injective.
(2) By Theorem (3.5) $\Psi_{L/F}(U_{n,F}N_{L/F}L^*) = \mathrm{Gal}(L/F)(n)$ for every finite abelian extension L/F. We deduce that $\Psi_F(U_{n,F})$ is a dense subset of $G(n)$. Since in our case $U_{n,F}$ is compact, we conclude that $\Psi_F(U_{n,F}) = G(n)$.
(3) Due to the definition of the upper ramification filtration in (3.5) Ch. III for an abelian extension L/F we know that $\mathrm{Gal}(L/F)(n)$ is the image of $G(n)$ in $\mathrm{Gal}(L/F)$. Since every $G(n)$ has finite index in $G(0)$ by (2), we deduce that every $\mathrm{Gal}(L/F)(x)$ has finite index in $\mathrm{Gal}(L/F)$. Thus, L/F is arithmetically profinite by Remark 1 in (5.1) Ch. III.
(4) For an upper ramification jump x of L/F from (3) we know that $\mathrm{Gal}(L/F)(x+1)$ is an open subgroup of $\mathrm{Gal}(L/F)$. Therefore, the fixed field E of $\mathrm{Gal}(L/F)(x+1)$ is a finite abelian extension of F. The jump x corresponds to the jump x of $\mathrm{Gal}(E/F)$ and therefore is integer by Theorem (3.5). □

REMARKS.

1. Lemma (1.4) implies that one may omit the word "open" in the Theorem if $\mathrm{char}(F) = 0$, but not if $\mathrm{char}(F) \neq 0$.

2. Theorems (3.3) and (6.2) can be reformulated as the existence of a canonical isomorphism between the group $X\big(\mathrm{Gal}(F^{\mathrm{sep}}/F)\big)$ of all continuous characters of the profinite group $\mathrm{Gal}(F^{\mathrm{sep}}/F)$ and the group $X(F^*)$ of all continuous characters of finite order of the abelian group F^*. As a generalization, a part of the local *Langlands programme* predicts existence of a certain bijection, satisfying some properties, between isomorphism classes of complex irreducible smooth representations of $GL_n(F)$ and isomorphism classes of complex n-dimensional semi-simple *Weil–Deligne* representations of the so called *Weil* group (closely related to $\mathrm{Gal}(F^{\mathrm{sep}}/F)$). This approach is often called a nonabelian class field theory. For introductory texts to the programme see Bibliography. Efforts of many mathematicians culminated in two proofs of this part of the Langlands programme by *G. Henniart* [Henn3] and *M. Harris–R. Taylor* [HT] in characteristic zero and in positive characteristic by *G. Laumon–M. Rapoport–U. Stuhler* [LRS], *L. Lafforgue* [L]. The proofs are quite difficult, and it is likely to take many years before the subject reaches the state of relative completion. In section 8 one can find an arithmetically oriented noncommutative generalization of the local reciprocity map.

DEFINITION. The field L, which is an abelian extension of finite degree over F, with the property $N_{L/F}L^* = N$ is called the *class field* of the subgroup $N \subset F^*$.

(6.3). Now we will generalize Theorem (6.2) for abelian (not necessarily finite) extensions of F. For an abelian extension L/F, we put

$$N_{L/F}L^* = \bigcap_M N_{M/F}M^*,$$

where M runs through all finite subextensions of F in L. Then the norm subgroup $N_{L/F}L^*$, as the intersection of closed subgroups, is closed in F^*. Theorem (4.2) implies that $N_{L/F}L^* = \bigcap_M \Psi_F^{-1}\big(\mathrm{Gal}(F^{\mathrm{ab}}/M)\big) = \Psi_F^{-1}\big(\mathrm{Gal}(F^{\mathrm{ab}}/L)\big)$. Moreover, for a closed subgroup N in F^* denote the topological closure of $\Psi_F(N)$ in $\mathrm{Gal}(F^{\mathrm{ab}}/F)$ by $G_{(N)}$. In other words, $G_{(N)}$ coincides with the intersection of all open subgroups H in $\mathrm{Gal}(F^{\mathrm{ab}}/F)$ with $H \supset \Psi_F(N)$. If an element $\alpha \in F^*$ belongs to $\Psi_F^{-1}(G_{(N)})$, then the automorphism $\Psi_F(\alpha)$ acts trivially on the fixed field of an open subgroup H with $H \supset \Psi_F(N)$. From Theorem (4.2) we deduce that $\alpha \in \bigcap_M N_{M/F}M^*$, where M corresponds to H. We conclude that $N = N_{L/F}L^*$ for the fixed field L of $G_{(N)}$ (or of $\Psi_F(N)$).

THEOREM. *The correspondence $L \to N_{L/F}L^*$ is an order reversing bijection between the lattice of abelian extensions of F and the lattice of closed subgroups in F^*. The quotient group $F^*/N_{L/F}L^*$ is isomorphic to a dense subgroup in* $\mathrm{Gal}(L/F)$.

Proof. It remains to use the injectivity of Ψ_F and the arguments in the proof of Proposition (4.2) and Theorem (6.2) (replacing the word "open" by "closed"). □

(6.4). Let L/F be a finite abelian extension, and L_0 be the maximal unramified subextension of F in L. Theorem (4.2) shows that $\Psi_F(U_F)|_L \subset \mathrm{Gal}(L/L_0)$. Conversely, if $\sigma \in \mathrm{Gal}(L/L_0)$ and $\sigma = \Psi_F(\alpha)|_L$ for $\alpha \in F^*$, then Theorem (4.2) implies that $v_F(\alpha) = 0$, i.e., $\alpha \in U_F$. Hence $\Psi_F(U_F)|_L = \mathrm{Gal}(L/L_0)$. The extension L^{ur}/F is abelian, and we similarly deduce that $\Psi_F(U_F)|_{L^{\mathrm{ur}}} = \mathrm{Gal}(L^{\mathrm{ur}}/F^{\mathrm{ur}})$. Since U_F is compact and Ψ_F is continuous, the group $\Psi_F(U_F)$ is closed and equal to $\mathrm{Gal}(F^{\mathrm{ab}}/F^{\mathrm{ur}})$.

Let π be a prime element in F and $\Psi_F(\pi) = \varphi$. Then $\varphi|_{F^{\mathrm{ur}}} = \varphi_F$, and for the fixed field F_π of φ we get

$$F_\pi \cap F^{\mathrm{ur}} = F, \quad F_\pi F^{\mathrm{ur}} = F^{\mathrm{ab}}$$

(the second equality can be deduced by the same arguments as in the proof of Proposition (2.1). The prime element π belongs to the norm group of every finite subextension L/F of F_π/F. The group $\mathrm{Gal}(F^{\mathrm{ab}}/F_\pi)$ is mapped isomorphically onto $\mathrm{Gal}(F^{\mathrm{ur}}/F)$) and the group $\mathrm{Gal}(F_\pi/F)$ is isomorphic $\mathrm{Gal}(F^{\mathrm{ab}}/F^{\mathrm{ur}})$. The latter group is often denoted by I_F and called the *inertia subgroup* of $G_F^{\mathrm{ab}} = \mathrm{Gal}(F^{\mathrm{ab}}/F)$.

We have

$$\mathrm{Gal}(F^{\mathrm{ab}}/F) \simeq \mathrm{Gal}(F_\pi/F) \times \mathrm{Gal}(F^{\mathrm{ur}}/F), \qquad \mathrm{Gal}(F_\pi/F) \simeq U_F, \mathrm{Gal}(F^{\mathrm{ur}}/F) \simeq \widehat{\mathbb{Z}}$$

and

$$\Psi_F(F^*) = \langle \varphi \rangle \times \mathrm{Gal}(F^{\mathrm{ab}}/F^{\mathrm{ur}}),$$

where $\langle \varphi \rangle$ is the cyclic group generated by φ. We observe that the distinction between F^* and $\mathrm{Gal}(F^{\mathrm{ab}}/F)$ is the same as that between $\mathbb{Z}$ and $\widehat{\mathbb{Z}}$. So if we define the group $\widehat{F}^*$ as $\varprojlim F^*/U$ where U runs over all open subgroups of finite index in F^*, then $\widehat{F}^* = U_F \times \widehat{\mathbb{Z}}$ and the reciprocity map Ψ_F extends to the isomorphism (and homeomorphism of topological spaces)

$$\widehat{\Psi}_F : \widehat{F}^* \longrightarrow \mathrm{Gal}(F^{\mathrm{ab}}/F) = G_F^{\mathrm{ab}}.$$

Define

$$\Upsilon_F = \widehat{\Psi}_F^{-1} : \mathrm{Gal}(F^{\mathrm{ab}}/F) \longrightarrow \widehat{F}^*.$$

Then Υ_F maps I_F homeomorphically onto U_F.

The field F_π can be explicitly generated by roots of iterated powers of the isogeny of a formal Lubin–Tate group associated to π. For this and other properties of F_π see Exercise 6 of this section and Exercises 5–7 section 1 Ch. VIII.

(6.5). Choose a prime element π. Then the surjective homomorphism $\mathrm{Gal}(F^{\mathrm{ab}}/F) \to \mathrm{Gal}(F_\pi/F)$ induces the epimorphism $G_F \to I_F$. Its composition with the restriction of the reciprocity homomorphism $\Upsilon_F: I_F \longrightarrow U_F$ defined in the previous subsection is a surjective homomorphism $\Phi_F = \Phi_{F,\pi}: G_F \longrightarrow U_F$. Certainly, $\Phi_{F,\pi}$ is just a modification of Υ_F: $\varphi = \Psi(\pi)$ instead of being sent to π is sent to 1, and $\Phi_{F,\pi}|_{I_F} = \Upsilon_F|_{I_F}$.

The homomorphism Φ_F can be viewed as an element of the group $H^1_c(G_F, F^*)$ of continuous cochains from G_F to F^* modulo coborders. Extend the target group F^* replacing it with the multiplicative group C^* of the completion C of F^{sep} with respect to the valuation on F^{sep}.

Now assume that F is of characteristic zero. *J. Tate* proved [T2] that the group $H^1_c(G_F, C^*)$ is isomorphic to $H^1_c(G(E/F), E^*)$ where E/F is any abelian extension with finite residue field extension and $\mathrm{Gal}(E/F) \simeq \mathbb{Z}_p$. He proved that $H^0_c(G_F, C) = F$ (for a simpler proof see [Ax]) and that $H^1_c(G_F, C)$ is a one-dimensional vector space over F generated by the class of $\log \circ \Phi_F: G_F \to F$. His work shows that if F is a finite Galois extension of $\mathbb{Q}_p$ and τ is a nontrivial element of $\mathrm{Gal}(F/\mathbb{Q}_p)$, then there is a non-zero element α_τ in the completion of F_π such that

$$\Upsilon_F(\sigma) = (\alpha_\tau^{\sigma-1})^\tau \qquad \text{for every } \sigma \in I_F \ .$$

For another proof which uses differential forms see [Fo3], see also Exercise 9. For a direct proof of the assertions of this paragraph see Exercise 8.

(6.6). Consider some generalizations of local class field theory (see also section 8 and the next chapter).

EXAMPLE 1. The completion $\mathcal{F} = \widehat{F^{\mathrm{ur}}}$ is a local field with the residue field $\mathbb{F}_q^{\mathrm{sep}}$, which is algebraically closed. As F^{ur} is Henselian, Theorem (2.8) Ch. II implies that the group $\mathrm{Gal}\big((F^{\mathrm{ur}})^{\mathrm{ab}}/F^{\mathrm{ur}}\big)$ is embedded isomorphically onto the group $\mathrm{Gal}\big((\widehat{F^{\mathrm{ur}}})^{\mathrm{ab}}/\widehat{F^{\mathrm{ur}}}\big)$. Let π be prime in F. Proposition (4.2) Ch. II and (6.4) show that the former group can be identified with the projective limit $\varprojlim \mathrm{Gal}(F_{n,\pi}/F_n)$ where F_n is the unramified extension of F of degree n. The preceding considerations and Theorem (4.2) now imply the existence of the isomorphism

$$\Psi_{\mathcal{F}}: \varprojlim U_{F_n} \longrightarrow \mathrm{Gal}(\mathcal{F}^{\mathrm{ab}}/\mathcal{F}),$$

where the projective limit is taken with respect to the norm maps. For $\mathcal{U}_{\mathcal{F}} = \varprojlim U_{F_n}$ and for a finite separable extension $\mathcal{L}/\mathcal{F}$ one can introduce the norm map $\mathcal{N}_{\mathcal{L}/\mathcal{F}}: \mathcal{U}_{\mathcal{L}} \to \mathcal{U}_{\mathcal{F}}$. For a finite abelian extension $\mathcal{L}/\mathcal{F}$

$$\mathcal{N}_{\mathcal{L}/\mathcal{F}}\mathcal{U}_{\mathcal{L}} = \Psi_{\mathcal{F}}^{-1}\big(\mathrm{Gal}(\mathcal{F}^{\mathrm{ab}}/\mathcal{L})\big), \qquad \mathcal{U}_{\mathcal{F}}/\mathcal{N}_{\mathcal{L}/\mathcal{F}}\mathcal{U}_{\mathcal{L}} \simeq \mathrm{Gal}(\mathcal{L}/\mathcal{F}).$$

Moreover, open subgroups in $\mathcal{U}_{\mathcal{F}}$ are in one-to-one correspondence with finite abelian extensions.

In the general case of a local field $\mathcal{F}$ with algebraically closed residue field k *J.-P. Serre*'s *geometric class field theory* describes the group $\mathrm{Gal}(\mathcal{F}^{\mathrm{ab}}/\mathcal{F})$ via the fundamental group $\pi_1(U_{\mathcal{F}})$ of $U_{\mathcal{F}}$ viewed as a proalgebraic group over k. For a finite Galois extension $\mathcal{L}/\mathcal{F}$ there is an exact sequence

$$\dots \to \pi_1(U_{\mathcal{L}}) \xrightarrow{N_{\mathcal{L}/\mathcal{F}}} \pi_1(U_{\mathcal{F}}) \xrightarrow{\partial} \pi_0(V_{\mathcal{L}}) \to \pi_0(U_{\mathcal{L}}) \to \dots$$

where $V_{\mathcal{L}}$ is the kernel of the norm map $N_{\mathcal{L}/\mathcal{F}}: U_{\mathcal{L}} \to U_{\mathcal{F}}$. Since $U_{\mathcal{L}}$ is connected and the connected component of $V_{\mathcal{L}}$ is $U(\mathcal{L}/\mathcal{F})$ defined in (1.7), Proposition (1.7) and the previous sequence induce the reciprocity map

$$\pi_1(U_{\mathcal{F}})/N_{\mathcal{L}/\mathcal{F}}\pi_1(U_{\mathcal{L}}) \to \mathrm{Gal}(\mathcal{L}/\mathcal{F})^{\mathrm{ab}}.$$

One shows that the corresponding reciprocity map $\pi_1(U_{\mathcal{F}}) \to \mathrm{Gal}(\mathcal{F}^{\mathrm{ab}}/\mathcal{F})$ is an isomorphism [Se2].

This theory can be also deduced from the approach discussed above for the field $\mathcal{F}$ with residue field $\mathbb{F}_q^{\mathrm{sep}}$, and for the general case see Exercise 4 section 3 of the next chapter.

EXAMPLE 2. Let $\mathcal{F}$ be an infinite separable extension of a complete discrete valuation field F with residue field $\mathbb{F}_q$. Put $\mathcal{F}^\times = \varprojlim M^*$, where M runs all finite subextensions of F in $\mathcal{F}$ and the projective limit is taken with respect to the norm maps. Assume that the residue field of $\mathcal{F}$ is finite. Then for an element $\mathrm{A} = (\alpha_M) \in \mathcal{F}^\times$ we put $v(\mathrm{A}) = v_M(\alpha_M)$ for M containing $\mathcal{F} \cap F^{\mathrm{ur}}$; v is a homomorphism of $\mathcal{F}^\times$ onto $\mathbb{Z}$. If $\mathcal{L}/\mathcal{F}$ is a finite separable extension, then it is a straightforward exercise to define the norm map $\mathcal{N}_{\mathcal{L}/\mathcal{F}}: \mathcal{L}^\times \to \mathcal{F}^\times$. It can be shown that $v(\mathcal{N}_{\mathcal{L}/\mathcal{F}}\mathcal{L}^\times) = f(\mathcal{L}/\mathcal{F})\mathbb{Z}$. If $\mathcal{L}/\mathcal{F}$ is a finite Galois extension, then $\mathrm{Gal}(\mathcal{L}/\mathcal{F})$ acts on $\mathcal{L}^\times$, and the set of fixed elements with respect to this action coincides with $\mathcal{F}^\times$. One can verify the assertions analogous to those of sections 2–4 and show that there is the isomorphism

$$\Upsilon_{\mathcal{L}/\mathcal{F}}: \mathrm{Gal}(\mathcal{L}/\mathcal{F})^{\mathrm{ab}} \longrightarrow \mathcal{F}^\times/\mathcal{N}_{\mathcal{L}/\mathcal{F}}\mathcal{L}^\times$$

(for more details see [Sch], [Kaw2], [N3, Ch. II, section 5]).

In the particular case of arithmetically profinite extension $\mathcal{F}/F$, the group $\mathcal{F}^\times$ is identified with $N(\mathcal{F}|F)^*$, and $\mathrm{Gal}(\mathcal{L}/\mathcal{F})^{\mathrm{ab}}$ is identified with $\mathrm{Gal}\big(N(\mathcal{L}|F)/N(\mathcal{F}|F)\big)^{\mathrm{ab}}$. We obtain isomorphisms $\mathrm{Gal}\big(N(\mathcal{L}|F)/N(\mathcal{F}|F)\big)^{\mathrm{ab}} \xrightarrow{\sim} \mathrm{Gal}(\mathcal{L}/\mathcal{F})^{\mathrm{ab}} \xrightarrow{\sim} \mathcal{F}^\times/\mathcal{N}_{\mathcal{L}/\mathcal{F}}\mathcal{L}^\times \xrightarrow{\sim} N(\mathcal{F}|F)/N_{N(\mathcal{L}|F)/N(\mathcal{F}|F)}N(\mathcal{L}|F)^*$. Thus, the reciprocity $\Upsilon_{\mathcal{L}/\mathcal{F}}$ in characteristic p or zero is connected with the reciprocity map $\Upsilon_{N(\mathcal{L}|F)/N(\mathcal{F}|F)}$ in characteristic p. See also Exercise 7.

Exercises.

1. Show that the map $(\cdot,\cdot]_n: F^* \times W_n(F) \to W_n(\mathbb{F}_p)$ (see Exercise 6 section 5) is continuous with respect to the discrete topologies on $W_n(F)$, $W_n(\mathbb{F}_p)$.
2. Prove Theorem (6.2) using Artin–Schreier extensions and the considerations of (1.5) instead of the pairings of F^* of section 5.

3. By using Exercise 2a) section 4 find another proof of the local Kronecker–Weber Theorem, different from the proof in Exercise 6 section 1. Show that the assertion the theorem does not hold if $\mathbb{Q}_p$ is replaced by $\mathbb{Q}_p(\zeta_p)$.
4. Show that the closed subgroups in U_F are in one-to-one correspondence with the abelian extensions L/F such that $F^{\text{ur}} \subset L$.
5. Prove the existence Theorem for a Henselian discrete valuation field of characteristic 0 with finite residue field (see Exercise 4 section 4 and Exercise 9 section 5). For the case of characteristic p see p. 160 of [Mi].
6. A field $E \subset F^{\text{ab}}$ is said to be a frame field if $E \cap F^{\text{ur}} = F$ and $EF^{\text{ur}} = F^{\text{ab}}$.
   a) Show that $\text{Gal}(F^{\text{ab}}/F) \simeq \text{Gal}(F^{\text{ab}}/E) \times \text{Gal}(F^{\text{ab}}/F^{\text{ur}})$ for a frame field E.
   b) Let $\varphi \in \text{Gal}(F^{\text{ab}}/F)$ be an extension of the Frobenius automorphism φ_F. Show that the fixed field E_φ of φ is a frame field and that the correspondence $\varphi \to E_\varphi$ is a one-to-one correspondence between extensions of φ_F and frame fields.
   c) Show that the correspondence $\pi \to \Psi_F(\pi)$ is a one-to-one correspondence between prime elements in F and extensions of φ_F. Therefore, the correspondence $\pi \to F_\pi$ is a one-to-one correspondence between prime elements in F and frame fields.
   d) $\pi \in N_{L/F}L^*$ for some prime element π if and only if $L \cap F^{\text{ur}} = F$.

   Further information on the field F_π can be deduced using Lubin–Tate formal groups, see Exercises 5–7 section 1 Ch. VIII.
7. ($\diamond$) Let F be a local field with finite residue field, and let L be a totally ramified infinite arithmetically profinite extension of F. Let $N = N(L|F)$. Show that there is a homomorphism $\Psi\colon N^* \to \text{Gal}(L^{\text{ab}}/L)$ induced by the reciprocity maps $\Psi_E\colon E^* \mapsto \text{Gal}(E^{\text{ab}}/E)$ for finite subextensions E/F in L/F. Show that $\chi \circ \Psi = \Psi_N$, where the homomorphism $\chi\colon \text{Gal}(L^{\text{ab}}/L) \to \text{Gal}(N^{\text{ab}}/N)$ is defined similarly to the homomorphism $\tau \mapsto \mathrm{T}$ of (5.6) Ch. III. For further details see [Lau4].
8. ($\diamond$) Let F be of characteristic zero. Let E/F be a totally ramified Galois extension with the group isomorphic to $\widehat{\mathbb{Z}}$. Let σ be a generator of $\text{Gal}(E/F)$. Denote by $\widehat{E}$ the completion of E. Denote by E_n the subextension of degree p^n over F.
   a) Let $\varepsilon \in N_{E/F}U_E$. Using properties of the Hasse–Herbrand function and Exercise 2 section 5 Ch. III show that there exist $\eta_n \in E_n^*$ such that $\varepsilon^{p^n} = N_{E_n/F}\eta_n$ and $v_{E_n}(\eta_n - 1) \geqslant p^{n-n_0}e(F|\mathbb{Q}_p)(n - n_0)$ for some n_0 and all sufficiently large n. Deduce that η_n tends to 1 when n tends to infinity. Write $\varepsilon = \eta_n\gamma_n^{\sigma-1}$ with $\gamma_n \in E_n^*$. Show that the limit γ of γ_n exists in $\widehat{E}^*$ and $\varepsilon = \gamma^{\sigma-1}$. Deduce that $N_{E/F}U_E \subset F^* \cap \widehat{E}^{*\sigma-1}$.
   b) Using a) show that the operator $N_{E_n/F}\colon \widehat{E}^* \to \widehat{E}^*$ is surjective. Then using the description of the reciprocity map in section 2 show that every $\varepsilon \in F^* \cap \widehat{E}^{*\sigma-1}$ belongs to $N_{E/F}U_E$.
   c) Deduce that $N_{E/F}U_E = F^* \cap \widehat{E}^{*\sigma-1}$ and $\widehat{E}^{*\sigma-1}$ coincides with the closure of $\cup_n N_{E/E_n}E^*$.
   d) Deduce from c) that if $F/\mathbb{Q}_p$ is a finite Galois extension and τ is a nontrivial element of $\text{Gal}(F/\mathbb{Q}_p)$, then there is a non-zero element α_τ in the completion of F_π such that $\tau^{-1}\Upsilon_F(\sigma) = \alpha_\tau^{\sigma-1}$ for every $\sigma \in I_F$.
   e) Show that the class of Φ_F in $H^1_c(G_F, C^*)$ is nontrivial.

Now, using the logarithm and two linear algebra–Galois theory exercises in [S6, Exercises 1–2 Appendix to Ch. III] one easily deduces avoiding *Hodge–Tate theory* that $H_c^1(G_F, C)$ is a one-dimensional vector space over F generated by the class of $\log \circ \Phi_F\colon G_F \to F$. It follows from c) that the class of $\log \circ \Phi_{F,\pi}\colon G_F \to F$ in $H_c^1(G_F, C)$ coincides with the class of $\log \circ \Phi_{\mathbb{Q}_p, N_{F/\mathbb{Q}_p}\pi}\colon G_F \to F$.

9. ($\diamond$) (*J.-M. Fontaine* [Fo3]) Let F be a local field of characteristic zero. Let v be the valuation on C normalized by $v(\pi_F) = 1$ where π_F is a prime element of F. Denote by Ω the module of relative differential forms $\Omega_{\mathcal{O}_{F^{\mathrm{sep}}}/\mathcal{O}_F}$. For a G_F-module M put $T_p(M) = \varprojlim_n {}_{p^n}M$ where ${}_{p^n}M$ stands for the p^n-torsion of M. For example, $T_p(F^{\mathrm{sep}*})$ is a free $\mathbb{Z}_p$-module of rank 1 with generator ζ and G_F-action given by $\sigma(\zeta) = \chi(\sigma)\zeta$ where $\chi\colon G_F \to \mathbb{Z}_p^*$ is the so called cyclotomic character: choose for every n a primitive p^n th root ζ_{p^n} of unity such that $\zeta_{p^{n+1}}^p = \zeta_{p^n}$, then $\sigma(\zeta_{p^n}) = \zeta_{p^n}^{\chi(\sigma)}$. Denote $M(1) = M \otimes_{\mathbb{Z}_p} T_p(F^{\mathrm{sep}*})$.
   a) Show that $\Omega_{\mathcal{O}_L/\mathcal{O}_F} = \mathcal{O}_L d\pi_L$ for a finite extension L/F, where π_L is a prime element of L. Denote by d_F the non-negative integer such that the ideal $\{\alpha \in \mathcal{O}_F : \alpha d\pi_F = 0 \text{ in } \Omega_{\mathcal{O}_F/\mathbb{Z}_p}\}$ is equal to $\mathcal{M}_F^{d_F}$.
   b) Show that if E/F is a subextension of a finite extension L/F, then the sequence
   $$0 \to \Omega_{\mathcal{O}_E/\mathcal{O}_F} \otimes_{\mathcal{O}_E} \mathcal{O}_L \to \Omega_{\mathcal{O}_L/\mathcal{O}_F} \to \Omega_{\mathcal{O}_L/\mathcal{O}_E} \to 0$$
   is exact.
   c) Define $g\colon F^{\mathrm{sep}}(1) \longrightarrow \Omega$, $\quad \alpha/p^n \otimes \zeta \mapsto \alpha \dfrac{d\zeta_{p^n}}{\zeta_{p^n}}$ where $\alpha \in \mathcal{O}_{F^{\mathrm{sep}}}$. Show that this map is a well defined surjective G_F-homomorphism. Show that its kernel equals to $A(1)$ where
   $$A = \{\alpha \in F^{\mathrm{sep}} : v(\alpha\pi_F^{1/(p-1)}) + d_F/e(F|\mathbb{Q}_p) \geqslant 0\}.$$
   d) Deduce that there is an isomorphism of G_F-modules ${}_{p^n}\Omega \simeq (A/p^n A)(1)$ and
   $$T_p(\Omega) \otimes_{\mathbb{Z}_p} \mathbb{Q}_p \simeq C(1).$$

10. Let M be the maximal abelian extension of the maximal abelian extension of the maximal abelian extension of F. Show, using the notations of Exercise 3 sect. 3 Ch. III, that $B(M/F)$ is dense in $[0, +\infty)$ and deduce that every nonnegative real number is an upper ramification jump of M/F. Therefore, every nonnegative real number is an upper ramification jump of F^{sep}/F.

7. Other approaches to the local reciprocity map

In this section we just briefly review other approaches to local class field theory.

We keep the conventions on F.

(7.1). The approaches of Hazewinkel and Neukirch for local fields with finite residue field can be developed without using each other, see [Haz1–2], [Iw5], [N4–5]; but each of them has to go through some "unpleasant" lemmas.

In characteristic p there is a very elegant elementary approach by *Y. Kawada* and *I. Satake* [KwS] which employs Artin–Schreier–Witt theory, see Exercise 8 section 5.

(7.2). The maximal abelian totally ramified extension of $\mathbb{Q}_p$ coincides with $\mathbb{Q}_p(\mu_{p^\infty})$ where μ_{p^∞} is the group of all roots of order a power of p (see Exercise 3 section 6). By using formal Lubin–Tate groups associated to a prime element π one can similarly construct the field F_π of (6.5). Due to explicit results on the extensions generated by roots of iterated powers of the isogeny of the formal group (see Exercises 5–7 section 1 Ch. VIII), one can develop an explicit class field theory for local fields with finite residue field, see for instance [Iw6]. Disadvantage of this approach is that it is not apparently generalizable to local fields with infinite residue field.

(7.3). All other approaches prove and use the fact (or its equivalent) that for the *Brauer group* of a local field F there is a (canonical) isomorphism

$$\operatorname{inv}_F\colon \operatorname{Br}(F) \longrightarrow \mathbb{Q}/\mathbb{Z}.$$

Historically this is the first approach [Schm], [Ch1].

Recall that the Brauer group of a field K is the group of equivalence classes of central simple algebras over K. A finite dimensional algebra A over K is called central simple if there exists a finite Galois extension L/K such that the algebra viewed over L isomorphic to a matrix algebra over L (in this case A is said to split over L). A central simple algebra A over K is isomorphic to $M_m(D)$ where D is a division algebra with centre K, $m \geqslant 1$. Two central simple algebras A, A' are said to be equivalent if the associated division algebras are isomorphic over K. The group structure of $\operatorname{Br}(K)$ is given by the class of the tensor product of representatives.

A standard way to prove the assertion about $\operatorname{Br}(F)$ is the show that every central simple algebra over F splits over some finite unramified extension of F, and then using $\operatorname{Gal}(F^{\mathrm{ur}}/F) \simeq \operatorname{Gal}(\mathbb{F}_q^{\mathrm{sep}}/\mathbb{F}_q)$ reduce the calculation to the fact that the group of continuous characters $\mathrm{X}_{\mathbb{F}_q}$ of $G_{\mathbb{F}_q}$ is canonically (due to the canonical Frobenius automorphism) isomorphic with $\mathbb{Q}/\mathbb{Z}$. For proofs of the existence of the isomorphism inv_F see for instance [W, Ch. XII] or a cohomological calculation in [Se3, Ch. XII] or a review of the latter in [Iw6, Appendix].

Now let a character $\chi \in \mathrm{X}_F = \operatorname{Hom}_c(G_F, \mathbb{Q}/\mathbb{Z})$ correspond to a cyclic extension L/F of degree n with generator σ such that $\chi(\sigma) = 1/n$. For every element $\alpha \in F^*$ there is a so called cyclic algebra $A_{\alpha,\chi}$ defined as $\oplus_{i=0}^{n-1} L\beta^i$ where $\beta^n = \alpha$, $a\beta = \beta \cdot \sigma(a)$ for every $a \in L$. We have a pairing

$$F^* \times \mathrm{X}_F \longrightarrow \mathbb{Q}/\mathbb{Z}, \quad (\alpha, \chi) \mapsto \operatorname{inv}_F([A_{\alpha,\chi}]).$$

This pairing induces then a homomorphism

$$F^* \longrightarrow \mathrm{Gal}(F^{\mathrm{ab}}/F) = \mathrm{Hom}(\mathrm{X}_F, \mathbb{Q}/\mathbb{Z}).$$

Then one proves that this homomorphism possesses all nice properties, i.e. establishes local class field theory.

If the field F contains a primitive n th root of unity, then Kummer theory supplies a homomorphism from F^*/F^{*n} to the n-torsion subgroup ${}_n\mathrm{X}_F$ and the resulting pairing $F^*/F^{*n} \times F^*/F^{*n} \to \frac{1}{n}\mathbb{Z}/\mathbb{Z}$ after identifications coincides with the Hilbert symbol in (5.1)–(5.3). Similarly, if F is of characteristic p then Artin–Schreier theory supplies a homomorphism $F/\wp(F) \to {}_p\mathrm{X}_F$ which then induces a pairing $F^*/F^{*p} \times F/\wp(F) \to \frac{1}{p}\mathbb{Z}/\mathbb{Z}$ which after identifications coincides with the pairing of (5.4)–(5.5).

The just described approach does not require cohomological tools and was known before the invention of those.

(7.4). Using cohomology groups one can perhaps simplify the proofs in the approach described in (7.3). From our point of view the exposition of class field theory for local fields with finite residue field given in this chapter is the most appropriate for a beginner; at a later stage the cohomological approach can be mastered. The real disadvantage of the cohomological approach is its unexplicitness whereas the approach in this chapter in addition to quite an explicit nature can be easily extended to many other situations.

If L is a finite Galois extension of F then one has an exact sequence

$$1 \to H^2(\mathrm{Gal}(L/F), L^*) \to \mathrm{Br}(F) \to \mathrm{Br}(L) \to 1$$

and $\mathrm{Br}(F)$ is the union of classes of algebras which split over L (i.e. the image of all $H^2(\mathrm{Gal}(L/F), L^*)$ for all finite Galois extensions L/F). So inv_F induces a canonical isomorphism

$$\mathrm{inv}_{L/F}\colon H^2(\mathrm{Gal}(L/F), L^*) \xrightarrow{\sim} \frac{1}{|L:F|}\mathbb{Z}/\mathbb{Z}.$$

Denote the element which is mapped to $1/|L:F|$ by $u_{L/F}$. If $\widehat{H}^r$ stands for the modified Tate's cohomology group, see [Se3, sect. 1 Ch. VIII], then the cup product with $u_{L/F}$ induces an isomorphism

$$\widehat{H}^r(\mathrm{Gal}(L/F), \mathbb{Z}) \xrightarrow{\sim} \widehat{H}^{r+2}(\mathrm{Gal}(L/F), L^*).$$

For $r = 0$ we have

$$\mathrm{Gal}(L/F)^{\mathrm{ab}} = \widehat{H}^0(\mathrm{Gal}(L/F), \mathbb{Z}) \xrightarrow{\sim} \widehat{H}^2(\mathrm{Gal}(L/F), L^*) = F^*/N_{L/F}L^*$$

which leads to the analog of Theorems (3.3) and (4.2). Certainly the last isomorphism in much more explicit form is $\Upsilon^{\mathrm{ab}}_{L/F}$.

Using cohomology groups one can interpret the pairing $F^* \times \mathrm{X}_F \longrightarrow \mathbb{Q}/\mathbb{Z}$ of the previous subsection as arising from the cup product

$$H^0(\mathrm{Gal}(L/F), L^*) \times H^2(\mathrm{Gal}(L/F), \mathbb{Z}) \to H^2(\mathrm{Gal}(L/F), L^*)$$

and the border homomorphism $H^1(\mathrm{Gal}(L/F), \mathbb{Q}/\mathbb{Z}) \to H^2(\mathrm{Gal}(L/F), \mathbb{Z})$ associated to the exact sequence

$$0 \to \mathbb{Z} \to \mathbb{Q} \to \mathbb{Q}/\mathbb{Z} \to 0.$$

For a field K one can try to axiomatize those properties of its cohomology groups which are sufficient to get a reciprocity map from K^* to G_K^{ab}, as it is well known this leads to the notion of class formation, see for example [Se3, Ch. XI].

(7.5). Assume that F is of characteristic zero with finite residue field of characteristic p. For $n \geqslant 1$ the p^n-component of the pairing $F^* \times \mathrm{X}_F \longrightarrow \mathbb{Q}/\mathbb{Z}$ defined in (7.3) is a pairing

$$H^1(G_F, \mu_{p^n}) \times H^1(G_F, \mathbb{Z}/p^n\mathbb{Z}) \to H^2(G_F, \mu_{p^n}).$$

If for every n one knows that this pairing is a perfect pairing, and the right hand side is a cyclic group of order p^n, then one deduces the p-part of class field theory of the field F.

More generally, for a finitely generated $\mathbb{Z}_p$-module M equipped with the action of G_F and annihilated by p^n define $M^*(1) = \mathrm{Hom}(M, \mu_{p^n})$. The previous pairing can be generalized to the pairing given by the cup product

$$H^i(G_F, M) \times H^{2-i}(G_F, M^*(1)) \to H^2(G_F, \mu_{p^n}).$$

By *Tate local duality* it is a perfect pairing of finite groups. So, if one can establish Tate local duality independently of local class field theory, then one obtains another approach to the p-part of local class field theory in characteristic zero.

J.-M. Fontaine's theory of $\Phi - \Gamma$-modules [Fo5] was used by *L. Herr* to relate $H^i(G_F, M)$ with cohomology groups of a simple complex of $\Phi - \Gamma$-modules.

Namely, let L be the cyclotomic $\mathbb{Z}_p$-extension of F, i.e. the only subfield of $F(\mu_{p^\infty})$ such that $\mathrm{Gal}(L/F) \simeq \mathbb{Z}_p$. It follows from Exercise 2 section 5 Ch. III that the Hasse–Herbrand function of L/F grows sufficiently fast as in Exercise 7 of the same section, so we have a continuous field homomorphism $N(L|F) \longrightarrow R = R(C)$ where C and $R(C)$ are defined in the same exercise. Denote by $X \in W(R)$ the multiplicative representative in $W(R)$ of the image in R of a prime element of $N(L|F)$. We have also a continuous ring homomorphism $W(\overline{F}) \longrightarrow W(R)$, denote by W its image. The action of elements of G_F is naturally extended on $W(R)$. One can show that the ring $O_L = W\{\{X\}\}$ is contained in $W(R)$, which means that the series of Example 4 of (4.5) Ch. I converge in $W(R)$.

The module $S = D(M) = (O_L \otimes_{\mathbb{Z}_p} M)^{G_L}$ is a finitely generated O_L-module endowed with an action of a generator γ of $\mathrm{Gal}(L/F)$ and an action of Frobenius automorphism φ. It is shown in [Fo5] and [Herr1] that $H^i(G_F, M)$ is equal to the i th cohomology group of the complex

$$0 \longrightarrow S \xrightarrow{\ f\ } S \oplus S \xrightarrow{\ g\ } S \longrightarrow 0$$

where $f(s) = ((\varphi - 1)s, (\gamma - 1)s)$ and $g(s,t) = (\gamma - 1)s - (\varphi - 1)t$ (for a review see [Herr2]).

Then Tate local duality can be established by working with the complex above and this provides another approach to the p-part of local class field theory [Herr1].

This approach is just a small application of the theory of Galois representations over local fields, see [A], [Colm] and references there.

8. Nonabelian Extensions

In (8.1) we shall introduce a description of totally ramified Galois extensions of a local field with finite residue field (extensions have to satisfy certain arithmetical restrictions if they are infinite) in terms of subquotients of formal power series $\mathbb{F}_p^{\text{sep}}[[X]]^*$. This description can be viewed as a non-commutative local reciprocity map (which is not in general a homomorphism but a cocycle) describing the Galois group in terms of certain objects related to the ground field. It can be viewed as a generalization of the reciprocity map of the previous sections.

In subsections (8.2)–(8.3) we review results on the absolute Galois group of local fields with finite residue field.

(8.1). Let F be a local field with finite residue field $\mathbb{F}_q$. Let φ in the absolute Galois group G_F of F be an extension of the Frobenius automorphism φ_F. Let F_φ be the fixed field of φ. It is a totally ramified extension of F and its compositum with F^{ur} coincides with the maximal separable extension of F. In this subsection *we shall work with Galois extensions of F inside F_φ*. For every finite subextension E/F of F_φ/F put $\pi_E = \Upsilon_E(\varphi|_{E^{\text{ab}}})$, see (6.4). Then π_E is a prime element of E and from functorial properties of the reciprocity maps we deduce that $\pi_M = N_{E/M}\pi_E$ for every subextension M/F of E/F.

Let $L \subset F_\varphi$ be a Galois totally ramified arithmetically profinite extension (see section 5 Ch. III) of F. If L/F is infinite, then the prime elements (π_E) in finite subextensions E of F_φ/F supply the sequence of norm-compatible prime elements (π_E) in finite subextensions of L/F and therefore by the theory of fields of norms (section 5 Ch. III) a prime element X of the local field $N = N(L|F)$. Denote by φ the automorphism of N^{ur} and of its completion $\widehat{N^{\text{ur}}}$ (which can be identified with $N(\widehat{L^{\text{ur}}}/\widehat{F^{\text{ur}}})$) corresponding to φ. Note that N and $\widehat{N^{\text{ur}}}$ are G_F-modules. If L/F is finite then we view N^* as just the group of norm compatible non-zero elements in subextensions of F in L.

DEFINITION. Define a *noncommutative local reciprocity map* [Fe13–14]

$$\Theta_{L/F}\colon \mathrm{Gal}(L/F) \longrightarrow U_{\widehat{N^{\mathrm{ur}}}}/U_N$$

by

$$\Theta_{L/F}(\sigma) = U \mod U_N,$$

where $U \in U_{\widehat{N^{\mathrm{ur}}}}$ satisfies the equation

$$U^{\varphi-1} = X^{1-\sigma}.$$

The element U exists by Proposition (1.8) applied to the local field $\widehat{N^{\mathrm{ur}}}$. It is uniquely determined modulo U_N due to the same Proposition.

A link between the reciprocity maps studied in the previous sections and the map $\Theta_{L/F}$ is supplied by the following

LEMMA.
(1) *The ground component* $u_{\widehat{F^{\mathrm{ur}}}}$ *of* $U = (u_{\widehat{M^{\mathrm{ur}}}})$ *belongs to* F.
(2) $\Theta_{L/F}(\sigma)_{\widehat{F^{\mathrm{ur}}}} = u_{\widehat{F^{\mathrm{ur}}}} = \Upsilon_F(\sigma) \mod N_{L/F}L^*$ *where* Υ_F *is defined in* (6.4).
(3) $\Theta_{L/F}$ *is injective.*
(4) $\Theta_{L/F}(\sigma\tau) = \Theta_{L/F}(\sigma)\,\sigma(\Theta_{L/F}(\tau))$.

Proof. The unit $u_{\widehat{F^{\mathrm{ur}}}}$ belongs to F, since $u_{\widehat{F^{\mathrm{ur}}}}^{\varphi-1} = 1$. The second assertion follows from Corollary in (3.2).

To show the third assertion assume that $\Theta_{L/F}(\sigma) = 1$. Then σ acts trivially on the prime elements π_M of finite subextension M/F in L/F, therefore $\sigma = 1$.

Finally, $\mathrm{X}^{1-\sigma\tau} = \mathrm{X}^{1-\sigma}(\mathrm{X}^{1-\tau})^{\sigma}$. □

This lemma shows that the ground component of Θ is the abelian reciprocity map Υ. We see that the reciprocity map $\Theta_{L/F}$ is not a homomorphism in general, but a Galois cocycle. The map $\Theta_{L/F}$ satisfies functorial properties which generalize those in (3.4).

Denote by $U^{\diamond}_{\widehat{N^{\mathrm{ur}}}}$ the subgroup of the group $U_{\widehat{N^{\mathrm{ur}}}}$ of those elements whose $\widehat{F^{\mathrm{ur}}}$-component belongs to U_F. From the previous Lemma we know that $\Theta(\sigma)$ belongs to $U^{\diamond}_{\widehat{N^{\mathrm{ur}}}}$.

Note that $\widehat{N^{\mathrm{ur}}} = \mathbb{F}_q^{\mathrm{sep}}((X))$ and so $U_{\widehat{N^{\mathrm{ur}}}} = \mathbb{F}_q^{\mathrm{sep}}[[X]]^*$. Hence the quotient group $U^{\diamond}_{\widehat{N^{\mathrm{ur}}}}/U_N$, where the image of the reciprocity map Θ is contained, is a subquotient of the invertible power series over $\mathbb{F}_q^{\mathrm{sep}}$.

The image of $\Theta_{L/F}$ is not in general closed with respect to the multiplication. Due to the Lemma the set $\mathrm{im}(\Theta_{L/F})$ endowed with new operation $x \star y = x\Theta_{L/F}^{-1}(x)(y)$ is a group isomorphic to $\mathrm{Gal}(L/F)$.

In order to describe the image of $\Theta_{L/F}$ one introduces another reciprocity map which is a generalization of the Hazewinkel map.

Denote by $U^1_{\widehat{N^{\mathrm{ur}}}}$ the subgroup of the group $U_{\widehat{N^{\mathrm{ur}}}}$ of those elements whose $\widehat{F^{\mathrm{ur}}}$-component is 1. This subgroup correspongs to the kernel of the norm map $N_{\widehat{L^{\mathrm{ur}}}/\widehat{F^{\mathrm{ur}}}}$. Instead of the subgroup $U(\mathcal{L}/\mathcal{F})$ as in Proposition (1.7), we introduce another subgroup Z of $U^1_{\widehat{N^{\mathrm{ur}}}}$. Assume, for simplicity, that there is only one root of order p in $\widehat{L^{\mathrm{ur}}}$. Let $F = E_0 - E_1 - E_2 - \dots$ be a tower of subfields, such that $L = \cup E_i$, E_i/F is a Galois extension, and E_i/E_{i-1} is cyclic of prime degree with generator σ_i. Let Z_i be a homomorphic image of $U^{\sigma_i - 1}_{\widehat{E_i^{\mathrm{ur}}}}$ in $U_{N(\widehat{L^{\mathrm{ur}}}/\widehat{E_i^{\mathrm{ur}}})}$, so that at the level of $\widehat{E_i^{\mathrm{ur}}}$-component it is the indentity map. The group Z_i can be viewed as a subgroup of $U_{\widehat{N^{\mathrm{ur}}}}$ and one can show that $\prod z_i,\ z_i \in Z_i$ converges in $U_{\widehat{N^{\mathrm{ur}}}}$. Denote by Z the subgroup generated by all such products. For the general case see [Fe13].

As a generalization of Proposition (1.7) one can show that the map

$$\ell\colon \mathrm{Gal}(L/F) \longrightarrow U^1_{\widehat{N^{\mathrm{ur}}}}/Z, \quad \sigma \mapsto X^{\sigma-1}$$

is a bijection. Using this result, one defines a generalization of the Hazewinkel map

$$U^{\diamond}_{\widehat{N^{\mathrm{ur}}}}/Y \longrightarrow \mathrm{Gal}(L/F)$$

where $Y = \{y \in U^{\diamond}_{\widehat{N^{\mathrm{ur}}}} : y^{\varphi-1} \in Z\}$. Using both reciprocity maps one verifies that $\mathrm{Gal}(L/F) \longrightarrow U^{\diamond}_{\widehat{N^{\mathrm{ur}}}}/Y$ is a bijection. For details see [Fe13].

REMARK. *H. Koch* and *E. de Shalit* [Ko7], [KdS] constructed a so called metabelian local class field theory which describes metabelian extensions of F (metabelian means that the second derived group of the Galois group is trivial). For totally ramified metabelian extensions their description is given in terms of the group

$$\mathfrak{n}(F) = \big\{(u \in U_F, \xi(X) \in \mathbb{F}_q^{\mathrm{sep}}[[X]]^*) : \xi(X)^{\varphi-1} = \{u\}(X)/X\big\}$$

with certain group structure. Here $\{u\}(X)$ is the residue series in $\mathbb{F}_q^{\mathrm{sep}}[[X]]^*$ of the endomorphism $[u](X) \in \mathcal{O}_F[[X]]$ of the formal Lubin–Tate group corresponding to π_F, q, u (see section 1 Ch. VIII).

Let M/F be the maximal totally ramified metabelian subextension of F_φ/F. Let R/F be the maximal abelian subextension of M/F. Note that the extension M/F is arithmetically profinite (apply Exercise 5 section 5 Ch. III and Corollary of (6.2) to $M/R/F$).

Send an element $U = (u_{\widehat{Q^{\mathrm{ur}}}}) \in U^{\diamond}_{N\widehat{(M|F)^{\mathrm{ur}}}}$ ($F \subset Q \subset M$, $|Q : F| < \infty$) satisfying $(u_{\widehat{Q^{\mathrm{ur}}}})^{\varphi-1} = (\pi_Q)^{1-\tau}$, $\tau \in \mathrm{Gal}(M/F)$, to

$$\big(u^{-1}_{\widehat{F^{\mathrm{ur}}}}, (u_{\widehat{E^{\mathrm{ur}}}}) \in U^{\diamond}_{N\widehat{(R|F)^{\mathrm{ur}}}}\big) \quad (F \subset E \subset R, |E : F| < \infty).$$

So we forget about the components of U lying above the level of R (like in abelian class field theory we don't need components lying above the ground level).

The element $\left(u^{-1}_{\widehat{F^{\mathrm{ur}}}}, (u_{\widehat{E^{\mathrm{ur}}}})\right)$ can be viewed as an element of $\mathfrak{n}(F)$, and we get a map

$$g\colon U^{\diamond}_{N\widehat{(M|F)}^{\mathrm{ur}}} \to \mathfrak{n}(F).$$

One can prove [Fe13] that the composite of this map with $\Theta_{L/F}$ is an isomorphism which makes Koch–de Shalit's theory a partial case of the theory of this subsection.

REMARK. A theorem of *I.R. Shafarevich* says that for every finite Galois extension F/K and abelian extension L/F the image of $\mathrm{inv}_{F/K} \in H^2(\mathrm{Gal}(F/K), F^*)$ (defined in (7.4)) with respect to

$$H^2(\mathrm{Gal}(F/K), F^*) \to H^2(\mathrm{Gal}(F/K), F^*/N_{L/F}L^*) \to H^2(\mathrm{Gal}(F/K), \mathrm{Gal}(L/F))$$

(where the last homomorphism is induced by $\Psi_{L/F}$) is equal to the cohomology class corresponding to the extension of groups

$$1 \to \mathrm{Gal}(L/F) \to \mathrm{Gal}(L/K) \to \mathrm{Gal}(F/K) \to 1.$$

This theorem (being appropriately reformulated) is used and reproved in metabelian local class field theory where its meaning becomes clearer.

(8.2). In this and next subsection we review results on the absolute Galois group G_F of a local field F with finite residue field. Let F^{ur} be the maximal unramified extension of F in F^{sep}, F^{tr} the maximal tamely ramified extension. Then $\mathrm{Gal}(F^{\mathrm{ur}}/F) \simeq \widehat{\mathbb{Z}}$ and $F^{\mathrm{ur}} = \bigcup_{(l,p)=1} F(\zeta_l)$, where ζ_l is a primitive lth root of unity. In addition, $F^{\mathrm{tr}} = \bigcup_{(l,p)=1} F^{\mathrm{ur}}(\sqrt[l]{\pi})$, where π is a prime element in F.

Let $n_1 < n_2 < \dots$ be a sequence of natural numbers, such that n_{i+1} is divisible by n_i and for every positive integer m there exists an index i for which n_i is divisible by m. Put $l_i = q^{n_i} - 1$. Choose primitive l_ith roots of unity ζ_{l_i} and $\sqrt[l_i]{\pi}$ so that $\zeta_{l_j}^{l_j l_i^{-1}} = \zeta_{l_i}$, $(\sqrt[l_j]{\pi})^{l_j l_i^{-1}} = \sqrt[l_i]{\pi}$ for $j > i$. Take $\sigma \in \mathrm{Gal}(F^{\mathrm{tr}}/F)$ such that $\sigma(\sqrt[l_i]{\pi}) = \sqrt[l_i]{\pi}$, $\sigma(\zeta_{l_i}) = \zeta_{l_i}^q$, and $\tau \in \mathrm{Gal}(F^{\mathrm{tr}}/F)$ such that $\tau(\sqrt[l_i]{\pi}) = \zeta_{l_i}\sqrt[l_i]{\pi}$, $\tau(\zeta_{l_i}) = \zeta_{l_i}$. Then $\sigma|_{F^{\mathrm{ur}}}$ coincides with the Frobenius automorphism of F and $\sigma\tau\sigma^{-1} = \tau^q$. A theorem of *H. Hasse–K. Iwasawa* ([Has12], [Iw1]) asserts that $G_{\mathrm{tr}} = \mathrm{Gal}(F^{\mathrm{tr}}/F)$ is topologically generated by σ and τ with the relation $\sigma\tau\sigma^{-1} = \tau^q$.

(8.3). Now let I be an index-set and let F_I be a free profinite group with a basis z_i, $i \in I$. Let $F_I * G_{\mathrm{tr}}$ be the free profinite product of F_I and G_{tr} (see [N2], [BNW]). Let H be the normal closed subgroup of $F_I * G_{\mathrm{tr}}$ generated by $(z_i)_{i\in I}$, and let K be the normal closed subgroup of H such that the factor group H/K is the maximal p-factor group of H. Put $F(I, G_{\mathrm{tr}}) = (F_I * G_{\mathrm{tr}})/K$ and denote the image of z_i in $F(I, G_{\mathrm{tr}})$ by x_i. The group $F(I, G_{\mathrm{tr}})$ has topological generators σ, τ, x_i, $i \in I$ with the relation $\sigma\tau\sigma^{-1} = \tau^q$.

Assume first that char(F) = p (the functional case). Then a theorem of *H. Koch* (see [Ko3]) says that the group G_F is topologically isomorphic to $F(\mathbb{N}, G_{\text{tr}})$. Recall that $U_{1,F}$ is a free $\mathbb{Z}_p$-module of rank $\mathbb{N}$ in this case.

Assume next that char(F) = 0, i.e., F is a local number field. If there is no p-torsion in F^*, then a theorem of *I.R. Shafarevich* (see [Sha1], [JW]) implies that the group G_F is topologically isomorphic to $F(n, G_{\text{tr}})$, where $n = |F : \mathbb{Q}_p|$. See also [Se4, II], [Mik1], [Mar2] for the case of a perfect residue field. Recall that $U_{1,F}$ is a free $\mathbb{Z}_p$-module of rank n in this case.

Assume, finally, that char(F) = 0 and $\mu_p \subset F^*$. Let $r \geqslant 1$ be the maximal integer such that $\mu_{p^r} \subset F^{\text{tr}^*}$. This is the most complicated case. Let χ_0 be a homomorphism of G_{tr} onto $(\mathbb{Z}/p^r\mathbb{Z})^*$ such that $\rho(\zeta_{p^r}) = \zeta_{p^r}^{\chi_0(\rho)}$ for $\rho \in G_{\text{tr}}$, where ζ_{p^r} is a primitive p^rth root of unity. Let $\chi\colon G_{\text{tr}} \to \mathbb{Z}_p^*$ be a lifting of χ_0. Let l be prime, $\{p_1, p_2, \dots\}$ the set of all primes $\neq l$. For $m \geqslant 1$ there exist integers a_m, b_m such that $1 = a_m l^m + b_m p_1^m p_2^m \dots p_m^m$. Put $\pi_l = \lim b_m p_1^m p_2^m \dots p_m^m \in \widehat{\mathbb{Z}}$. For elements $\rho \in G_{\text{tr}}$, $\xi \in F(I, G_{\text{tr}})$ put

$$(\xi, \rho) = \left(\xi^{\chi(1)} \rho \xi^{\chi(\rho)} \rho \dots \xi^{\chi(\rho^{p-2})} \rho\right)^{\pi_p/(p-1)},$$

$$\{\xi, \rho\} = \left(\xi^{\chi(1)} \rho^2 \xi^{\chi(\rho)} \rho^2 \dots \xi^{\chi(\rho^{p-2})} \rho^2\right)^{\pi_p/(p-1)}.$$

If $n = |F : \mathbb{Q}_p|$ is even, put

$$\lambda = \sigma x_0^{-1} \sigma^{-1} (x_0, \tau)^{\chi(\sigma)^{-1}} x_1^{p^n} x_1 x_2 x_1^{-1} x_2^{-1} x_3 x_4 x_3^{-1} x_4^{-1} \dots x_{n-1} x_n x_{n-1}^{-1} x_n^{-1}.$$

If $n = |F : \mathbb{Q}_p|$ is odd, let a, b be integers such that $-\chi_0(\sigma\tau^a)$ is a square $\mod p$ and $-\chi_0(\sigma\tau^b)$ is not a square $\mod p$. Put

$$\lambda_1 = \tau_2^{p+1} x_1 \tau_2^{-(p+1)} \sigma_2 \tau_2^a \{x_1, \tau_2^{p+1}\} \tau_2^{-a+b} \{\{x_1, \tau_2^{p+1}\}, \sigma_2 \tau_2^a\} \tau_2^{-b} \sigma_2^{-1}$$
$$\times \tau_2^{(p+1)/2} \{\{x_1, \tau_2^{p+1}\}, \sigma_2 \tau_2^a\} \tau_2^{-(p+1)/2},$$

where $\sigma_2 = \sigma^{\pi_2}$, $\tau_2 = \tau^{\pi_2}$. Put

$$\lambda = \sigma x_0^{-1} \sigma^{-1} (x_0, \tau)^{\chi(\sigma)^{-1}} x_1^{p^r} x_1 \lambda_1 x_1^{-1} \lambda_1^{-1} x_2 x_3 x_2^{-1} x_3^{-1} \dots x_{n-1} x_n x_{n-1}^{-1} x_n^{-1}.$$

For $n + 1$ we choose the indexset $I = \{0, 1, \dots, n\}$.

Theorems of *H. Koch* [Ko1–5]), *S.P. Demushkin* [Dem1–2], *A.V. Yakovlev* (see [Yak1–5], *J.-P. Labute* [Lab]) and theorems of *U. Jannsen–K. Wingberg* (see [Jan], [Wig1], [JW]) assert that if $p > 2$, then the absolute Galois group G_F is topologically isomorphic to $F(n + 1, G_{\text{tr}})/(\lambda)$, where (λ) is the closed normal subgroup of $F(n + 1, G_{\text{tr}})$ generated by λ. Recall that $U_{1,F}$ is a $\mathbb{Z}_p$-module of rank $n + 1$ with one relation. The case $p = 2$, $\sqrt{-1} \in F$ was considered in [Di], [Ze]; see also [Gor], [JR2], and [Mik2], [Kom] for a brief discussion of the proofs.

Unfortunately, the description of the absolute Galois groups does not provide arithmetical information on their generators.

REMARKS.

1. *M. Jarden* and *J. Ritter* ([JR1], [Rit1]) proved that two absolute Galois groups G_F and G_L for local number fields F and L are topologically isomorphic if and only if $|F : \mathbb{Q}_p| = |L : \mathbb{Q}_p|$ and $F \cap \mathbb{Q}_p^{\mathrm{ab}} = L \cap \mathbb{Q}_p^{\mathrm{ab}}$ (for $p > 2$ or $p = 2$, $\sqrt{-1} \in F, L$).

2. Recall that a theorem first proved by *F. Pop* [Po2] states that if two absolute Galois groups of finitely generated fields over $\mathbb{Q}$ are isomorphic, then so are the fields. The previous Remark shows that this is not true in the local situation. One can ask which additional conditions should be imposed on an isomorphism between two absolute Galois groups of local fields so that one can deduce that the fields are isomorphic. *Sh. Mochizuki* (in the case of characteristic zero, [Moc1]) and *V.A. Abrashkin* (in the general case [Ab8]) proved that if the isomorphism translates upper ramification subgroups onto each other, then the fields are isomorphic.

3. A formally p-adic field (defined in Exercise 6 sect. 2 Ch. I) K is said to be a *p-adically closed field* if for every proper algebraic extension L/K of valuation fields the quotient of the ring of integers of L modulo p is strictly larger than the quotient of the ring of integers of K modulo p. Certainly, finite extensions of $\mathbb{Q}_p$ are p-adically closed fields.

The works of *I. Efrat* [Ef1] (odd p) and *J. Koenigsman* [Koen2], extending earlier results of *J. Neukirch* [N1] and *F. Pop* [Po1], prove that every field F with the absolute Galois group G_F isomorphic to an open subgroup of $G_{\mathbb{Q}_p}$ is p-adically closed. The proof involves a construction of Henselian valuations using only Galois theoretic data. For the situation in positive characteristic see [EF].

CHAPTER 5

Local Class Field Theory II

In this chapter we consider various generalizations of local class field theory established in the previous chapter. In sections 1–3 we study the question for which complete discrete valuation fields their abelian extensions are described by their multiplicative group in the way similar to the theory of the previous chapter. We shall see in section 1 that such fields must have a quasi-finite residue field, i.e. a perfect field with absolute Galois group isomorphic to $\widehat{\mathbb{Z}}$. Then we indicate which results of the previous chapter (except sections 6-8) indeed take place for local fields with quasi-finite residue field. If the residue field is infinite of positive characteristic, it is not true that every open subgroup of finite index is the norm group of an abelian extension. To prove the existence theorem for local fields with quasi-finite residue field we study additive polynomials over quasi-finite fields of positive characteristic in section 2. Then in section 3 we state and prove the existence theorem for local fields with quasi-finite residue field.

In section 4 we describe abelian totally ramified p-extensions of a local field with arbitrary perfect residue field of characteristic p which is not separably p-closed. The corresponding reciprocity maps are a generalization of those in sections 2 and 3 of the previous chapter. Finally, in section 5 we review other generalizations of local class field theory: for complete discrete valuation fields with imperfect residue field and for certain abelian varieties over local fields.

1. The Multiplicative Group and Abelian Extensions

In this section we discuss to which local fields one can generalize class field theory of the previous chapter so that still the multiplicative group essentially describes abelian extensions of the fields. We shall show that except the existence theorem, all other ingredients of the theory of the previous chapter can be extended to local fields with quasi-finite residue field.

(1.1). For which complete discrete valuation fields their abelian extensions correspond to subgroups in the multiplicative group? The answer is as follows.

PROPOSITION. *Let F be a complete discrete valuation field. Assume that for every finite separable extension M of F and every cyclic extension L of M of prime* *degree the index of the norm group* $N_{L/M}L^*$ *in* M^* *coincides with the degree of* L/M. *Then the residue field* $K = \overline{F}$ *is perfect, and for any* $n \geqslant 1$ *there exists exactly one separable extension of* K *of degree* n. *Moreover, such an extension is cyclic. Conversely, if the residue field* $\overline{F}$ *is perfect and there exists exactly one Galois extension of degree* n *over* $\overline{F}$ *for* $n \geqslant 1$ *and it is cyclic, then for the fields* M *and* L *as above* $|M^*/N_{L/M}L^*| = |L : M|$.

Proof. To verify the first part of the Proposition we use the computations of norm subgroups in section 1 Ch. III. Note that the assertions which will be proved for the field F hold also for every finite separable extension of F. Proposition (1.2) Ch. III shows that the norm map and the trace map must be surjective for every finite residue extension. Let l be a prime, different from char($\overline{F}$). If a primitive lth root of unity belongs to $\overline{F}$, then by Hensel's Lemma, this is also true for F. The extension $F(\sqrt[l]{\pi})/F$ is a totally and tamely ramified Galois extension for a prime element π in F. Proposition (1.3) Ch. III shows that the subgroup $\overline{F}^{*l}$ is of index l in $\overline{F}^*$. Next, Proposition (1.5) Ch. III shows that if char($\overline{F}$) $= p > 0$, then $\overline{F}^p = \overline{F}$, and the image of the right vertical homomorphism in the fourth diagram is of index p in $\overline{F}$. In terms of those Propositions this image can be written as $\overline{\eta}^p \wp\left(\overline{F}\right)$. Thus, we deduce that the subgroup $\wp\left(\overline{F}\right)$ is of index p in $\overline{F}$.

Kummer theory and Artin–Schreier theory imply that there is exactly one cyclic extension of prime degree l (char($\overline{F}$) $\nmid l, \mu_l \subset F$) over $\overline{F}$, and that there is exactly one cyclic extension of degree p (if char($\overline{F}$) $= p$) over $\overline{F}$. This assertion also holds for a finite extension of F. In particular, putting $L = F(\mu_l)$ if $\mu_l \not\subset F, \operatorname{char}(\overline{F}) \nmid l$, we get exactly one cyclic extension of degree l over $\overline{L}$. The Galois theory immediately implies that there exists exactly one cyclic extension of degree l over $\overline{F}$ (note that $F(\mu_l)/F$ is a cyclic extension of degree $< l$).

Now we verify that there is exactly one cyclic extension of degree n over $K = \overline{F}$, $n \geqslant 1$. The uniqueness is shown easily: if K_1/K, K_2/K are cyclic extensions of degree n and l is a prime divisor of n, $l < n$, then K_1 and K_2 are cyclic extensions of degree n/l over the field K_3 that is the cyclic extension of degree l over K. Then induction arguments show that $K_1 = K_2$.

For the existence of cyclic extensions it suffices to construct cyclic extensions of degree l^n for a prime l, $n \geqslant 1$. If $l = p$, then, as it has been shown, $K/\wp(K)$ is of order p; therefore $W_n(K)/\wp W_n(K)$ is of order $\geqslant p^n$ and by the Witt theory (see also Exercise 6 in section 5 Ch. IV) there exists a cyclic extension of degree p^n over K. If $l \neq p$, then denote $d = |K(\mu_l) : K|$. It suffices to construct a cyclic extension of degree dl^n over K. Put $K_1 = \cup_{i \geqslant 1} K(\mu_{l^i})$. If $|K_1 : K| \geqslant dl^n$, then the desired extension can be chosen as a subextension in K_1/K. If $dl^m = |K_1 : K| < dl^n$, then one can find an element $a \in K_1$ such that a is not an l^{n-m}-power in K_1^*. Indeed, otherwise $K_1^{*l^i} = K_1^{*l^{i+1}}$ for some $1 \leqslant i < n - m$ and then $K_1^* = K_1^{*l}$, which is impossible by the previous considerations. Now $K_2 = K_1(\sqrt[l^{n-m}]{a})$ is a cyclic extension of K_1

and the unique cyclic extension of degree l^{n-m} over K_1. Let τ be a generator of $\mathrm{Gal}(K_1/K)$. Then by Kummer theory we deduce that for a root ζ of order equal to a power of l there exists some j such that $\tau(a\zeta) \equiv (a\zeta)^j \mod K_1^{*l^{n-m}}$. This congruence implies that K_2/K is cyclic of degree dl^n.

Note that the existence and uniqueness of cyclic extensions imply that if K''/K', K'/K are cyclic extensions, then K''/K is cyclic. Let K_1/K be a finite Galois extension, let $\sigma \in \mathrm{Gal}(K_1/K)$ be of prime order l, and let K_2 be the fixed field of σ. Then for the cyclic extension K'/K of degree l we get $K'K_2 \subset K_1$ and $K' \subset K_1$. Now, by induction arguments we may assume that K_1/K' is cyclic. Since K'/K is also cyclic, we deduce that K_1/K is cyclic as well. Finally, every finite separable extension of K is a subextension in a finite Galois extension, which is cyclic. Thus, every finite separable extension is cyclic.

To verify the second part of the Proposition, assume that there is exactly one Galois extension of degree n over $\overline{F}$ and it is cyclic, $n \geqslant 1$. Then, by the same arguments as just above, every finite separable extension of $\overline{F}$ is cyclic. Hence, if K'/K is a cyclic extension of prime degree n, then the uniqueness of K' implies that the polynomial $X^n - \alpha$ splits completely in $K'[X]$ for every $\alpha \in K$. We deduce that $-\alpha = N_{K'/K}(-\gamma)$, where γ is a root of this polynomial. This shows that the norm map is surjective for every finite residue extension. This is also true for the trace map.

Kummer and Artin–Schreier theories imply that $\overline{F}^{*l}$ is of index l in $\overline{F}^*$ for a prime l, $\mathrm{char}(\overline{F}) \nmid l$, $\mu_l \subset \overline{F}^*$; $F^{*l} = F^*$ if $\mu_l \cap \overline{F} = \{1\}$, and $\wp\left(\overline{F}\right)$ is of index p in $\overline{F}$ if $\mathrm{char}(\overline{F}) = p$. Now Propositions (1.2), (1.3), (1.5) Ch. III show that the index of the norm subgroup $N_{L/F}L^*$ in F^* is equal to the degree of the Galois extension L/F when this degree is prime.

The same assertion holds for a finite separable extension M/F. This completes the proof. □

(1.2). A field K satisfying the conditions of the Proposition (1.1) is called *quasi-finite*. From the previous Proposition we conclude that $\mathrm{Gal}(K^{\mathrm{sep}}/K)$ is isomorphic to $\widehat{\mathbb{Z}}$. This explains the name, since $\mathrm{Gal}(\mathbb{F}_q^{\mathrm{sep}}/\mathbb{F}_q) \simeq \widehat{\mathbb{Z}}$. In particular, the arguments in the proof of the Proposition (1.1) show that the norm and trace maps are surjective for every finite extension of a quasi-finite field. Below we shall show that class field theory of the previous chapter can be generalized to a local field with quasi-finite residue field. This generalization was developed by *M. Moriya, O.F.G. Schilling, G. Whaples, J.-P. Serre* and *K. Sekiguchi*.

EXAMPLES.

1. Let K be a quasi-finite field, and let L be its extension in K^{sep}. Let $\deg(L/K) = \prod_l l^{n(l)}$ be the Steinitz degree, which defines the degree of L over K: the formal product taken over all primes l, $n(l) \in \mathbb{N} \cup \{+\infty\}$, such that K has an extension of degree l^n in L if and only if $n \leqslant n(l)$. Then L is a quasi-finite field if and only if

$n(l) \neq +\infty$ for all prime l. In particular, an extension L over $\mathbb{F}_p$ with all $n(l) \neq +\infty$ is a quasi-finite field.

2. Let $\mathbb{Q}^{\text{cycl}}$ denote the field generated by all the roots of unity over $\mathbb{Q}$. Then $\mathbb{Q}^{\text{cycl}} = \bigcup_{n\geqslant 1} \mathbb{Q}(\mu_n)$ and

$$\text{Gal}(\mathbb{Q}^{\text{cycl}}/\mathbb{Q}) = \varprojlim \text{Gal}(\mathbb{Q}(\mu_n)/\mathbb{Q}) = \varprojlim (\mathbb{Z}/n\mathbb{Z})^* = \widehat{\mathbb{Z}}^*$$

(the group $\text{Gal}(\mathbb{Q}(\mu_n)/\mathbb{Q})$ is isomorphic to the multiplicative group of invertible elements in $\mathbb{Z}/n\mathbb{Z}$, see [La1, Ch. VIII]). As $\widehat{\mathbb{Z}} = \prod_p \mathbb{Z}_p$, we get

$$\widehat{\mathbb{Z}}^* \simeq \prod_p \mathbb{Z}_p^* \simeq \prod_p \mathbb{Z}_p \times \mathbb{Z}/2\mathbb{Z} \times \prod_{p\neq 2} \mathbb{Z}/(p-1)\mathbb{Z}.$$

Hence, the fixed field F of the subgroup $\mathbb{Z}/2\mathbb{Z} \times \prod_{p\neq 2} \mathbb{Z}/(p-1)\mathbb{Z}$ in $\widehat{\mathbb{Z}}^*$ is a $\widehat{\mathbb{Z}}$-extension of $\mathbb{Q}$ (it plays an important role in global class field theory [N3–5]).

3. Let E be an algebraically closed field, and let $\{x_i\}_{i\in I}$ be a basis of transcendental elements in E over the prime field E_0 in E (see [La1, Ch. X]). Put $M = E_0(\{x_i\}_{i\in I})$. Since the prime field E_0 has a $\widehat{\mathbb{Z}}$-extension E_1 ($\mathbb{F}_p^{\text{sep}}$ or F, as above), we deduce that M has the $\widehat{\mathbb{Z}}$-extension $M_1 = E_1(\{x_i\}_{i\in I})$. The field E is algebraic over the field M and is its algebraic closure. Let L be the fixed field of all automorphisms of E over M. Then L/M is purely inseparable and E/L is separable (see [La1, Ch. VII]). Let $\widetilde{\sigma} \in \text{Gal}(E/L)$ denote an automorphism, such that its restriction $\widetilde{\sigma}|_{LM_1} \in \text{Gal}(LM_1/L)$ is a topological generator of $\text{Gal}(LM_1/L)$. Then, applying the same arguments as in the proof of Proposition (2.1) Ch. IV, we conclude that the fixed field K of $\widetilde{\sigma}$ satisfies $\text{Gal}(E/K) \simeq \widehat{\mathbb{Z}}$, i.e., K is quasi-finite. We have shown that every algebraically closed field E has a subfield K which is quasi-finite.

4. Let E be an algebraically closed field of characteristic 0, $K = E((X))$. Then there is the unique extension $E((X^{1/n}))$ of degree n over K, and K is a quasi-finite field of characteristic 0.

(1.3). Now we will give a brief review of the previous chapter from the standpoint of a generalization of its assertions to a local field F with quasi-finite residue field.

Section 1

(1.1) There are three types of local fields with quasi-finite residue field, the additional third class is that of $\text{char}(F) = \text{char}(\overline{F}) = 0$. Note that in this case Corollary (5.5) Ch. I shows that $U_{1,F}$ is uniquely divisible. This means that the group $U_{1,F}$ of such a field is not interesting from the standpoint of class field theory. Since abelian extensions of a local field with quasi-finite residue field of characteristic zero are tamely ramified, it is relatively easy to describe them without using the method of the previous chapter, see Exercise 9.

Denote by $\mathcal{R}$ the set of multiplicative representatives if $\operatorname{char}(\overline{F})$ is positive and a coefficient field if $\operatorname{char}(\overline{F}) = 0$.
Further, F is not locally compact and U_F is not compact if $\overline{F}$ is not finite (see Exercise 1 in section 1 Ch. IV).

(1.2) The Galois group of a finite Galois extension L/F is solvable, since the absolute Galois group of the residue field is abelian.
As for an analog of the Frobenius automorphism, the problem is that there is no canonical choice of a generator of $\operatorname{Gal}(F^{\mathrm{ur}}/F)$ unless the residue field is finite. Therefore, *from now on we fix an isomorphism of* $\operatorname{Gal}(\overline{F}^{\mathrm{sep}}/\overline{F})$ *onto* $\widehat{\mathbb{Z}}$ and let $\overline{\varphi}$ denote the element of $\operatorname{Gal}(\overline{F}^{\mathrm{sep}}/\overline{F})$ which is mapped to 1 under this isomorphism $\operatorname{Gal}(\overline{F}^{\mathrm{sep}}/\overline{F}) \longrightarrow \widehat{\mathbb{Z}}$.
Propositions (3.2) and (3.3) Ch. II show that for the maximal unramified extension F^{ur} of F its Galois group is isomorphic to $\widehat{\mathbb{Z}}$. Let φ_F denote the automorphism in $\operatorname{Gal}(F^{\mathrm{ur}}/F)$, such that φ_F is mapped to $\overline{\varphi}$. Then the group $\operatorname{Gal}(F^{\mathrm{ur}}/F)$ is topologically generated by φ_F. We get $U_F \simeq \mathcal{R}^* \times U_{1,F}$ due to section 5 Ch. I.

(1.4) If $\operatorname{char}(\overline{F}) = p$, then there are analogs of the expansions in (1.4) Ch. IV. Namely, the index-set J numerates now elements in $R_0 \subset \mathcal{O}_F$ such that their residues form a basis of $\overline{F}$ over $\mathbb{F}_p$.
In the case of $\operatorname{char}(F) = p$ an element $\alpha \in U_{1,F}$ can be uniquely expressed as convergent product

$$\alpha = \prod_{\substack{p \nmid i \\ i>0}} \prod_{j \in J} \left(1 + \theta_j \pi_i\right)^{a_{ij}}$$

with $\theta_j \in R_0, a_{ij} \in \mathbb{Z}_p$ and the sets $J_{i,c} = \{j \in J : v_p(a_{ij}) \leqslant c\}$ finite for all $c \geqslant 0, p \nmid i, i > 0$, where v_p is the p-adic valuation.
In the case of $\operatorname{char}(F) = 0$ we know from the proof of Proposition (1.1) that $\wp\left(\overline{F}\right)$ is of index p in $\overline{F}$. Hence by (6.3), (6.4) Ch. I an element $\alpha \in U_{1,F}$ can be expressed as convergent product

$$\alpha = \prod_{i \in I} \prod_{j \in J} \left(1 + \theta_j \pi_i\right)^{a_{ij}} \omega_*^a$$

with $I = \{1 \leqslant i < \frac{pe}{p-1}, p \nmid i\}$, the absolute index of ramification $e = e(F)$, and the index-set J as above, $a_{ij} \in \mathbb{Z}_p$. Conditions on ω_*^a are the same as in (1.4) Ch. IV.
If $\operatorname{char}(\overline{F}) = 0$, then F^{*n} is an open subgroup of finite index in F^*, since according to the proof of Proposition (1.1) $\overline{F}^{*n}$ is of finite index in $\overline{F}^*$. If $\operatorname{char}(F) = 0$, $\operatorname{char}(\overline{F}) = p$, then F^{*n} is an open subgroup in F^* but not of finite index if $\overline{F}$ is infinite and $p|n$. If $\operatorname{char}(F) = p$, then F^{*n} is an open subgroup in F^* only if $p \nmid n$; and in this case it is of finite index.

(1.5) We have seen in Proposition (1.1) that if L/F is a cyclic extension of prime degree, then $|F^*/N_{L/F}L^*| = |L:F|$. The assertions of (1.5) Ch. IV for unramified and tamely ramified extensions of local fields with quasi-finite residue field are valid. We shall show below (see (3.6)) that an open subgroup N of finite index in F^* is not in general a norm subgroup if $\operatorname{char}(\overline{F}) \neq 0$. This may explain why we need to study some additional topics in section 2 to follow.

(1.6)–(1.9) Everything works for local fields with quasi-finite residue field.

Section 2 The definition of the Neukirch map is exactly the same. All the assertions hold for F.

Section 3 The definition of the Hazewinkel homomorphism for a finite Galois totally ramified extension is exactly the same. All results of section 3 remains valid.

Section 4 Everything remains valid. Thus, we have the *reciprocity map* $\Psi_F\colon F^* \longrightarrow \operatorname{Gal}(F^{\mathrm{ab}}/F)$.

Section 5

(5.1) The definition of the Hilbert norm residue symbol is valid for F, and all its properties described in Proposition (5.1) Ch. IV remian valid.

(5.2) The Theorem is not true if $\overline{F}$ is infinite, since not every open subgroup of finite index is the norm subgroup of a finite abelian extension (see Corollary 2 in (3.6)).

(5.3) The Theorem must be formulated as follows. Let $\operatorname{char}(\overline{F}) \nmid n$ and $\mu_n \subset F^*$. From the proof of Proposition (1.1) we know that $\overline{F}^*/\overline{F}^{*n}$ is a cyclic group of order n. Define a homomorphism

$$\nu_n\colon \overline{F}^*/\overline{F}^{*n} \longrightarrow \mu_n, \qquad \theta \mapsto \rho^{-1}\overline{\varphi}(\rho),$$

where an element $\rho \in \overline{F}^{\mathrm{sep}}$ with $\rho^n = \theta$. It is easy to show that ν_n is an isomorphism. Then for $\alpha, \beta \in F^*$ we obtain

$$(\alpha,\beta)_n = \nu_n d(\alpha,\beta), \qquad d(\alpha,\beta) = \overline{\gamma} \mod \overline{F}^{*n},$$
$$\gamma = \beta^{v_F(\alpha)}\alpha^{-v_F(\beta)}(-1)^{v_F(\alpha)v_F(\beta)}.$$

The proof of this assertion is carried out in the same way as that of Theorem (5.3) Ch. IV. In particular, for an element $\theta \in \mathcal{R}^*$ we get

$$(\pi,\theta)_n = \rho^{\varphi_F - 1}, \qquad \text{where } \rho^n = \theta.$$

(5.4)–(5.6) These assertions except Corollary (5.6) Ch. IV (see Exercise 7) can be appropriately reformulated to remain valid.

(5.7) Not true in general.

Section 6 We shall consider the Existence Theorem below in section 3.

Exercises.

1. (*G. Whaples*) Let K be an algebraic extension of $\mathbb{F}_p$, and S the set of primes l such that $n(l) = +\infty$ in $\deg(K/\mathbb{F}_p) = \prod l^{n(l)}$. Assume that $p \notin S$ and $\mu_{l^n} \subset K$ for every $n \geqslant 1$, $l \in S$ (e.g., $K = \bigcup_{n\geqslant 1} \mathbb{F}_3(\mu_{2^n})$). Let I be the additive subgroup of rational numbers m/n with integer m, n, n relatively prime to any $l \in S$. Let K' be the formal power series field $\sum_{\substack{i\in I\\ i\geqslant i_0}} a_i X^i$, $a_i \in K$. Show that K' is quasi-finite and that K is the algebraic closure of $\mathbb{F}_p$ in K'.
2. Let K be a field, and let G be the group of all automorphisms of K^{alg} over K. There is a natural continuous map
$$\widehat{\mathbb{Z}} \times G \to G, \quad (a, \sigma) \mapsto \sigma^a.$$
An element $\sigma \in G$ has a period $a \in \widehat{\mathbb{Z}}$ that is a generator of the ideal $A \subset \widehat{\mathbb{Z}}$ of those elements $b \in \widehat{\mathbb{Z}}$ for which $\sigma^b = 1$. Show that
   a) K has an algebraic extension, which is a quasi-finite field, if and only if there is an element of period 0 in G.
   b) If for every $n \geqslant 1$ there is a cyclic extension over K of degree n, then there is an element of period 0 in G.
3. ($\diamond$) ([Wh4], [Wen])
   a) Let n be any positive integer. Show that there exists a field K with no extensions of degree $\leqslant n$, but with algebraic extensions of degree divisible by n.
   b) Show that if a field K has a cyclic extension of degree l, where l is an odd prime, then K has cyclic extensions K_n of degree l^n over K for every $n \geqslant 1$, such that $K_n \subset K_{n+1}$ (then for $K' = \bigcup_{n\geqslant 1} K_n$ the group $\mathrm{Gal}(K'/K)$ is isomorphic to $\mathbb{Z}_l$; such an extension is called a $\mathbb{Z}_l$-extension). Show that if a field K has a cyclic extension of degree 4, then K has a $\mathbb{Z}_2$-extension. Show that if a field has a cyclic extension of degree 2 but not of degree 4, then K is a formally real field (see [La1, Ch. XI]) of characteristic 0.
4. (*G. Whaples* [Wh3]) A field K is said to be a *Brauer field* if it is perfect and there is at most one extension over K in K^{alg} of degree n for every $n \geqslant 1$.
   a) Show that every finite extension of a Brauer field K is cyclic.
   b) Let K be a Brauer field and $\deg(K^{\mathrm{sep}}/K) = \prod l^{d(l)}$. Show that if l is an odd prime, then $d(l) = 0$ or $d(l) = +\infty$. Show that $d(2) = 0$, or $d(2) = 1$, or $d(2) = +\infty$. Prove that for a finite extension E/K the norm map is surjective if $d(2) \neq 1$, and
$$N_{E/K}E^* = \begin{cases} K^*, & \text{if } |E:K| \text{ is odd,} \\ K^{*2} \neq K^*, & \text{if } |E:K| \text{ is even,} \end{cases}$$
   if $d(2) = 1$.
5. Let F be a complete discrete valuation field and let its residue field $\overline{F}$ be a Brauer field. Define the Neukirch map $\Upsilon_{L/F}$ and show that Theorem (4.2) Ch. IV holds for all finite

abelian extensions of degree dividing

$$\deg(\overline{F}^{\mathrm{sep}}/\overline{F}) = \prod_l l^{n(l)}$$

when $n(2) \neq 1$ and is, in addition, odd when $n(2) = 1$.

6. ($\diamond$) Let F be a local field with quasi-finite residue field. Let $(F_i)_{i\in\mathbb{Z}}$ be an increasing chain of separable finite extensions of F, $\mathcal{F} = \cup_i F_i$. Let S denote the set of primes l, such that $|F_{i+1} : F_i|$ is divisible by l for almost all i.
   a) Let $\mathcal{L}$ be a finite abelian extension of $\mathcal{F}$. Show that if $\mathrm{Gal}(\mathcal{L}/\mathcal{F})$ is isomorphic to $\mathcal{F}^*/N_{\mathcal{L}/\mathcal{F}}\mathcal{L}^*$, then the degree $|\mathcal{L} : F|$ is relatively prime with all $l \in S$.
   b) Show that Theorem (4.2) Ch. IV holds for all finite abelian extensions $\mathcal{L}/\mathcal{F}$ of degree relatively prime to all $l \in S$.
7. Let F be a local field of characteristic p with quasi-finite residue field.
   a) Show that for the map $(\cdot,\cdot]\colon F^* \times F \to \mathbb{F}_p$ defined by the formula

$$(\alpha, \beta] = \Psi_F(\alpha)(\gamma) - \gamma, \qquad \text{where } \wp(\gamma) = \beta,$$

   all the properties in Proposition (5.4) Ch. IV, except (6), hold.
   b) Let $\rho_n : \overline{F}/\wp\left(\overline{F}\right) \to \mathbb{F}_p$ be the homomorphism defined as

$$\theta \mod \wp\left(\overline{F}\right) \to \overline{\varphi}(\eta) - \eta$$

   with $\wp(\eta) = \theta$, where $\overline{\varphi}$ is as in (1.3). Show that ρ_n is an isomorphism. Show that

$$(\alpha, \beta] = \rho_n \operatorname{res}\left(\beta\alpha^{-1}\frac{\partial\alpha}{\partial\pi}\right).$$

8. ($\diamond$) (*Sh. Sen* [Sen1, 2], *E. Maus* [Mau2]) Let F be a local field of characteristic 0 with perfect residue field of characteristic p. Let L/F be a finite abelian p-extension, $G = \mathrm{Gal}(L/F)$, $h = h_{L/F}$, $e = e(F)$. Assertion:

$$\text{if } n \leqslant \frac{e}{p-1} \quad \text{then } G^p_{h(n)} \subset G_{h(pn)};$$
$$\text{if } n > \frac{e}{p-1} \quad \text{then } G^p_{h(n)} = G_{h(n+e)}.$$

   a) Using Proposition (5.7) Ch. I, show that the assertion is true when $\overline{F}$ is quasi-finite.
   b) Show that the assertion is true when $\overline{F}$ is algebraically closed.
   c) Show that the assertion is true when $\overline{F}$ is perfect.
9. Let F be a local field with perfect residue field.
   a) Let L/F be a finite abelian tamely ramified extension. Put $L_0 = L \cap F^{\mathrm{ur}}$ and denote $e = |L : L_0|$. Using (3.5) Ch. II show that F contains a primitive eth root of unity and there is a prime element $\pi \in F$ such that $L = L_0(\sqrt[e]{\pi})$.
   b) Denote by F^{abtr} the maximal abelian tamely ramified extension of F and by F^{abur} the maximal abelian unramified extension of F. Fix a prime element π if F and denote by E_π the subfield of F^{abtr} generated by $\sqrt[e]{\pi}$ where e runs over all integers not divisible by $\mathrm{char}(\overline{F})$ and such that $\mu_e \subset F$. Show that F^{abtr} is the compositum of linearly disjoint abelian extension F^{abur} and E_π.

c) Choose primitive roots ζ_e of unity of order e not divisible by $\operatorname{char}(\overline{F})$ in such a way that $\zeta_{ee'}^{e} = \zeta_e$ for all e, e'. The choice of the roots determine an isomorphism between the Galois group of a Kummer extension of F and the corresponding quotient of F^*. Show that with respect to this choice the Galois group $\operatorname{Gal}(E/F)$ is isomorphic to $\varprojlim_e F^*/F^{*e}$. Show that if $\mathcal{R}$ is the set of multiplicative representatives in F or a coefficient field (in the case $\operatorname{char}(\overline{F}) = 0$), then $F^*/F^{*e} \simeq \mathbb{Z}/e\mathbb{Z} \times \mathcal{R}^*/\mathcal{R}^{*e}$.

2. Additive Polynomials

In this section we consider the theory of additive polynomials which will be applied in the next section. This theory was developed by *O. Ore, H. Hasse* and *E. Witt* in the general case, and by *G. Whaples* in the case of quasi-finite fields.

(2.1). Let K be a field. A polynomial $f(X)$ over K is called *additive* if for every $\theta, \eta \in K$ the equality $f(\theta + \eta) = f(\theta) + f(\eta)$ holds.

LEMMA. *Let $q \leqslant +\infty$ be the cardinality of K. If q is finite, then assume that $\deg f(X) \leqslant q$. Then $f(X)$ is additive if and only if $f(X+Y) = f(X) + f(Y)$ in $K[X,Y]$. In this case $f(X) = aX$ with $a \in K$ if* $\operatorname{char}(K) = 0$, *and* $f(X) = \sum_{m=0}^{n} a_m X^{p^m}$ *with $a_m \in K$ if* $\operatorname{char}(K) = p$.

Proof. Assume that $f(X+Y) - f(X) - f(Y) = \sum h_i(Y)X^i \neq 0$ in $K[X,Y]$, where h_i are polynomials over K. Then there is an index i such that $h_i(Y) \neq 0$. Since $\deg h_i < q$, there exists an element $\theta \in K$ for which $h_i(\theta) \neq 0$. Then the polynomial $\sum h_i(\theta)X^i \in K[X]$ is not zero and its degree is less that q. Therefore, there exists an element $\eta \in K$ such that $\sum h_i(\theta)\eta^i \neq 0$. This is impossible because $f(\theta + \eta) = f(\theta) + f(\eta)$.

Now we deduce that the derivative $f'(X)$ is a constant and obtain the last assertion. □

From this point on, we assume that $\operatorname{char}(K) = p > 0$.

(2.2). The sum of two additive polynomials is additive, but the product, in general, is not. So we introduce another operation of composition and put $f \circ g = f\big(g(X)\big)$. The ring of additive polynomials with respect to $+, \circ$ is isomorphic to the ring of noncommutative polynomials $K[\Lambda]$ with multiplication defined as $(a\Lambda)(b\Lambda) = ab^p\Lambda^2$ for $a, b \in K$, under the map

$$\sum_{m=0}^{n} a_m X^{p^m} \mapsto \sum_{m=0}^{n} a_m \Lambda^m.$$

If a polynomial $f(X) \in K[X]$ is written as $g(X) \circ h(X)$, then $g(X)$ is called an *outer component* of $f(X)$ and $h(X)$ is called an *inner component* of $f(X)$.

LEMMA. *For additive polynomials $f(X), g(X) \in K[X]$, $g(X) \neq 0$, there exist additive polynomials $h(X), q(X)$ such that $f(X) = h(X) \circ g(X) + q(X)$ and the degree of $q(X)$ is smaller than the degree of $g(X)$. If K is perfect, then there exist additive polynomials $h_1(X), q_1(X)$ such that $f(X) = g(X) \circ h_1(X) + q_1(X)$ with* $\deg q_1(X) < \deg g(X)$.

Proof. Let $f(X) = \sum_{m=0}^{n} a_m X^{p^m}, g(X) = \sum_{m=0}^{k} b_m X^{p^m}, n \geqslant k$. Then

$$\deg\left(f(X) - a_n b_k^{-p^{n-k}} X^{p^{n-k}} \circ g(X)\right) < p^n,$$

$$\deg\left(f(X) - g(X) \circ \left(\left(a_n b_k^{-1}\right)^{p^{-k}} X^{p^{n-k}}\right)\right) < p^n.$$

Now the proof of the Lemma follows by induction. □

PROPOSITION. *The ring of additive polynomials under addition and composition is a left Euclidean principal ideal ring. If K is perfect, then it is also a right Euclidean principal ideal ring.*

Proof. It immediately follows from the previous Lemma. □

REMARK. If $f(X) = g(X) \circ h(X)$ for additive polynomials over K^{sep} and two of these polynomials have coefficients in K, then the coefficients of the third are also in K.

COROLLARY. *Let K be perfect, and let $f_1(X), f_2(X)$ be additive polynomials. If $f_3(X)$ is a least common outer multiple of $f_1(X), f_2(X)$ and $f_4(X)$ is a greatest common outer divisor of $f_1(X), f_2(X)$, then*

$$f_3(K) \subset f_1(K) \cap f_2(K), \qquad f_4(K) = f_1(K) + f_2(K).$$

Proof. Let $f_3(X)$ be a least common outer multiple of f_1, f_2, i.e., $f_3(X)$ is an additive polynomial of the minimal positive degree such that $f_3 = f_1 \circ g_1 = f_2 \circ g_2$, with additive polynomials g_1, g_2 (for the existence of $f_3(X)$ see Exercise 2). Then $f_3(K) \subset f_1(K) \cap f_2(K)$. Let $f_4(X)$ be a greatest common outer divisor of f_1, f_2, i.e., an additive polynomial of the maximal degree such that $f_1 = f_4 \circ h_1$, $f_2 = f_4 \circ h_2$, with additive polynomials h_1, h_2. The polynomial f_4 can be also presented in the form

$$f_4 = f_1 \circ p_1 + f_2 \circ p_2$$

with additive polynomials p_1, p_2. Therefore,

$$f_4(K) \subset f_1(K) + f_2(K) \subset f_4(K) + f_4(K) = f_4(K). \square$$

This Corollary shows a connection between additive polynomials and subgroups in K. Of great importance is the following assertion.

(2.3). PROPOSITION. *Any finite additive subgroup* $H \subset K$ *is the set of all roots of some additive polynomial* $f(X)$ *over* K *such that* $\deg(f) = |H|$.

Proof. Put $f(X) = \prod_{a_i \in H}(X - a_i)$. Assume that $g(X,Y) = f(X+Y) - f(X) - f(Y) \neq 0$ in $K[X,Y]$. Observing that $f(\theta) = f(\theta + a_i)$ for every $\theta \in K$, we obtain that the polynomial $g(X,\theta)$ of degree $< \deg(f)$ has roots a_i. This implies

$$g(X,\theta) = 0 \quad \text{and} \quad f(\eta + \theta) = f(\eta) + f(\theta) \qquad \text{for } \theta, \eta \in K,$$

as desired. □

COROLLARY 1. *Let* H *be any finite additive subgroup in* K^{sep}, *such that* $\sigma(H) = H$ *for every* $\sigma \in \text{Gal}(K^{\text{sep}}/K)$. *Then* H *is the set of all roots of some additive polynomial over* K.

COROLLARY 2. *Let* $\{a_i\} \subset K$ *be a set of* n *linearly independent elements over* $\mathbb{F}_p$, *and let* $\{b_i\}$ *be a set of* n *elements in* K. *Then there exists an additive polynomial* $f(X)$ *of degree* $\leqslant p^n$ *over* K *such that* $f(a_i) = b_i$.

Proof. It suffices to show that there exists an additive polynomial $f(X)$ such that $f(a_1) = \dots = f(a_{n-1}) = 0, f(a_n) \neq 0$. Let H be an additive group of order p^{n-1} generated by $a_1, \dots, a_{n-1}$. If f is an additive polynomial with H as the set of its roots, then $f(a_n) \neq 0$. □

(2.4). From this point on we assume that K is a quasi-finite field of characteristic p.

PROPOSITION. *Let* $f(X)$ *be a nonzero additive polynomial. Then the index of* $f(K)$ *in* K *coincides with the number of roots of* $f(X)$ *in* K.

Proof. Let H be the set of roots of $f(X)$ in K^{sep}. Let $\overline{\varphi}$ be a topological generator of $\text{Gal}(K^{\text{sep}}/K) \simeq \widehat{\mathbb{Z}}$, which is mapped to 1. As H is finite, the kernel and cokernel of the homomorphism $\overline{\varphi} - 1\colon H \to H$ are of the same order. Thus, it suffices to show that the index of $f(K)$ in K coincides with the order of $H/(\overline{\varphi} - 1)H$. We shall verify a more general assertion, namely, there is an isomorphism $\psi\colon K/f(K) \xrightarrow{\sim} H/(\overline{\varphi} - 1)H$.

Let $a \in K$; put $\psi(a \mod f(K)) = \overline{\varphi}(b) - b$, where $b \in K^{\text{sep}}, f(b) = a$. Then ψ is well defined and is an injective homomorphism. Any element $c \in H$ can be regarded as an element of a finite extension K_1 of K. Then $\text{Tr}_{K_2/K_1} c = 0$, where K_2 is the cyclic extension of K_1 of degree p. For the same reasons as in the proof of Proposition (1.8) Ch. IV, there exists an element $d \in K_2$ such that $\overline{\varphi}(d) - d = c$. Then $f(d) \in K$ and $\psi\big(f(d) \mod f(K)\big) = c$. This means that ψ is surjective, and the proof is completed. □

COROLLARY. *Let* $f(X)$ *be an additive polynomial over* K, $f'(0) \neq 0$, *and let all the roots of* f *belong to* K. *Let* $g(X)$ *be an additive polynomial over* K. *Then* $g(K) \subset f(K)$ *if and only if* $f(X)$ *is an outer component of* $g(X)$.

Proof. The "if" part is clear. Let $h(X)$ be a greatest common outer divisor of $f(X), g(X)$. If $g(K) \subset f(K)$, then by Corollary (2.2) $h(K) = f(K)$. Now the Proposition implies $\deg(h) = \deg(f)$. Therefore, $h(X) = af(X)$ for some $a \in K$, and $f(X)$ is an outer component of $g(X)$. □

(2.5). There is a close connection between inner components and the sets of roots of additive polynomials.

PROPOSITION. *Let $f(X)$ be an additive polynomial over K and $f'(0) \neq 0$. Let $g(X)$ be an additive polynomial over K. Then the set of roots of $f(X)$ in K^{sep} is a subset of the set of roots of $g(X)$ in K^{sep} if and only if $f(X)$ is an inner component of $g(X)$.*

Proof. The "if" part is clear. To prove the "only if" part, put $H' = f(H)$, where H is the set of roots of $g(X)$. By Proposition (2.3), there exists an additive polynomial $h(X)$ with H' as its set of roots. One may assume $h'(0) \neq 0$. Then the polynomials $h\big(f(X)\big)$ and $g(X)$ have the same roots. Since $h\big(f(X)\big)$ is simple, i.e., $(h \circ f)'(0) \neq 0$, we conclude that $g(X) = ah\big(f(X)\big)^{p^m}$ for some $a \in K, m \geqslant 0$. This means that $f(X)$ is an inner component of $g(X)$. □

REMARK. The Proposition holds also for perfect fields.

(2.6). PROPOSITION. *Let $f(X)$ be an additive polynomial over K. Then there exists an additive polynomial $g(X)$ over K with $g'(0) \neq 0$, such that $f = g \circ h$ for some additive polynomial $h(X)$ over K, $f(K) = g(K)$, and all roots of $g(X)$ belong to K.*

Proof. Let H be the set of roots of $f(X)$ in K^{sep}, $L = K(H)$. Since K is quasi-finite, one can choose a generator σ of $G = \mathrm{Gal}(L/K)$. Put $H_1 = \{a \in H : \sigma(a) = a\}$. The theory of linear operators in finite-dimensional spaces (see [La1, Ch. XV]) implies that there exists a decomposition of H into a direct sum of indecomposable $\mathbb{F}_p[G]$-submodules $H^{(i)}, 1 \leqslant i \leqslant m$. If $H^{(i)} \cap H_1 = 0$, then we put $H_2^{(i)} = H^{(i)}$. If $H^{(i)} \cap H_1 \neq 0$, then the minimal polynomial of the restriction of σ on $H^{(i)}$ is $(X-1)^n$, where $n = \dim_{\mathbb{F}_p} H^{(i)}$. In this case, there exists a Jordan basis of $H^{(i)}: a_1, \dots, a_n$, such that $\sigma(a_j) = a_j + a_{j+1}$ if $1 \leqslant j \leqslant n-1, \sigma(a_n) = a_n$. Then $H^{(i)} \cap H_1 = a_n\mathbb{F}_p$. Put $H_2^{(i)} = \bigoplus_{j=2}^{j=n} a_j\mathbb{F}_p$.

Now for $H_2 = \oplus H_2^{(i)}$ we get

$$\dim_{\mathbb{F}_p} H_2 = \dim_{\mathbb{F}_p} H - \dim_{\mathbb{F}_p} H_1,\ \sigma(H_2) = H_2,\ (\sigma - 1)H \subset H_2.$$

Let, by Corollary 1 of Proposition (2.3), $h(X)$ be an additive polynomial over K, such that H_2 is its set of roots. One may assume $h'(0) \neq 0$. Then, by Proposition (2.5), there exists an additive polynomial $g(X)$ over K^{sep} such that $f(X) = g\big(h(X)\big)$. The set of roots of g coincides with $h(H)$. In fact, the coefficients of $g(X)$ belong to K. Since the element $\sigma a - a$ belongs to H_2 for an element $a \in H$, we get $h(H) \subset K$. On the other hand, the order of $h(H)$ is equal to the index of H_2 in H, i.e., the order of H_1. Finally, $f(K) \subset g(K)$ and, by Proposition (2.4), $|K/f(K)| = |K/g(K)|$. Thus, $f(K) = g(K)$ and $g(X)$ is the required polynomial. □

COROLLARY. *Let $f(X)$ be a nonzero additive polynomial over K. The following conditions are equivalent:*
(i) $f(K) \neq K$,
(ii) *f has a root $\neq 0$ in K,*
(iii) *$\wp(aX)$ is an inner component of $f(X)$ for some $a \in K^*$,*
(iv) *$b\wp(X)$ is an outer component of $f(X)$ for some $b \in K^*$.*

Proof. Proposition (2.4) shows the equivalence of (i) and (ii), and proposition (2.5) that of (ii) and (iii). The implication (iv) $\Rightarrow$ (i) follows immediately, because $\wp(K)$ is of index p in K. To show that (i) $\Rightarrow$ (iv), we write $f = g \circ h$ as in the Proposition. As $g(K) = f(K) \neq K$, we get $g = g_1 \circ \wp(aX)$ for some additive polynomial $g_1(X)$ over K, and $a \in K$ by Proposition (2.5). If the polynomial $g_1(X)$ is not linear, then it has a root $c \neq 0$ in K^{sep}. Then an element $d \in K^{\mathrm{sep}}$, such that $\wp(ad) = c$, is a root of the polynomial $g(X)$. Since all roots of $g(X)$ belong to K, we obtain $d \in K$. Therefore, $c \in K$, and Proposition (2.4) shows that $g_1(K) \neq K$. Applying the previous arguments to $g_1(X)$, we deduce after a series of steps that $b\wp(X)$ is an outer component of $f(X)$ for some $b \in K^*$, as desired. □

(2.7). Let $f(X)$ be a nonzero additive polynomial over K, $S = f(K)$. Then S is a subgroup of finite index in K according to Proposition (2.4). Our first goal is to show that every intermediate subgroup between S and K is the set of values of some additive polynomial.

PROPOSITION. *The endomorphisms of the $\mathbb{F}_p$-space K/S are induced by additive polynomials.*

Proof. One may assume, by Proposition (2.6), that all roots of $f(X)$ belong to K and $f'(0) \neq 0$. Denote $H = \ker(f)$. By Corollary 2 of Proposition (2.3) endomorphisms of the set H of all roots of $f(X)$ are induced by additive polynomials. Let an additive polynomial $h(X)$ induce an endomorphism of H. This means that the set of roots of f is a subset of the set of roots of $f \circ h$. By Proposition (2.5) there exists an additive polynomial $g(X)$ over K such that $f \circ h = g \circ f$. Then $g(X)$ induces an endomorphism of K/S. Conversely, if $g(X)$ is an additive polynomial which induces an endomorphism of K/S, then $g\big(f(K)\big) \subset f(K)$. By Corollary (2.4), there exists an additive polynomial $h(X)$ over K such that $f \circ h = g \circ f$. Then $h(X)$ induces an endomorphism of H. Thus, there is the isomorphism $f \mapsto h$ between the ring of endomorphisms of H, which are induced by additive polynomials, and the ring of endomorphisms of K/S, which are induced by additive polynomials. Since the dimensions of $\operatorname{End}(H)$ and $\operatorname{End}(K/S)$ coincide by Proposition (2.4), we obtain the desired assertion. □

COROLLARY 1. *Any intermediate subgroup between $f(K)$ and K can be presented as $g(K)$ for some additive polynomial.*

COROLLARY 2. *The homomorphisms of $K/f_1(K)$ to $K/f_2(K)$, where f_1, f_2 are additive polynomials, are induced by additive polynomials.*

Proof. Let f_3 be as in Corollary (2.2). Then $\operatorname{Hom}(K/f_1(K), K/f_2(K))$ is a subfactor of the space $\operatorname{End}(K/f_3(K))$. □

COROLLARY 3. *$f(K)$ is the intersection of a suitable finite set of $b_i\wp(K)$, $b_i \in K$.*

Proof. The intersection of all intermediate subgroups of index p between $f(K)$ and K coincides with $f(K)$. Such a subgroup can be written as $h(K)$ by Corollary 1. Corollary (2.6) shows that $h(K) = b\wp(g(K))$ for some additive polynomial $g(X)$. As $h(K)$ is of index p in K, we conclude that $h(K) = b\wp(K)$. □

(2.8). The assertions of (2.7) and (2.2) show that the set of subgroups $f(K)$, where f runs through the set of additive polynomials over K, forms a basis of neighborhoods of a linear topology on K. This topology is said to be *additive*. Any neighborhood S of 0 can be written as $f(K)$ for some additive polynomial $f(X)$ by Corollary 1 of (2.7).

PROPOSITION. *Additive polynomials define continuous endomorphisms of K with respect to the additive topology. The subring of these endomorphisms is dense in the ring of all continuous endomorphisms of K.*

Proof. Let S be a neighborhood of 0 in K. Then $S = f(K)$ for some additive polynomial f. Let $g(X)$ be an additive polynomial and let $h(X)$ be a least common outer multiple of $f(X), h(X)$. Then $h = f \circ f_1 = g \circ g_1$ for some additive polynomials $f_1(X), g_1(X)$ over K and $g_1(K) \subset g^{-1}(S)$. This means that g induces a continuous endomorphism of K.

Let A be a continuous endomorphism of K. For a neighborhood $S_2 = f_2(K)$ of 0 in K there exists a neighborhood $S_1 = f_1(K)$ with $A(S_1) \subset S_2$. By Corollary 2 of (2.7) the induced homomorphism $A\colon K/S_1 \to K/S_2$ is induced by an additive polynomial $f(X)$ over K. Then $(A - f)(K) \subset S_2$ and we obtain the second assertion of the Proposition. □

(2.9). Finally, we show that every polynomial can be transformed to an additive polynomial.

PROPOSITION. *Let $f(X)$ be a nonzero polynomial over K, $f(0)=0$. Then there exists a finite set of elements $a_i \in K$, such that $g(X)=\sum f(a_iX)$ is an additive polynomial and $\sum a_i = 1$. Moreover, there exists a finite set of polynomials $h_i(X)$ over K, such that $h(X)=\sum f\big(h_i(X)\big)$ is a nonzero additive polynomial.*

Proof. Let q be the cardinality of K and $\deg(f) \geqslant q$. Then one can write $f(X) = p(X)(X^q - X) + r(X)$ with $p(X), r(X) \in K[X]$, $\deg(r) < q$. In this case $f(\theta) = r(\theta)$ for $\theta \in K$, and we may assume, without loss of generality, that $\deg(f) < q$. Now let $n < q$ and let n be relatively prime to p. Let $m \mid n$ be the maximal integer such that a primitive mth root of unity belongs to K. If $m > 1$, then putting $c_i = 1, 1 \leqslant i \leqslant p-1, c_p = \zeta$, where ζ is a primitive mth root of unity, we get $\sum c_i^n = 0, \sum c_i \neq 0$.

If $m = 1$, then let l be prime, $l \mid n$. Assume that $K^{*l} \neq K^*$. Then for $a \in K^*, a \notin K^{*l}$, the extension $K(\sqrt[l]{a})/K$ is cyclic of degree l since K is quasi-finite. Therefore, a primitive lth root of unity belongs to $K(\sqrt[l]{a})$ and does not belong to K, which is impossible. Thus, $K^{*l} = K^*$ and $K^{*n} = K^*$. The conditions on n imply that there exist elements $c_1, c_2 \in K$ such that $c_1 + c_2 \neq -1$, $c_1^n + c_2^n = -1$. Hence, for $c_3 = 1$ we get $c_1^n + c_2^n + c_3^n = 0$, $c_1 + c_2 + c_3 \neq 0$.

Thus, we conclude that the polynomial $\sum f(c_iX)$ has the coefficient 0 at X^n and $\sum c_i \neq 0$. After a series of steps of this kind we obtain the elements $a_i \in K$ indicated in the first assertion of the Proposition.

To prove the second assertion, we take a polynomial $h(X)$ such that the degree of $f\big(h(X)\big)$ is a power of p. As above, we find elements $a_1, a_2, \dots$ in K, such that $\sum a_i = 1$ and $g(X) = \sum f\big(h(a_iX)\big)$ is an additive polynomial. Then $g(X) \neq 0$, as required. □

COROLLARY 1. *Let $p(X)$ be a given nonzero additive polynomial, and let $f(X)$ be as in the Proposition. Then there exist polynomials $f_i(X), g_i(X)$ over K such that $\sum f_i(X)$ is a nonzero additive polynomial and $p(X)$ is an outer component of the additive polynomial $\sum f \circ f_i$ and of the nonzero additive polynomial $\sum f \circ g_i$ (0 is considered as having $p(X)$ as an outer component).*

Proof. Let $g(X), h(X)$, $a_i \in K$, $h_i(X)$ be as in the Proposition. Let $\widetilde{p}(X)$ be a least common outer multiple of $g(X), h(X), p(X)$. Then $\widetilde{p} = g \circ \widetilde{g} = h \circ \widetilde{h}$ for some additive polynomials $\widetilde{g}(X), \widetilde{h}(X)$ over K. Putting $f_i = a_i\widetilde{g}, g_i = h_i \circ \widetilde{h}$, we get the required assertion. □

COROLLARY 2. *A neighborhood of 0 in the additive topology in K can be redefined as a vector subspace over $\mathbb{F}_p$ that contains the set of values of some nonzero polynomial $f(X)$ over K with $f(0) = 0$.*

Proof. Let $h(X)$ for $f(X)$ be as in the Proposition. Then $h(K)$ is contained in every vector subspace over $\mathbb{F}_p$ containing the set $f(K)$. □

Exercises.

1. a) Show that $f = g \circ h$ in the ring of additive polynomials over K if and only if $h(X)$ divides $f(X)$ in $K[X]$.
   b) Show that for a polynomial $f(X)$ of degree n over K there exists an additive polynomial $g(X)$ of degree $\leqslant p^n$ over K, such that $f(X)$ divides $g(X)$ in $K[X]$.
2. a) Let $f_1(X), f_2(X)$ be nonzero additive polynomials and let
$$f_1(X) = q_1(X) \circ f_2(X) + f_3(X), \ \dots,$$
$$f_i(X) = q_i(X) \circ f_{i+1}(X) + f_{i+2}(X), \ \dots,$$
$$f_{n-1}(X) = q_{n-1}(X) \circ f_n(X)$$
be the Euclid algorithm for $f_1(X)$, $f_2(X)$ in the ring of additive polynomials. Show that
$$f_{n-1}(X) \circ f_n(X)^{-1} \circ f_{n-2}(X) \circ f_{n-1}(X)^{-1} \circ \dots \circ f_2(X) \circ f_3(X)^{-1} \circ f_1(X)$$
is an additive polynomial and a least common inner multiple of the polynomials $f_1(X), f_2(X)$.
   b) Show that if K is perfect and
$$g_1(X) = g_2(X) \circ r_1(X) + g_3(X), \ \dots,$$
$$g_i(X) = g_{i+1}(X) \circ r_i(X) + g_{i+2}(X), \ \dots, g_{m-1}(X) = g_m(X) \circ r_{m-1}(X)$$
is the Euclid algorithm for nonzero additive $g_1(X)$, $g_2(X)$, then
$$g_1(X) \circ g_3(X)^{-1} \circ g_2(X) \circ g_4(X)^{-1} \circ g_3(X) \circ \dots \circ g_m(X)^{-1} \circ g_{m-1}(X)$$
is an additive polynomial and a least common outer multiple of the polynomials $g_1(X), g_2(X)$.
3. Define a generalized additive polynomial as a finite sum of $a_i X^{p^i}$ with $i \in \mathbb{Z}$. Show that generalized additive polynomials form a ring under addition and composition. For a generalized additive polynomial $f(X) = \sum a_i X^{p^i}$ put $f^*(X) = \sum a_i^{p^{-i}} X^{p^{-i}}$.
   a) Show that $(f+g)^* = f^* + g^*, (f \circ g)^* = g^* \circ f^*, (f^*)^* = f$.
   b) Let K be a quasi-finite field of characteristic p. Show that an additive polynomial $f(X)$ over K has a nonzero root in K if and only if $f^*(X)$ does.
   c) Let K be quasi-finite, and let $f(X)$ be an additive polynomial over K. Show that the set $\{b \in K : b\wp(X)$ is an outer component of $f(X)\}$ is an additive group of order equal to the index of $f(K)$ in K.
   d) Let K be quasi-finite. Show that the number of roots in K of an additive polynomial $f(X)$ over K is equal to the number of roots in K of $f^*(X)$.
4. Let K be quasi-finite of characteristic p. Let $f(X)$ be an additive polynomial over K, and H the set of its roots in K^{sep}.
   a) Assume that there are no additive polynomials $h(X)$ of degree $< \deg(f)$, that are inner components of $f(X)$. Show that the degree of $K(H)/K$ is relatively prime to p.

b) Show that $f(X)$ is a composition of $\wp(X), X^p, aX$ with $a \in K$ if and only if $K(H)/K$ is a p-extension.

5. a) Show that Proposition (2.4) does not hold in general for a perfect field K of characteristic p.

b) Show that for the polynomial $h(X)$ of Proposition (2.6) it is not always true that $h(K) = K$.

6. ($\diamond$) (*V.G. Drinfeld* [Dr]) Let L be a finite extension of $\mathbb{F}_q((X))$, and let Γ be a finite discrete $\mathbb{F}_q[X]$-submodule of dimension d in L^{sep} such that $\text{Gal}(L^{\text{sep}}/L)$ acts trivially on Γ. Put

$$e_\Gamma(t) = t \prod_{\substack{a\in\Gamma \\ a\neq 0}} \left(1 - \frac{t}{a}\right).$$

Show that $e_\Gamma(t+u) = e_\Gamma(t) + e_\Gamma(u)$ and that e_Γ induces the isomorphism $L^{\text{alg}}/\Gamma \xrightarrow{\sim} L^{\text{alg}}$ of $\mathbb{F}_q[X]$-modules. Introduce a new structure of $\mathbb{F}_q[X]$-module on L^{alg}, putting $a * y = e_\Gamma(az)$ for $a \in \mathbb{F}_q[X]$, where $e_\Gamma(z) = y$, $z \in L^{\text{alg}}$. Show that

$$e_\Gamma(at) = a e_\Gamma(t) \prod_{\substack{b\in a^{-1}\Gamma/\Gamma \\ b\notin\Gamma}} \left(1 - \frac{e_\Gamma(t)}{e_\Gamma(b)}\right) = p_a\left(e_\Gamma(t)\right),$$

where $p_a(X) = \sum_{i=0}^n a_i X^{q^i}$, $n = d\deg a(X), a_0 = a$. The correspondence $a(X) \mapsto \sum_{i=0}^n a_i \Lambda^i$ determines an injective $\mathbb{F}_q$-homomorphism $\psi_\Gamma : \mathbb{F}_q[X] \to L^{\text{alg}}[\Lambda]$ (the ring of noncommutative polynomials, see (2.2)). This homomorphism is said to determine an elliptic $\mathbb{F}_q[X]$-module over $\mathbb{F}_q((X))$ of rank d.
Conversely, for a given ψ_Γ there uniquely exists a series

$$e_\Gamma(t) = t + \sum_{i\geqslant 1} b_i t^{q^i} \in L[[t]],$$

such that $e_\Gamma(Xt) = e_\Gamma(t)\psi_\Gamma(X)$ and $e_\Gamma(at) = e_\Gamma(t)\psi_\Gamma(a)$ for $a \in \mathbb{F}_q[X]$. The kernel of $e_\Gamma(t)$ is a finite discrete $\mathbb{F}_q[X]$-submodule Γ' of dimension d in L^{sep} and $\Gamma' = \Gamma$. (This construction is used to describe abelian ($d = 1$) and non-abelian ($d > 1$) extensions of $\mathbb{F}_q(X)$.)
For an introduction to Drinfeld modules see [Gos].

3. Normic Subgroups

In this section we apply the theory developed in the previous section to describe the norm subgroups in the case of a local field F with quasi-finite residue field. This theory was first obtained by *G. Whaples* [Wh1]. From our description it will follow that for infinite residue fields not every open subgroup of finite index is a norm subgroup (see Corollary 2 in (3.6)). We shall define the notion of a normic subgroup in (3.1) and the normic topology on F^* in (3.3). In (3.4) we prove the Existence Theorem which claims that there is a on-to-one correspondence between normic subgroups of finite index and norm subgroups of finite abelian extensions. Using the Existence Theorem we shall show in (3.6) that the kernel of the reciprocity map $\Psi_F\colon F^* \longrightarrow \operatorname{Gal}(F^{\mathrm{ab}}/F)$ is equal to the subgroup of divisible elements in F^*.

(3.1). Let π be a prime element in F.

DEFINITION. An open subgroup N in F^* is said to be *normic* if there exist polynomials $f_i(X) \in \mathcal{O}_F[X]$, such that the residue polynomials $\overline{f}_i(X) \in \overline{F}[X]$ are not constants and $1 + f_i(\alpha)\pi^i \in N$ for $\alpha \in \mathcal{O}_F, i > 0$.

This definition does not depend on the choice of a prime element π, because for $\pi' = \pi\varepsilon$ one can take $f_i'(X) = f_i(X)\varepsilon^{-i} \in \mathcal{O}_F[X]$. If $\overline{F} = \mathbb{F}_q$ is finite, then every open subgroup N in F^* is normic. Indeed, there exists an integer s such that $U_{s+1,F} \subset N$. Putting $f_i(X) = (X^q - X)^{p^s}$ for $1 \leqslant i \leqslant s$, we get $1 + f_i(\alpha)\pi^i \in U_{s+1,F}$ for $\alpha \in \mathcal{O}_F, i > 0$. If $\operatorname{char}(\overline{F}) = 0$, then the group $U_{1,F}$ is uniquely divisible and any open subgroup N of finite index in F^* contains $U_{1,F}$, and hence is normic. *From now on we shall assume that $\overline{F}$ is infinite of characteristic p.*

We may assume $f_i(0) = 0$, replacing $f_i(X)$ by $\widetilde{f}_i(X)$ otherwise, where $\widetilde{f}_i(X)\pi^i = \bigl(1+f_i(X)\pi^i\bigr)\bigl(1+f_i(0)\pi^i\bigr)^{-1} - 1$. By Proposition (2.9) there exist polynomials $g_{ij}(X) \in \overline{F}[X]$, such that $\sum_j \overline{f}_i\bigl(g_{ij}(X)\bigr)$ is a nonzero additive polynomial over $\overline{F}$. Then for polynomials $h_{ij}(X) \in \mathcal{O}_F[X]$, such that $\overline{h}_{ij} = g_{ij}$, and the polynomial $g_i(X) \in \mathcal{O}_F[X]$, such that

$$1 + g_i(X)\pi^i = \prod_j \Bigl(1 + f_i\bigl(h_{ij}(X)\pi^i\bigr)\Bigr),$$

we get $1+g_i(\alpha)\pi^i \in N$ for $i > 0, \alpha \in \mathcal{O}_F$, and $\overline{g}_i(X)$ is a nonzero additive polynomial over $\overline{F}$. Therefore, in the definition of a normic subgroup one can assume that the residue polynomial $\overline{f}_i(X)$ is nonzero additive over $\overline{F}$. In terms of the homomorphisms λ_i defined in section 5 Ch. I, we get

$$\lambda\bigl((N \cap U_{i,F})U_{i+1,F}/U_{i+1,F}\bigr) \supset \overline{f}_i(\overline{F}).$$

Since $\overline{f}_i(\overline{F})$ is of finite index in $\overline{F}$ by Proposition (2.4), we obtain that $N \cap U_{1,F}$ is of finite index in $U_{1,F}$.

(3.2). Now we show that the norm subgroups are normic.

PROPOSITION. *Let L be a finite Galois extension of F. Then $N_{L/F}L^*$ is a normic subgroup of finite index in F^*.*

Proof. Since the assertion holds in the case when $\overline{F}$ is finite, we assume that $\overline{F}$ is infinite. The arguments of Proposition (6.1) Ch. IV show that $N_{L/F}L^*$ is an open subgroup of finite index in F^*. Since the Galois group of L/F is solvable, it suffices to verify that for a cyclic extension L/F of prime degree the norm map $N_{L/F}$ transforms normic groups in L^* to normic groups in F^*.

Let L/F be unramified, and π a prime element in F. Let N be normic in L^*, $1 + f_i(\alpha)\pi^i \in N$ for $\alpha \in \mathcal{O}_L$, where $f_i(X) \in \mathcal{O}_L[X]$ is such that $\overline{f}_i$ is a nonzero additive polynomial over $\overline{L}$. Since the trace map $\operatorname{Tr}_{\overline{L}/\overline{F}}$ is surjective and $\overline{F}$ is infinite, the index of $\ker(\operatorname{Tr}_{\overline{L}/\overline{F}})$ in $\overline{L}$ is infinite. Proposition (2.4) implies that $\overline{f}_i(\overline{L})$ is of finite index in $\overline{L}$. Therefore, there exists an element $\beta \in \mathcal{O}_L$ with $\operatorname{Tr}_{\overline{L}/\overline{F}}\big(\overline{f}_i(\overline{\beta})\big) \neq 0$. Then Lemma (1.1) Ch. III shows that

$$N_{L/F}\big(1 + f_i(\beta\alpha)\pi^i\big) = 1 + g_i(\alpha)\pi^i \qquad \text{for } \alpha \in \mathcal{O}_F,$$

where $g_i(X) \in \mathcal{O}_F[X]$ with $\deg(\overline{g}_i) > 0$. Thus, $N_{L/F}(N)$ is normic in F^*.

Let L/F be a totally ramified Galois extension of prime degree n, π_L a prime element in L, $\pi_F = N_{L/F}\pi_L$. Let $n \nmid i$ and let $p(X) = X^n + \beta_{n-1}X^{n-1} + \dots + \beta_0$ be the monic irreducible polynomial of π_L^i over F. Then

$$N_{L/F}(1 - \alpha\pi_L^i) = \alpha^n p(\alpha^{-1}) = \beta_0\alpha^n + \beta_1\alpha^{n-1} + \dots + \beta_{n-1}\alpha + 1$$

for $\alpha \in \mathcal{O}_F$. Let $h_{L/F}$ be the Hasse-Herbrand function of L/F (see section 3 Ch. III). For $\alpha \in \mathcal{O}_F$ we get

$$N_{L/F}(1 + \alpha\pi_L^i) = 1 + g_i(\alpha)\pi_F^j$$

for a suitable polynomial $g_i(X)$ over $\mathcal{O}_F$ with $\bar{g}_i \neq 0$ and $j > 0$. The same assertion is trivially true also for $n \mid i$. Propositions (1.3), (1.5) Ch. III show that $j > h_{L/F}^{-1}(i)$ for $i \notin h_{L/F}(\mathbb{N})$, and if $i = h_{L/F}(j_0)$, then one may take $j = j_0$ and then $\deg\big(\bar{g}_i(X)\big) > 0$.

Let N be a normic subgroup in L^*, $U_{s+1,L} \subset N$, $1 + f_i(\alpha)\pi_L^i \in N$ for $\alpha \in \mathcal{O}_L$, where $f_i(X) \in \mathcal{O}_L[X]$ and $\deg\big(\overline{f}_i(X)\big) > 0$. Let $U_{r+1,F} \subset N_{L/F}(N)$. The previous arguments imply that for $i = h_{L/F}(j)$ there exists a polynomial $g(X)$ over $\mathcal{O}_F$, such that $\overline{g}(X) = \overline{g}_i\big(\overline{f}_i(X)\big)$ is not a constant and

$$N_{L/F}\big(1 + f_i(\alpha)\pi_L^i\big) \equiv 1 + g(\alpha)\pi_F^j \mod \pi_F^{r+1} \quad \text{for } \alpha \in \mathcal{O}_F.$$

Thus, $N_{L/F}(N)$ is normic. □

(3.3). PROPOSITION. *The normic subgroups in F^* determine in F^* a basis of neighborhoods in for the so called normic topology on F^*. If L/F is a finite Galois extension, then the norm map $N_{L/F}$ is continuous with respect to the normic topology.*

Proof. We must show that the intersection of two normic subgroups and the pre-image $N_{L/F}^{-1}$ of a normic subgroup is a normic subgroup. As $\operatorname{Gal}(L/F)$ is solvable, it suffices to verify the last assertion only for a cyclic extension of prime degree. Note that the pre-image of a normic subgroup is an open subgroup by the arguments in the proof of Proposition (6.1) Ch. IV.

Let L be either a totally ramified Galois extension of prime degree over F or $L = F$. Let N_1 be a normic subgroup in L^*, and N a normic subgroup in F^*. We shall verify that $N_1 \cap N_{L/F}^{-1}(N)$ is normic in L^*. This will complete the proof of the Proposition, except for the case of an unramified extension L/F. We leave the verification of the latter case to the reader. In fact, the case of an unramified extension will not be used in the sequel.

Let π_L be prime in L, $\pi_F = N_{L/F}\pi_L$. Let $f(X)$ be a polynomial over $\mathcal{O}_L$, such that $1 + f(\alpha)\pi_L^i \in N_1$ for $\alpha \in \mathcal{O}_L$ and $\overline{f}(X)$ is a nonzero additive polynomial over $\overline{L} = \overline{F}$. Then the arguments in the proof of the previous Proposition show that there exist a number j and a polynomial $g(X) \in \mathcal{O}_F[X]$, such that $N_{L/F}(1 + f(\alpha)\pi_L^i) \equiv 1 + g(\alpha)\pi_F^j \mod N$ for $\alpha \in \mathcal{O}_F$.

Let $q(X)$ be a polynomial over $\mathcal{O}_F$, such that $\overline{q}(X)$ is a nonzero additive polynomial over $\overline{F}$ and $1 + q(\alpha)\pi_F^j \in N$ for $\alpha \in \mathcal{O}_F$. Corollary 1 of (2.9) shows that there are polynomials $h_k(X) \in \mathcal{O}_F[X]$, such that $\sum_k \overline{h}_k(X)$ is a nonzero additive polynomial over $\overline{F}$, and

$$\sum \overline{g}(\overline{h}_k(X)) = \overline{q}(\overline{h}(X))$$

for some polynomial $h(X) \in \mathcal{O}_F[X]$, such that its residue polynomial $\overline{h}$ is additive. Define the polynomial $f_1(X)$ by the equality

$$1 + f_1(X)\pi_L^i = \prod_k \Big(1 + f\big(h_k(X)\big)\pi_L^i\Big).$$

Then $\overline{f}_1$ is a nonzero additive polynomial over $\overline{F}$,

$$1 + f_1(\alpha)\pi_L^i \in N_1 \qquad \text{for } \alpha \in \mathcal{O}_L,$$

and for $\alpha \in \mathcal{O}_F$

$$N_{L/F}(1 + f_1(\alpha)\pi_L^i) = \prod_k (1 + g\big(h_k(\alpha)\big)\pi_F^j) = (1 + q(h(\alpha))\pi_F^j)(1 + g_1(\alpha)\pi_F^{j+1})$$

for some polynomial $g_1(X) \in \mathcal{O}_F[X]$. Therefore,

$$N_{L/F}\big(1 + f_1(\alpha)\pi_L^i\big) \equiv 1 + g_1(\alpha)\pi_F^{j+1} \mod N \qquad \text{for } \alpha \in \mathcal{O}_F.$$

Proceeding in this way, one can find $f_m(X) \in \mathcal{O}_L[X]$ and $g_m(X) \in \mathcal{O}_F[X]$, such that $\overline{f}_m(X)$ is a nonzero additive polynomial over $\overline{F}$, $1 + f_m(\alpha)\pi_L^i \in N_1$ for $\alpha \in \mathcal{O}_L$, and

$$N_{L/F}\big(1 + f_m(\alpha)\pi_L^i\big) \equiv 1 + g_m(\alpha)\pi_F^{j+m} \mod N \qquad \text{for } \alpha \in \mathcal{O}_F.$$

Since $U_{r+1,F} \subset N$ for some integer $r > 0$, we obtain that $1 + f_m(\alpha)\pi_L^i \in N_1 \cap N_{L/F}^{-1}(N)$ for sufficiently large $m, \alpha \in \mathcal{O}_F$. Let $U_{s+1,L} \subset N_1$ and $N_{L/F}(U_{s+1,L}) \subset N$. Subsections (5.7) and (5.8) of Ch. I imply that $U_{1,L}^{p^t} \subset U_{s+1,L}$ for sufficiently large

t. As $\overline{L} = \overline{F}$, we deduce that $\mathcal{O}_L^{p^t} \subset \mathcal{O}_F U_{s+1,L}$, and hence

$$1 + f_m(\alpha^{p^t})\pi_L^i \in N_1 \cap N_{L/F}^{-1}(N) \qquad \text{for } \alpha \in \mathcal{O}_L.$$

This means that $N_1 \cap N_{L/F}^{-1}(N)$ is normic. □

(3.4). THEOREM ("EXISTENCE THEOREM"). *Let F be a local field with quasi-finite residue field. There is a one-to-one correspondence between normic subgroups of finite index in F^* and the norm subgroups of finite abelian extensions: $N \longleftrightarrow N_{L/F}L^*$. This correspondence is an order reversing bijection between the lattice of normic subgroups of finite index in F^* and the lattice of finite abelian extensions of F.*

Proof. Similarly to the proof of Theorem (6.2) Ch. IV, it suffices to verify that a normic subgroup of finite index contains $N_{L/F}L^*$ for some finite separable extension L/F.

Let N be a normic subgroup of index n in F^* and $\text{char}(\overline{F}) \nmid n$. Then $N \supset U_{1,F}$, and the arguments in (1.5) Ch. IV show that N coincides with $N_{L/F}L^*$ for some tamely ramified abelian extension of degree n.

Let $n = \text{char}(\overline{F}) = p$. If $U_{1,F} \subset N$, then $U_F \subset N$ and a prime element π of F does not belong to N. In this case $N = N_{L/F}L^*$, where L is the unramified extension of degree p over F. Let $U_{s,F} \not\subset N, U_{s+1,F} \subset N$ for $s \geqslant 1$. As N is normic, we get in terms of the homomorphism λ_i from section 5 Ch. I that

$$\lambda_i\big((N \cap U_{i,F})U_{i+1,F}/U_{i+1,F}\big) = \overline{F}, \qquad \text{if} \quad i \neq s$$
$$\lambda_i\big((N \cap U_{i,F})U_{i+1,F}/U_{i+1,F}\big) = \overline{\eta}\wp\left(\overline{F}\right), \qquad \text{if} \quad i = s,$$

where η is a proper element of $\mathcal{O}_F$. The arguments in (1.5) Ch. IV show that there exists a cyclic extension L/F of degree p (a Kummer extension or an Artin–Schreier extension), such that the $\lambda_i\big((N_{L/F}L^* \cap U_{i,F})U_{i+1,F}/U_{i+1,F}\big)$ are the same as those for N, and some prime element π in F is contained in $N \cap N_{L/F}L^*$.

If $s = 1$, then obviously $N = N_{L/F}L^*$. Otherwise we can proceed by induction on s. If $N \neq N_{L/F}L^*$, then the group $N \cap N_{L/F}L^*$ is normic of index p^2 in F^* by Propositions (3.2) and (3.3). Therefore, there exists an integer $s_1 < s$ such that

$$\lambda_{s_1}\big((N \cap N_{L/F}L^* \cap U_{s_1,F})U_{s_1+1,F}/U_{s_1+1,F}\big) \neq \overline{F}.$$

Then the group $(N \cap N_{L/F}L^*)U_{s_1+1,F}$ is normic of index p in F^*. By the induction assumption $(N \cap N_{L/F}L^*)U_{s_1+1,F} = N_{L_1/F}L_1^*$ for a cyclic extension L_1/F of degree p. Then N contains the group $N \cap N_{L/F}L^* = N_{L_1/F}L_1^* \cap N_{L/F}L^*$, which coincides with $N_{L_1L/F}(L_1L)^*$ and is a norm subgroup in F^*. Therefore, N is a norm subgroup.

The case $n = p^m$ can be considered in the same way as in the proof of Theorem (6.2) Ch. IV, using Propositions (3.2) and (3.3). Now, repeating the arguments in the proof of Theorem (6.2) Ch. IV, we obtain that every normic subgroup of finite index in F^* is a norm subgroup. The remaining assertions of the Theorem are proved similarly, as in the proof of Theorem (6.2). □

REMARK. Another proof of this Theorem can be carried out using pairings of the multiplicative group F^*, similarly to the proof of Theorem (6.2) Ch. IV. For the case of $\operatorname{char}(F) = p$ see [Sek1]; that paper also contains another description of normic subgroups.

(3.5). There exists another description of normic subgroups, more convenient in some cases. Let $\operatorname{char}(\overline{F}) = p$ and let

$$\mathcal{E}(\,\cdot\,, X)\colon W(\overline{F}) \longrightarrow 1 + X\mathcal{O}_F[[X]]$$

be the Artin–Hasse map (see (9.3) Ch. I). We keep the notations of section 9 Ch. I. For an element $\alpha \in W(\overline{F})$ and a prime element π in F we put $\mathcal{E}(\alpha, \pi^i) = \mathcal{E}(\alpha, X^i)|_{X=\pi}$. Note that $\mathcal{E}(\alpha, \pi^i) \equiv 1 + r(c_0)\pi^i \mod \pi^{i+1}$; this follows from Proposition (9.3) Ch. I, where $\alpha = \sum_{i\geqslant 0} r_0(c_i)p^i$ with $c_i \in \overline{F}$, r_0 is the Teichmüller map $\overline{F} \to W(\overline{F})$, and r is the Teichmüller map $\overline{F} \to \mathcal{O}_F$. An advantage of introducing the map $\mathcal{E}(\,\cdot\,, \pi^i)\colon W(\overline{F}) \to U_{i,F}$ is its linearity: $\mathcal{E}(\alpha + \beta, \pi^i) = \mathcal{E}(\alpha, \pi^i)\mathcal{E}(\beta, \pi^i)$.

Let $\mathcal{A}$ denote the ring of linear operators on $W(\overline{F})$ of the form

$$A = \sum_{m=0}^{n} \alpha_m \mathbf{F}^m, \qquad \alpha_m \in W(\overline{F}),$$

where $\mathbf{F}$ is the Frobenius map (see section 8 Ch. I). Then

$$A(\beta) = \sum_{m=0}^{n} \alpha_m \mathbf{F}^m(\beta)$$

for $\beta \in W(\overline{F})$. The ring $\mathcal{A}$ is isomorphic to the ring of noncommutative polynomials $W(\overline{F})[\Lambda]$ mentioned in (2.2):

$$\sum_{m=0}^{n} \alpha_m \mathbf{F}^m \mapsto \sum_{m=0}^{n} \alpha_m \Lambda^m.$$

Since $\overline{F}$ is perfect, arguments similar to those in (2.2) show that the ring $\mathcal{A}$ is a left and right Euclidean principal ideal ring under addition and composition.

There is also the natural homomorphism from the ring $\mathcal{A}$ to the ring of additive polynomials over $\overline{F}$:

$$A = \sum_{m=0}^{n} \alpha_m \mathbf{F}^m \mapsto \overline{A} = \sum_{m=0}^{n} \overline{\alpha}_m X^{p^m} \in \overline{F}[X].$$

PROPOSITION. *An open subgroup N in F^* is normic if and only if for a prime element π in F there exists a linear operator $A \in \mathcal{A}$, such that* $\deg(\overline{A}) > 0$ *and* $\mathcal{E}(A(\alpha), \pi^i) \in N$ *for all* $\alpha \in W(\overline{F}), i > 0$.

Proof. Let $U_{s+1,F} \subset N$. Suppose that there exists a linear operator $A \in \mathcal{A}$, with the properties indicated in the Proposition. Put

$$\mathcal{E}\big(Ar_0(a), \pi^i\big) \equiv 1 + f\big(r(a)\big)\pi^i \mod \pi^{s+1} \qquad \text{for } a \in \overline{F},$$

where $f(X) \in \mathcal{O}_F[X]$ and $\deg(\overline{f}) > 0$. If β is an element in $\mathcal{O}_F$ such that $\overline{\beta} = a$, then Lemma (7.2) Ch. I shows that $r(a^{p^s}) = r(a)^{p^s} \equiv \beta^{p^s} \mod \pi^{s+1}$. Therefore, $1 + f(\beta^{p^s})\pi^i \in N$ for $\beta \in \mathcal{O}_F$ and $\deg(\overline{f(X^{p^s})}) > 0$. Thus, N is normic.

Conversely, let N be a normic subgroup and $U_{s+1,F} \subset N$. We saw in (3.1) that $N \cap U_{1,F}$ is of finite index in $U_{1,F}$. Let p^m be this index. Then $\mathcal{E}\big(p^m A(\alpha), \pi^i\big) \in N$ for $\alpha \in W(\overline{F}), i \geqslant 0$, where A is any linear operator in $\mathcal{A}$.

Let $f(X) \in \mathcal{O}_F[X]$ be as above, $g(X) = f(X^{p^{m-1}})$. Then writing $\alpha \equiv r_0(a)$ mod $pW(\overline{F})$ for elements $\alpha \in W(\overline{F})$ and $a \in \overline{F}$, we obtain

$$\mathcal{E}\big(p^{m-1}A(\alpha), \pi^i\big) \equiv \mathcal{E}\Big(p^{m-1}A\big(r_0(a)\big), \pi^i\Big) \equiv 1 + g\big(r(a)\big)\pi^i \mod N,$$

Proposition (2.9) shows that there are elements $\alpha_i \in W(\overline{F})$, such that $\sum \alpha_j = 1$ and $\sum \overline{g}(\overline{\alpha}_j X)$ is an additive polynomial over $\overline{F}$. Then

$$\mathcal{E}\big(p^{m-1}A(\alpha), \pi^i\big) \equiv \prod\Big(1 + g\big(r(\alpha_j a)\big)\Big)\pi^i \mod N,$$

and we may assume, without loss of generality, that $\overline{g}$ is an additive polynomial over $\overline{F}$.

Let $h_i(X)$ be a polynomial over $\mathcal{O}_F$, such that $\overline{h}_i$ is a nonzero additive polynomial and

$$1 + h_i(\alpha)\pi^i \in N \qquad \text{for } \alpha \in \mathcal{O}_F.$$

Choose an operator $A_1 \in \mathcal{A}$ such that the polynomial $\overline{g} \circ \overline{A}_1$ has an outer component $\overline{h}_i(X)$. Then

$$\mathcal{E}\big(p^{m-1}AA_1(\alpha), \pi^i\big) \equiv 1 + g\big(A_1 r(a)\big)\pi^i \equiv 1 + g_1\big(r(a)\big)\pi^{i+1} \mod N$$

for some polynomial $g_1(X) \in \mathcal{O}_F[X]$. Continuing in this way, one can find operators $A_2, \dots \in \mathcal{A}$ such that for $B_1^{(i)} = AA_1A_2\dots$

$$\mathcal{E}\big(p^{m-1}B_1^{(i)}(\alpha), \pi^i\big) \in N \qquad \text{for } \alpha \in W(\overline{F}).$$

Proceeding by induction on m, we conclude that there exist operators $B_m^{(i)} \in \mathcal{A}$ such that

$$\mathcal{E}\big(B_m^{(i)}(\alpha), \pi^i\big) \in N \qquad \text{for } \alpha \in W(\overline{F}), 0 < i \leqslant s.$$

Now let $B \in \mathcal{A}$ be a least common outer multiple of the $B_m^{(i)}$. Then we deduce that $\mathcal{E}\big(B(\alpha), \pi^i\big) \in N$, as desired. □

COROLLARY. *An open subgroup N in F^* is normic if and only if there exist polynomials $p_i(X) \in \mathcal{O}_F[X]$, such that the polynomial $\overline{p}_i$ is of positive degree and $\mathcal{E}\big(p_i(\alpha), \pi^i\big) \in N$ for $\alpha \in W(\overline{F})$, $i > 0$.*

(3.6). Finally, we shall find another characterization of normic subgroups.

Let N be an open subgroup of index p in $U_{1,F}$. Let $U_{s+1,F} \subset N$ and $U_{s,F} \not\subset N$. Then the group

$$H = \lambda_s\big((N \cap U_{s,F})U_{s+1,F}/U_{s+1,F}\big)$$

is of index p in $\overline{F}$. For every i, $0 < i < s$, and $\alpha \in W(\overline{F})$ there exists an element $f_i(\alpha) \in W(\overline{F})$ such that

$$\mathcal{E}(\alpha, \pi^i)\mathcal{E}\big(f_i(\alpha), \pi^s\big) \in N.$$

Then $\mathcal{E}\big(f_i(\alpha+\beta), \pi^s\big) \equiv \mathcal{E}(\alpha+\beta, \pi^i)^{-1} = \mathcal{E}(\alpha, \pi^i)^{-1}\mathcal{E}(\beta, \pi^i)^{-1} \mod N$. We obtain that $\mathcal{E}\big(f_i(\alpha+\beta) - f_i(\alpha) - f_i(\beta), \pi^s\big) \in N$ and

$$f_i(\alpha+\beta) \equiv f_i(\alpha) + f_i(\beta) \mod H.$$

Since $\mathcal{E}(p\alpha, \pi^i) \in N$, we deduce that f_i, in fact, depends on the residue classes of α mod $pW(\overline{F})$. Hence, f_i induces the linear homomorphism $\overline{f}_i : \overline{F} \to \overline{F}/H$.

PROPOSITION. *Let N be a subgroup of index p in $U_{1,F}$ such that*

$$U_{s+1,F} \subset N \qquad \text{and} \qquad U_{s,F} \not\subset N$$

for some $s \geqslant 1$. Then $N \leftrightarrow (H, \overline{f}_1, \dots, \overline{f}_{s-1})$ is a one-to-one correspondence between such subgroups and sequences of a subgroup H of index p in $\overline{F}$, and homomorphisms $\overline{f}_i : \overline{F} \to \overline{F}/H$. A subgroup N is normic if and only if H is open in the additive topology on $\overline{F}$ and the homomorphisms $\overline{f}_i$ are induced by additive polynomials.

Proof. Assume that $(H, \overline{f}_1, \dots, \overline{f}_{s-1}) \neq (H', \overline{f}'_1, \dots, \overline{f}'_{s-1})$. If $H \neq H'$, then clearly $N \neq N'$. If $\overline{f}_i \neq \overline{f}'_i$ then $NN' = U_{1,F}$ and $N \neq N'$.

If N is normic then H is open. Let $g_i(X)$ be a polynomial over $\mathcal{O}_F$, such that $\overline{g}_i$ is a nonzero additive polynomial over $\overline{F}$ and $1 + g_i(\alpha)\pi^i \in N$ for $\alpha \in \mathcal{O}_F$. Then $(\overline{f}_i\overline{g}_i)(\overline{F}) = 0$, and Proposition (2.4) shows that $\overline{g}_i(\overline{F})$ is of finite index in $\overline{F}$. Therefore, by Corollary 2 of (2.7) the homomorphism $\overline{f}_i : \overline{F}/\overline{g}_i(\overline{F}) \to \overline{F}/H$ is induced by an additive polynomial. Conversely, let H be open in the additive topology on $\overline{F}$ and let $g(X) \in \mathcal{O}_F[X]$ be such that $\overline{g}(\overline{F}) = H$ and $\overline{g}$ is an additive polynomial. Let $g_i(X) \in \mathcal{O}_F[X]$ be such that $\overline{f}_i$ is induced by an additive polynomial $\overline{g}_i(X)$ over $\overline{F}$. Let $\overline{h}_i(X) \in \mathcal{O}_F[X]$ be such that $\overline{h}_i$ is a least common outer multiple of $\overline{g}, \overline{g}_i$. Put

$\overline{h}_i = \overline{g}_i \circ \overline{p}_i$ with $p_i(X) \in \mathcal{O}_F[X]$. Then

$$\mathcal{E}\big(g(\alpha), \pi^s\big) \in N$$
$$\mathcal{E}\big(p_i(\alpha), \pi^i\big) \equiv \mathcal{E}\big(g_i p_i(\alpha), \pi^s\big) \equiv 1 \mod N,$$

for $\alpha \in W(\overline{F})$. Now Corollary (3.5) shows that N is normic. □

COROLLARY 1. *The reciprocity map Ψ_F is continuous with respect to the normic topology on F^*. Its kernel coincides with the subgroup of divisible elements in F^*.*

Proof. Denote $\Lambda_F = \cap N_{L/F} L^*$. The intersection of all normic subgroups of index l coincides with F^{*l}. Hence, $\Lambda_F = \cap F^{*l}$. Fix l. For every $a \in \Lambda_F$ and every L there is $b \in F^*$ such that $a = b^l$ and $b \in N_{L/F} L^*$. Therefore, the intersection of finitely many closed subgroups $N_{L_i/F} L_i^*$ with the finite discrete set $\sqrt[l]{a}$ is nonempty. Then there is $c \in \sqrt[l]{a}$ which belongs to Λ_F. Thus, Λ_F is l-divisible. It coincides with the subgroup of multiplicative representatives of $\overline{F}$ in F which are in the image of the subgroup of divisible elements of $\overline{F}$ in F. □

COROLLARY 2. *Let* $\operatorname{char}(\overline{F}) = p$. *Suppose that the cardinality of F is q. The set of all subgroups $N \cap U_{1,F}$ for normic subgroups N of finite index has the cardinality q (we assume that q is not finite). The set of all open subgroups N in F^* of finite index in $U_{1,F}$, such that*

$$\lambda_i\big(U_{i+1,F}(N \cap U_{i,F})/U_{i+1,F}\big)$$

is open in $\overline{F}$ with respect to the additive topology for $i > 0$, has the cardinality 2^q. The set of all open subgroups of finite index in $U_{1,F}$ has the cardinality 2^q.

Proof. For every normic subgroup N of finite index in F^* there is a totally ramified extension L/F such that $N \cap U_{1,F} = N_{L/F} L^* \cap U_{1,F}$. This extension is obtained by adjoining a root of a polynomial, such that its coefficients may be written as polynomials in a prime element π of F with coefficients in $r(\overline{F})$ (see Exercise 5 in section 3 Ch. II). Therefore, there are at most q such extensions. By the previous Proposition there are q normic subgroups N of index p in $U_{1,F}$ such that $U_{2,F} \subset N$. We conclude that there are q normic subgroups of finite index. This Proposition also shows that there are 2^q open subgroups of index p in $U_{1,F}$, since there are 2^q subgroups H of index p in $\overline{F}$ and 2^q homomorphisms of $\overline{F}$ to $\overline{F}/H$. Therefore, there are 2^q open subgroups of finite index in $U_{1,F}$.

Assume that if $\operatorname{char}(F) = 0, p > 2$, then the absolute index of ramification $e(F) \neq 1$. Then Corollary 2 of (5.8) Ch. I shows that there exists an index $s > 1, p \nmid s$, such that $U_{s+1,F} \subset U_F^p, U_{s,F} \not\subset U_F^p$. Choose a subgroup H of index p in $\overline{F}$ open in the additive topology, such that

$$\lambda_s\big(U_{s+1,F}(U_{1,F}^p \cap U_{s,F})/U_{s+1,F}\big) \subset H.$$

As there are 2^q homomorphisms of $\overline{F}$ to $\overline{F}/H$, using the previous Proposition we conclude that there are 2^q open subgroups N of index p in $U_{1,F}$, such that

$$\lambda_s\big(U_{s+1,F}(N\cap U_{s,F})/U_{s+1,F}\big)=H \quad \text{and} \quad \lambda_i\big(U_{i+1,F}(N\cap U_{i,F})/U_{i+1,F}\big)=\overline{F}$$

for $i\neq s$. If $\operatorname{char}(F)=0, p>2, e(F)=1$, then it is straightforward to show that there are 2^q open subgroups N of index p^2 in $U_{1,F}$, such that their images $\lambda_i\big(U_{i+1,F}(N\cap U_{i,F})/U_{i+1,F}\big)$ are open in the additive topology of $\overline{F}$. Thus, in the general case there are 2^q such open subgroups of finite index. □

REMARK. Another description of normic groups, using the language of algebraic groups over $\overline{F}$, can be found in [Se3, sect. 2 Ch. XV].

Exercises.

1. Show that for a normic group $N\subset U_{1,F}$, such that $U_{s+1,F}\subset N$, there exists a sequence of linear operators $A_{ij}\in\mathcal{A}, 1\leqslant i\leqslant s, i\leqslant j\leqslant s$, such that $\overline{A}_{ii}(\overline{F})=\lambda_i\big(U_{i+1,F}(N\cap U_{i,F})/U_{i+1,F}\big)$ and N is generated by $U_{s+1,F}$ and the elements
$$\beta_i(\alpha)=\prod_{j=i}^{s}\mathcal{E}\big(A_{ij}(\alpha),\pi^j\big),\qquad \alpha\in W(\overline{F}), 1\leqslant i\leqslant s.$$
2. An open subgroup N of finite index in $U_{1,F}$, such that $\lambda_i\big(U_{i+1,F}(N\cap U_{i,F})/U_{i+1,F}\big)$ are open in the additive topology of $\overline{F}$ for all $i>0$, is called pseudonormic.
Show that the intersection of two pseudonormic subgroups is not always pseudonormic when $\overline{F}$ is an infinite field of characteristic p.
3. Generalize the arguments of (6.4) Ch. IV to a local field with quasi-finite residue field.
4. Using Example 2 in section 1 and Exercise 3, describe abelian extensions of a local field with an algebraically closed residue field.
5. (◇) Let F be a local field such that its residue field is a Brauer field (see Exercises 4, 5 in section 1). The notion of a normic group in F^* is the same as in the previous section. Show that normic groups of index n that divides $\deg(\overline{F}^{\mathrm{sep}}/\overline{F})=\prod l^{n(l)}$ (n is odd when $n(2)=1$) are in one-to-one correspondence with finite abelian extensions of degree n.

4. Local p-Class Field Theory

In this section we consider a local field F with perfect residue field of characteristic p and describe its abelian totally ramified p-extensions by using the group of principal units $U_{1,F}$. This theory is a generalization of the theory of Chapter IV, and the methods of section 2 and 3 of that chapter. Note that abelian totally tamely ramified extensions are described by Kummer theory (see Exercise 9 section 1) and unramified extensions just correspond to separable extensions of the residue field.

Let $\widetilde{F}$ denote the maximal abelian unramified p-extension of F and let L/F be a finite Galois totally ramified p-extension. We shall show in (4.5) that a generalization $\Upsilon_{L/F}$ of the Neukirch map of section 2 Ch. IV induces an isomorphism

$$\operatorname{Hom}_{\mathbb{Z}_p}\big(\operatorname{Gal}(\widetilde{F}/F), \operatorname{Gal}(L/F)^{\mathrm{ab}}\big) \xrightarrow{\sim} U_F/N_{L/F}U_L,$$

where $\operatorname{Hom}_{\mathbb{Z}_p}$ denotes continuous $\mathbb{Z}_p$-homomorphisms from the group $\operatorname{Gal}(\widetilde{F}/F)$ endowed with the topology of profinite group to the discrete finite group $\operatorname{Gal}(L/F)^{\mathrm{ab}}$. In (4.6)–(4.9) we show how various results of class field theory for local fields with (quasi-)finite residue field can be generalized in p-class field theory.

(4.1). Let F be a complete (or Henselian) discrete valuation field with perfect residue field $\overline{F}$ of characteristic $p > 0$. Let $\wp(X)$ denote as usually the polynomial $X^p - X$. Let

$$\kappa = \dim_{\mathbb{F}_p} \overline{F}/\wp(\overline{F}).$$

Further we will assume that $\kappa \neq 0$. This means that the field $\overline{F}$ is not separably p-closed, i.e., it has nontrivial separable extensions of degree p. If $\overline{F}$ is quasi-finite, then $\kappa = 1$.

REMARK. If $\kappa = 0$ then the field $\overline{F}$ is separably p-closed. By choosing nontrivial perfect subfields of it and local fields $F_i \subset F$ having them as residue fields and containing a prime element of F for sufficiently large i one can view extensions of F as coming from extensions of local fields F_i. Then one can describe abelian totally ramified p-extensions of F using the description for F_i similarly to Example 1 of (6.6) Ch. IV.

Denote by $\widetilde{F}$ the maximal abelian unramified p-extension of F. Due to Witt theory (see Exercise 6 section 5 Ch. IV) there is a canonical isomorphism

$$\operatorname{Gal}(\widetilde{F}/F) \simeq \operatorname{Hom}(W(\overline{F})/\wp W(\overline{F}) \otimes \mathbb{Q}_p/\mathbb{Z}_p, \mathbb{Q}_p/\mathbb{Z}_p).$$

Non-canonically $\operatorname{Gal}(\widetilde{F}/F)$ is isomorphic to $\prod_\kappa \mathbb{Z}_p$ (we have a canonically defined generator of this group, the Frobenius automorphism, only when the residue field is finite).

Let L/F be a Galois totally ramified p-extension. Then $\operatorname{Gal}(L/F)$ can be identified with $\operatorname{Gal}(\widetilde{L}/\widetilde{F})$, and $\operatorname{Gal}(\widetilde{L}/F)$ is isomorphic to $\operatorname{Gal}(\widetilde{L}/\widetilde{F}) \times \operatorname{Gal}(\widetilde{L}/L)$.

DEFINITION. Denote by $\operatorname{Gal}(L/F)^{\sim} = \operatorname{Hom}_{\mathbb{Z}_p}\big(\operatorname{Gal}(\widetilde{F}/F), \operatorname{Gal}(L/F)\big)$ the group of continuous homomorphisms from the profinite group $\operatorname{Gal}(\widetilde{F}/F)$ which is a $\mathbb{Z}_p$-module ($a \cdot \sigma = \sigma^a$, $a \in \mathbb{Z}_p$) to the discrete $\mathbb{Z}_p$-module $\operatorname{Gal}(L/F)$.

By Witt theory there is a canonical isomorphism

$$\operatorname{Hom}_{\mathbb{Z}_p}(\operatorname{Gal}(\widetilde{F}/F), \mathbb{Z}/p^n\mathbb{Z}) \simeq W_n(\overline{F})/\wp(W_n(\overline{F})).$$

Hence if $\operatorname{Gal}(L/F)^{p^n} = \{1\}$ then $\operatorname{Gal}(L/F)^{\sim}$ is canonically isomorphic to $\operatorname{Gal}(L/F)\otimes W_n(\overline{F})/\wp(W_n(\overline{F}))$. This group is non-canonically isomorphic to $\oplus_\kappa \operatorname{Gal}(L/F)$.

Let in addition the degree of L/F be finite. For $\chi \in \operatorname{Gal}(L/F)^{\sim}$ denote by Σ_χ the fixed field of all $\sigma_\varphi \in \operatorname{Gal}(\widetilde{L}/F)$, where $\sigma_\varphi\big|_{\widetilde{F}} = \varphi, \sigma_\varphi\big|_L = \chi(\varphi)$ and φ runs over all elements (or just a topological $\mathbb{Z}_p$-basis) of $\operatorname{Gal}(\widetilde{F}/F)$.

Then $\Sigma_\chi \cap \widetilde{F} = F$, i.e., Σ_χ/F is a totally ramified p-extension.

DEFINITION. For $\chi \in \operatorname{Gal}(L/F)^{\sim}$ let π_χ be a prime element of Σ_χ. Put

$$\Upsilon_{L/F}(\chi) = N_{\Sigma_\chi/F}\pi_\chi N_{L/F}\pi_L^{-1} \mod N_{L/F}U_L,$$

where π_L is a prime element in L.

(4.2). LEMMA. *The map* $\Upsilon_{L/F}\colon \operatorname{Gal}(L/F)^{\sim} \longrightarrow U_F/N_{L/F}U_L$ *is well defined.*

Proof. $\Upsilon_{L/F}$ does not depend on the choice of π_L. Let M be the compositum of Σ_χ and L. Then M/Σ_χ is unramified and a prime element in Σ_χ can be written as $\pi_\chi N_{M/\Sigma_\chi}\varepsilon$ for a suitable $\varepsilon \in U_M$. Since $N_{M/F}\varepsilon = N_{L/F}(N_{M/L}\varepsilon) \in N_{L/F}U_L$, we complete the proof. □

Since we are working with p-extensions, the inclusion $U_{1,F} \to U_F$ induces an isomorphism

$$U_{1,F}/N_{L/F}U_{1,L} \simeq U_F/N_{L/F}U_L.$$

From now on we assume that the image of $\Upsilon_{L/F}$ is in $U_{1,F}/N_{L/F}U_{1,L}$.

(4.3). Denote by $\widehat{F}$ the completion of $\widetilde{F}$. We always assume that $\widehat{F} \subset \widehat{L}$. We also denote an extension of $\varphi \in \operatorname{Gal}(\widetilde{F}/F)$ on $\widetilde{L}$ by φ, and use the same notation for its continuous extension on $\widehat{L}$.

For every finite Galois totally ramified p-extension L/F in accordance with Remark in (1.6) Ch. IV the norm map $N_{\widehat{L}/\widehat{F}}\colon U_{1,\widehat{L}} \to U_{1,\widehat{F}}$ is surjective (working with totally ramified p-extensions and principal units we can use $\widetilde{F}$ instead of the maximal unramified extension of F).

Now we introduce the map inverse to $\Upsilon_{L/F}$.

Let L/F be a finite Galois totally ramified p-extension. Let $\varepsilon \in U_{1,F}$ and $\varphi \in \operatorname{Gal}(\widetilde{F}/F)$. Let $\eta \in U_{1,\widehat{L}}$ be such that $N_{\widehat{L}/\widehat{F}}\eta = \varepsilon$. Let π_L be a prime element in L. Since $N_{\widehat{L}/\widehat{F}}(\eta^{\varphi-1}) = 1$, we deduce from Proposition and Remark in (1.7) Ch. IV that $\eta^{\varphi-1} \equiv \pi_L^{1-\sigma} \mod U(\widehat{L}/\widehat{F})$ for a $\sigma \in \operatorname{Gal}(\widehat{L}/\widehat{F})$ which is uniquely determined as an element of $\operatorname{Gal}(\widehat{L}/\widehat{F})^{\mathrm{ab}}$. Similarly to Lemma (3.1) Ch. IV the element σ does not depend on the choice of η. Set $\chi(\varphi) = \sigma\big|_L$. Then $\chi(\varphi_1\varphi_2) = \sigma_1\sigma_2$, since

$$\eta^{\varphi_1\varphi_2-1} \equiv \eta^{\varphi_1-1}(\eta^{\varphi_2-1})^{\varphi_1} \equiv \pi_L^{1-\sigma_1}\pi_L^{1-\sigma_2} \equiv \pi_L^{1-\sigma_1\sigma_2} \mod U(\widehat{L}/\widehat{F}).$$

This means $\chi \in (\mathrm{Gal}(L/F)^{\mathrm{ab}})^{\sim}$.

Similarly to the proof of Lemma (3.1) Ch. IV we deduce that the map $\Psi_{L/F}\colon U_{1,F}/N_{L/F}U_{1,L} \longrightarrow (\mathrm{Gal}(L/F)^{\mathrm{ab}})^{\sim}$, $\varepsilon \mapsto \chi$ is a homomorphism.

(4.4). We need the following auxiliary assertions.

LEMMA. *Let L/F be a finite Galois totally ramified p-extension. Let φ_i, $i \in I$, be $\mathbb{Z}_p$-linearly independent elements of $\mathrm{Gal}(\widetilde{F}/F)$. Let $\psi \in \mathrm{Gal}(\widetilde{L}/F)$ be such that its restriction on $\widetilde{F}$ and φ_i, $i \in I$, are $\mathbb{Z}_p$-linearly independent. Let E be the fixed subfield of $\{\varphi_i\}_{i\in I}$ in $\widehat{L}$, and let $\varepsilon \in U_{n,E}$, $n \geqslant 1$. Then there exists $\eta \in U_{n,E}$ such that $\varepsilon = \eta^{\psi-1}$.*

Proof. Note that E is a complete field and one can construct the desired element η similarly to the proof of Proposition (1.8) Ch. IV. □

PROPOSITION. *Let L/F be a finite Galois totally ramified p-extension, and M/F be a Galois subextension in L/F, $M \neq L$. Let $\psi_i \in \mathrm{Gal}(\widetilde{L}/F)$, $i \in I$, be such that $\psi_i|_{\widetilde{F}}$ are $\mathbb{Z}_p$-linearly independent in $\mathrm{Gal}(\widetilde{F}/F)$. For a set $S \subset \widehat{L}$ let $S^{\langle\psi_i\rangle}$ denote the set of the fixed elements under the action of all ψ_i.*

Let $\alpha \in \widehat{L}^*$. Then

(1) *if $\alpha^{\psi_i-1} \in U(\widehat{L}/\widehat{M})$ for all $i \in I$, then $\alpha \in U(\widehat{L}/\widehat{M})\widehat{L}^{\langle\psi_i\rangle}$.*

Assume that there is $\psi \in \mathrm{Gal}(\widetilde{L}/F)$ such that $\psi|_{\widetilde{F}}$ and $\{\psi_i|_{\widetilde{F}}\}_{i\in I}$ are $\mathbb{Z}_p$-linearly independent. Then

(2) $U(\widehat{L}/\widehat{M})^{\langle\psi_i\rangle} = \left(U(\widehat{L}/\widehat{M})^{\langle\psi_i\rangle}\right)^{\psi-1}$.

(3) $U(\widehat{M}/\widehat{F})^{\langle\psi_i\rangle} = N_{\widehat{L}/\widehat{M}}\left(U(\widehat{L}/\widehat{F})^{\langle\psi_i\rangle}\right)$.

Proof. In the first half of the proof assume that $|L{:}M| = p$. Let $s = s(L|M)$ be as in (1.4) Ch. III.

First we prove (2). Let $E = \widehat{L}^{\langle\psi_i\rangle}$ be the fixed field of ψ_i, $i \in I$. Let σ be a generator of $\mathrm{Gal}(\widehat{L}/\widehat{M})$. If $\varepsilon \in U(\widehat{L}/\widehat{M})^{\langle\psi_i\rangle}$, then it follows from (1.5) Ch. III that $\varepsilon \in U_{s+1,E}$. By Lemma there exists an element $\eta \in U_{s+1,E}$ such that $\varepsilon = \eta^{\psi-1}$. Then for $\rho = N_{\widehat{L}/\widehat{M}}\eta$ we obtain $\rho \in U_{s+1,\widehat{M}} \cap E^{\langle\psi\rangle}$. Applying (1.5) Ch. III once again we deduce that $\rho = N_{\widehat{L}/\widehat{M}}\xi$ for some $\xi \in E^{\langle\psi\rangle} \cap U_{s+1,E}$. Then $N_{\widehat{L}/\widehat{M}}(\eta\xi^{-1}) = 1$ and $\eta\xi^{-1} \in U_{s+1,E}$. Therefore, $\eta = \xi\mu^{\sigma-1}$ for some $\mu \in U_{1,E}$. Thus, $\varepsilon = \mu^{(\sigma-1)(\psi-1)} \in \left(U(\widehat{L}/\widehat{M})^{\langle\psi_i\rangle}\right)^{\psi-1}$.

To prove (1) we use induction on the cardinality of I. Let $\alpha = \alpha_1\alpha_2$ with $\alpha_1 \in U(\widehat{L}/\widehat{M})$, $\alpha_2 \in \widehat{L}^{\langle\psi_i\rangle}, i \in J$, $J = I - \{j\}$. Then $\alpha_2^{\psi_j-1} \in U(\widehat{L}/\widehat{M})^{\langle\psi_i\rangle}$, $i \in J$, and by (2) we deduce $\alpha_2^{\psi_j-1} = \alpha_3^{\psi_j-1}$ for a suitable $\alpha_3 \in U(\widehat{L}/\widehat{M})^{\langle\psi_i\rangle}, i \in J$. Then $\alpha_2\alpha_3^{-1} \in \widehat{L}^{\langle\psi_i\rangle}, i \in I$, and $\alpha = (\alpha_1\alpha_3)(\alpha_2\alpha_3^{-1}) \in U(\widehat{L}/\widehat{M})\widehat{L}^{\langle\psi_i\rangle}$.

To establish (3) let $\alpha \in U(\widehat{M}/\widehat{F})^{\langle\psi_i\rangle}$. Then by (1) there is $\alpha_1 \in U(\widehat{M}/\widehat{F})^{\langle\psi_i\rangle}$ such that $\alpha = \alpha_1^{\psi-1}$. Let $\alpha_1 = N_{\widehat{L}/\widehat{M}}\beta_1$ with $\beta_1 \in U(\widehat{L}/\widehat{F})$. Then $\beta_1^{\psi_i-1} \in \ker N_{\widehat{L}/\widehat{M}}$ and $(\beta_1^{\psi-1})^{\psi_i-1} \in U(\widehat{L}/\widehat{M})$ for $i \in I$. Hence by (1) we obtain that the element $\beta_1^{\psi-1}$ lies in $U(\widehat{L}/\widehat{M})U(\widehat{L}/\widehat{F})^{\langle\psi_i\rangle}$, $\alpha \in N_{\widehat{L}/\widehat{M}}U(\widehat{L}/\widehat{F})^{\langle\psi_i\rangle}$, as desired.

In the general case we proceed by induction on $|L{:}M|$ using the solvability of Galois groups of totally ramified extensions.

(3) follows immediately.

To prove (2) let $\alpha \in U(\widehat{L}/\widehat{M})^{\langle\psi_i\rangle}, i \in I$, and let $J = I - \{j\}$. We may assume that $\alpha = \alpha_1^{\psi-1}$ with $\alpha_1 \in U(\widehat{L}/\widehat{M})^{\langle\psi_i\rangle}, i \in J$. Suppose that $\alpha_1^{\psi_j-1} \neq 1$. By the induction assumption for a Galois subextension K/M in L/M, $K \neq L$, we get $N_{\widehat{L}/\widehat{K}}\alpha = \beta^{\psi-1}$ with $\beta \in U(\widehat{K}/\widehat{M})^{\langle\psi_i\rangle}, i \in I$. Assume that $0 \notin I$ and denote $\psi_0 = \psi$. Put $J' = J \cup \{0\}$. Then $\rho = \beta^{-1}N_{\widehat{L}/\widehat{K}}\alpha_1 \in U(\widehat{K}/\widehat{M})^{\langle\psi_i\rangle}, i \in J'$. Now for elements $\gamma \in U(\widehat{L}/\widehat{M})^{\langle\psi_i\rangle}, i \in I$, and $\mu \in U(\widehat{L}/\widehat{M})^{\langle\psi_i\rangle}, i \in J'$, with the properties $N_{\widehat{L}/\widehat{K}}\gamma = \beta$, $N_{\widehat{L}/\widehat{K}}\mu = \rho$ we obtain $\alpha_1 = \gamma\mu\delta$ with $\delta \in (\ker N_{\widehat{L}/\widehat{K}})^{\langle\psi_i\rangle}, i \in J$. Then $\delta^{\psi-1} \in U(\widehat{L}/\widehat{K})^{\langle\psi_i\rangle}, i \in I$, and $\delta^{\psi-1} = \lambda^{\psi-1}$ with some $\lambda \in U(\widehat{L}/\widehat{K})^{\langle\psi_i\rangle}, i \in I$. Thus $\alpha = (\gamma\lambda)^{\psi-1}$ and $\gamma\lambda \in U(\widehat{L}/\widehat{M})^{\langle\psi_i\rangle}, i \in I$, as required.

(1) follows formally from (2) as above. □

(4.5). THEOREM. *Let L/F be a finite Galois totally ramified p-extension. The map $\Upsilon^{\text{ab}}_{L/F}\colon \big(\operatorname{Gal}(L/F)^{\text{ab}}\big)^{\sim} \longrightarrow U_{1,F}/N_{L/F}U_{1,L}$ is an isomorphism, and the map $\Psi_{L/F}$ is the inverse one.*

Proof. First we verify that $\Psi_{L/F} \circ \Upsilon^{\text{ab}}_{L/F} = \text{id}$. Indeed, let $\pi_\chi = \pi_L\eta$ with $\eta \in U_{\widehat{L}}$. Let $\varphi \in \operatorname{Gal}(\widetilde{L}/L)$ and $\sigma_\varphi \in \operatorname{Gal}(\widetilde{L}/F)$, where $\sigma_\varphi|_{\widetilde{F}} = \varphi$, $\sigma_\varphi|_L = \sigma = \chi(\varphi)$. Then

$$\pi_L^{1-\sigma} = \eta^{\sigma_\varphi-1} \equiv \eta^{\varphi-1} \mod U(\widehat{L}/\widehat{F})$$

and $N_{\widehat{L}/\widehat{F}}\eta = N_{\Sigma_\chi/F}\pi_\chi N_{L/F}\pi_L^{-1}$. Therefore, $\chi = \Psi_{L/F}(\Upsilon^{\text{ab}}_{L/F}(\chi))$.

Next we show that $\Upsilon^{\text{ab}}_{L/F} \circ \Psi_{L/F} = \text{id}$. Let $\varepsilon \in U_{1,F}$ and $\varepsilon = N_{\widehat{L}/\widehat{F}}\eta$ for some $\eta \in U_{1,\widehat{L}}$. Suppose that $\eta^{\varphi_i-1} \equiv \pi_L^{1-\sigma_i} \mod U(\widehat{L}/\widehat{F})$ for $\varphi_i \in \operatorname{Gal}(\widetilde{L}/L)$, $\sigma_i \in \operatorname{Gal}(\widetilde{L}/\widetilde{F})$. Put $\psi_i = \varphi_i\sigma_i$ and apply (1) of the previous Proposition. Then, as $(\pi_L\eta)^{\psi_i-1} \in U(\widehat{L}/\widehat{F})$, we obtain that $\pi_L\eta = \eta_1\eta_2$ with $\eta_1 \in U(\widehat{L}/\widehat{F})$, $\eta_2 \in \widehat{L}^{\langle\psi_i\rangle}$. This means that $\pi_L\eta\eta_1^{-1} \in \Sigma_\chi$, where $\chi(\varphi_i|_{\widetilde{F}}) = \sigma_i|_L$, and
$\varepsilon \equiv N_{\Sigma_\chi/F}\big(\pi_L\eta\eta_1^{-1}\big)N_{L/F}\pi_L^{-1} \mod N_{L/F}U_{1,L}$. Thus, $\Upsilon^{\text{ab}}_{L/F} \circ \Psi_{L/F} = \text{id}$. □

COROLLARY. *Let M/F be the maximal abelian subextension in a Galois totally ramified p-extension L/F. Then $N_{M/F}U_{1,M} = N_{L/F}U_{1,L}$.*

(4.6). The proof of the following Proposition is similar to the proof of Propositions (3.4) and (3.6) Ch. IV.

PROPOSITION.

(1) *Let L/F, L'/F' be finite Galois totally ramified p-extensions, and let F'/F, L'/L be finite totally ramified extensions. Then the diagram*

$$\begin{array}{ccc} \mathrm{Gal}(L'/F')^{\sim} & \longrightarrow & U_{1,F'}/N_{L'/F'}U_{1,L'} \\ \downarrow & & \downarrow{\scriptstyle N_{F'/F}} \\ \mathrm{Gal}(L/F)^{\sim} & \longrightarrow & U_{1,F}/N_{L/F}U_{1,L} \end{array}$$

is commutative, where the left vertical homomorphism is induced by the natural restriction $\mathrm{Gal}(L'/F') \to \mathrm{Gal}(L/F)$ *and the isomorphism* $\mathrm{Gal}(\widetilde{F'}/F') \xrightarrow{\sim} \mathrm{Gal}(\widetilde{F}/F)$.

(2) *Let L/F be a Galois totally ramified p-extension, and let σ be an automorphism. Then the diagram*

$$\begin{array}{ccc} \mathrm{Gal}(L/F)^{\sim} & \longrightarrow & U_{1,F}/N_{L/F}U_{1,L} \\ {\scriptstyle \sigma^{\sim}}\downarrow & & \downarrow \\ \mathrm{Gal}(\sigma L/\sigma F)^{\sim} & \longrightarrow & U_{1,\sigma F}/N_{\sigma L/\sigma F}U_{1,\sigma L} \end{array}$$

is commutative, where $(\sigma^{\sim}\chi)(\sigma\varphi\sigma^{-1}) = \sigma\chi(\varphi)\sigma^{-1}$.

(3) *Let L/F be a Galois totally ramified p-extension and M/F be its subextension. Then the diagram*

$$\begin{array}{ccc} \left(\mathrm{Gal}(L/F)^{\mathrm{ab}}\right)^{\sim} & \longrightarrow & U_{1,F}/N_{L/F}U_{1,L} \\ {\scriptstyle \mathrm{Ver}^{\sim}}\downarrow & & \downarrow \\ \left(\mathrm{Gal}(L/M)^{\mathrm{ab}}\right)^{\sim} & \longrightarrow & U_{1,M}/N_{L/M}U_{1,L} \end{array}$$

is commutative, where $\mathrm{Ver}^{\sim}$ *is induced by* $\mathrm{Ver}\colon \mathrm{Gal}(L/F)^{\mathrm{ab}} \to \mathrm{Gal}(L/M)^{\mathrm{ab}}$.

Similarly to section 4 Ch. IV one proves

COROLLARY. *Let L_1/F, L_2/F, L_1L_2/F be abelian totally ramified p-extensions. Put $L_3 = L_1L_2$, $L_4 = L_1 \cap L_2$. Then*

$$N_{L_3/F}U_{1,L_3} = N_{L_1/F}U_{1,L_1} \cap N_{L_2/F}U_{1,L_2}, \quad N_{L_4/F}U_{1,L_4} = N_{L_1/F}U_{1,L_1}N_{L_2/F}U_{1,L_2}.$$

Moreover,

$$N_{L_1/F}U_{1,L_1} \subset N_{L_2/F}U_{1,L_2} \quad \text{if and only if} \quad L_1 \supset L_2;$$
$$N_{L_1/F}U_{1,L_1} = N_{L_2/F}U_{1,L_2} \quad \text{if and only if} \quad L_1 = L_2.$$

(4.7). The following assertion is proved in a similar way to Theorem (3.5) Ch. IV.

THEOREM. *Assume that L/F is a finite abelian totally ramified p-extension and $G = \mathrm{Gal}(L/F)$. Let $h = h_{L/F}$ be the Hasse–Herbrand function of L/F Then for $n \geqslant 1$ the reciprocity isomorphism $\Psi_{L/F}$ maps the quotient group $U_{n,F}N_{L/F}L^*/N_{L/F}L^*$ isomorphically onto the group $G_{h(n)}\widetilde{\ }$, and the reciprocity isomorphism $\Upsilon_{L/F}$ maps the group $G_{h(n)+1}\widetilde{\ }$ isomorphically onto $U_{n+1,F}N_{L/F}L^*/N_{L/F}L^*$.*

Therefore, $G_{h(n)+1} = G_{h(n+1)}$, i.e., upper ramification jumps of L/F are integers.

REMARK. Since for a local field F with separably p-closed residue field of characteristic p its finite abelian totally ramified extension L/F is generated by an element which is defined over a local field $E \subset F$ with non-separably-p-closed residue field, we can apply the previous Theorem to deduce the validity of the Hasse–Arf Theorem in the general case.

(4.8). Let F^{abp}/F be the maximal p-subextension in F^{ab}/F. Let $\{\psi_i\}$ be a set of automorphisms in $\mathrm{Gal}(F^{\mathrm{abp}}/F)$ such that $\psi_i|_{\widetilde{F}}$ are linearly independent and generate $\mathrm{Gal}(\widetilde{F}/F)$. Then the group $\mathrm{Gal}(\Sigma/F)$ for the fixed field Σ of ψ_i is isomorphic to $\mathrm{Gal}(F^{\mathrm{abp}}/\widetilde{F})$. Passing to the projective limit we obtain the *p-class reciprocity map*

$$\Psi_F\colon U_{1,F} \longrightarrow \mathrm{Hom}_{\mathbb{Z}_p}\big(\mathrm{Gal}(\widetilde{F}/F), \mathrm{Gal}(F^{\mathrm{abp}}/\widetilde{F})\big).$$

This map possesses functional properties analogous to stated in Proposition. The kernel of Ψ_F coincides with the intersection of all norm groups $N_{L/F}U_{1,L}$ for abelian totally ramified p-extensions L/F, $L \subset \Sigma$.

Similarly to the case of quasi-finite residue field the Existence Theorem requires an additional study of additive polynomials over perfect fields of characteristic p. We refer to [Fe6] for details.

The Existence Theorem implies that the reciprocity map Ψ_F is injective. It is not surjective unless the residue field of F is finite.

Another Corollary of the Existence Theorem is the following assertion [Fe6, sect.3]:

Let π be a prime element in F. Let F_π be the compositum of all finite abelian extensions L of F such that $\pi \in N_{L/F}L^*$. Then F_π is a maximal abelian totally ramified p-extension of F and the maximal abelian p-extension F^{abp} of F is the compositum of linearly disjoint extensions F_π and $\widetilde{F}$.

It is an open problem to generate the field F_π over F explicitly (similar to how Lubin–Tate formal groups do in the case of finite residue field, see section 1 Ch. VIII).

REMARK. There is another (historically the first) approach to class field theory of local fields with infinite perfect residue field due *M. Hazewinkel* [Haz1]. It provides a description of abelian extensions in terms of maximal constant quotients of the fundamental group of the group of units of a local field viewed with respect to its pro-quasi-algebraic structure. This is a generalization of *J.-P. Serre*'s geometric class field theory [Se2] (see

Example 1 section 6 Ch. IV). The method is to use a generalization of the Hazewinkel map and to go from the case of algebraically closed residue field to the situation of perfect residue field. Unfortunately, we know almost nothing about the structure of the fundamental groups involved.

5. Generalizations

In this section we discuss two further generalizations of local class field theory: imperfect residue field case in (5.1) and abelian varieties with ordinary good reduction over local fields with finite residue field in (5.2).

(5.1). Let F be a complete (Henselian) discrete valuation field with residue field $\overline{F}$ of characteristic p. We assume that $\overline{F}$ is not necessarily perfect and that $\kappa = \dim_{\mathbb{F}_p} \overline{F}/\wp(\overline{F})$ is not zero.

Denote by $\widetilde{F}$ be the maximal abelian unramified p-extension of F.

Let L be a totally ramified Galois p-extension of F. Similarly to the previous section put

$$\operatorname{Gal}(L/F)^{\sim} = \operatorname{Hom}_{\mathbb{Z}_p}(\operatorname{Gal}(\widetilde{F}/F), \operatorname{Gal}(L/F)).$$

In a similar way to the previous section define the map

$$\Upsilon_{L/F}\colon \operatorname{Gal}(L/F)^{\sim} \to U_{1,F}/N_{L/F}U_{1,L}.$$

In fact the image lies in $(U_{1,F} \cap N_{\widetilde{L}/\widetilde{F}}U_{1,\widetilde{L}})/N_{L/F}U_{1,L}$ and we denote this new map by the same notation.

DEFINITION. Let $\mathbf{F}$ be complete discrete valuation field such that $\mathbf{F} \supset \widetilde{F}$, $e(\mathbf{F}|\widetilde{F}) = 1$ and the residue field of $\mathbf{F}$ is the perfection of the residue field K of $\widetilde{F}$, i.e., $\cup_{n\geqslant 0} K^{p^{-n}}$. Such a field $\mathbf{F}$ exists by (5.3) Ch. II. So the residue field of $\mathbf{F}$ does not have algebraic p-extensions.

Put $\mathbf{L} = L\mathbf{F}$.

Denote by $U(\widetilde{L}/\widetilde{F})$ the subgroup of $U_{1,\widetilde{L}}$ generated by elements $\varepsilon^{\sigma-1} \in U_{1,\widetilde{L}}$ where ε runs through $U_{1,\mathbf{L}}$, and σ runs through $\operatorname{Gal}(L/F)$.

Then similarly to Proposition (1.7) Ch. IV one shows that the sequence

$$1 \to \operatorname{Gal}(L/F)^{\mathrm{ab}} \xrightarrow{\ell} U_{1,\widetilde{L}}/U(\widetilde{L}/\widetilde{F}) \xrightarrow{N_{\widetilde{L}/\widetilde{F}}} N_{\widetilde{L}/\widetilde{F}}U_{1,\widetilde{L}} \to 1$$

is exact (note that $N_{\widetilde{L}/\widetilde{F}}U_{1,\widetilde{L}} \neq U_{1,\widetilde{F}}$ in general).

Generalizing the Hazewinkel homomorphism introduce

DEFINITION. Define a homomorphism

$$\Psi_{L/F}\colon (U_{1,F} \cap N_{\widetilde{L}/\widetilde{F}} U_{1,\widetilde{L}})/N_{L/F}U_{1,L} \to \operatorname{Gal}(L \cap F^{\mathrm{ab}}/F)^{\sim}, \qquad \varepsilon \mapsto \chi$$

where $\chi(\varphi) = \ell^{-1}(\eta^{1-\varphi})$ and $\eta \in U_{1,\widetilde{L}}$ is such that $\varepsilon = N_{\widetilde{L}/\widetilde{F}}\eta$. This homomorphism is well defined.

PROPERTIES OF $\Upsilon_{L/F}, \Psi_{L/F}$ [Fe9].

(1) $\Psi_{L/F} \circ \Upsilon_{L/F} = \mathrm{id}$ on $\operatorname{Gal}(L \cap F^{\mathrm{ab}}/F)^{\sim}$, so $\Psi_{L/F}$ is surjective.

(2) Let F be a complete discrete valuation field such that $\mathrm{F} \supset F$, $e(\mathrm{F}|F) = 1$ and the residue field of F is the perfection of the residue field of F, i.e., is equal to $\cup_{n \geqslant 0} \overline{F}^{p^{-n}}$. Such a field exists by (5.3) Ch. II. Put $\mathrm{L} = L\mathrm{F}$. The embedding $F \to \mathrm{F}$ induces the homomorphism

$$\lambda_{L/F}\colon (U_{1,F} \cap N_{\widetilde{L}/\widetilde{F}} U_{1,\widetilde{L}})/N_{L/F}U_{1,L} \to U_{1,\mathrm{F}}/N_{\mathrm{L}/\mathrm{F}}U_{1,\mathrm{L}}.$$

Then the diagram

$$\begin{array}{ccccc}
\operatorname{Gal}(L/F)^{\sim} & \xrightarrow{\Upsilon_{L/F}} & (U_{1,F} \cap N_{\widetilde{L}/\widetilde{F}} U_{1,\widetilde{L}})/N_{L/F}U_{1,L} & \xrightarrow{\Psi_{L/F}} & \operatorname{Gal}(L \cap F^{\mathrm{ab}}/F)^{\sim} \\
\downarrow & & \lambda_{L/F}\downarrow & & \mathrm{iso}\downarrow \\
\operatorname{Gal}(\mathrm{L}/\mathrm{F})^{\sim} & \xrightarrow{\Upsilon_{\mathrm{L}/\mathrm{F}}} & U_{1,\mathrm{F}}/N_{\mathrm{L}/\mathrm{F}}U_{1,\mathrm{L}} & \xrightarrow{\Psi_{\mathrm{L}/\mathrm{F}}} & \operatorname{Gal}(\mathrm{L} \cap \mathrm{F}^{\mathrm{ab}}/\mathrm{F})^{\sim}
\end{array}$$

is commutative.

(3) Since $\Psi_{\mathrm{L}/\mathrm{F}}$ is an isomorphism by the previous section, we deduce that $\lambda_{L/F}$ is surjective and $\ker(\Psi_{L/F}) = \ker(\lambda_{L/F})$, and therefore we have an isomorphism

$$(U_{1,F} \cap N_{\widetilde{L}/\widetilde{F}} U_{1,\widetilde{L}})/N_*(L/F) \xrightarrow{\sim} \operatorname{Gal}(L \cap F^{\mathrm{ab}}/F)^{\sim}$$

where $N_*(L/F) = U_{1,F} \cap N_{\widetilde{L}/\widetilde{F}} U_{1,\widetilde{L}} \cap N_{\mathrm{L}/\mathrm{F}}U_{1,\mathrm{L}}$ is the group of elements of $U_{1,F}$ which are norms at the level of the maximal abelian unramified p-extension (where the residue field is separably p-closed) and at the level of F (where the residue field is perfect).

THEOREM. *Let L/F be a finite cyclic totally ramified p-extension. Then*

$$\Upsilon_{L/F}\colon \operatorname{Gal}(L/F)^{\sim} \to (U_{1,F} \cap N_{\widetilde{L}/\widetilde{F}} U_{1,\widetilde{L}})/N_{L/F}U_{1,L}$$

is an isomorphism.

Proof. Since L/F is cyclic, we get $U(\widetilde{L}/\widetilde{F}) = U_{1,\widetilde{L}}^{\sigma-1}$ where σ is a generator of the Galois group.

Let $\Psi_{L/F}(\varepsilon) = 1$ for $\varepsilon = N_{\widetilde{L}/\widetilde{F}}\eta \in U_{1,F}$. Then $\eta^{\varphi-1} \in U(\widetilde{L}/\widetilde{F}) \cap U_{1,\widetilde{L}}^{\varphi-1}$. Similarly to Proposition (4.3) we deduce that $\eta \in U(\widetilde{L}/\widetilde{F})L_{\varphi}$ where L_{φ} is the fixed subfield of $\widetilde{L}$ with respect to φ. Hence $\varepsilon \in N_{L_\varphi/F\cap L_\varphi} U_{1,L_\varphi}$. By induction on κ we deduce that $\varepsilon \in N_{L/F} U_{1,L}$ and $\Psi_{L/F}$ is injective. □

REMARKS.

1. *H. Miki* proved this theorem in a different setting [Mik4] which does not mention class field theory.

2. It is an open problem what is the kernel of $\Psi_{L/F}$ for an arbitrary finite abelian extension L/F, in other words how different is $N_*(L/F)$ from $N_{L/F} U_{1,L}$.

COROLLARY.

(1) *Let F be a complete discrete valuation field with residue field of characteristic p. Let L_1/F and L_2/F be finite abelian totally ramified p-extensions. Let $N_{L_1/F} L_1^* \cap N_{L_2/F} L_2^*$ contain a prime element of F. Then $L_1 L_2/F$ is totally ramified.*

(2) *Assume that $\overline{F} \neq \wp(\overline{F})$. Let L_1/F, L_2/F be finite abelian totally ramified p-extensions. Then*

$$N_{L_1/F} L_1^* = N_{L_2/F} L_2^* \iff L_1 = L_2.$$

For the proof see [Fe9]. The second assertion can be viewed as an extension of the similar assertion of Proposition (4.1) Ch. IV to the most general case.

REMARK. Let F be of characteristic zero with absolute ramification index equal to 1. Let π be a prime element of F.

Define a homomorphism

$$\mathcal{E}_{n,\pi}: W_n(\overline{F}) \to U_{1,F}/U_{1,F}^{p^n}, \quad \mathcal{E}_{n,\pi}((a_0, \dots, a_{n-1})) = \prod_{0 \leqslant i \leqslant n-1} E(\widetilde{a_i}^{p^{n-i}} \pi)^{p^i}$$

where $E(X)$ is the Artin–Hasse function of (9.1) Ch. I and $\widetilde{a_i} \in \mathcal{O}_F$ is a lifting of $a_i \in \overline{F}$. This homomorphism is injective and if $\overline{F}$ is perfect then $\mathcal{E}_{n,\pi}$ is an isomorphism.

Assume that $\overline{F} \neq \wp(\overline{F})$. Then one can prove using the theory of this subsection and the theory of fields of norms of section 5 Ch. III that cyclic totally ramified extensions L/F of degree p^n such that $\pi \in N_{L/F} L^*$ are in one-to-one correspondence with subgroups

$$\mathcal{E}_{n,\pi}\big(\mathbf{F}(w)\wp(W_n(\overline{F}))\big) U_{1,F}^{p^n}$$

of $U_{1,F}/U_{1,F}^{p^n}$ where w runs over elements of $W_n(\overline{F})^*$, see [Fe9]. This is a variant of the existence theorem for the absolutely unramified field F.

This theorem in a stronger form was first discovered by *M. Kurihara* [Ku2].

For a generalization to higher dimensional local fields see (4.13) Ch. IX.

(5.2). In this short subsection we discuss a class field theoretical interpretation of a result of *B. Mazur* [Maz] on abelian varieties with good ordinary reduction over a local field K with finite residue field, which was given another proof in *J. Lubin* and *M. Rosen* [LR].

Let $\mathcal{K}$ be the completion of the maximal unramified extension of K. Let A be an abelian variety over K of dimension d with good ordinary reduction $\widetilde{A}$. Let F be the formal group of dimension d corresponding to the Neron model A° of A, then

$$F(\mathcal{M}_K) = A^\circ(K) = \ker(A(K) \to \widetilde{A}(\overline{K})).$$

Over the field $\mathcal{K}$ the formal group of A due to the assumption on the type of its reduction is isomorphic to the torus

$$\lambda\colon F(\mathcal{M}_{\mathcal{K}}) \xrightarrow{\sim} U_{1,\mathcal{K}}^{\oplus d}.$$

In fact, the theory below can be slightly extended to any formal group over a local field which is isomorphic to a torus over $\mathcal{K}$.

Let φ (the continuous extension of φ_K) act on series coefficientwise. Then $\varphi\lambda$ is an isomorphism as well, so $(\varphi\lambda)^{-1}\lambda$ as an element of $\operatorname{Aut}(U_{1,\mathcal{K}}^{\oplus d}) = \mathrm{GL}_d(\mathbb{Z}_p)$ corresponding to an invertible matrix $M \in \mathrm{GL}_d(\mathbb{Z}_p)$ which is called the twist matrix of F.

Let L/K be a finite Galois totally ramified p-extension. The norm map

$$N_{L/K}^A\colon A(L) \to A(K)$$

induces the norm map $N_{L/K}^A\colon A^\circ(L) \to A^\circ(K)$.

For $a \in A^\circ(K) = F(\mathcal{M}_K)$ let $\lambda(a) = (\varepsilon_1, \dots \varepsilon_d)$ and $\varepsilon_i = N_{\mathcal{L}/\mathcal{K}}\eta_i$, $\eta_i \in U_{1,\mathcal{L}}$ in accordance with (1.6) Ch. IV. Then $N_{\mathcal{L}/\mathcal{K}}(\gamma_i) = 1$ where $(\gamma_1, \dots, \gamma_d) = (\eta_1, \dots, \eta_d)^{\varphi - M}$ and therefore by (1.7) Ch. IV we deduce that

$$\gamma_i \equiv \pi_L^{1-\sigma_i} \mod U(\mathcal{L}/\mathcal{K}) \quad \text{with } \sigma_i \in \operatorname{Gal}(L/K).$$

Define the *twisted reciprocity homomorphism*

$$\Psi_{L/K}\colon A^\circ(K)/N_{L/K}^A A^\circ(L) \longrightarrow \operatorname{Gal}(L \cap K^{\mathrm{ab}}/K)^{\oplus d}/(\operatorname{Gal}(L \cap K^{\mathrm{ab}}/K)^{\oplus d})^{E-M},$$

$$a \mapsto (\sigma_1, \dots, \sigma_d) \mod (\operatorname{Gal}(L \cap K^{\mathrm{ab}}/K)^{\oplus d})^{E-M}.$$

This homomorphism is an isomorphism as was first proved in [Maz] and more explicitly in [LR] without using the language of class field theory.

Certainly, using the methods of this and the previous chapters this result is easily established in the framework of class field theory. In fact the twisted reciprocity homomorphism can be defined for every formal group which is isomorphic to a torus over the maximal unramified extension.

This result has applications to Iwasawa theory of abelian varieties, see [Maz], [Man], [CG]. The norm groups $N_{L/K}^A A^\circ(L)$ have been intensively studied, see [CG] and references there.

CHAPTER 6

The Group of Units of Local Number Fields

In this chapter we assume that F is a local field of characteristic 0 with finite residue field of characteristic p, i.e., F is a local number field. We extend the investigation of the multiplicative structure of the group of principal units, in particular for applications in the next chapter.

Section 1 presents power series and some issues of their convergence in the non-Archimedean case. Section 2 introduces a generalization E_X of Artin–Hasse maps, defined in section 9 Ch. I. This time E_X acts as an operator map on power series by using an operator $\triangle$. The inverse map to it is a generalization l_X of the logarithm map; and the map l_X can be extended to a larger domain as in subsection (2.3). In section 3 we associate to a p^n th root of unity several series whose various properties are studied in detail. Subsections (3.5)–(3.6) contain auxiliary results important for Ch. VII. Section 4 discusses p^n-primary elements and their explicit presentation in terms of power series. Finally, section 5 presents a specific basis of the group of principal units of F, called the *Shafarevich basis*. The latter is very useful for the understanding of explicit formulas for the Hilbert pairing of Ch. VII.

1. Formal Power Series

Local fields of characteristic zero are in many senses similar to power series fields. It is convenient to use power series when working with elements of local fields as we have seen in section 6 Ch. I. In this section we discuss elementary properties of power series, including their convergence.

(1.1). Let K be a field. Then the field $K((X))$ of formal power series over K is a complete discrete valuation field (with respect to the valuation v_X; see section 2 Ch. I). Its residue field can be identified with K. Besides addition and multiplication, there is the operation of composition in some cases. Let $f(X) \in K((X))$ and $g(X) \in XK[[X]]$. Writing $f(X) = \sum_{n\geqslant n_0} \alpha_n X^n$, $g(X) = \sum_{n\geqslant 1} \beta_n X^n$, put

$$(f \circ g)(X) = f(g(X)) = \sum_{n\geqslant n_0} \alpha_n g(X)^n, \quad n \geqslant n_0,$$

where

$$g(X)^n = \sum_{i_1+\dots+i_n=m} \beta_{i_1}\dots\beta_{i_n}X^m \quad \text{for} \quad n>0$$

(there is only a finite number of addends defining the coefficient of X^m), $g(X)^n = (1/g(X))^{-n}$ for $n<0$, and

$$1/g(X) = \beta_i^{-1}X^{-i}(1+\sum_{n\geqslant i+1}\beta_i^{-1}\beta_nX^{n-i})^{-1},$$

where β_i is the first nonzero coefficient. Then $v_X(\alpha_n g(X)^n) \to +\infty$ as $n\to+\infty$ and $f\circ g$ is well defined.

(1.2). EXAMPLE. Let K be of characteristic 0. Consider the formal power series

$$\exp(X) = 1+X+\frac{X^2}{2!}+\frac{X^3}{3!}+\dots,$$
$$\log(1+X) = X-\frac{X^2}{2}+\frac{X^3}{3}-\dots,$$

Then $\log(1+(\exp(X)-1)) = X$, $\exp(\log(1+X)) = 1+X$ and $\exp(X+Y) = \exp(X)\cdot\exp(Y)$, $\log((1+X)(1+Y)) = \log(1+X)+\log(1+Y)$ in the field $K((X))((Y))$. (These equalities hold for $K=\mathbb{Q}$, and therefore for an arbitrary K of characteristic 0). In particular, for series $f(X)$, $g(X)\in XK[[X]]$ we obtain

$$\exp(f(X)+g(X)) = \exp(f(X))\exp(g(X)),$$
$$\log\big((1+f(X))(1+g(X))\big) = \log(1+f(X))+\log(1+g(X)).$$

Suppose that $v_X(f_n(X)) \to +\infty$ as $n\to+\infty$ for formal power series $f_n(X)\in XK[[X]]$. Then

$$\exp(\sum_{n\geqslant1} f_n(X)) = \prod_{n\geqslant1}\exp(f_n(X)),$$
$$\log\Big(\prod_{n\geqslant1}(1+f_n(X))\Big) = \sum_{n\geqslant1}\log(1+f_n(X)).$$

Finally, if $K=F$ is a local number field and $a\in\mathbb{Z}_p$, then put

$$(1+X)^a = \lim_{n\to+\infty}(1+X)^{a_n},$$

where $a_n\in\mathbb{Z}$, $\lim a_n = a$. For a formal power series $f(X)\in XF[[X]]$ put, similarly,

$$(1+f(X))^a = \lim_{n\to+\infty}(1+f(X))^{a_n} = \exp\big(a\log(1+f(X))\big).$$

The series $(1+X)^a$, $(1+f(X))^a$ so defined do not depend on the choice of (a_n) (see (6.1) Ch. I), and

$$(1+X)^{a+b} = (1+X)^a(1+X)^b,$$
$$(1+X)^a(1+Y)^a = (1+(X+Y+XY))^a,$$
$$((1+X)^a)^b = (1+X)^{ab}.$$

(1.3). Let F be a local number field with the discrete valuation v and a prime element π. In Example 4 of (4.5) Ch. I we introduced the field $F\{\{X\}\}$ of formal series $\sum_{-\infty}^{+\infty} \alpha_n X^n$, such that $v(\alpha_n) \to +\infty$ as $n \to -\infty$ and, for some integer c, $v(\alpha_n) \geqslant c$ for all integer n (here F coincides with its completion). This field is a complete discrete valuation field with a prime element π, and its residue field is isomorphic to $\overline{F}((X))$. Let $\mathcal{O}$ be the ring of integers of F. For $f(X), g(X) \in \mathcal{O}\{\{X\}\} = \{\sum \alpha_n X^n \in F\{\{X\}\} : \alpha_n \in \mathcal{O}\}$ we shall write

$$\operatorname{res}_X(f) = \operatorname{res}(f) = \alpha_{-1},$$
$$f(X) \equiv g(X) \mod \deg m \qquad \text{if } f(X) - g(X) \in X^m\mathcal{O}[[X]],$$
$$f(X) \equiv g(X) \mod (\pi^n, \deg m) \qquad \text{if } f(X) - g(X) \in \pi^n\mathcal{O}\{\{X\}\} + X^m\mathcal{O}[[X]].$$

By the way, subgroups $\pi^n\mathcal{O}\{\{X\}\} + X^m\mathcal{O}[[X]]$ with $n \geqslant 0$, $m \in \mathbb{Z}$, form a basis of neighborhoods of 0 in the additive group $\mathcal{O}\{\{X\}\}$ for the topology induced by the discrete valuation $v_*\colon F\{\{X\}\} \to \mathbb{Z} \oplus \mathbb{Z}$ of rank 2:

$$v_*\left(\sum_{-\infty}^{+\infty} \alpha_n X^n\right) = \min_n(v(\alpha_n), n).$$

LEMMA. *A series $f(X) \in \mathcal{O}\{\{X\}\}$ is invertible in $\mathcal{O}\{\{X\}\}$ if and only if $f(X) \notin \pi\mathcal{O}\{\{X\}\}$.*

Proof. Let $f(X) = \sum_{-\infty}^{+\infty} \alpha_n X^n$, and let m be the minimal integer such that α_m belongs to the unit group U. Then

$$f(X) = \alpha_m X^m(1+g(X)),$$

where $g(X) = \sum_{-\infty}^{+\infty} \beta_n X^n$, $\beta_0 = 0$, and $\beta_n \in \pi\mathcal{O}$ for $n < 0$. Hence

$$1/f(X) = \alpha_m^{-1} X^{-m}\left(1 - g(X) + g(X)^2 - g(X)^3 + \dots\right).$$

The sum converges , because $g(X) \in \pi\mathcal{O}\{\{X\}\} + X\mathcal{O}[[X]]$, and for fixed r, s we get $g(X)^n \in \pi^r\mathcal{O}\{\{X\}\} + X^s\mathcal{O}[[X]]$, where $n \geqslant 2\max(r, s)$. Thus, we deduce that $f(X)$ is invertible in $\mathcal{O}\{\{X\}\}$. The converse assertion is clear. □

Let U be the group of units of $\mathcal{O}$, $\mathcal{M}$ be its maximal ideal.

PROPOSITION ("WEIERSTRASS PREPARATION THEOREM"). *Let* $f(X) = \sum_{n\geqslant 0} \alpha_n X^n$ *be a series of* $\mathcal{O}[[X]]$ *invertible in* $\mathcal{O}\{\{X\}\}$. *Let* $m > 0$ *be the minimal integer such that* $\alpha_m \in U$. *Then there exists a series* $h(X) \in \mathcal{O}[[X]]$, *uniquely determined and invertible in* $\mathcal{O}[[X]]$, *and a monic polynomial* $g(X)$ *of degree* m *over* $\mathcal{O}$, *such that* $f(X) = g(X)h(X)$.

Proof. Put $g(X) = \beta_0 + \cdots + \beta_{m-1}X^{m-1} + X^m$, $h(X) = \gamma_0 + \gamma_1 X + \gamma_2 X^2 + \dots$, with $\beta_i \in \mathcal{M}$, $\gamma_i \in \mathcal{O}$, $\gamma_0 \in U$. The equality to be proved is equivalent to the system of equations

$$\begin{aligned} \alpha_0 &= \beta_0\gamma_0 \\ \alpha_1 &= \beta_0\gamma_1 + \beta_1\gamma_0 \\ &\dots \\ \alpha_m &= \beta_0\gamma_m + \cdots + \beta_{m-1}\gamma_1 + \gamma_0 \\ \alpha_{m+1} &= \beta_0\gamma_{m+1} + \cdots + \beta_{m-1}\gamma_2 + \gamma_1 \\ &\dots \end{aligned}$$

We shall show by induction on n that this system of equations has a unique solution $\gamma_0^{(n)} \in U$, $\gamma_i^{(n)} \in \mathcal{O}$ modulo π^n, $\beta_i^{(n)} \in \mathcal{M}$ modulo π^{n+1}, and that the limits $\gamma_i = \lim_n \gamma_i^{(n)}$, $\beta_i = \lim_n \beta_i^{(n)}$ exist; the latter then form the unique solution of the system.

Assume first that $n = 1$. Using the $(m+1+i)$th equation, put $\gamma_i^{(1)} \equiv \alpha_{m+i} \mod \pi$ for $i \geqslant 0$; then $\gamma_0^{(1)} \in U$. Using the ith equation, we get

$$\alpha_{i-1} \equiv \beta_0^{(1)}\gamma_{i-1}^{(1)} + \cdots + \beta_{i-1}^{(1)}\gamma_0^{(1)} \mod \pi^2,$$

and we find $\beta_0^{(1)}, \beta_1^{(1)}, \dots, \beta_{m-1}^{(1)}$ from the first m equations.

Furthermore, put $\beta_i^{(n)} = \beta_i^{(n-1)} + \pi^n\delta_i$, $\gamma_i^{(n)} = \gamma_i^{(n-1)} + \pi^{n-1}\varepsilon_i$ for $n \geqslant 2$. Then the $(m+1+i)$th equation implies that

$$\pi^{n-1}\varepsilon_i \equiv \alpha_{m+i} - \beta_0^{(n-1)}\gamma_{m+i}^{(n-1)} - \cdots - \beta_{m-1}^{(n-1)}\gamma_{i+1}^{(n-1)} - \gamma_i^{(n-1)} \mod \pi^n,$$

because $\beta_i^{(n-1)} \in \mathcal{M}$. Therefore, as the right-hand expression is divisible by π^{n-1} by the induction assumption, ε_i is uniquely determined modulo π. The first m equations imply the congruences

$$\begin{aligned} &\pi^n\gamma_0^{(n)}\delta_0 \equiv \alpha_0 - \gamma_0^{(n)}\beta_0^{(n-1)} \mod \pi^{n+1} \\ &\pi^n\gamma_0^{(n)}\delta_i \equiv \alpha_i - \beta_i^{(n-1)}\gamma_0^{(n)} - \beta_0^{(n)}\gamma_i^{(n)} - \cdots - \beta_{i-1}^{(n)}\gamma_1^{(n)} \mod \pi^{n+1}, \quad i \geqslant 1. \end{aligned}$$

Since the expressions on the right-hand sides are divisible by π^n by the induction assumption, $\delta_0, \delta_1, \dots$ are uniquely determined modulo π. This completes the proof. □

Note that for $f(X) \in \mathcal{O}[[X]]$, $\alpha \in \mathcal{M}_F$, the expression $f(\alpha)$ is well defined.

COROLLARY. *Let $f_1(X), f_2(X) \in \mathcal{O}[[X]]$, and let the free coefficient $f_1(0)$ of f_1 be a prime element in F. Then, $f_2(X)$ is divisible by $f_1(X)$ in the ring $\mathcal{O}[[X]]$ if and only if $f_1(X)$ and $f_2(X)$ have a common root α in the maximal ideal $\mathcal{M}_L$ of some finite extension L over F.*

Proof. By the Proposition, one can write $f_1 = g_1 h_1$, $f_2 = g_2 h_2$, where $g_1(X)$, $g_2(X)$ are monic polynomials over $\mathcal{O}$ and $h_i(X)$ are invertible elements in $\mathcal{O}[[X]]$. We obtain that $g_1(X) = X^n + \beta_{n-1}X^{n-1} + \dots + \beta_0$ with $n > 0$ and $v(\beta_0) = 1$, $v(\beta_i) \geqslant 1$ for $i \geqslant 0$. Therefore, $g_1(X)$ is an Eisenstein polynomial over F (see (3.6) Ch. II). We also deduce that the set of roots of $f_i(X)$ in the maximal ideal $\mathcal{M}_L$ of any finite extension L/F coincides with the set of roots of $g_i(X)$ in $\mathcal{M}_L$. Recall that all roots of $g_1(X)$ in L belong to $\mathcal{M}_L$.

If f_2 is divisible by f_1, then a root α of $g_1(X)$ in $\mathcal{M}_L$ is a root of $g_2(X)$, where $L = F(\alpha)$. If $f_1(\alpha) = f_2(\alpha) = 0$ for $\alpha \in \mathcal{M}_L$ and some finite extension L/F, then α is a root of the Eisenstein polynomial $g_1(X)$, which is irreducible. As $g_2(\alpha) = 0$, $f_2(X)$ is divisible by $f_1(X)$. □

REMARK. For other proofs of the Proposition see [Cas, Ch. VI], [Man].

(1.4). Let $f(X) = \sum_{n \geqslant 0} \alpha_n X^n \in F((X))$. Put

$$c = -\varliminf_{n \geqslant 1} \frac{v(\alpha_n)}{n}.$$

Then for an element $\alpha \in F$ with $v(\alpha) > c$ we get $v(\alpha_n \alpha^n) \to +\infty$ as $n \to +\infty$. This means that the sum $\sum_{n \geqslant 0} \alpha_n \alpha^n$ is convergent in F. We put $f(\alpha) = \sum_{n \geqslant 0} \alpha_n \alpha^n$. The series $f(X)$ is then said to converge at $\alpha \in F$. It is easy to show that $f(X)$ converges on the set

$$\mathcal{O}^c = \{\alpha : v(\alpha) > c\},$$

and does not converge on the set $\{\alpha : v(\alpha) < c\}$. If we pass to the absolute values $\|\cdot\|$, the constant c should be replaced by the radius of convergence. A special feature of formal series over local number fields is that the necessary condition $v(\alpha_n \alpha^n) \to +\infty$ for convergence is also sufficient. It immediately follows that $f(X)$ determines a continuous function $f\colon \mathcal{O}^c \to F$.

EXAMPLE. The series $\exp(X)$ converges on $\mathcal{O}^c$ with $c = e/(p-1)$, $e = v(p)$. Indeed,

$$\frac{v(n!)}{n} = \frac{e}{n}\left(\left[\frac{n}{p}\right] + \left[\frac{n}{p^2}\right] + \dots\right) < e \sum_{m \geqslant 1} p^{-m} = \frac{e}{p-1}.$$

On the other hand, for $n = p^m$ we get

$$\frac{v(n!)}{n} = e\frac{1-p^{-m}}{p-1};$$

therefore, $\exp(X)$ does not converge at $\alpha \in F$ with $v(\alpha) \leqslant e/(p-1)$. The series $\log(1+X)$ converges on $\mathcal{M}$, because

$$\underline{\lim}\, \frac{v(n^{-1})}{n} = 0.$$

Note that $\exp(X)$ induces an isomorphism of the additive group $\mathcal{O}^c$ onto the multiplicative group $1+\mathcal{O}^c$ and $\log(X)$ induces the inverse isomorphism.

(1.5). If $*$ denotes one of the operations $+, \times$, and formal power series $f(X)$, $g(X)$ converge at $\alpha \in F$, then the formal power series $h(X) = f(X) * g(X)$ converges at $\alpha \in F$ and $h(\alpha) = f(\alpha) * g(\alpha)$. The operation of composition is more complicated (see Exercise 3). The following assertion will be useful below:

PROPOSITION. *Let $f(X) = \sum_{n\geqslant 0} \alpha_n X^n$, $g(X) = \sum_{n\geqslant 1} \beta_n X^n$ be formal power series over F. Let $\widehat{\mathcal{O}}$ be the ring of integers in the complete discrete valuation field $\widehat{F^{\mathrm{ur}}}$. Assume that $f(X)$ converges on $\widehat{\mathcal{O}}^c$, $g(X)$ converges on $\widehat{\mathcal{O}}^d$. Let $g(\alpha) \in \widehat{\mathcal{O}}^c$ for all $\alpha \in \widehat{\mathcal{O}}^b \subset \widehat{\mathcal{O}}^d$. Then the formal power series $h(X) = (f\circ g)(X)$ converges on $\widehat{\mathcal{O}}^b$ and $h(\alpha) = f(g(\alpha))$ for $\alpha \in \widehat{\mathcal{O}}^b$.*

Proof. Denote the discrete valuation on $\widehat{F^{\mathrm{ur}}}$ by $\hat{v}$. Assume that for some $\alpha \in \widehat{\mathcal{O}}^b$ the element $\beta_n\alpha^n$ does not belong to $\widehat{\mathcal{O}}^c$ for some n. Put $a = \min_{n\geqslant 1} \hat{v}(\beta_n\alpha^n)$. Let S denote the finite set of those indices n, for which $\hat{v}(\beta_n\alpha^n) = a$. For a prime element π in F there exists an element $\theta \in U_{\widehat{F^{\mathrm{ur}}}}$, such that the residues of $\beta_n\alpha^n\theta^n\pi^{-a}$, $n\in S$, are linearly independent over $\mathbb{F}_p$ (because the residue field of $\widehat{F^{\mathrm{ur}}}$ is infinite). Then $\alpha\theta \in \widehat{\mathcal{O}}^b$, $f(\alpha\theta) \notin \widehat{\mathcal{O}}^c$, and we get a contradiction. Thus, $\beta_n\alpha^n \in \widehat{\mathcal{O}}^c$ for $n \geqslant 1$.

Put $\kappa_n = \hat{v}(\beta_n\alpha^n)$, then $\kappa_n \to +\infty$. Since $f(X)$ converges on $\widehat{\mathcal{O}}^c$, we obtain

$$\hat{v}(\alpha_n\beta_{i_1}\dots\beta_{i_n}\alpha^m) = v(\alpha_n) + \hat{v}(\beta_{i_1}\alpha^{i_1}) + \dots + \hat{v}(\beta_{i_n}\alpha^{i_n}) \to +\infty$$

as $n\to+\infty$, for any $i_1,\dots,i_n \geqslant 1$ with $i_1+\dots+i_n = m$. This means that for a fixed s there exists an index n_0 such that $\hat{v}(\alpha_n g(\alpha)^n) > s$ for $n > n_0$. There exists also an index m_0, such that $\kappa_m > s - \min(v(\alpha_1),\dots,v(\alpha_{n_0})) - n_0\cdot t$, $\kappa_m > 0$ for $m > m_0n_0^{-1}$, where $t = \min(0, \inf_n \kappa_n)$. Then $\hat{v}(\alpha_n\beta_{i_1}\dots\beta_{i_n}\alpha^m) > s$ for $1\leqslant n\leqslant n_0$, $i_1+\dots+i_n = m > m_0$. Putting $h(X) = \sum_{m\geqslant 0}\gamma_m X^m$ we conclude that $\hat{v}(\gamma_m\alpha^m) > s$ for $m > m_0$. Therefore, $h(X)$ converges at α. As $g(X) \in XF[[X]]$, we get

$$\sum_{m=0}^{m_0} \alpha_m g(X)^m = \sum_{m=0}^{m_0}\gamma_m X^m \mod \deg m_0+1.$$

Hence

$$\hat{v}\left(\sum_{m=0}^{n} \alpha_m g(\alpha)^m - \sum_{m\geqslant 0} \gamma_m \alpha^m\right) > s \qquad \text{for} \quad n \geqslant m_0$$

and $\sum_{m=0}^{n} \alpha_m g(\alpha)^m \to h(\alpha)$ as $n \to +\infty$. This means that $f(g(\alpha)) = h(\alpha)$. □

Exercises.

1. a) Let $f(X) \in \mathcal{O}\{\{X\}\}$. Show that

$$f(X) \equiv X^r \mod \pi$$

if and only if

$$1/f(X) \equiv X^{-r} \mod \pi.$$

b) Let $f(X), g(X)$ be invertible in $\mathcal{O}\{\{X\}\}$. Show that

$$f(X) \equiv g(X) \mod \pi^m$$

if and only if

$$1/f(X) \equiv 1/g(X) \mod \pi^m.$$

c) Let $f(X), g(X) \in \mathcal{O}\{\{X\}\}$. Let $h(X)$ be invertible in $\mathcal{O}\{\{X\}\}$. Show that

$$f(X) \equiv g(X) \mod \pi^m$$

if and only if

$$f(X)/h(X) \equiv g(X)/h(X) \mod \pi^m.$$

2. Let $g(X)$ be an element of $\mathcal{O}[[X]]$ invertible in $\mathcal{O}\{\{X\}\}$. Show that for an element $f(X) \in \mathcal{O}[[X]]$ there exist uniquely determined series $q(X) \in \mathcal{O}[[X]]$ and polynomial $r(X)$ of degree $< v_X(\overline{g}(X))$ over $\mathcal{O}$, such that $f = gq+r$ ($\overline{g}(X) \in \overline{F}((X))$ is the residue of the polynomial $g(X)$).
3. (*G. Henniart* [Henn2]) Let $\operatorname{char}(\overline{F}) = p$.
   a) Let $f(X) = \sum_{n\geqslant 0} X^n$, $g(X) = p^{-2}X - p^{-3}X^2$. Show that g converges at $\alpha = p$, f converges at $g(\alpha)$, but $f \circ g$ does not converge at $\alpha = p$.
   b) Let $p = 2$, $f(X) = \exp(X)$, $g(X) = \log(1+X)$. Show that $g, f \circ g$ converge at $\alpha = \sqrt{2}$, but f does not converge at $g(\alpha)$.
   c) Let $f(X) = \exp(X)$, $g(X) = \log(1+X)$, $\alpha = \zeta - 1$, where ζ is a primitive pth root of unity. Show that $g(\alpha) = 0$, $f(g(\alpha)) = 1$, but $(f \circ g)(\alpha) = \zeta$.
   d) Let $\exp(X) = \prod_{n\geqslant 1}(1 + a_n X^n)$; put $f_n(X) = a_n(\log(1+X))^n$. Show that $\prod_{n\geqslant 1}(1 + f_n(X)) = 1 + X$. For α as in c) show that $f_n(\alpha) = 0$ and check that $\prod_{n\geqslant 1}(1 + f_n(\alpha)) \neq 1 + \alpha$.
4. (*G. Henniart* [Henn2]) Let $f(X) = \sum_{n\geqslant 0} \alpha_n X^n$, $g(X) = \sum_{n\geqslant 1} \beta_n X^n$ be formal power series over F, $h(X) = f(X) \circ g(X)$. Put

$$a_m = \inf_{i_1+\dots+i_n=m} v(\alpha_n \beta_{i_1} \dots \beta_{i_n}) \qquad \text{for} \quad m > 0.$$

Let $a_m + mv(\alpha) \to +\infty$ as $m \to +\infty$ and let $g(X)$ converge at α. Show that f converges at $g(\alpha)$, h converges at α, and $f(g(\alpha)) = h(\alpha)$.

5. Let $f(X) \in F[[X]]$ and let $f'(X) \in F[[X]]$ be its formal derivative. Show that if $f(X)$ converges on $\mathcal{O}^c$, then $f'(X)$ converges on $\mathcal{O}^c$ and

$$f'(\alpha) = \lim_{\substack{v(\beta)\to+\infty \\ \beta\in\mathcal{O}^c}} \frac{f(\alpha+\beta)-f(\alpha)}{\beta}, \qquad \alpha \in \mathcal{O}^c.$$

6. Show that $\log(1+\alpha) = \lim_{n\to+\infty} \dfrac{(1+\alpha)^{p^n}-1}{p^n}$ for $\alpha \in \mathcal{M}$.

7. Show that
   a) The series $(1+X)^a$ converges on $\mathcal{O}^c$ with $c = e/(p-1)$ if $a \in \mathbb{Q}_p$, and on $\mathcal{O}^0$ if $a \in \mathbb{Z}_p$.
   b) $(1+\alpha)^a = \exp(a\log(1+\alpha))$ for $a \in \mathbb{Z}_p$, $\alpha \in \mathcal{O}^c$.
   c) The function $(1+\alpha)^a$ depends continuously on $a \in \mathbb{Q}_p$ for $\alpha \in \mathcal{O}^c$.

8. ($\diamond$) Let $\widehat{\mathcal{O}}$, $\hat{v}$ be as in (1.5), and let $f(X)$ be an element of $\mathcal{O}[[X]]$, convergent on $\widehat{\mathcal{O}}$.
   a) Show that if $f(X) = \sum_{n\geqslant 1} \alpha_n X^n$, then $\inf \hat{v}(\alpha_n) = \inf_{\alpha\in\widehat{\mathcal{O}}} v(f(\alpha))$.
   b) Show that if $f(X)$ vanishes on some non-empty open set $A \subset \mathcal{O}$ then $f = 0$.
   c) Show that the maximum of $f(X)$ on any set of the form

$$\{\alpha \in \widehat{\mathcal{O}} : a \leqslant \hat{v}(\alpha) \leqslant b\}, \quad a \geqslant 0,$$

   is attained on the set $\{\alpha \in \widehat{\mathcal{O}} : \hat{v}(\alpha) = a \text{ or } \hat{v}(\alpha) = b\}$.

(For other properties of analytic functions in non-Archimedean setting see [Kr2], [T4], [BGR], [Cas], [Kob1], [Kob2]).

2. The Artin–Hasse–Shafarevich Map

The Artin–Hasse maps, discussed in section 9 Ch. I, play an important role in the arithmetics of local fields. *E. Artin* and *H. Hasse* used these maps in computations of the values of the Hilbert norm residue symbol in cyclotomic extensions of $\mathbb{Q}_p$ ([AH2], 1928). Later *H. Hasse* used them for establishing an explicit form of p-primary elements ([Has8], 1936). *I.R. Shafarevich* generalized and applied these maps to the construction of a canonical basis of a local number field ([Sha2], 1950). This construction allows one to derive explicit formulas for the Hilbert norm residue symbol. In this section we consider a generalization of the Artin–Hasse maps as linear operators on $\mathbb{Z}_p$-modules.

(2.1). As usual, we denote by $\mathbb{Q}_p^{\mathrm{ur}}$ the maximal unramified extension of $\mathbb{Q}_p$. Recall that $\mathrm{Gal}(\mathbb{Q}_p^{\mathrm{ur}}/\mathbb{Q}_p)$ is topologically generated by the Frobenius automorphism which will be denoted by φ (see (1.2) Ch. IV).

Let $\widehat{\mathcal{O}}$ denote the ring of integers of the completion $\widehat{\mathbb{Q}_p^{\mathrm{ur}}}$ of $\mathbb{Q}_p^{\mathrm{ur}}$ and φ the continuous extension of φ to $\widehat{\mathbb{Q}_p^{\mathrm{ur}}}$.

For a formal power series $f(X) = \sum \alpha_n X^n$ over $\widehat{\mathcal{O}}$, define the Frobenius operator $\triangle_X$ as follows:

$$\triangle_X (f) = f^{\triangle_X} = \sum \varphi(\alpha_n) X^{pn}.$$

Then $\triangle_X$ is a $\mathbb{Z}_p$-endomorphism of $\widehat{\mathcal{O}}[[X]]$:

$$\triangle_X (f+g) = \triangle_X (f) + \triangle_X (g), \qquad \triangle_X (fg) = \triangle_X (f) \triangle_X (g),$$
$$\triangle_X (af) = a \triangle_X (f) \quad \text{for} \quad a \in \mathbb{Z}_p.$$

Note that $\triangle_X$ depends on X. We will often write $\triangle$ instead of $\triangle_X$.

Put

$$\left(1 - \frac{\triangle}{p}\right)^{-1} = 1 + \frac{\triangle}{p} + \frac{\triangle^2}{p^2} + \dots .$$

For a formal power series $g(X) \in X\widehat{\mathcal{O}}[[X]]$

$$\left(1 - \frac{\triangle}{p}\right)^{-1} (g(X)) = g(X) + \frac{\triangle\, g(X)}{p} + \frac{\triangle^2\, g(X)}{p^2} + \dots$$

is an element of $X\widehat{\mathcal{O}}[[X]]$, because $v_X(\triangle^n g(X)) \to +\infty$ as $n \to +\infty$.

(2.2). Regarding the additive group $X\widehat{\mathcal{O}}[[X]]$ as a $\mathbb{Z}_p$-module ($a \circ f(X) = af(X)$ for $a \in \mathbb{Z}_p$, $f(X) \in \widehat{\mathcal{O}}[[X]]$) and the multiplicative group $1 + X\widehat{\mathcal{O}}[[X]]$ as a $\mathbb{Z}_p$-module ($a \bullet g(X) = g(X)^a$ for $a \in \mathbb{Z}_p$, $g(X) \in 1 + X\widehat{\mathcal{O}}[[X]]$), we introduce the *Artin–Hasse–Shafarevich map*

$$E_X \colon X\widehat{\mathcal{O}}[[X]] \to 1 + X\widehat{\mathcal{O}}[[X]]$$

by the formula

$$E_X(f(X)) = \exp\left(\left(1 - \frac{\triangle_X}{p}\right)^{-1} f(X)\right).$$

Then $E_X(X) = E(X)$, where $E(X)$ is the Artin–Hasse function (see (9.1) Ch. I).

Introduce also the map $l_X \colon 1 + X\widehat{\mathcal{O}}[[X]] \to X\widehat{\mathcal{O}}[[X]]$ by the formula

$$l_X(1 + f(X)) = \left(1 - \frac{\triangle_X}{p}\right) \left(\log(1 + f(X))\right) = \left(1 - \frac{\triangle_X}{p}\right) \left(\sum_{i \geqslant 1} \frac{-(-f)^i}{i}\right).$$

PROPOSITION. *E_X induces a $\mathbb{Z}_p$-isomorphism of $X\widehat{\mathcal{O}}[[X]]$ onto $1 + X\widehat{\mathcal{O}}[[X]]$, and the map l_X is the inverse isomorphism. If $\alpha \in \widehat{\mathcal{O}} = W(\mathbb{F}_p^{\text{sep}})$ then $E_X(\alpha X) = \mathcal{E}(\alpha, X)$, where $\mathcal{E}$ was defined in* (9.3) Ch. I.

Proof. For the arguments below, it is convenient to put in evidence the following result.

LEMMA. *Let* $f(X) \in \widehat{\mathcal{O}}[[X]]$. *Then*

$$f(X)^{mp} \equiv f(X)^{m\triangle} \mod pm.$$

Proof. One can assume $m = p^i$. If $i = 0$ then the congruence

$$\triangle f(X) \equiv f(X)^p \mod p$$

follows from the definition of $\triangle$ and the congruence $\varphi(\alpha) \equiv \alpha^p \mod p$. It remains to use Lemma (7.2) Ch. I. □

It is clear that

$$E_X(f+g) = E_X(f)E_X(g), \quad E_X(af) = E_X(f)^a$$

for $a \in \mathbb{Z}_p$, $f, g \in X\widehat{\mathcal{O}}[[X]]$, and

$$l_X((1+f)(1+g)) = l_X(1+f) + l_X(1+g), \quad l_X((1+f)^a) = a l_X(1+f)$$

for $a \in \mathbb{Z}_p$, $f, g \in X\widehat{\mathcal{O}}[[X]]$ (see (1.2)).

First we show that $E_X(f) \in 1 + X\widehat{\mathcal{O}}[[X]]$ for $f(X) \in X\widehat{\mathcal{O}}[[X]]$. By linearity, one can assume $f(X) = \alpha X^n$ with $\alpha \in \widehat{\mathcal{O}}$. By Proposition (1.2) Ch. IV the ring $\widetilde{\mathcal{O}}$ is generated over $\mathbb{Z}_p$ by mth roots of unity with $(m,p) = 1$. Therefore, by linearity and continuity one can assume that $f(X) = \theta X^n$ with θ an mth root of unity, $(m,p) = 1$. Then $\varphi(\theta) = \theta^p$ (see (1.2) Ch. IV) and $E_X(\theta X^n) = E(\theta X^n)$, where E is the Artin–Hasse function, defined in (9.1) Ch. I. Lemma (9.1) Ch. I implies that $E(\theta X^n) \in 1 + \theta X^n \mathbb{Z}_p[[\theta X^n]] \subset 1 + X\widehat{\mathcal{O}}[[X]]$, and we conclude that $E_X(f) \in 1 + X\widehat{\mathcal{O}}[[X]]$.

Furthermore, for $f \in X\widehat{\mathcal{O}}[[X]]$

$$\begin{aligned} l_X(1+f) &= \left(1 - \frac{\triangle}{p}\right)\left(\sum_{n\geqslant 1} \frac{(-1)^{n-1}}{n} f^n\right) \\ &= \sum_{n\geqslant 1} \frac{(-1)^{n-1}}{n} f^n - \sum_{n\geqslant 1} \frac{(-1)^{n-1}}{np} f^{n\triangle} \\ &= \sum_{\substack{(n,p)=1\\ n\geqslant 1}} \frac{(-1)^{n-1}}{n} f^n - \sum_{n\geqslant 1} \frac{(-f)^{np} - (-f)^{n\triangle}}{np}. \end{aligned}$$

The first sum is an element of $\widehat{\mathcal{O}}[[X]]$ because $n^{-1} \in \mathbb{Z}_p$ for $(n,p) = 1$. The terms in the second sum belong to $\widehat{\mathcal{O}}[[X]]$ by the Lemma. Therefore, $l_X(1+f) \in X\widehat{\mathcal{O}}[[X]]$.

From the definitions we deduce

$$(l_X \circ E_X)(f) = \left(1 - \frac{\triangle}{p}\right)(\log \circ \exp)\left(\left(1 - \frac{\triangle}{p}\right)^{-1}(f)\right) = f$$

and, similarly, $(E_X \circ l_X)(1+f) = 1+f$. Therefore, E_X and l_X are inverse to each other and are $\mathbb{Z}_p$-isomorphisms. □

(2.3). LEMMA. *The map* l_X

$$l_X(f) = \log(f) - \frac{\Delta}{p}\log(f) = \frac{1}{p}\log(f^p/f^{\Delta})$$

is a homomorphism from the group $1 + 2p\widehat{\mathcal{O}} + X\widehat{\mathcal{O}}[[X]]$ *to* $2\widehat{\mathcal{O}} + X\widehat{\mathcal{O}}[[X]]$; *the group* $1 + (2p, X)\mathcal{O}[[X]]$ *is mapped to* $2\mathcal{O}[[X]] + X\mathcal{O}[[X]]$.

Proof. The group $1+2p\widehat{\mathcal{O}}+X\widehat{\mathcal{O}}[[X]]$ is the product of its subgroups $1+2p\widehat{\mathcal{O}}$ and $1+X\widehat{\mathcal{O}}[[X]]$. We have already seen in (2.2) that l_X maps $1+X\widehat{\mathcal{O}}[[X]]$ isomorphically to $X\widehat{\mathcal{O}}[[X]]$. It remains to use (1.4) according to which exp and log induce isomorphisms between $2p\widehat{\mathcal{O}}$ and $1+2p\widehat{\mathcal{O}}$. □

We can extend the map l_X even further.

Put

$R = \widehat{\mathcal{O}}((X))^*$ if $p > 2$,

$R = \left\{X^m a\varepsilon(X) : \varepsilon(X) \in 1 + X\widehat{\mathcal{O}}[[X]], a \in \widehat{\mathcal{O}}^*, a^{\varphi} \equiv a^2 \bmod 4, m \in \mathbb{Z}\right\}$ if $p = 2$.

For $f \in R$ we get $f^p/f^{\Delta} \in 1 + (2p, X)\widehat{\mathcal{O}}[[X]]$. Extend l_X to R by the formula

$$l_X(f) = \frac{1}{p}\log(f^p/f^{\Delta}).$$

Then $l_X(f) \in (2, X)\widehat{\mathcal{O}}[[X]]$.

(2.4). At the end of this section we make several remarks. First, if $f \equiv \alpha X^n \bmod \deg n+1$, $\alpha \in \widehat{\mathcal{O}}$, then

$$E_X(f) \equiv 1 + \alpha X^n \mod \deg n + 1.$$

Similarly, if $f \equiv \alpha X^n \mod (p^k, \deg n+1)$, $\alpha \in \widehat{\mathcal{O}}$, $k \geqslant 1$, then

$$E_F(f) \equiv (1+\alpha X^n)(1+g)^{p^k} \mod \deg n+1$$

for some $g \in X\widehat{\mathcal{O}}[[X]]$.

For a formal power series $f \in X\widehat{\mathcal{O}}[[X]]$ one has

$$E_X(f)^p = \exp(pf)E_X(f^{\Delta}),$$

since $E_X(f)^p E_X(\Delta\, f)^{-1} = E_X((p-\Delta)f)$.

Exercises.

1. Let π be a prime element of a finite extension F over $\mathbb{Q}_p$. Let $f = \sum_{n\geqslant m} \alpha_n X^n \in X\mathcal{O}_0[[X]]$, where $\mathcal{O}_0$ is the ring of integers of $F_0 = F \cap \mathbb{Q}_p^{\mathrm{ur}}$. Show that

$$E_X(f(X))|_{X=\pi} \equiv 1 + \alpha_m \pi^m \mod \pi^{m+1}.$$

Let α be a nonzero root of the polynomial $f = pX - X^p$. Show that

$$E_X(f)|_{X=\alpha} \neq 1 = E_X(f(\alpha)).$$

2. Let $l_X : R \to \widehat{\mathcal{O}}[[X]$ be the map defined in the end of (2.3).
   a) Let $f \in R$. Show that the free coefficient of $l_X(f)$ belongs to $(\varphi - 1)\widehat{\mathcal{O}} + 2\widehat{\mathcal{O}} + X\widehat{\mathcal{O}}[[X]]$.
   b) Show that the kernel of l_X is equal to $\langle X \rangle \times \mu$, where μ is the group generated by roots of unity of order relatively prime to p and -1, and the image of l_X is equal to $(\varphi - 1)\widehat{\mathcal{O}} + 2\widehat{\mathcal{O}} + X\widehat{\mathcal{O}}[[X]]$.
   c) Let $\mathcal{L}_X : R \to \widehat{\mathcal{O}}[[X]]$ be the map defined by

$$\mathcal{L}_X(f) = \frac{f^p - f^{\Delta_X}}{p f^{\Delta_X}}$$

   Show that $\mathcal{L}_X(f) + \mathcal{L}_X(\Delta_X\ f) \equiv l_X(f) \mod 2$ if $p = 2$, and $\mathcal{L}_X(f) \equiv l_X(f) \mod p$ if $p \neq 2$.

3. ($\diamond$) Let F be a local number field, $\mathcal{O}_0$ the ring of integers of $F_0 = F \cap \mathbb{Q}_p^{\mathrm{ur}}$. Let L/F be a totally and tamely ramified finite Galois extension, $G = \mathrm{Gal}(L/F)$. Let π be a prime element in L such that π^n is a prime element in F, $n = |L : F|$. For $\sigma \in G$, the element $\varepsilon_\sigma = \pi^{-1}\sigma(\pi)$ belongs to the set of multiplicative representatives in F (see section 4 Ch. II). Define the action of G on $\mathcal{O}_0[[X]]$ by $\sigma(X) = \varepsilon_\sigma X$.
   a) Let I be a set of n integers such that all their residues modulo n are distinct. Let A_I denote the $\mathcal{O}_0[G]$-module generated by X^i, $i \in I$. Show that A_I is a free $\mathcal{O}_0[G]$-module of rank 1, and an element $\alpha = \sum_{i\in I} \alpha_i X^i$ with $\alpha_I \in \mathcal{O}_0$ is a generator of A_I if and only if all α_i are invertible elements in $\mathcal{O}_0$.
   b) Let $e = e(F|\mathbb{Q}_p)$. Denote

$$I_m = \left\{ \frac{pn(m-1)}{p-1} \leqslant i < \frac{pnm}{p-1}, (i,p) = 1 \right\}, \qquad 1 \leqslant m \leqslant e,$$

$$I^{(p)} = \cup_{1\leqslant m\leqslant e} I_m.$$

   Let A_m denote the $\mathcal{O}_0[G]$-submodule in $\mathcal{O}_0[[X]]$ generated by X^i with $i \in I_m$, $A^{(p)} = \bigoplus_{1\leqslant m\leqslant e} A_m$. Choose in A_m a $\mathcal{O}_0[G]$-generator $\alpha_m = \alpha_m(X)$ as in a). Let $\beta_1, \dots, \beta_f$ be a basis of $\mathcal{O}_0$ over $\mathbb{Z}_p$. Show that $A^{(p)}$ is a free $\mathbb{Z}_p[G]$-module of rank fe with generators $\beta_i\alpha_m$, $1 \leqslant i \leqslant f$, $1 \leqslant m \leqslant e$.
   c) Let the field L contain no nontrivial pth roots of unity. Prove that the map $f(X) \to E_X(f(X))|_{X=\pi}$ induces an isomorphism of $\mathbb{Z}_p[G]$-module $A^{(p)}$ onto $\mathbb{Z}_p[G]$-module $U_{1,L}$. This means that $U_{1,L}$ is a free $\mathbb{Z}_p[G]$-module of rank ef with generators $E_X(\beta_i\alpha_m(X))|_{X=\pi}$, $1 \leqslant i \leqslant f$, $1 \leqslant m \leqslant e$.

4. Let F be a complete discrete valuation field of characteristic 0 with perfect residue field $\overline{F}$ of characteristic p. Let $\mathbf{F}$ be the Frobenius map on the field of fractions of $W(\overline{F}^{\text{sep}})$ (see (8.2) Ch. I). For a formal power series $f(X) = \sum \alpha_n X^n$ over $W(\overline{F}^{\text{sep}})$ define

$$\triangle_X f(X) = \sum \mathbf{F}(\alpha_n) X^{pn}.$$

Show that the maps E_X, l_X, defined similarly to (2.2)–(2.3) have properties similar to the assertions of this section.

3. Series Associated to Roots

In this section we consider various formal power series associated to a p^n th primitive root of unity; these will be applied in the next two sections and Chapter VII. We also state and prove several auxiliary results in (3.5)–(3.6) which will be in use in Chapter VII when we study explicit pairings. The reader may omit subsections (3.3)–(3.6) in the first reading.

(3.1). Suppose that F contains nontrivial pth roots of unity and let $n \geqslant 1$ be the maximal integer such that a p^n th primitive root ζ of unity is contained in F. Let π be a prime element in F. Denote the ring of integers of the inertia subfield $F_0 = F \cap \mathbb{Q}_p^{\text{ur}}$ by $\mathcal{O}_0$. By Corollary 2 of (2.9) Ch. II we get an expansion

$$\zeta = 1 + c_1\pi + c_2\pi^2 + \dots, \qquad c_i \in \mathcal{O}_0.$$

Let $z(X) = 1 + z_0(X)$ denote the following formal power series:

$$z(X) = 1 + c_1 X + c_2 X^2 + \dots \quad .$$

Then $z(X) \in \mathcal{O}_0[[X]]$ and $z(\pi) = \zeta$. The formal power series $z(X)$ depends on the choice of the prime element π and the expansion of ζ as power series in the prime element π.

Put

$$s_m(X) = z(X)^{p^m} - 1, \qquad s(X) = s_n(X),$$
$$u_m(X) = \frac{s_m(X)}{s_{m-1}(X)}, \qquad u(X) = u_n(X).$$

Then $s_m \in \mathcal{O}_0[[X]]$ and $u_m \in \mathcal{O}_0[[X]]$, because

$$u_m = \frac{(1 + s_{m-1})^p - 1}{s_{m-1}} = p + \sum_{i=1}^{p-1} \binom{p}{i+1} s_{m-1}^i .$$

We have also $s(\pi) = u(\pi) = 0$. The series $u(X)$ belongs to $p\mathcal{O}_0 + X\mathcal{O}_0[[X]]$. Hence for every $g(X) \in \mathcal{O}_0[[X]]$ and $p > 2$ we get in accordance with (2.3)

$$l_X(1+ug) = \sum_{i\geqslant 1} \frac{(-1)^{i-1}u^i g^i}{i} - \sum_{i\geqslant 1} \frac{(-1)^{i-1}u^{i\Delta} g^{i\Delta}}{pi}.$$

To have a similar expression for $p = 2$ we need to impose an additional restriction that $ug \in 2p\mathcal{O}_0 + X\mathcal{O}_0[[X]]$.

Let $e = e(F|\mathbb{Q}_p)$ and $e_m = e/(p-1)p^{m-1}$ for $m \geqslant 1$. If v denotes the discrete valuation on F, then $v(\zeta - 1) = e_n$ by Proposition (5.7) Ch. I.

PROPOSITION. *The formal power series s_m, u_m are invertible in the ring $\mathcal{O}_0\{\{X\}\}$. Moreover, $v_X(\overline{s}_m) = p^m e_n$, $v_X(\overline{u}_m) = p^{m-n}e$, where v_X is the discrete valuation of $\overline{F}((X))$. The following congruences hold for $m \geqslant 1$:*

a) $s_m \equiv z_0^{p^m} \mod p,$

b) $s_m \equiv\Delta\, s_{m-1} \mod p^m,$

c) $\dfrac{1}{s_m} \equiv \dfrac{1}{\Delta\, s_{m-1}} \mod p^m,$

d) $s'_m \equiv \left(\dfrac{1}{s_m}\right)' \equiv 0 \mod p^m,$

where $s'_m(X)$ is the formal derivative of $s_m(X)$.

Proof. The first congruence follows from the definition of s_m and Lemma (7.2) Ch. I.

If all coefficients c_i of $z_0(X)$ were divisible by p, then $v(\zeta - 1) \geqslant v(\pi p) = e + 1$ which contradicts $v(\zeta-1) = e_n$. So let $v_X(\overline{z}_0(X)) = i$. Then for $z_0(X) = c_1 X + c_2 X^2 + \dots$ we obtain $c_1 \equiv \dots \equiv c_{i-1} \equiv 0 \mod p$. Therefore, $v(z_0(\pi)) \geqslant \min(i, e+1)$. But $v(z_0(\pi)) = v(\zeta - 1) = e_n \leqslant e$, and hence $v_X(\overline{z}_0) = e_n$. The first congruence implies now that $v_X(\overline{s}_m) = p^m e_n$, $v_X(\overline{u}_m) = p^{m-n}e$. Lemma (1.3) shows that s_m, u_m are invertible in $\mathcal{O}_0\{\{X\}\}$.

We shall verify the second congruence by induction on m. If $m = 1$ then

$$s_1 = (1+z_0)^p - 1 \equiv z_0^p \mod p$$

and

$$z_0^p \equiv\Delta\, z_0 =\Delta\, s_0 \mod p.$$

Hence, $s_1 \equiv\Delta\, s_0 \mod p$. Further, if $m > 1$ and $s_{m-1}(X) \equiv\Delta\, s_{m-2} \mod p^{m-1}$, then by Lemma (7.1) Ch. I

$$s_{m-1}^p \equiv\Delta\, s_{m-2}^p \mod p^m,$$

and

$$s_m \equiv\Delta\, ((1+s_{m-2})^p - 1) =\Delta\, s_{m-1} \mod p^m.$$

The third congruence follows from the second one, because s_m is invertible in $\mathcal{O}_0\{\{X\}\}$.

The last congruence follows from the definition of s_m and the equality

$$(1/s_m)' = -s_m'/s_m^2.$$

□

COROLLARY. *Let the expansion of* ζ *be the following:*

$$\zeta = 1 + c_{e_n}\pi^{e_n} + c_{e_n+1}\pi^{e_n+1} + \dots \quad \textit{with } c_i \in \mathcal{O}_0.$$

Then $E_X(a\,s(X)) \equiv (1 + a\,c\,X^{pe_1})(1 + g(X))^p \mod \deg pe_1 + 1$ *for* $a \in \mathcal{O}_0$ *and some* $c \in \mathcal{O}_0$, $g(X) \in X\mathcal{O}_0[[X]]$.

Proof. In this case $s(X) \equiv c_{e_n}^{p^n} X^{pe_1} \mod (p, \deg pe_1 + 1)$. It remains to apply (2.4). □

LEMMA. *Let the expansion of* ζ *be the following:*

$$\zeta = 1 + c_{e_n}\pi^{e_n} + c_{e_n+1}\pi^{e_n+1} + \dots \quad \textit{with } c_i \in \mathcal{O}_0.$$

Then $v((\triangle_X^m s(X))|_{X=\pi}) > e(1 + \max(m, n))$ *and, in addition, for* $p = 2$

$$v((\triangle_X^m s(X))|_{X=\pi}) \geqslant e(2 + m).$$

Proof. By b) of the Proposition we get

$$\triangle\, s_{n+k-1} = s_{n+k} + p^{n+k} f_k, \quad k \geqslant 1$$

with $f_k \in \mathcal{O}_0[[X]]$. As $z_0 \equiv 0 \mod \deg e_n$, we deduce

$$s_k \equiv 0 \mod \deg e_n \quad \text{for } k \geqslant 1, \qquad \text{and} \quad f_m \equiv 0 \mod \deg e_n.$$

Acting by $\triangle^{m-k-1}$ on the equality and summing for $1 \leqslant k \leqslant m$, we obtain

$$\triangle^m\, s = s_{n+m} + p^{n+m} f_m + p^{n+m-1} \triangle\, f_{m-1} + \dots + p^{n+1} \triangle^{m-1} f_1.$$

One has

$$v(p^{n+k} \triangle^{m-k} f_k(X)|_{X=\pi}) \geqslant (n + k)e + p^{m-k} e_n.$$

Now,

if $n \geqslant m$, then $(n + k)e + p^{m-k}e_n > e(1 + n)$ and $\geqslant e(2 + m)$ for $p = 2$.
if $n < m$ and $n + k \geqslant m + 2$, then $(n + k)e + p^{m-k}e_n \geqslant e(2 + m)$.
if $n < m$ and $n + k \leqslant m + 1$, $k \geqslant 1$, then $(n + k)e + p^{m-k}e_n > e(1 + m)$ and $(n + k)e + p^{m-k}e_n \geqslant e(2 + m)$ for $p = 2$.

This proves the Lemma. □

(3.2). PROPOSITION. *There is an invertible formal power series* $g(X) \in \mathcal{O}_0[[X]]$ *such that* $u(X)g(X)$ *is the Eisenstein polynomial of* π *over* F_0. *Any formal power series* $f(X) \in \mathcal{O}_0[[X]]$ *with* $f(\pi) = 0$ *is divisible by* $u(X)$ *in* $\mathcal{O}_0[[X]]$.

Proof. The first assertion follows from the previous Proposition and Proposition (1.3), the second from Corollary (1.3). □

(3.3). Now we compare distinct formal power series corresponding to distinct expansions of ζ in a power series in π.

PROPOSITION. *Let* $s(X), s^{(1)}(X)$ *be two formal power series over* $\mathcal{O}_0$ *which correspond to two expansions of* ζ *in a power series in* π. *Then*

$$s^{(1)} = s + p^n g_1 + p^{n-1} s^{p-1} g_2 + s^p g_3$$

for some $g_1 \in X\mathcal{O}_0[[X]]$, $g_2, g_3 \in X^2\mathcal{O}_0[[X]]$.

Proof. Let $z(X), z^{(1)}(X)$ be two elements of $1 + X\mathcal{O}_0[[X]]$ with $z(\pi) = z^{(1)}(\pi) = \zeta$. Then, by Proposition (3.2) the series $z^{(1)}(X)/z(X) - 1$ is divisible by $u(X)$. Put $z^{(1)} = z(1 + u\psi)$ where $\psi \in \mathcal{O}_0[[X]]$. Since $u(0) = p$, we obtain that $\psi \in X\mathcal{O}_0[[X]]$. According to (3.1), we can write

$$z^{(1)} = z + p\psi_1 + s_{n-1}^{p-1}\psi_2$$

for some formal power series ψ_1, ψ_2 in $X\mathcal{O}_0[[X]]$. By induction on m one can obtain that

$$s_m^{(1)} = s_m + p^{m+1}\psi_{1,m} + s_{n-1}^{p-1} p^m \psi_{2,m} + s_{n-1}^p \psi_{3,m}$$

for some $\psi_{i,m} \in X\mathcal{O}_0[[X]]$. Then

$$\begin{aligned} s^{(1)} &\equiv \triangle\ s_{n-1}^{(1)} \equiv \triangle\ s_{n-1} + p^{n-1}\ \triangle\ (s_{n-1}^{p-1}\psi_{2,n-1}) + \triangle\ (s_{n-1}^p\psi_{3,n-1}) \\ &\equiv s + p^{n-1}s^{p-1}\ \triangle\ (\psi_{2,n-1}) + s^p\ \triangle\ (\psi_{3,n-1}) \mod p^n \end{aligned}$$

by Proposition (3.1), b). This completes the proof. □

COROLLARY. *If* $p > 2$, *then*

$$1/s(X) \equiv 1/s^{(1)}(X) \mod (p^n, \deg 0).$$

Proof. By Proposition

$$1/s^{(1)} \equiv 1/s \cdot 1/(1 + p^{n-1}s^{p-2}g_2 + s^{p-1}g_3) \mod p^n.$$

Since $p > 2$, we deduce

$$1/(1 + p^{n-1}s^{p-2}g_2 + s^{p-1}g_3) = 1/(1 + sg_4) = 1 + \sum_{m\geqslant 1} (-1)^m s^m g_4^m$$

for some $g_4 \in \mathcal{O}_0[[X]]$. Therefore,

$$1/s^{(1)} \equiv 1/s + \sum_{m \geqslant 1} (-1)^m s^{m-1} g_4^m \equiv 1/s \mod (p^n, \deg 0).$$

□

REMARK. If $p = 2$ then $r(X)/s(X) \equiv r^{(1)}(X)/s^{(1)}(X) \mod (p^n, \deg 0)$ where the polynomial $r(X)$ depends on the series $s(X)$ and is defined in (3.4).

(3.4). In this subsection $p = 2$. We introduce a series $h(X)$ and polynomial $r(X)$ introduced by G. Henniart in the case $p = 2$.

Define

$$h(X) = \frac{\triangle\,(s_{n-1}(X)) - s(X)}{2^n}.$$

Then, by Proposition (3.1), b) the series h belongs to $\mathcal{O}_0[X]]$. Let $r_0(X) \in X\mathcal{O}_0[X]$ be a polynomial of degree $e - 1$, satisfying the condition:

$$\triangle^2\, r_0 + (1 + (2^{n-1} - 1)s_{n-1}) \triangle\, r_0 + s_{n-1} r_0 \equiv h \operatorname{modev}(2, \deg 2e),$$

where we introduced the notation

$$\sum_{m \geqslant 0} \alpha_m X^m \equiv 0 \operatorname{modev}(2, \deg 2e)$$

if $\alpha_{2m} \equiv 0 \mod 2$ for $0 \leqslant m < e$. Put

$$r(X) = 1 + 2^{n-1} \triangle_X\, r_0(X).$$

Observing that Proposition (3.1) implies

$$s_{n-1} \equiv \alpha_e X^e \mod (2, \deg e + 1),$$

we get

$$s_{n-1} \equiv \sum_{m=e}^{2e-1} \alpha_m X^m \mod (2, \deg 2e), \quad \alpha_m \in \mathcal{O}_0.$$

Let

$$h \equiv \sum_{m=1}^{2e-1} \beta_m X^m \mod (2, \deg 2e), \quad r_0 = \sum_{m=1}^{e-1} \rho_m X^m$$

with $\beta_m, \rho_m \in \mathcal{O}_0$. Then the condition on r_0 is equivalent to the following one: for $m < e$ the coefficient of X^{2m} in the expression

$$\sum_{m=1}^{e-1} \varphi^2(\rho_m)X^{4m} + \bigl(1 + (2^{n-1} - 1)\sum_{m=e}^{2e-1} \alpha_m X^m\bigr)\bigl(\sum_{m=1}^{e-1} \varphi(\rho_m)X^{2m}\bigr)$$
$$+ \bigl(\sum_{m=e}^{2e-1} \alpha_m X^m\bigr)\bigl(\sum_{m=1}^{e-1} \rho_m X^m\bigr)$$

is congruent modulo 2 to the coefficient β_{2m}. Thus, every subsequent coefficient ρ_m linearly depends on $\varphi^i(\rho_1), \dots, \varphi^i(\rho_{m-1})$, $i = 0, 1, -1$. This linear system of equations has the unique solution when $\beta_{2m} = 0$ for $1 \leqslant m < e$. Therefore, the polynomials $r_0(X)$ and $r(X)$ are uniquely determined by the conditions indicated. From Proposition (3.1) one deduces that the condition on r_0 is equivalent to the following one: for $m < 0$ the coefficient of X^{4m} in the series

$$H(r) = \frac{\triangle^2 r - \triangle (1 + 2^{n-1}h) \triangle r}{\triangle s} + \frac{\triangle r - \triangle (1 + 2^{2n-2}h)r}{s}$$

is divisible by 2^n.

(3.5). This subsection and the following one contain several auxiliary assertions which will be applied in Ch. VII.

LEMMA.

a) *For* $i \geqslant 1$,

$$\frac{u^i}{s} \equiv \frac{p^{i-1}}{s_{n-1}} + \frac{(i-1)p^{i-1}(p-1)}{2} \quad \text{mod deg } 1.$$

In particular,

$$\frac{u^i}{s} \equiv \frac{p^{i-1}}{s_{n-1}} \quad \text{mod deg } 0.$$

b) *For* $p > 2$, $i \geqslant 1$,

$$\frac{u^{i\triangle}}{s} \equiv \frac{p^i}{s} + \frac{ip^i(p-1)}{2} \quad \text{mod } (p^{i+n-1}, \text{deg } 1).$$

In particular,

$$\frac{u^{i\triangle}}{s} \equiv \frac{p^i}{s} \quad \text{mod } (p^{i+n-1}, \text{deg } 0).$$

Moreover,

$$\frac{u^{\triangle}}{s} \equiv \frac{p}{s} + \frac{p(p-1)}{2} \quad \text{mod } (p^{n+1}, \text{deg } 1).$$

c) *For* $p = 2,\ i \geqslant 1$,

$$\frac{u^{i\triangle}}{s} \equiv \frac{(2 + 2^n h)^i}{s} + i2^{i-1} \mod \deg 1$$

where the series h *is defined in* (3.4).

d) *For* $p > 2,\ i \geqslant 1$,

$$\frac{(u^i)'}{i^2 s} \equiv \frac{(u^i)'}{is} \equiv \frac{p^{i-1}}{i}\Big(\frac{1}{s_{n-1}}\Big)' \mod (p^n, \deg 0).$$

Proof. a) Since $u = p + \sum_{j=2}^{p} \binom{p}{j} s_{n-1}^{j-1}$ we have

$$\frac{u^i}{s} = \frac{u^{i-1}}{s_{n-1}} = \frac{1}{s_{n-1}}\left(p + \binom{p}{2} s_{n-1} + \dots\right)^{i-1}$$
$$\equiv p^{i-1}/s_{n-1} + (i-1)p^{i-2}\binom{p}{2} \mod \deg 1.$$

b) By Proposition (3.1), b) we get $\triangle\ u = u_{n+1} + p^n g$ for some $g(X) \in \mathcal{O}_0[[X]]$. Hence

$$\frac{u^{i\triangle}}{s} = \sum_{j=0}^{i} \binom{i}{j} p^{nj} g^j u_{n+1}^{i-j}/s = \frac{1}{s}\sum_{j=0}^{i} \binom{i}{j} p^{nj} g^j \left(p + \binom{p}{2} s + \dots\right)^{i-j}$$
$$\equiv \sum_{j=0}^{i} \binom{i}{j} p^{nj} g^j \left(p^{i-j}/s + (i-j)p^{i-j}(p-1)/2\right) \mod \deg 1$$

which for $p > 2$ is congruent to $p^i/s + ip^i(p-1)/2 \mod (\deg 1, p^{i+n-1})$.

For $i = 1,\ p > 2$ we deduce

$$\frac{u^{\triangle}}{s} = \frac{\left(p + \binom{p}{2} s_{n-1} + \dots + s_{n-1}^{p-1}\right)^{\triangle}}{s}$$
$$\equiv \frac{p + \binom{p}{2} s + \dots + s^{p-1}}{s} \equiv \frac{p}{s} + \frac{p(p-1)}{2} \mod (p^{n+1}, \deg 1),$$

since $ps_{n-1}^{i\triangle} \equiv ps^i \mod p^{n+1}$ from the congruence b) of Proposition (3.1).

c) Next, for $p = 2$ we get $\triangle\ u = 2 + 2^n h + s$, hence

$$u^{i\triangle}/s \equiv (2 + 2^n h)^i/s + i2^{i-1} \mod \deg 1.$$

d) Finally, Proposition (3.1) implies that

$$u' \equiv (s_{n-1}^{p-1})' \equiv -s'_{n-1} s_{n-1}^{p-2} \mod p^n.$$

Then

$$u'/s \equiv -s'_{n-1}/s_{n-1}^2 \mod p^n.$$

By Proposition (3.1) d) we know that $(1/s_{n-1})' \equiv 0 \mod p^{n-1}$. Then

$$\frac{(u^i)'}{i^2 s} = \frac{u^{i-1}u'}{is} \equiv \frac{p^{i-1}s'_{n-1}}{is^2_{n-1}} \mod (p^n, \deg 0)$$

since $p^{i-1}s'_{n-1}/i \equiv 0 \mod p^n$ for $i \geqslant 2,\ p > 2$. The latter is $\equiv 0 \mod (p^n, \deg 0)$ unless $i = 1$, in which case $\dfrac{(u^i)'}{i^2 s} = \dfrac{(u^i)'}{is}$. □

(3.6). And, finally, another three lemmas.
Put $V(X) = 1/2 + 1/s(X)$.

LEMMA 1. *Let* $f(X) \in \mathcal{O}_0\{\{X\}\}$. *Then*

$$\operatorname{res} f'/s \equiv f'V \equiv 0 \mod p^n.$$

Proof. By Proposition (3.1), d)

$$(f/s)' = f'/s + f(1/s)' \equiv f'/s \mod p^n.$$

Since $\operatorname{res} g' = 0$ for every $g \in F_0\{\{X\}\}$, the assertion follows. □

LEMMA 2. *Let* $f(X)$ *belong to* R *defined in* (2.3). *Let* i *be divisible by* $p^k,\ k \geqslant 0$. *Then for* $p > 2$

$$f(X)^{ip} - f(X)^{i\Delta} \equiv ipl_X(f(X))f(X)^{i\Delta} \mod p^{2(k+1)}.$$

Proof. We have

$$f^{ip} - f^{i\Delta} = f^{i\Delta}(f^{ip}/f^{i\Delta} - 1) = f^{i\Delta}\big(\exp(ipl_X(f)) - 1\big).$$

This means that

$$f^{ip} - f^{i\Delta} = f^{i\Delta} ipl_X(f) + f^{i\Delta}\sum_{j\geqslant 2}\frac{(ip)^j}{j!}l_X(f)^j.$$

Since $(ip)^{j-2}/j! \in \mathbb{Z}_p$ for $p > 2,\ j \geqslant 2$, the assertion follows. □

LEMMA 3. *Let* $f(X) \in \mathcal{O}_0((X)),\ g(X) \in \mathcal{O}_0((X))^*,\ h(X) \in \mathcal{O}_0\{\{X\}\}$. *Then*

a) $(\Delta\, f)' = pX^{p-1} \Delta\, (f') = pX^{-1} \Delta\, (Xf')$,

b) $g'/g = l_X(g)' + X^{p-1} \Delta\, (g'/g)$,

c) $\operatorname{Tr}_{F_0/\mathbb{Q}_p} \operatorname{res} X^{-1}h = \operatorname{Tr}_{F_0/\mathbb{Q}_p} \operatorname{res} X^{-1} \Delta\, h$.

Proof.

a) Let $f = \sum \alpha_i X^i,\ \alpha_i \in \mathcal{O}_0$. Then

$$(\Delta\, f)' = \Big(\sum \varphi(\alpha_i)X^{pi}\Big)' = \sum pi\varphi(\alpha_i)X^{pi-1} = pX^{p-1} \Delta\, (f').$$

b) Let $g = \alpha X^m \varepsilon(X)$ with $\alpha \in \mathcal{O}_0^*$, $\varepsilon = 1 + \sum_{i\geqslant 1} \beta_i X^i$, $\beta_i \in \mathcal{O}_0$. Then using (2.3) we get $g'/g = mX^{-1} + \varepsilon'/\varepsilon$ and $l_X(g) - l_X(\varepsilon) \in \mathcal{O}_0$. Now

$$l_X(g)' = l_X(\varepsilon)' = \left(\left(1 - \frac{\Delta}{p}\right)\log\varepsilon\right)' = (\log\varepsilon)' - X^{p-1} \Delta\, (\log\varepsilon)'$$
$$= \varepsilon'/\varepsilon - X^{p-1} \Delta\, (\varepsilon'/\varepsilon) = g'/g - X^{p-1} \Delta\, (g'/g).$$

c) Let $h(X) = \sum \alpha_i X^i$ with $\alpha_i \in \mathcal{O}_0$. Then

$$\mathrm{Tr}_{F_0/\mathbb{Q}_p} \operatorname{res} X^{-1} h = \mathrm{Tr}_{F_0/\mathbb{Q}_p} \alpha_0 = \mathrm{Tr}_{F_0/\mathbb{Q}_p} \varphi(\alpha_0) = \mathrm{Tr}_{F_0/\mathbb{Q}_p} \operatorname{res} X^{-1} \Delta\, h.$$

□

Exercises.

1. a) Show that if

$$\psi_1/s \equiv \psi_2/s \mod (p^n, \deg 1)$$

for $\psi_1, \psi_2 \in \mathcal{O}_0\{\{X\}\}$, then

$$\Delta\, (\psi_1)/s \equiv \Delta\, (\psi_2)/s \mod (p^n, \deg 1).$$

b) Show that for $m \geqslant 1$ there exists a series $g_m \in -1 + X\mathcal{O}_0[[X]]$, such that

$$s_n^{pm} = s_{n+1}^m + pms_n^m g_m.$$

c) Show that $l_X(s_n) \equiv s_n g \mod p^n$ for some $g \in X\mathcal{O}_0[[X]]$.

2. Show that for $p = 2$, $m \geqslant 2$, $k \geqslant 1$
   a) $\Delta^k\, (u^m)/s \equiv (2 + 2^n \Delta^{k-1}\, (h))^m/s \mod (p^{n+m}, \deg 0)$,
   b) $(\Delta^k\, (u^m)/m2^k)'/s \equiv X^{2^k-1} \Delta^k\, (s')/s \mod (2^n, \deg 0)$.
   c) $\Delta\, (u^m)/s \equiv 2^m/s \mod (2^{n+m}, \deg 0)$ if m is a power of 2, $m \geqslant 3$.
   d) $2(1/s_{n-1})' + s'/s + s'/s_{n-1} \equiv 0 \mod (2^{n+1}, \deg 1)$
   e) $(\Delta\, s_{n-1})'/2 \equiv s'/2 + 2^{n-1}h' \mod 2^n$.
3. Let $p = 2$.
   a) Show that for $f \in R$ (R is defined in (2.3)), $g \in \widehat{\mathcal{O}}\{\{X\}\}$

$$\operatorname{res} f' \Delta\, (g)/f = \operatorname{res} X \Delta\, (f'g/f).$$

   b) Show that if $f = X^m a\varepsilon \in R$ with $a \in \widehat{\mathcal{O}}^*$, $\varepsilon \in 1 + X\widehat{\mathcal{O}}[[X]]$, then

$$\left(\frac{f^2 - f^{\Delta}}{2f^{\Delta}}\right)' \equiv l_X(f)' \equiv \varepsilon'(X\varepsilon)'/\varepsilon^2 \mod 2.$$

   c) Show that for $g \in \widehat{\mathcal{O}}\{\{X\}\}$

$$\operatorname{res} g' r/s \equiv 0 \mod 2^n.$$

4. Let $p = 2$. Using Exercise 2 show that for $g \in \mathcal{O}_0[[X]]$ and $i, m \geqslant 1$

$$\mathrm{Tr}_{F_0/\mathbb{Q}_2} \operatorname{res} \Big(\frac{\Delta^i\, (ug)^m}{2^i m}\Big)' r/s \equiv 0 \mod 2.$$

5. ($\diamond$) Let $p = 2$. Let $g \in R$. Put

$$f_j = \frac{g^{i2^j} - g^{i2^{j-1}\triangle}}{i2^j} - l_X(g)g^{i2^{j-1}\triangle}.$$

a) Show that the series

$$\sum_{j\geqslant 1} \frac{f_j}{2^j} - \frac{i}{2} \triangle (g^i L_X(g)) - \frac{i}{2} \sum_{j\geqslant 1} \triangle \left(g^{2^{j-1}i} l_X(g)\right)$$

belongs to $\widehat{\mathcal{O}}[[X]]$.

b) Show that

$$\operatorname{res}\Big(\sum_{j\geqslant 1} f_j/2^j\Big)' r/s \equiv \operatorname{res}\Big(\triangle \Big(ig^i L_X(g) + il_X(g) \sum_{j\geqslant 1} g^{2^{j-1}i}\Big)\Big)' r/(2s) \mod 2^n.$$

6. Let $p > 2$.

a) Show that V' belongs to $p^n X^{-2pe/(p-1)}\mathcal{O}_0[[X]]\,[[pX^{-e}]]$.

b) Let $g \in \mathcal{O}_0[[X]]$. Show that $\log(1 + ug)$ belongs to $\mathcal{O}_0[[X]]\,[[p^{-1}X^{pe}]]$.

c) Deduce that for every $\alpha \in \mathcal{O}_0((X))^*$

$$l_X(\alpha)\log(1 + ug)V' \equiv 0 \mod (p^n, \deg 1).$$

7. Let $p > 2$. Deduce from (3.5) and (3.6) that for every $f \in \mathcal{O}_0[[X]]$

$$\operatorname{res}(1 - p\,\triangle)(V)\, f\, \frac{\triangle}{p} \log(1 + ug) \equiv 0 \mod p^n.$$

8. Let $f(X) \in X\mathcal{O}_0[[X]]$. Show using Proposition (3.2) that $E_X(f(X))|_{X=\pi}$ belongs to F^{*p^n} if and only if $f(X) - l_X(1 + u(X)g(X)) = p^n t(X)$ for some $g \in X\mathcal{O}_0[[X]]$, $t \in X\mathcal{O}_0[[X]]$.

4. Primary Elements

In this section we shall construct primary elements of a local number field F which contains a primitive p^n th root ζ of unity. F_0 denotes, as usually, the inertia subfield $F_0 = F \cap \mathbb{Q}_p^{\mathrm{ur}}$ of F, $\mathcal{O}_0$ denotes its ring of integers. The continuous extension of the Frobenius automorphism $\varphi \in \mathrm{Gal}(\mathbb{Q}_p^{\mathrm{ur}}/\mathbb{Q}_p)$ on the completion $\widehat{\mathbb{Q}_p^{\mathrm{ur}}}$ will be denoted by the same notation. Let $\varphi_F \in \mathrm{Gal}(F^{\mathrm{ur}}/F)$ be the Frobenius automorphism of F, and let its continuous extension to $\widehat{F^{\mathrm{ur}}}$ be denoted by the same notation.

From now on we denote the trace map $\mathrm{Tr}_{F_0/\mathbb{Q}_p}$ by Tr.

(4.1). An element $\omega \in F^*$ is said to be p^n*-primary* if $F(\sqrt[p^n]{\omega})/F$ is an unramified extension (see Exercise 7 section 1 Ch. IV). According to Proposition (1.8) Ch. IV, for an element $a \in \mathcal{O}_0$ there exists an element κ in the ring of integers of $\widehat{\mathbb{Q}_p^{\mathrm{ur}}}$ such that $\varphi(\kappa) - \kappa = a$. Let π be a prime element in F and let $z(X)$ be as in section 3.

PROPOSITION.

(1) *The element*

$$H(a) = E_X\bigl(p^n\varphi(\kappa)l_X(z(X))\bigr)|_{X=\pi}$$

is p^n*-primary. Let* $\gamma \in F^{\mathrm{ur}}$ *be a* p^n *th root of* $H(a)$*. Then*

$$\gamma^{\varphi_F - 1} = \zeta^{\operatorname{Tr} a}.$$

(2) *The element* $H(a)$ *does not depend, up to* p^n *th powers, on the choice of* κ *and prime element* π *and on the choice of expansion of* ζ *in a series in* π.

Proof. Let $f = f(F|\mathbb{Q}_p)$. Then $\varphi^f|_{\mathbb{Q}_p^{\mathrm{ur}}} \in \operatorname{Gal}(\mathbb{Q}_p^{\mathrm{ur}}/F_0)$ is the Frobenius automorphism of F_0. We get

$$\varphi^{f+1}(\kappa) - \varphi(\kappa) = \varphi(1 + \varphi + \cdots + \varphi^{f-1})(\varphi(\kappa) - \kappa) = \operatorname{Tr} a = b,$$

where $b \in \mathbb{Z}_p$. Observing that φ^f commutes with $\triangle$, we deduce

$$\varphi^f E_X\bigl(\varphi(\kappa)l_X(z)\bigr) = E_X\bigl((\varphi(\kappa) + b)l_X(z)\bigr) = E_X\bigl(\varphi(\kappa)l_X(z)\bigr)z^b, \qquad (*)$$

by Proposition (2.2). Hence for the element $\gamma = E_X\bigl(\varphi(\kappa)l_X(z(X))\bigr)|_{X=\pi}$ we get $\gamma^{\varphi_F - 1} = \zeta^{\operatorname{Tr} a}$. The element $H(a)$ belongs to $\widehat{F^{\mathrm{ur}}} = F\widehat{\mathbb{Q}_p^{\mathrm{ur}}}$. We have $\varphi_F|_{\widehat{\mathbb{Q}_p^{\mathrm{ur}}}} = \varphi^f$ and

$$\varphi_F H(a) = H(a)z(\pi)^{p^n b} = H(a)\zeta^{p^n b} = H(a).$$

Therefore, $H(a) \in F$ by Proposition (1.8) Ch. IV. Thus, $H(a)$ is a p^n-primary element in F.

If for a κ_1 in the ring of integers in $\widehat{\mathbb{Q}_p^{\mathrm{ur}}}$ we have $\varphi(\kappa_1) - \kappa_1 = a$, then $\varphi(\kappa_1 - \kappa) = \kappa_1 - \kappa$, and by Proposition (1.8) Ch. IV we get $\kappa_1 = \kappa + c$ for some $c \in \mathbb{Z}_p$. The same arguments as above show that $E_X(p^n\varphi(\kappa_1)l_X(z(X)))|_{X=\pi}$ coincides with $H(a)$ up to p^n th powers.

Let $H_1(a)$ be an element constructed in the same way as $H(a)$ but for another prime element π or for another series $z^{(1)}(X) \in 1 + X\mathcal{O}_0[[X]]$ with $z^{(1)}(\pi) = \zeta$. Then as above we deduce that $\gamma_1^{\varphi_F - 1} = \zeta^{\operatorname{Tr} a}$. Since both elements γ and γ_1 belong to F^{ur}, we deduce that $\gamma = \gamma_1 c$ with $c \in F$. Thus, $H_1(a) = H(a)c^{p^n}$ as required. □

REMARK. The elements $H(a)$ were constructed by *H. Hasse* in [Has8]. Hasse's elements $H(a)$ are not suitable for our purposes, because they involve elements which do not belong to the base field F. Later in (4.2) we shall obtain other forms of primary elements.

LEMMA. $H(a) = E_X(p^n\, a \log z(X))|_{X=\pi}$

Proof. One has

$$\left(1 - \frac{\Delta}{p}\right)(\kappa \log z) = \varphi(\kappa) l_X(z) - a \log z,$$

$$E_X\left(p^n\left(1 - \frac{\Delta}{p}\right)(\kappa \log z)\right) = \exp(p^n \kappa \log z).$$

Hence

$$E_X(p^n \varphi(\kappa) l_X(z)) = E_X(p^n\, a \log z) \exp(p^n \kappa \log z).$$

To apply Proposition (1.5), let $f(X) = \exp(X)$, $g(X) = p^n \kappa \log z(X)$, $c = e/(p-1)$, where $e = e(F|\mathbb{Q}_p)$, and $d = 0$. Therefore, putting $b = 0$, we get $g(\widehat{\mathcal{O}}^0) \subset \widehat{\mathcal{O}}^c$. Now Proposition (1.5) shows that

$$\exp(p^n \kappa \log z(X))|_{X=\pi} = \exp(p^n \kappa \log(\zeta)).$$

Since $\zeta^{p^n} = 1$, we obtain $\log(\zeta) = 0$. Thus,

$$E_X(p^n \varphi(\kappa) l_X(z))|_{X=\pi} = E_X(p^n\, a \log(z))|_{X=\pi},$$

as required. □

(4.2). Our next goal is to replace the formal power series $p^n\, a \log(z)$ in the previous Lemma with another series over $\mathcal{O}_0[[X]]$.

THEOREM. *The element*

$$\omega(a) = E_X(a\, s(X))|_{X=\pi}, \quad a \in \mathcal{O}_0,$$

coincides with $H(a)$ up to the elements of the p^n th power in F.

Thus, $\omega(a)$ is a p^n-primary element in F and does not depend, up to the p^n th powers in F, on the choice of prime element π and on the choice of expansion of ζ in a series in π.

Proof. First we verify the assertion of the Theorem for the series $z = 1 + c_{e_n} X^{e^n} + c^{e_n+1} X^{e_n+1} + \dots$ with $c_i \in \mathcal{O}_0$ (see (3.1)). The equality $p^n \log(z) = \log(1+s)$ implies

$$E_X(p^n\, a \log(z)) = E_X(a\, s) E_X(a(\log(1+s) - s)).$$

Put $\psi = \log(1 + s(X)) - s(X)$. We shall show that $E_X(a\psi)|_{X=\pi} = \varepsilon^{p^n}$ for some $\varepsilon \in F^*$. Then $H(a) = \omega(a)\varepsilon^{p^n}$, as desired. We get

$$E_X(a\psi) = \exp(a\psi) \exp\left(\sum_{i \geqslant 1} \Delta^i\, (a\psi)/p^i\right).$$

Let v be the discrete valuation of $\widehat{F^{\mathrm{ur}}}$. Since $s(\alpha) = (1 + z_0(\alpha))^{p^n} - 1$ for an element $\alpha \in \widehat{F^{\mathrm{ur}}}$ with $v(\alpha) \geqslant 1$, we deduce $v(s(\alpha)) > e$. Then Proposition (1.5) and Example (1.4) show that $\log(1 + s(X))|_{X=\alpha} = \log(1 + s(\alpha))$ and $v(\psi(\alpha)) > e$. Therefore, by that Proposition,

$$\exp(a\psi(X))|_{X=\pi} = \exp(a\log(1 + s(\pi)) - a\, s(\pi)) = 1.$$

Further, $\psi = \sum_{m\geqslant 2}(-1)^{m-1}s(X)^m/m$; consequently

$$\exp\left(\sum_{i\geqslant 1} \Delta^i\,(a\psi)/p^i\right) = \exp\left(\sum_{i\geqslant 1}\sum_{m\geqslant 2} \varphi^i(a)\psi_{m,i}\right)^{p^n}$$

where $\psi_{m,i} = (-1)^{m-1}\,\Delta^i\, s^m/mp^{i+n}$. For an element $\alpha \in \widehat{F^{\mathrm{ur}}}$ with $v(\alpha) \geqslant 1$, Lemma (3.1) shows that

$$v(\psi_{m,i}(\alpha)) > -(m-1)e + me(1 + \max(i, n)) - (i + n)e \geqslant e,$$

because $v(m) \leqslant (m-1)e$ for $m \geqslant 1$. We also obtain that $v(\psi_{m,i}(\alpha)) \to +\infty$ as $m \to +\infty$ or $i \to +\infty$. Therefore, Proposition (1.5) implies that

$$\exp\left(\sum_{i\geqslant 1} \Delta^i\,(a\psi(X))/p^i\right)\Bigg|_{X=\pi} = \exp\left(\sum_{i\geqslant 1}\sum_{m\geqslant 2} \varphi^i(a)\psi_{m,i}(\pi)\right)^{p^n}.$$

Thus, $H(a)$ coincides with $\omega(a)$ up to F^{*p^n}.

Now we verify the assertion of the Theorem for an arbitrary expansion of ζ in a series in π. Let $z^{(1)}(X)$, $s^{(1)}(X)$ be the corresponding series. By Proposition (3.3) we get

$$s^{(1)} = s + p^n g_1 + p^{n-1}s^{p-1}g_2 + s^p g_3$$

with $g_i \in X\mathcal{O}_0[[X]]$. Then

$$E_X(p^n g_1(X))|_{X=\pi} = E_X(g_1(X))^{p^n}|_{X=\pi} \in F^{*p^n}.$$

In the same way as above, we deduce that

$$\exp(p^{n-1}s^{p-1}g_2)|_{X=\pi} = 1, \quad \exp(s^p g_3)|_{X=\pi} = 1.$$

Finally, for an element $\alpha \in \widehat{F^{\mathrm{ur}}}$ with $v(\alpha) \geqslant 1$ we obtain, by Lemma (3.1), that

$$v(p^{n-1}\,\Delta^i\,(s^{p-1}g_2)/p^i|_{X=\alpha}) = v_i > e,\ v(\Delta^i\,(s^p g_3)/p^i|_{X=\alpha}) = w_i > e$$

and $v_i, w_i \to +\infty$ as $i \to +\infty$. Therefore, Proposition (1.5) implies

$$E_X(p^{n-1}s^{p-1}g_2)|_{X=\pi} \in F^{*p^n},\ E_X(s^p g_3)|_{X=\pi} \in F^{*p^n}$$

and $E_X(a\,s^{(1)}(X))|_{X=\pi}$ coincides with $\omega(a)$ up to F^{*p^n}.

The last assertion of the theorem follows from Proposition (4.1). □

(4.3). PROPOSITION. *A primary element $\omega(a)$, $a \in \mathcal{O}_0$, is a p^n th power in F if and only if* $\operatorname{Tr} a \equiv 0 \mod p^n$, *where* $\operatorname{Tr} = \operatorname{Tr}_{F_0/\mathbb{Q}_p}$.

Proof. From the previous theorem and $(*)$ in the proof of Proposition (4.1) we deduce that $\omega(a) \in F^{*p^n}$ if and only if $H(a) \in F^{*p^n}$ if and only if $z(X)^{\operatorname{Tr} a}|_{X=\pi} = 1$ which is equivalent to $\operatorname{Tr} a \equiv 0 \mod p^n$. □

COROLLARY. *Let Ω be the group of all p^n-primary elements in F^*. Then the quotient group Ω/F^{*p^n} is a cyclic group of order p^n and is generated by $\omega(a_0)$ with $a_0 \in \mathcal{O}_0$,* $\operatorname{Tr} a_0 \not\equiv 0 \mod p$.

Proof. Since F has unique unramified extension of degree p^n, Kummer theory implies that Ω/F^{*p^n} is a cyclic group of order p^n. Let Ω_1 be the subgroup in Ω generated by $\omega(a)$ with $a \in \mathcal{O}_0$. The kernel of the surjective homomorphism

$$\chi\colon \Omega_1 \to \mu, \quad \omega(a) \to \zeta^{\operatorname{Tr} a},$$

where μ is the group of p^n th roots of unity in F, is equal to F^{*p^n}. Therefore, $\Omega = \Omega_1$. □

Exercises.

1. Show that $H(a) \equiv 1 + a(\zeta - 1)^{p^n} \mod \pi^{pe_1+1}$.
2. Let $f(X)$ be an invertible series in $\mathcal{O}_0((X))$ and $f(\pi) = 1$. Show that

$$f(X) = (1 - \alpha u)(1 - ug)$$

   for some $\alpha \in \mathcal{O}_0$ and $g \in \mathcal{O}_0((X))$, $g(0) = 0$.
3. Let $F, \mathbf{F}, l_X, E_X$ be as in Exercise 4 section 2. Let a primitive p^n th root ζ of unity belong to F, π prime in F and $z(X) \in 1 + XW(\overline{F})[[X]]$, $s(X)$ as in (3.1).
   a) Show that for an element $a \in W(\overline{F})$ there exists an element $\kappa \in W(\overline{F}^{\text{sep}})$ such that $\mathbf{F}(\kappa) - \kappa = a$. Show that for every $\sigma \in \operatorname{Gal}(F_0^{\text{ur}}/F_0)$ the element $\sigma(\kappa) - \kappa$ belongs to $\mathbb{Z}_p$.
   b) Show that the element

$$H(a) = E_X\left(p^n \mathbf{F}(\kappa) l_X(z(X))\right)|_{X=\pi}$$

   is p^n-primary and does not depend, up to the p^n th powers in F, on the choice of κ, π and $z(X)$.
   c) Show that the element

$$\omega(a) = E_X(a\, s(X))|_{X=\pi}$$

   coincides with $H(a)$ up to a p^n th power in F, and, thus, it is a p^n-primary element of F.
   d) Show that $\omega(a) \in F^{*p^n}$ if and only if

$$a \equiv b^p - b \mod p^n$$

   for some $b \in W(\overline{F})$

e) Show that $\omega(a)$, $a \in W(\overline{F})$, generate the group Ω of p^n-primary elements of F and

$$\Omega/F^{*p^n} \simeq W(\overline{F})/(p^n W(\overline{F}) + \wp W(\overline{F})),$$

where $\wp(b) = b^p - b$ for $b \in W(\overline{F})$.

5. The Shafarevich Basis

We keep the notations of the preceding sections. In particular, we fix a prime element π of F. We shall construct a special system of generators of the multiplicative $\mathbb{Z}_p$-module $U_1 = U_{1,F}$ of a local number field F by using the Artin–Hasse–Shafarevich map E_X.

(5.1). PROPOSITION. *Let a local number field F contain no nontrivial pth roots of unity. Then for a unit $\varepsilon \in U_1$ there exists a unique polynomial $w(X) = \sum_{\substack{1 \leqslant i < pe_1 \\ (i,p)=1}} \alpha_i X^i$, $\alpha_i \in \mathcal{O}_0$ $(e_1 = e/(p-1), e = e(F|\mathbb{Q}_p))$, such that*

$$\varepsilon = E_X(w(X))|_{X=\pi}.$$

Proof. Let $\theta_1, \dots, \theta_f$ be a set of representatives in F of a basis of $\overline{F}$ over $\mathbb{F}_p$. Put

$$\varepsilon_{ij} = E_X(\theta_j X^i)|_{X=\pi}, \quad 1 \leqslant j \leqslant f, \quad 1 \leqslant i < pe_1, \quad (i,p) = 1.$$

Then by (2.4)

$$\varepsilon_{ij} \equiv 1 + \theta_j \pi^i \mod \pi^{i+1}.$$

Proposition (6.4) and Corollary (6.5) Ch. I show that ε_{ij} form a basis of the $\mathbb{Z}_p$-module U_1. This means that for some $a_{ij} \in \mathbb{Z}_p$

$$\varepsilon = \prod_{i,j} \varepsilon_{ij}^{a_{ij}} = E_X\left(\sum_{i,j} a_{ij}\theta_j X^i\right)\Bigg|_{X=\pi}.$$

Putting $w(X) = \sum \alpha_i X^i$ with $\alpha_i = \sum_j a_{ij}\theta_j$, we get the required assertion. □

(5.2). PROPOSITION (THE SHAFAREVICH BASIS). *Let $n \geqslant 1$ be the maximal integer such that a primitive p^n th root of unity ζ belongs to F. Then for a unit $\varepsilon \in U_1$ there exists an element a of $\mathcal{O}_0$ and a polynomial $w(X) = \sum_{\substack{1 \leqslant i < pe_1 \\ (i,p)=1}} \alpha_i X^i$, $\alpha_i \in \mathcal{O}_0$, such that*

$$\varepsilon = E_X(w(X))|_{X=\pi}\omega(a).$$

$\varepsilon \in F^{*p^n}$ *if and only if* $\operatorname{Tr} a \equiv 0 \mod p^n$ *and* $w(X) \equiv 0 \mod p^n$.

Proof. Let

$$\zeta = 1 + c_{e_n}\pi^{e_n} + c_{e_n+1}\pi^{e_n+1} + \dots, \qquad c_i \in \mathcal{O}_0,$$

and let $z(X)$ be the corresponding series over $\mathcal{O}_0$. Then Corollary (3.1) shows that

$$\omega(a) \equiv 1 + a\,c\,\pi^{pe_1} \mod \pi^{pe_1+1}$$

for some $c \in \mathcal{O}_0$. If $\operatorname{Tr} a \not\equiv 0 \mod p$, then by Corollary (4.3), $\omega(a) \notin F^{*p}$. Let θ_j, $1 \leqslant j \leqslant f$, be as in the proof of the previous Proposition. Now (6.4) and (6.5) Ch. I imply (take $\pi_i = \pi^i$) that the elements

$$\begin{array}{lll} \varepsilon_{ij} = E_X(\theta_j X^i)|_{X=\pi} & \text{with} & 1 \leqslant j \leqslant f, \quad 1 \leqslant i < pe_1, \quad (i,p) = 1, \\ \omega(a_0) & \text{for a fixed } a_0 \in \mathcal{O}_0 & \text{with} \quad \operatorname{Tr} a_0 \not\equiv 0 \mod p \end{array}$$

form a basis of the $\mathbb{Z}_p/p^n\mathbb{Z}_p$-module $U_1/U_1^{p^n}$. Thus, by the same arguments as in the proof of Proposition (5.1), we obtain the required decomposition.

Further, if $\operatorname{Tr} a \equiv 0 \mod p^n$, $w(X) \equiv 0 \mod p^n$, then, by Proposition (4.3), $\varepsilon \in F^{*p^n}$.

Conversely, assume that $\varepsilon \in F^{*p^n}$. Then, since the elements $\varepsilon_{ij}, \omega(a) \mod U_1^{*p^n}$ form a basis of $U_1/U_1^{p^n}$, we deduce that $w(X) \equiv 0 \mod p^n$ and $\omega(a) \in F^{*p^n}$. Now Proposition (4.3) implies that $\operatorname{Tr} a \equiv 0 \mod p^n$. This completes the proof. □

COROLLARY. *Instead of the Shafarevich basis one can take as a basis of $U_1/U_1^{p^n}$ the elements $1 - \theta_j\pi^i$, $\omega(a_0)$ where θ_j, i and a_0 are as in the proof of the Proposition.*

Exercise.

1. Let $n \geqslant 1$ be the maximal integer such that a primitive p^n th root of unity ζ belongs to F. In notations of Exercise 5 section 2 and Exercise 3 section 4, show that for every unit $\varepsilon \in U_{1,F}$ there exist $a \in W(\overline{F})$ and a polynomial

$$w(X) = \sum_{\substack{1 \leqslant i < pe_1 \\ (i,p)=1}} \alpha_i X^i, \quad \alpha_i \in W(\overline{F}),$$

such that

$$\varepsilon = E_X(w(X))|_{X=\pi}\,\omega(a)$$

and $\varepsilon \in F^{*p^n}$ if and only if $a \in \wp W(\overline{F}) + p^n W(\overline{F})$ and $\alpha_i \in p^n W(\overline{F})$.

CHAPTER 7

Explicit Formulas for the Hilbert Symbol

This chapter presents comprehensive explicit formulas for the (p^n th) Hilbert symbol defined on a local number field. Origin of the formulas is discussed in section 1. Section 2 introduces a pairing $\langle \cdot, \cdot \rangle_X$ on formal power series which satisfies the *Steinberg property*. This pairing specializes to a pairing $\langle \cdot, \cdot \rangle_\pi$ on F^* in (2.2); it is well defined and does not depend on the choice of a prime element π. Subsection (2.5) presents the technically more difficult case of even p. We apply results of section 2 to construct an explicit class field theory for Kummer extension in section 3; the latter does not depend on results of Ch. IV. In section 4 we prove the equality of the Hilbert symbol and the pairing of section 2, thus establishing an explicit formula for the Hilbert symbol. Several other types of explicit formulas are discussed in section 5. There we also comment on the explicit formula and its generalizations to local fields and n-dimensional local fields.

We keep the notations of Ch. VI.

1. Origin of Formulas

In subsection (1.1) we calculate values of the Hilbert symbol using the Shafarevich basis introduced in Chapter VI. Then, as a motivation for the explicit formulas to come in sections 2 and 4, we treat the case of $\mathbb{Q}_p(\zeta)$ in subsections (1.2)–(1.4).

(1.1). Let a primitive p^n th root of unity ζ belong to F, and let π be a prime element in F. Let $(\cdot, \cdot)_{p^n}$ be the p^n th Hilbert symbol (see section 5 Ch. IV).

First, we compute the values of $(\pi, \varepsilon)_{p^n}$ for $\varepsilon \in U_{1,F}$. Let

$$\varepsilon = E_X(w(X))|_{X=\pi}\omega(a), \quad w(X) = \sum_{\substack{1 \leqslant i < pe_1 \\ (i,p)=1}} \alpha_i X^i,$$

with $\alpha_i, a \in \mathcal{O}_0$, be the Shafarevich basis as in section 5 Ch. VI. Applying Theorem (4.2) Ch. VI we get

$$(\pi, \omega(a))_{p^n} = (\pi, H(a))_{p^n}.$$

Since $H(a)$ is a p^n-primary element in F, using the definition of the Hilbert symbol we deduce that

$$(\pi, H(a))_{p^n} = \varphi_F E_X(\varphi(\kappa) l_X(z(X)))|_{X=\pi} E_X(-\varphi(\kappa) l_X(z(X)))|_{X=\pi},$$

where $\varphi_F = \Psi_F(\pi)|_{F^{\mathrm{ur}}}$ is the Frobenius automorphism (see (1.2) Ch. IV), $\kappa \in \widehat{F^{\mathrm{ur}}}$ and $\varphi(\kappa) - \kappa = a$ as in (4.1) Ch. VI. The computation in the proof of Proposition (4.1) Ch. VI shows that

$$(\pi, \omega(a))_{p^n} = E_X\left(l_X(z(X))\right)^{\operatorname{Tr} a}|_{X=\pi} = \zeta^{\operatorname{Tr} a}.$$

Now let θ be any nonzero multiplicative representative of $\overline{F}$ in F. Then, as noted in the proof of Proposition (2.2) Ch. VI, $E_X(\theta X^i)|_{X=\pi} = E(\theta X^i)|_{X=\pi} = E(\theta \pi^i)$, where $E(X)$ is the Artin-Hasse function (see (9.1) Ch. I). Lemma (9.1) Ch. I implies that

$$(\pi, E(\theta\pi^i))_{p^n} = \prod_{\substack{j\geqslant 1\\(j,p)=1}} \left(\pi, (1-\theta^j\pi^{ij})^{-\mu(j)/j}\right)_{p^n},$$

where μ is the Möbius function. Then

$$\begin{aligned}\left(\pi, (1-\theta^j\pi^{ij})^{-\mu(j)/j}\right)_{p^n}^{ij^2} &= (\pi^{ij}, 1-\theta^j\pi^{ij})_{p^n}^{-\mu(j)}\\ &= (\pi^{ij}\theta^j, 1-\theta^j\pi^{ij})_{p^n}^{-\mu(j)} = 1,\end{aligned}$$

by Proposition (5.1), (2) Ch. IV and Exercise 1 b) section 5 Ch. IV. If i is relatively prime to p, then

$$(\pi, (1-\theta^j\pi^{ij})^{-\mu(j)/j})_{p^n} = 1, \qquad (\pi, E(\theta\pi^i))_{p^n} = 1.$$

Since the set of multiplicative representatives generates $\mathcal{O}_0$ over $\mathbb{Z}_p$ and E_X is $\mathbb{Z}_p$-linear, we conclude that

$$(\pi, E_X(w(X))|_{X=\pi})_{p^n} = 1.$$

Therefore,

$$(\pi, \varepsilon)_{p^n} = (\pi, E_X(w(X))|_{X=\pi}\, \omega(a))_{p^n} = \zeta^{\operatorname{Tr} a}.$$

(1.2). Let $F = \mathbb{Q}_p(\zeta)$, where ζ_p is a primitive pth root of unity. Then F is a totally ramified extension of degree $p-1$ over $\mathbb{Q}_p$. By (1.3) Ch. IV $\pi = \zeta_p - 1$ is prime in F. The corresponding series $z(X) = 1 + X$ and $s(X) = s_1(X) = (1+X)^p - 1 \equiv X^p \bmod p$.

For a principal unit $\varepsilon = E_X(w(X))|_{X=\pi}\omega(a)$ in F, $w(X) = \sum_{1\leqslant i\leqslant p-1} a_i X^i$, $a_i \in \mathbb{Z}_p$, $a \in \mathbb{Z}_p$, we get

$$(\pi, \varepsilon)_p = \zeta^a.$$

In the case under consideration

$$\operatorname{res} X^{-1} l_X E_X(w(X))/s(X) \equiv 0 \mod p$$

and

$$\operatorname{res} X^{-1} l_X E_X(a\, s(X))/s(X) = a.$$

This means that for $\psi(X) = E_X(w(X) + a\, s(X))$ with $\psi(\pi) = \varepsilon$, one can write

$$(\pi, \varepsilon)_p = \zeta_p^{\operatorname{res} X^{-1} l_X(\psi(X))/s(X)} \tag{$*$}$$

In other words, for computing $(\pi, \psi(\pi))_p$ we must find the residue of the series

$$X^{-1} l_X(\psi(X))/s(X).$$

Note that π is a pole of this series.

Now let $\psi(X) \in 1 + X\mathbb{Z}_p[[X]]$ be an arbitrary series with $\psi(\pi) = \varepsilon$. By (2.4) Ch. VI we can express $\psi(X)$ as

$$\psi(X) = \prod_{i \geqslant 1} E_X(a_i X^i), \quad a_i \in \mathbb{Z}_p.$$

Then $\psi(X) = \prod_{i \geqslant 1} E(X^i)^{a_i}$ and $\varepsilon = \prod_{i \geqslant 1} E(\pi^i)^{a_i}$. The arguments of (1.1) show that $(\pi, E(\pi^i))_p = 1$ for $(i, p) = 1$. Subsections (5.7) and (5.8) Ch. I imply $U_{p+1,F} \subset F^{*p}$, hence

$$(\pi, \varepsilon)_p = (\pi, E(\pi^p))_p^{a_p}.$$

But, according to (2.4) Ch. VI

$$\omega(a_p) = E_X(a_p s(X))|_{X=\pi} = E(\pi^p)^{a_p} \eta^p \qquad \text{for some} \quad \eta \in U_{1,F}.$$

Therefore,

$$(\pi, \varepsilon)_p = (\pi, \omega(a_p))_p = \zeta_p^{a_p}.$$

On the other hand,

$$a_p \equiv \operatorname{res} X^{-1} l_X(\psi(X))/s(X) \mod p.$$

Thus, we conclude that formula $(*)$ holds for an arbitrary expansion of ε in a series in π.

(1.3). We next compute the values of $(\varepsilon, \rho)_p$ for $\varepsilon, \rho \in U_{1,F}$.

Let θ, η belong to the set of nonzero multiplicative representatives of $\mathbb{F}_p$ in $F = \mathbb{Q}_p(\zeta)$ (i.e., $\theta^{p-1} = \eta^{p-1} = 1$). By Exercise 1, f) below,

$$(E(\theta\pi^i), E(\eta\pi^j))_p = \prod_{n \geqslant 0} (-\eta\pi^j, E(\theta\eta^{p^n} \pi^{i+p^n j}))_p^{-1} \prod_{m \geqslant 1} (-\theta\pi^i, E(\theta^{p^m} \eta \pi^{p^m i + j}))_p.$$

Exercise 1 in section 5 Ch. IV and the equality $\varphi(\theta) = \theta^p$ for the Frobenius automorphism of F imply that for $p > 2$

$$\begin{aligned}&(E_X(\theta X^i)|_{X=\pi}, E_X(\eta X^j)|_{X=\pi})_p\\&= \prod_{n\geqslant 0}(\pi, E(\theta\pi^i j\varphi^n(\eta)\pi^{p^n j}))_p^{-1}\prod_{m\geqslant 1}(\pi, E(\eta\pi^j i\varphi^m(\theta)\pi^{p^m i}))_p\\&= (\pi, E_X(-\theta X^i(1+\triangle+\triangle^2+\dots)(j\eta X^j)+\eta X^j(\triangle+\triangle^2+\dots)(i\theta X^i))|_{X=\pi})_p,\end{aligned}$$

where $\triangle$ is defined in (2.1) Ch. VI. Note that this formula holds for every $\theta, \eta \in \mathbb{Z}_p$, due to the $\mathbb{Z}_p$-linearity of E_X.

Let $\varepsilon = \varepsilon(X)|_{X=\pi}$, $\rho = \rho(X)|_{X=\pi}$ with $\varepsilon(X), \rho(X) \in 1 + X\mathbb{Z}_p[[X]]$. Let

$$l_X(\varepsilon(X)) = \sum_{i\geqslant 1} a_i X^i, \qquad l_X(\rho(X)) = \sum_{i\geqslant 1} b_i X^i, \qquad \text{with} \quad a_i, b_i \in \mathbb{Z}_p.$$

Then we get

$$(\varepsilon,\rho)_p = (E_X(l_X(\varepsilon(X)))|_{X=\pi}, E_X(l_X(\rho(X)))|_{X=\pi})_p = (\pi, E_X(\nu(X))|_{X=\pi})_p$$

where

$$\begin{aligned}\nu(X) = &-l_X(\varepsilon(X))(1+\triangle+\triangle^2+\dots)\Big(\sum_{i\geqslant 1} ib_iX^i\Big)\\&+l_X(\rho(X))(\triangle+\triangle^2+\dots)\Big(\sum_{i\geqslant 1} ia_iX^i\Big).\end{aligned}$$

Since $\triangle^j \left(\sum_{i\geqslant 1} ib_iX^i\right) = \triangle^j \left(X\left(l_X(\rho(X))\right)'\right) = X\left(p^{-j}\ \triangle^j\ l_X(\rho(X))\right)'$, we obtain

$$(1+\triangle+\triangle^2+\dots)\Big(\sum_{i\geqslant 1} ib_iX^i\Big) = X\Big(\sum_{j\geqslant 0}\frac{\triangle^j}{p^j} l_X(\rho(X))\Big)' = X\left(\log(\rho(X))\right)',$$

$$(\triangle+\triangle^2+\dots)\Big(\sum_{i\geqslant 1} ia_iX^i\Big) = X\left(\log(\varepsilon(X)) - l_X((\varepsilon(X))\right)'.$$

Thus,

$$(\varepsilon,\rho)_p = \left(\pi, E_X\left(X(-l_X(\rho)l_X(\varepsilon)' + l_X(\rho)\log\varepsilon(X))' - l_X(\varepsilon)\log(\rho(X))')\right)|_{X=\pi}\right)_p$$

and, by the formula $(*)$ of (1.2), $(\varepsilon,\rho)_p = \zeta_p^c$ with $c = \operatorname{res}\widetilde{\Phi}_{\varepsilon,\rho}(X)/s(X)$, where

$$\widetilde{\Phi}_{\varepsilon,\rho}(X) = -l_X(\rho(X))l_X(\varepsilon(X))' + l_X(\rho(X))\varepsilon(X)^{-1}\varepsilon(X)' - l_X(\varepsilon(X))\rho(X)^{-1}\rho(X)'.$$

Here we used the equality $\log(\rho(X))' = \rho(X)^{-1}\rho(X)'$. Since $\operatorname{res}\left((l_X(\rho)l_X(\varepsilon))'/X^p\right) \equiv 0 \mod p$ one can replace $\widetilde{\Phi}_{\varepsilon,\rho}(X)$ with

$$\Phi_{\varepsilon,\rho}(X) = l_X(\varepsilon(X))l_X(\rho(X))' - l_X(\varepsilon(X))\rho(X)^{-1}\rho(X)' + l_X(\rho(X))\varepsilon(X)^{-1}\varepsilon(X)'.$$

(1.4). Now we treat the case of $(\alpha,\beta)_p$ with arbitrary $\alpha,\beta\in\mathbb{Q}_p(\zeta)^*$. Let $\alpha=\pi^i\theta\varepsilon$, $\beta=\pi^j\eta\rho$ with $\varepsilon,\rho\in U_{1,F}$, $i,j\in\mathbb{Z}$, $\theta^{p-1}=\eta^{p-1}=1$. Let $\varepsilon(X)$, $\rho(X)$ be as in (1.3). By Exercise 1 in section 5 Ch. IV we get

$$(\alpha,\beta)_p=(\pi^i,\rho)_p(\pi^j,\varepsilon)_p^{-1}(\varepsilon,\rho)_p=(\pi,\rho^i\varepsilon^{-j})_p(\varepsilon,\rho)_p.$$

Therefore,

$$(\alpha,\beta)_p=\zeta_p^{\operatorname{res}\Phi_{\alpha,\beta}(X)/s(X)}, \qquad (**),$$

where

$$\begin{aligned}\Phi_{\alpha,\beta}(X)&=\Phi_{\varepsilon,\rho}(X)+X^{-1}l_X(\rho(X)^i\varepsilon(X)^{-j}\\ &=\Phi_{\varepsilon,\rho}(X)+iX^{-1}l_X(\rho(X))-jX^{-1}l_X(\varepsilon(X))\\ &=l_X(\varepsilon)l_X(\rho)'-l_X(\varepsilon)(X^j\eta\rho(X))'/(X^j\eta\rho(X))+l_X(\rho)(X^i\theta\varepsilon(X))'/(X^i\theta\varepsilon(X)),\end{aligned}$$

because

$$(X^i\theta\varepsilon(X))'(X^i\theta\varepsilon(X))^{-1}=iX^{-1}+\varepsilon(X)'\varepsilon(X)^{-1}.$$

The same formula holds for $p=2$ (see Exercise 3).

The series $\Phi_{\alpha,\beta}(X)$ on the right-hand side of $(**)$ does depend on the choice of expansion of α,β in series in π and the choice of a prime element π.

REMARKS.

1. Note that for the field $F=\mathbb{Q}_p(\zeta_p)$ the inertia subfield $F_0=F\cap\mathbb{Q}_p^{\mathrm{ur}}$ coincides with $\mathbb{Q}_p$. In the general case we shall add the trace operator $\operatorname{Tr}=\operatorname{Tr}_{F_0/\mathbb{Q}_p}$ in front of res.

2. If we allow the series $\varepsilon(X),\rho(X)$ be arbitrary invertible series in $\mathbb{Z}_p((X))^*$ which give the elements ε,ρ when X is replaced with π, then we must slightly modify $1/s(X)$ by replacing it with $1/2+1/s(X)$.

3. Although for the field F the structure of the formulas for the Hilbert symbol $(\cdot,\cdot)_p$ is the same for $p=2$ and $p>2$, in the general case of a local number field the formulas for the Hilbert symbol differ for $p>2$ and $p=2$. When $p=2$, we shall add series $\Phi^{(1)}_{\alpha,\beta}$, $\Phi^{(2)}_{\alpha,\beta}$ defined in (2.5) below to $\Phi_{\alpha,\beta}$ and the factor $r(X)$ defined in (3.4) Ch. VI.

Exercises.

1. Let F be a local number field, let a primitive p^n th root of unity ζ belong to F, and let π be a prime element in F, $\alpha,\beta\in F$. Let $(\cdot,\cdot)$ be the p^n th Hilbert symbol in F.
   a) Show that for a sufficiently large integer c
   $$(1-\alpha,1-\beta)=1$$
   for all $\alpha,\beta\in\mathcal{M}_F{}^c$.
   b) Show that for $\alpha,\beta\in F^*$, $\alpha\neq1$, $\beta\neq1$,
   $$(1-\alpha,1-\beta)=(1-\alpha,1-\alpha\beta)(-\beta,1-\alpha\beta)(1-\alpha\beta,1-\beta).$$

c) Prove that for $\alpha, \beta \in \mathcal{M}_F$

$$(1-\alpha, 1-\beta) = \prod(-\alpha^{i_0}\beta^{j_0}, 1-\alpha^i\beta^j),$$

where $i, j \geqslant 1$, $(i,j) = 1$, $i_0, j_0 \geqslant 0$ run through all pairs of integer numbers with $ij_0 - i_0j = 1$. (Hint: Use a) and b) for $\alpha, \alpha\beta$ and $\alpha\beta, \beta$).

d) Prove that for $\alpha, \beta \in \mathcal{M}_F$

$$(1-\alpha, 1-\beta) = \prod(-\alpha^{i_0}\beta^{j_0}, E(\alpha^i\beta^j))^{-1},$$

where $i, j \geqslant 1$, g.c.d.$(i, j, p) = 1$, $i_0, j_0 \geqslant 0$ with $ij_0 - i_0j =$ g.c.d.(i, j). (Hint: Use Exercise 2 in section 9 Ch. I).

e) Prove that for $\alpha, \beta \in \mathcal{M}_F$

$$(1-\alpha, E(\beta)) = \prod(-\alpha^{i_0}\beta^{j_0}, E(\alpha^i\beta^{p^s})),$$

where $i \geqslant 1$ is relatively prime to p, $s \geqslant 0$, $i_0, j_0 \geqslant 0$ with $ij_0 - i_0p^s = 1$. (Hint: Use Lemma (9.1) Ch. I).

f) (*M. Kneser*) Prove that for $\alpha, \beta \in \mathcal{M}_F$

$$(E(\alpha), E(\beta)) = \prod_{i\geqslant 1}(-\alpha, E(\alpha^{p^i}\beta)) \prod_{j\geqslant 0}(-\beta^{-1}, E(\alpha\beta^{p^j})).$$

g) (*M. Kneser*) Show that for $p = 2$, $\alpha \in \mathcal{M}_F$

$$(-1, E(\alpha)) = \prod_{i\geqslant 0}(\alpha^{2^i}, E(\alpha^{2^{i+1}})).$$

2. ($\diamond$) Let $\alpha, \beta \in \mathcal{M}_F$, $\varepsilon = E(\alpha)$, $\rho = E(\beta)$, $\Phi_{\varepsilon,\rho}$ as in (1.4), $(\cdot,\cdot)$ as in Exercise 1. Show that for $p > 2$

$$(E(\alpha), E(\beta)) = (\pi, E_X(X\Phi_{\varepsilon,\rho}(X))|_{X=\pi}),$$

for $p = 2$

$$(E(\alpha), E(\beta)) = (\pi, E_X(X\Phi_{\varepsilon,\rho}(X) + X\Phi^{(1)}_{\varepsilon,\rho}(X))|_{X=\pi}),$$

where

$$\Phi^{(1)}_{\varepsilon,\rho}(X) = \left(\frac{\triangle}{2}(L_X(\psi(X))L_X(\varphi(X)))\right)', \quad L_X(\psi(X)) = (1 + \triangle + \triangle^2 + \dots)l_X(\psi(X))$$

with $\varepsilon = \psi(\pi), \rho = \varphi(\pi), \psi, \varphi \in 1 + X\mathbb{Z}[[X]]$.

3. Using Exercise 3 in section 5 Ch. IV show that for the Hilbert symbol $(\cdot,\cdot)_2$ in $\mathbb{Q}_2$, $\alpha, \beta \in \mathbb{Q}_2^*$

$$(\alpha, \beta)_2 = (-1)^{\operatorname{res}\Phi_{\alpha,\beta}(X)/s(X)},$$

where $\Phi_{\alpha,\beta}$ is as in (1.4).

2. The Pairing $\langle \cdot, \cdot \rangle$

We introduce a pairing $\langle \cdot, \cdot \rangle_X$ on formal power series in subsection (2.1) and study its properties. Then in subsection (2.2) we define a pairing $\langle \cdot, \cdot \rangle_\pi$ on the multiplicative group of a local number field F and study its properties. We show that $\langle \cdot, \cdot \rangle_\pi$ is well defined in (2.2) and that it does not depend on the choice of prime element π in (2.4). For the case of $p = 2$ see subsection (2.5). Later in section 4 we shall prove that $\langle \cdot, \cdot \rangle_\pi$ coincides with the Hilbert symbol.

(2.1). From this point until (2.5) we assume that $p > 2$.

Recall that $\mathcal{O}_0$ is the ring of integers of $F_0 = F \cap \mathbb{Q}_p^{\mathrm{ur}}$ and $\mathcal{R}$ is the group of multiplicative representatives of the residue field of F in $\mathcal{O}_0$. Recall that in (3.6) of Ch. VI we defined the series $V(X) = 1/2 + 1/s(X)$ of $\mathcal{O}_0\{\{X\}\}$ where $s(X) = z(X)^{p^n} - 1$ and $z(X) \in 1 + X\mathcal{O}_0[[X]]$ is such that $z(\pi) = \zeta$ is a p^n th primitive root of unity in F. We denote by l the map l_X (which is a sort of a special logarithm) on $\mathcal{O}_0((X))^*$ defined in (2.3) of Ch. VI, so

$$l(\alpha) = \frac{1}{p} \log(\alpha^p / \alpha^\Delta).$$

Introduce a pairing

$$\langle \cdot, \cdot \rangle_X : \mathcal{O}_0((X))^* \times \mathcal{O}_0((X))^* \to \langle \zeta \rangle$$

as

$$\boxed{\langle \alpha, \beta \rangle_X = \zeta^{\operatorname{Tr} \operatorname{res} \Phi_{\alpha,\beta} V}}$$

where $\operatorname{Tr} = \operatorname{Tr}_{F_0/\mathbb{Q}_p}$,

$$\boxed{\Phi_{\alpha,\beta} = \alpha^{-1} \alpha' \, l(\beta) - l(\alpha) \, \beta^{-1} \beta' + l(\alpha) \, l(\beta)'}$$

Note that $\Phi_{\alpha,\beta}$ belongs to $\mathcal{O}_0((X))$.

Using the equality

$$l(\beta)' = \beta^{-1}\beta' - \frac{1}{p} \beta^{-\Delta} (\beta^\Delta)'$$

(see Lemma 3 b) of (3.6) Ch. VI) we can rewrite $\Phi_{\alpha,\beta}$ as

$$\Phi_{\alpha,\beta} = \frac{\alpha'}{\alpha} \, l(\beta) - l(\alpha) \, \frac{1}{p} \, \frac{(\beta^\Delta)'}{\beta^\Delta}.$$

REMARK. If $\alpha, \beta \in \mathcal{O}_0^*$, then $\alpha' = \beta' = (l(\beta))' = 0$ and so $\langle \alpha, \beta \rangle_X = 1$.

PROPOSITION.

a) *The pairing* $\langle \cdot, \cdot \rangle_X$ *is bilinear*

$$\langle \alpha_1\alpha_2, \beta \rangle_X = \langle \alpha_1, \beta \rangle_X \langle \alpha_2, \beta \rangle_X,$$
$$\langle \alpha, \beta_1\beta_2 \rangle_X = \langle \alpha, \beta_1 \rangle_X \langle \alpha, \beta_2 \rangle_X$$

and antisymmetric

$$\langle \alpha, \beta \rangle_X \langle \beta, \alpha \rangle_X = 1.$$

b) $\langle \alpha, \alpha \rangle_X = 1, \qquad \langle \theta, \alpha \rangle_X = 1$ *for* $\theta \in \mathcal{R}^*$.

c) *Steinberg property*

$$\langle \alpha, 1 - \alpha \rangle_X = 1 \quad \textit{for every } \alpha \neq 1.$$

Proof. Bilinearity of $\langle \cdot, \cdot \rangle$ follows from the properties of l ((2.2) and (2.3) Ch. VI).

Furthermore,

$$\Phi_{\alpha,\beta} + \Phi_{\beta,\alpha} = l(\alpha)l(\beta)' + l(\beta)l'(\alpha) = (l(\alpha)l(\beta))'.$$

Lemma 1 of (3.6) Ch. VI now implies that $\langle \alpha, \beta \rangle = \langle \beta, \alpha \rangle^{-1}$.

An element $\theta \in \mathcal{R}^*$ can be written as η^{p^n} with $\eta \in \mathcal{R}^*$; therefore $\langle \theta, \alpha \rangle_X = \langle \eta, \alpha \rangle_X^{p^n} = 1$.

Since $p > 2$, the equality $\langle \alpha, \alpha \rangle^2 = 1$ implies $\langle \alpha, \alpha \rangle = 1$.

To prove the Steinberg property (which take some time) we *first assume that* $\alpha \in X\mathcal{O}_0[[X]]$. Then

$$\begin{aligned}\Phi_{\alpha,1-\alpha} &= l(1-\alpha)\alpha^{-1}\alpha' - l(\alpha)\frac{1}{p}(1-\alpha)^{-\Delta}\big((1-\alpha)^{\Delta}\big)' \\ &= -\alpha^{-1}\alpha'\Big(1 - \frac{\Delta}{p}\Big)\sum_{i\geqslant 1}\frac{\alpha^i}{i} + l(\alpha)\Big(\sum_{i\geqslant 1}\frac{\alpha^{i\Delta}}{p^i}\Big)' \\ &= -\alpha' - \sum_{p\nmid i\geqslant 1}\frac{\alpha^i}{i}\alpha^{-1}\alpha' - \sum_{i\geqslant 1}\Big(\frac{\alpha^{ip} - \alpha^{i\Delta}}{ip}\alpha^{-1}\alpha' - l(\alpha)\frac{\alpha^{i\Delta}}{ip}\Big).\end{aligned}$$

Using Lemma 3 of (3.6) Ch. VI we deduce that

$$\frac{\alpha^{ip} - \alpha^{i\Delta}}{ip}\alpha^{-1}\alpha' - l(\alpha)\frac{\alpha^{i\Delta}}{ip} = g_i', \quad \text{where } g_i = \frac{\alpha^{ip} - \alpha^{i\Delta}}{(ip)^2} - l(\alpha)\frac{\alpha^{i\Delta}}{ip}.$$

Thus, in this case

$$\Phi_{\alpha,1-\alpha} = -\Big(\alpha + \sum_{p\nmid i\geqslant 1}\frac{\alpha^i}{i^2} + \sum_{i\geqslant 1} g_i\Big)'.$$

By Lemma 2 of (3.6) Ch. VI we have $g_i \in \mathcal{O}_0[[X]]$, so by Lemma 1 in the same section we get

$$\operatorname{res}\Phi_{\alpha,1-\alpha}V \equiv 0 \mod p^n, \qquad \langle \alpha, 1 - \alpha \rangle = 1.$$

Now suppose that $\alpha^{-1} \in X\mathcal{O}_0[[X]]$. Then

$$1 = \langle \alpha^{-1}, 1-\alpha^{-1} \rangle = \langle \alpha, (1-\alpha)/(-\alpha) \rangle^{-1} = \langle \alpha, -\alpha \rangle \langle \alpha, 1-\alpha \rangle^{-1} = \langle \alpha, 1-\alpha \rangle^{-1},$$

so $\langle \alpha, 1-\alpha \rangle = 1$.

Finally, in the remaining case $\alpha = a\beta$ with $a \in \mathcal{O}_0^*$, $1-a \in \mathcal{O}_0^*$ and $\beta \in 1 + X\mathcal{O}_0[[X]]$. The element $\gamma = (1-\beta)/(1-a\beta)$ belongs to $X\mathcal{O}_0[[X]]$, so from the previous we get

$$\begin{aligned} 1 &= \langle 1-\gamma, \gamma \rangle = \langle -(a-1)\beta/(1-a\beta), (1-\beta)/(1-a\beta) \rangle \\ &= \langle a-1, 1-\beta \rangle \langle a-1, 1-a\beta \rangle^{-1} \langle \beta, 1-a\beta \rangle^{-1} \langle 1-a\beta, 1-\beta \rangle^{-1}, \end{aligned}$$

since $\langle \beta, 1-\beta \rangle = \langle 1-a\beta, 1-a\beta \rangle = 1$. The element $\gamma' = a(1-\beta)/(a-1)$ belongs to $X\mathcal{O}_0[[X]]$, so similarly

$$\begin{aligned} 1 &= \langle 1-\gamma', \gamma' \rangle = \langle (1-a\beta)/(1-a), a(1-\beta)/(a-1) \rangle \\ &= \langle 1-a\beta, a \rangle \langle 1-a\beta, a-1 \rangle^{-1} \langle 1-a\beta, 1-\beta \rangle \langle 1-a, 1-\beta \rangle^{-1}, \end{aligned}$$

since $\langle a/(a-1), 1-a \rangle = 1$ due to the Remark above. So we deduce that

$$1 = \langle 1-\gamma, \gamma \rangle \langle 1-\gamma', \gamma' \rangle = \langle a\beta, 1-a\beta \rangle^{-1}.$$

Thus, the Steinberg property is proved. □

(2.2). Now let π be a prime element in F, $\alpha, \beta \in F^*$, and let $\alpha(X), \beta(X)$ be any series in $\mathcal{O}_0((X))^*$ such that $\alpha(\pi) = \alpha$, $\beta(\pi) = \beta$. Put

$$\boxed{\langle \alpha, \beta \rangle_\pi = \langle \alpha(X), \beta(X) \rangle_X}$$

PROPOSITION. *The value* $\langle \alpha, \beta \rangle_\pi$ *does not depend of the way the elements* α, β, ζ *are expanded in power series in* π. *Thus, the pairing* $\langle \cdot, \cdot \rangle_\pi \colon F^* \times F^* \to \langle \zeta \rangle$ *is well defined.*

It is bilinear, and antisymmetric. Moreover,

$$\langle \alpha, \alpha \rangle_\pi = 1, \quad \langle \theta, \alpha \rangle_\pi = 1 \qquad \textit{for} \quad \alpha \in F^*, \theta \in \mathcal{R}^*$$

and $\langle \alpha, 1-\alpha \rangle_\pi = 1$ *for every* α *different from 0 and 1.*

Proof. Let $s(X), s^{(1)}(X)$ be two distinct series corresponding to ζ. Then Corollary (3.3) Ch. VI shows that

$$\operatorname{res} \Phi_{\alpha,\beta}/s(X) \equiv \operatorname{res} \Phi_{\alpha,\beta}/s^{(1)}(X) \mod p^n.$$

Therefore, $\langle \alpha, \beta \rangle_\pi$ does not depend on the choice of an expansion of ζ in a power series in π.

Due to antisymmetry it is sufficient to show that if $\alpha_1(X), \alpha_2(X) \in \mathcal{O}_0((X))^*$ with $\alpha_1(\pi) = \alpha_2(\pi) = \alpha$, then

$$\langle \alpha_1(X), \beta(X)\rangle_X = \langle \alpha_2(X), \beta(X)\rangle_X.$$

The series $\alpha_1(X)/\alpha_2(X)$ is equal to 1 at $X = \pi$. Proposition (3.2) Ch. VI shows now that $\alpha_1(X)/\alpha_2(X) = 1 - ug$ with some $g \in \mathcal{O}_0[[X]]$.

Using the bilinearity of $\langle \cdot, \cdot \rangle_X$, we need to verify that

$$\langle 1 - ug, \beta\rangle_X = 1.$$

One has

$$\Phi_{1-ug,\beta} = \big(\log(1-ug)\big)' l(\beta) - \big(1 - \frac{\Delta}{p}\big)\big(\log(1-ug)\big)\frac{1}{p}\frac{(\beta^\Delta)'}{\beta^\Delta}.$$

First assume that $\beta(X) \in \mathcal{O}_0^*(1 + X\mathcal{O}_0[[X]])$. Then $\beta'/\beta \in \mathcal{O}_0[[X]]$ and $l(\beta) \in \mathcal{O}_0[[X]]$. So $\operatorname{res}\Phi_{1-ug,\beta}/2 = 0$. We can now apply Lemma (3.5) of Ch. VI (those parts of it which contain mod deg 0 congruences). Then

$$-\operatorname{res}\Phi_{1-ug,\beta}/s = \operatorname{res}\sum_{i\geqslant 1}\Big(\frac{(u^i g^i)'}{is} l(\beta) - \big(\frac{u^i g^i}{is} - \frac{u^{i\Delta} g^{i\Delta}}{pis}\big)\frac{1}{p}\frac{(\beta^\Delta)'}{\beta^\Delta}\Big) = \operatorname{res}\sum_{i\geqslant 1} f_i$$

where f_i is congruent modulo p^n to

$$g^i\big(\frac{1}{s_{n-1}}\big)'\frac{p^{i-1}}{i} l(\beta) + (g^i)'\frac{1}{s_{n-1}}\frac{p^{i-1}}{i} l(\beta) - g^i\frac{1}{s_{n-1}}\frac{p^{i-1}}{i}\frac{1}{p}\frac{(\beta^\Delta)'}{\beta^\Delta} + g^{i\Delta}\frac{1}{s}\frac{p^{i-1}}{i}\frac{1}{p}\frac{(\beta^\Delta)'}{\beta^\Delta}.$$

Note that $p^{i-1}/i \in \mathbb{Z}$ and $(\beta^\Delta)'/(p\beta^\Delta) = X^{p-1}(\beta'/\beta)^\Delta = -l(\beta)' + \beta'/\beta$ belongs to $\mathcal{O}_0((X))$ by Lemma 3 (3.6) Ch. VI. By the same Lemma

$$\operatorname{Tr}\operatorname{res} g^{i\Delta}\frac{1}{s_{n-1}^\Delta}\frac{p^{i-1}}{i}\frac{1}{p}\frac{(\beta^\Delta)'}{\beta^\Delta} = \operatorname{Tr}\operatorname{res} g^i\frac{1}{s_{n-1}}\frac{p^{i-1}}{i}\frac{\beta'}{\beta}.$$

Thus,

$$-\operatorname{Tr}\operatorname{res}\Phi_{1-ug,\beta}V \equiv \operatorname{Tr}\operatorname{res}\Big(\sum_{i\geqslant 1} g^i\frac{1}{s_{n-1}}\frac{p^{i-1}}{i} l(\beta)\Big)' = 0 \mod p^n,$$

i.e., $\langle 1 - ug, \beta\rangle_X = 1$.

Now, *in the general case of* $\beta = aX^m\beta_1(X)$ with $\beta_1 \in 1 + X\mathcal{O}_0[[X]]$ and $a \in \mathcal{O}_0^*$ due to bilinearity of $\langle\cdot,\cdot\rangle_X$ it remains to treat the case $\beta(X) = X$. Then $\Phi_{1-ug,X} = -X^{-1}l(1-ug)$. Similarly to the previous arguments using mod deg 1 congruences of Lemma (3.5) of Ch. VI we deduce that

$$\operatorname{res}\Phi_{1-ug,X}/s \equiv \operatorname{res} X^{-1}\Big(\sum_{i\geqslant 1}\frac{p^{i-1}g^i}{is_{n-1}} - \sum_{i\geqslant 1}\frac{p^{i-1}g^{i\Delta}}{is}\Big)$$
$$+ X^{-1}\Big(\sum_{i\geqslant 1}\frac{p^{i-1}(i-1)(p-1)g^i}{2i} - \sum_{i\geqslant 1}\frac{p^{i-1}(p-1)g^{i\Delta}}{2}\Big) \mod p^n$$

(the first two terms annihilate each other due to the previous discussions). We also have

$$\operatorname{res}\Phi_{1-ug,X}/2 \equiv \operatorname{res} X^{-1}\Big(\sum_{i\geqslant 1}\frac{p^{i-1}pg^i}{2i} - \sum_{i\geqslant 1}\frac{p^{i-1}g^{i\Delta}}{2i}\Big)$$

(note that $\Delta\ u \equiv u \equiv p \mod \deg 1$). Using Lemma 3 c) of (3.6) Ch. VI we conclude that

$$\operatorname{res}\Phi_{1-ug,X}V \equiv \operatorname{res} X^{-1}\Big(\sum_{i\geqslant 1} g^i p^{i-1}\Big(\frac{(i-1)(p-1)}{2i} - \frac{p-1}{2} + \frac{p}{2i} - \frac{1}{2i}\Big)\Big)$$

which is zero.

The other properties follow from Proposition (2.1). □

REMARKS.

1. If $\alpha(X)$ and $\beta(X)$ are chosen from the subgroup

$$P = \{X^m\theta\varepsilon(X) : m \in \mathbb{Z}, \theta \in \mathcal{R}^*, \varepsilon(X) \in 1 + X\mathcal{O}_0[[X]]\}$$

then the quotient α_1/α_2, which is considered in the proof of the previous Proposition, belongs to $1 + X\mathcal{O}_0[[X]]$ and therefore the series g belongs to $X\mathcal{O}_0[[X]]$. Then

$$\langle\alpha_1(X), \beta(X)\rangle' = \langle\alpha_2(X), \beta(X)\rangle'$$

where by $\langle\cdot,\cdot\rangle'$ we denoted the pairing with $1/s(X)$ instead of $V(X)$. The pairing $\langle\cdot,\cdot\rangle' \colon P\times P \to \langle\zeta\rangle$ is therefore well defined. It can be used instead of the pairing $\langle\cdot,\cdot\rangle$ in the following sections of this chapter (as it was in the first edition of this book).

2. For another proof of independence see Exercise 5.

(2.3). FIRST PROPERTIES OF $\langle\cdot,\cdot\rangle_\pi$.

1. *If* $\varepsilon = E_X\big(\sum_{\substack{1\leqslant i<pe_1\\(i,p)=1}}\alpha_i X^i\big)\big|_{X=\pi}$, $\alpha_i \in \mathcal{O}_0$, *is as in section 5* Ch. VI, *then*

$$\langle\pi, \varepsilon\rangle_\pi = 1.$$

This follows from

$$\Phi = X^{-1}\big(\sum \alpha_i X^i\big) = \big(\sum i^{-1}\alpha_i X^i\big)'$$

and Lemma 1 of (3.6) Ch. VI.

2. *If* $\omega(a) = E_X(a\,s(X))|_{X=\pi}$ *is a* p^n*-primary element* ($a \in \mathcal{O}_0$) *as in section 5 of* Ch. VI, *then*

$$\langle \pi, \omega(a)\rangle_\pi = \zeta^{\operatorname{Tr} a}.$$

Indeed, $s(0) = 0$ and so

$$\operatorname{res}\Phi V = \operatorname{res} X^{-1} as(X)(1/s(X) + 1/2) = a.$$

3. *If* $\varepsilon \in U_{1,F}$ *then*

$$\langle \varepsilon, \omega(a)\rangle_\pi = 1.$$

Indeed, if $\varepsilon = \varepsilon(X)|_{X=\pi}$ with $\varepsilon(X) \in \mathcal{O}_0((X))^*$ then

$$\Phi = as(\log\varepsilon(X))' - l(\varepsilon(X))\Big(\sum_{i\geqslant 1}(as)^{\Delta^i}/p^i\Big)'.$$

From Lemma 3 a) of (3.6) Ch. VI we deduce by induction on i that

$$(s^{\Delta^i})'/p^i = X^{-1}\ \Delta^i\ (Xs')$$

which is congruent to 0 $\mod p^n$ due to Proposition (3.1) d) of Ch. VI. Thus, $\operatorname{res}\Phi V \equiv 0 \mod p^n$.

(2.4). The next property of $\langle\cdot,\cdot\rangle_\pi$ to be verified is its invariance with respect to the choice of a prime element π in F. In other words, we will show that for $\alpha,\ \beta \in F^*$

$$\langle \alpha, \beta\rangle_\pi = \langle \alpha, \beta\rangle_\tau,$$

for prime elements π, τ in F.

PROPOSITION. *The pairing* $\langle\cdot,\cdot\rangle_\pi$ *is invariant with respect to the choice of a prime element* π *in* F.

Proof. Assume that for any prime elements π, τ in F and $\beta \in U_{1,F}$

$$\langle \pi, \beta\rangle_\pi = \langle \pi, \beta\rangle_\tau. \qquad (*)$$

Let ε be a principal unit in F, then $\pi\varepsilon$ is prime in F. We then deduce

$$\langle \pi\varepsilon, \beta\rangle_{\pi\varepsilon} = \langle \pi\varepsilon, \beta\rangle_\pi = \langle \pi\varepsilon, \beta\rangle_\tau.$$

Therefore

$$\langle \varepsilon, \beta\rangle_\pi = \langle \pi\varepsilon, \beta\rangle_\pi \langle \pi, \beta\rangle_\pi^{-1} = \langle \pi\varepsilon, \beta\rangle_\tau \langle \pi, \beta\rangle_\tau^{-1} = \langle \varepsilon, \beta\rangle_\tau.$$

Now the linear property of $\langle\cdot,\cdot\rangle_\pi$ of Proposition (2.2) implies the invariance of $\langle\cdot,\cdot\rangle_\pi$.

To prove $(*)$ first note that due to Proposition (2.2) we get

$$\langle \pi, \pi^i\theta\varepsilon\rangle_\pi = \langle \pi, \varepsilon\rangle_\pi$$

and similarly for $\langle\pi,\pi^i\theta\varepsilon\rangle_\tau$, where $\theta\in\mathcal{R}^*$, $\varepsilon\in U_1$. Now by Corollary of (5.2) Ch. VI in its notations we can express

$$\varepsilon=\prod_{(i,p)=1}(1-\theta_j\pi^i)^{a_{ij}}\,\omega(a)$$

with some $\theta_j\in\mathcal{R}$, $a\in\mathcal{O}_0$, $a_{ij}\in\mathbb{Z}_p$. Then Proposition (2.2) implies

$$\langle\pi,1-\theta_j\pi^i\rangle_\rho^i=\langle\pi^i,1-\theta_j\pi^i\rangle_\rho=\langle\theta_j\pi^i,1-\theta_j\pi^i\rangle_\rho=1$$

for $\rho=\pi$ or $=\tau$. Therefore, since i is prime to p we deduce that

$$\langle\pi,\beta\rangle_\pi=\langle\pi,\omega(a)\rangle_\pi,\quad \langle\pi,\beta\rangle_\tau=\langle\pi,\omega(a)\rangle_\tau.$$

Let $\pi=\tau\eta$ with $\eta\in\mathcal{O}^*$. By Property 3 and 2 in (2.3) we get

$$\langle\pi,\omega(a)\rangle_\tau=\langle\tau,\omega(a)\rangle_\tau=\zeta^{\operatorname{Tr}a}=\langle\pi,\omega(a)\rangle_\pi.$$

Thus, due to the independence of the choice of power expansion in a prime element in Proposition (2.2) we conclude that

$$\langle\pi,\varepsilon\rangle_\pi=\prod\langle\pi,1-\theta_j\pi^i\rangle_\pi^{a_{ij}}\langle\pi,\omega(a)\rangle_\pi=\langle\pi,\omega(a)\rangle_\tau\prod\langle\pi,1-\theta_j\pi^i\rangle_\tau^{a_{ij}}=\langle\pi,\varepsilon\rangle_\tau.$$

□

REMARK. For another proof see Exercise 6.

(2.5). In this subsection we treat the special case of $p=2$.

The first essential difference with the case $p>2$ is that the pairing for the formal series is defined not for all invertible series in $\mathcal{O}_0((X))$ but for series which belong to $Q=R\cap\mathcal{O}_0((X))=$

$$\{X^m a\varepsilon(X):\varepsilon(X)\in 1+X\mathcal{O}_0[[X]],a\in\mathcal{O}_0^*,a^\varphi\equiv a^2\mod 4,m\in\mathbb{Z}\}$$

(R is defined in (2.3) of Ch. VI).

Certainly, the group of series P defined in Remark of (2.2) is a subgroup of Q. The reason why we have to work with Q is that for $p=2$ the formula $l_X(f)=\frac{1}{p}\log(f^p/f^{\Delta_X})$ for the map l_X of (2.3) Ch. VI is defined for $f\in Q$ and not for an arbitrary invertible series of $\mathcal{O}_0((X))$.

For $\alpha,\beta\in Q$ put

$$\Phi^{(1)}_{\alpha,\beta}=\left(\frac{\Delta}{2}\left(\frac{\alpha^2-\alpha^\Delta}{2\alpha^\Delta}\,\frac{\beta^2-\beta^\Delta}{2\beta^\Delta}\right)\right)'$$

and

$$\Phi^{(2)}_{\alpha,\beta}=X^{-1}v_X(\alpha)v_X(\beta)\,l_X(1+s_{n-1}(X))$$

where v_X is the discrete valuation of $\mathcal{O}_0((X))$ corresponding to X. The series $1+s_{n-1}(X)\in Q$ corresponds to -1, since $1+s_{n-1}(\pi)=z(\pi)^{2^{n-1}}=-1$. The series

$\Phi^{(2)}_{\alpha(X),\beta(X)}$ takes care of the fact that $\langle \pi, \pi\rangle_\pi = \langle \pi, -1\rangle_\pi$ is not necessarily equal to 1 in the case $p = 2$.

Introduce the pairing

$$\langle \cdot, \cdot\rangle_X : Q \times Q \to \langle \zeta \rangle$$

by the formula

$$\boxed{\langle \alpha, \beta\rangle_X = \zeta^{\operatorname{Tr}\operatorname{res}\left(\Phi_{\alpha,\beta} + \Phi^{(1)}_{\alpha,\beta} + \Phi^{(2)}_{\alpha,\beta}\right) r(X) V(X)}}$$

where $V(X), \Phi_{\alpha,\beta}$ are as in (2.1), and $r(X)$ as in (3.4) Ch. VI.

One can show that Proposition (2.1) holds for $\langle \cdot, \cdot\rangle_X$. Then for elements $\alpha, \beta \in F^*$ let $\alpha(X), \beta(X) \in Q$ be such that $\alpha(\pi) = \alpha$ and $\beta(\pi) = \beta$. Put

$$\boxed{\langle \alpha, \beta\rangle_\pi = \langle \alpha(X), \beta(X)\rangle_X}$$

One can show that Propositions (2.2) and (2.4) hold for the pairing $\langle \cdot, \cdot\rangle_\pi$. The proofs can be carried in the same way as above, but with longer calculations. For details see Exercises 2–4.

Exercises.

1. Show that $\langle a, b\rangle_\pi = 1$ for $a, b \in \mathcal{O}_0^*$.
2. ($\diamond$) Let $p = 2$. We use the definitions of (2.5).
   a) Show that $\Phi_{\alpha,\beta} + \Phi^{(1)}_{\alpha,\beta} + \Phi^{(2)}_{\alpha,\beta} \in \mathcal{O}_0[[X]]$.
   b) Show that $\operatorname{Tr}\operatorname{res}(\Phi^{(1)}_{\alpha,\beta} + \Phi^{(2)}_{\alpha,\beta}) rV \equiv 0 \mod 2^{n-1}$.
   c) Show that for $p = 2$ and $\alpha, \beta \in 1 + X\mathcal{O}_0[[X]]$
   $$\frac{\alpha^2 - \alpha^\Delta}{2\alpha^\Delta} \equiv (1 + \Delta + \Delta^2 + \dots) l_X(\alpha) \mod \deg 2$$
   and therefore
   $$\Phi^{(1)}_{\alpha(X),\beta(X)} = \left(\Delta\ (L_X(\alpha(X)) L_X(\beta(X))/2\right)',$$
   where $L_X(\alpha) = (1 + \Delta + \Delta^2 + \dots) l_X(\alpha)$.
   d) Show using section 3 (and its exercises) of Ch. VI that the pairing $\langle \cdot, \cdot\rangle_\pi$ is bilinear, antisymmetric and satisfies the Steinberg property.
3. ($\diamond$) Let $p = 2$. Using section 3 (and its exercises) of Ch. VI show that $\langle 1 + ug, \beta\rangle_X = 1$ for $\beta(X) \in Q$, $g(X) \in \mathcal{O}_0[[X]]$, $1 + ug \in Q$. Deduce that the pairing $\langle \cdot, \cdot\rangle_\pi$ is well defined.
4. Let $p = 2$. Using Exercises 2 and 3 show that $\langle \cdot, \cdot\rangle_\pi$ is invariant with respect to the choice of a prime element π.
5. Let $p > 2$. Prove independence of $\langle \alpha, \beta\rangle_\pi$ of the power series expansion of α, β in π following the steps below.
   a) Similarly to the beginning of the proof of Proposition (2.2) it suffices to show that $\langle 1 - ug, \beta\rangle_X = 1$.

b) Using Lemma 3 of (3.6) Ch. VI and Exercise 7 section 3 Ch. VI show that

$$\operatorname{res}\frac{1}{p}\frac{(\beta^{\Delta})'}{\beta^{\Delta}}\frac{\Delta}{p}\log(1-ug)\,V\equiv\operatorname{res}X^{-1}f^{\Delta}$$

where $f = X\beta^{-1}\beta'\log(1-ug)\,V$.

c) Using Exercise 6 section 3 Ch. VI show that

$$\operatorname{res}l(\beta)(\log(1-ug))'\,V\equiv-\operatorname{res}(l(\beta))'\log(1-ug)\,V\mod p^n.$$

d) Deduce from the previous congruences that

$$\operatorname{res}\Phi_{1-ug,\beta}\equiv\operatorname{res}X^{-1}(f-f^{\Delta})\mod p^n.$$

and using Lemma 3 of (3.6) Ch. VI conclude that

$$\operatorname{Tr}\operatorname{res}\Phi_{1-ug,\beta}\equiv 0\mod p^n$$

and therefore $\langle 1-ug,\beta\rangle_X = 1$.

6. Let $p>2$. Let π,τ be prime elements of F and $\pi = g(\tau)$ with $g(X)\in X\mathcal{O}_0[[X]]$. Let $\beta=\beta(\pi)$ with $\beta(X)\in 1+X\mathcal{O}_0[[X]]$. Show that

$$\langle\pi,\beta\rangle_\pi = \langle\pi,\beta\rangle_\tau$$

following the steps below.

a) Show that it suffices to check the equality for $\beta = E(\theta X^j)|_{X=\pi}$ with $\theta\in\mathcal{R}$.

b) Show that

$$E_X(\theta X^j) = E_Y\left(\left(1-\frac{\Delta_Y}{p}\right)f(Y)\right)$$

where $f(Y)=\sum_{i\geqslant 0}\theta^{p^i}g(Y)^{jp^i}/p^i$. Using Lemma (2.2) Ch. VI show that $f(Y)\in\mathcal{O}_0((Y))$.

c) Using arguments similar to the proof of Proposition (2.1) show that

$$\Phi_{g(Y),\beta(g(Y))} = \theta g(Y)^{i-1}g(Y)' + \left(\sum_{i\geqslant 1}\theta^{p^i}f_i\right)'$$

where

$$f_i = \frac{g^{p^ij}-g^{p^{i-1}j\Delta_Y}}{p^{2i}j} - \frac{g^{p^{i-1}j\Delta_Y}l_Y(g)}{p^i}.$$

d) Deduce that

$$\operatorname{Tr}\operatorname{res}_X\Phi_{X,\beta(X)}V(X)\equiv\operatorname{Tr}\operatorname{res}_Y\Phi_{g(Y),\beta(g(Y))}V(Y)\mod p^n.$$

3. Explicit Class Field Theory for Kummer Extensions

In this section we will show, without employing local class field theory of Ch. IV, that the norm subgroups of abelian extensions of exponent p^n of a local number field F, which contains a primitive p^n th root ζ of unity, are in one-to-one correspondence with subgroups in F^* of exponent p^n. This relation is described by means of the pairing $\langle \cdot, \cdot \rangle_\pi$.

(3.1). We shall use the following

PROPOSITION (CHEVALLEY). *Let M be an arbitrary field, and let*

$$M_1/M, M_2/M$$

be cyclic extensions of degree m. Assume that $M_1 \cap M_2 = M$ and M_3/M is a cyclic subextension of degree m in M_1M_2/M such that $M_1 \cap M_3 = M$. Then an element $\alpha \in M^*$ belongs to the subgroups $N_{M_1/M}M_1^*$ and $N_{M_2/M}M_2^*$ if and only if it belongs to $N_{M_1/M}M_1^*$ and $N_{M_3/M}M_3^*$.

Proof. Let $M_3 \neq M_1$, $M_3 \neq M_2$. Then the Galois group $\text{Gal}(M_1M_2/M)$ is isomorphic to $\text{Gal}(M_1/M) \times \text{Gal}(M_2/M)$. Let σ_1 and σ_2 be elements of the group $\text{Gal}(M_1M_2/M)$ such that $\sigma_1|_{M_2}$, $\sigma_2|_{M_1}$ are trivial automorphisms, $\sigma_1|_{M_1}$ is a generator of $\text{Gal}(M_1/M)$, $\sigma_2|_{M_2}$ is a generator of $\text{Gal}(M_2/M)$, and M_3 is the fixed field of $\sigma_1\sigma_2$.

Let $\alpha \in N_{M_1/M}M_1^* \cap N_{M_2/M}M_2^*$. Write

$$\alpha = \prod_{i=0}^{m-1} \sigma_1^i(\beta) = \prod_{i=0}^{m-1} \sigma_2^i(\gamma),$$

with $\beta \in M_1$, $\gamma \in M_2$. Then we deduce that $\prod_{i=0}^{m-1}(\sigma_1\sigma_2)^i(\beta\gamma^{-1}) = 1$, i.e., $N_{M_1M_2/M_3}(\beta\gamma^{-1}) = 1$. By Proposition (4.1) Ch. III, we get $\beta\gamma^{-1} = \lambda^{-1}(\sigma_1\sigma_2)(\lambda)$ for some $\lambda \in M_1M_2$. Now we put $\kappa = \beta\lambda\sigma_1(\lambda^{-1})$. Then

$$\sigma_2^{-1}\sigma_1^{-1}(\kappa) = \sigma_2^{-1}(\gamma\lambda^{-1}\sigma_2(\lambda)) = \sigma_2^{-1}(\beta\sigma_2(\lambda)(\sigma_1\sigma_2)(\lambda^{-1})) = \kappa,$$

i.e., $\kappa \in M_3$. We also obtain that

$$N_{M_3/M}(\kappa) = \prod_{i=0}^{m-1} \sigma_1^i(\kappa) = \prod_{i=0}^{m-1} \sigma_1^i(\beta) = N_{M_1/M}\beta. \square$$

(3.2). We verify the following assertion for local number fields without employing class field theory.

PROPOSITION. *Let F be a complete discrete valuation field with finite residue field. Let L/F be a cyclic extension of degree n. Then the quotient group $F^*/N_{L/F}L^*$ is a cyclic group of order n.*

Proof. If n is prime, then the required assertion follows from (1.4) Ch. IV.

Let σ be a generator of $\mathrm{Gal}(L/F)$, and let M/F be a subextension in L/F. Denote the set $\{\alpha^{-1}\sigma(\alpha) : \alpha \in M^*\}$ denote by $M^{*\sigma-1}$. We claim that

$$M^{*\sigma-1} \subset N_{L/M}(L^{*\sigma-1}), \quad M^* \subset F^* N_{L/M}L^*.$$

Indeed, if L/M is of prime degree, then, by (1.4) Ch. IV, $M^*/N_{L/M}L^*$ is a cyclic group of the same order. It is generated by $\alpha N_{L/M}L^*$ for some $\alpha \in M^*$. Then (1.4) Ch. IV implies that $\alpha^{-1}\sigma(\alpha) \in N_{L/M}L^*$; therefore $\alpha^{-1}\sigma(\alpha) = N_{L/M}\beta$ for some $\beta \in L^*$. We get $N_{L/F}\beta = 1$, and Proposition (4.1) Ch. III shows that $\beta = \gamma^{-1}\sigma(\gamma)$ for some $\gamma \in L^*$. Thus, $M^{*\sigma-1} \subset N_{L/M}(L^{*\sigma-1})$. In general, we proceed by induction on the degree $|L : M|$. Let M_1/M be a proper subextension in L/M. Then, by the induction assumption, $M^{*\sigma-1} \subset N_{M_1/M}(M_1^{*\sigma-1})$ and $M_1^{*\sigma-1} \subset N_{L/M_1}(L^{*\sigma-1})$, hence $M^{*\sigma-1} \subset N_{L/M}(L^{*\sigma-1})$. Now for $\alpha \in M^*$ there exists $\beta \in N_{L/M}L^*$ with $\alpha^{-1}\sigma(\alpha) = \beta^{-1}\sigma(\beta)$. Then $\sigma(\alpha\beta^{-1}) = \alpha\beta^{-1}$ and $M^* \subset F^* N_{L/M}L^*$.

Assume that there exists a proper divisor m of n, such that $F^{*m} \subset N_{L/F}L^*$. Let M/F be a subextension in L/F of degree m. Then $N_{M/F}F^* \subset N_{L/F}L^*$ and by Proposition (4.1) Ch. III we deduce $F^* \subset (N_{L/M}L^*)M^{*\sigma-1} \subset N_{L/M}L^*$. Then $M^* \subset F^* N_{L/M}L^* \subset N_{L/M}L^*$, which is impossible because $M^* \neq N_{L/M}L^*$ ($M^*/N_{L/M}L^*$ is of order $\geqslant l$, where l is a prime divisor of nm^{-1}). Thus, $F^{*m} \not\subset N_{L/F}L^*$.

On the other hand,

$$\begin{aligned} |F^* : N_{L/F}L^*| &= |F^* : N_{M/F}M^*||N_{M/F}M^* : N_{M/F}(N_{L/M}L^*)| \\ &\leqslant |F^* : N_{M/F}M^*||M^* : N_{L/M}L^*| = n, \end{aligned}$$

and we conclude that $F^*/N_{L/F}L^*$ is cyclic of order n. □

COROLLARY. *Let L/F be a cyclic extension of degree l^n, where l is prime, $n \geqslant 1$. Let M/F be a subextension of degree l^{n-1} in L/F. Let $\alpha \in F^*$. Then the condition $\alpha^l \in N_{L/F}L^*$ is equivalent to $\alpha \in N_{M/F}M^*$.*

Proof. If $\alpha = N_{M/F}\beta$, then $\alpha^l = N_{M/F}\beta^l = N_{L/F}\beta$. If $\alpha^l \in N_{L/F}L^*$, then, by the Proposition, $\alpha \in N_{M/F}M^*$. □

(3.3). PROPOSITION. *Let F be a complete discrete valuation field with a finite residue field of characteristic p. Let $\alpha, \beta \in F^*$. Let a primitive p^nth root of unity belong to F.*

Then the conditions $\alpha \in N_{F(\sqrt[p^n]{\beta})/F} F(\sqrt[p^n]{\beta})^*$ and $\beta \in N_{F(\sqrt[p^n]{\alpha})/F} F(\sqrt[p^n]{\alpha})$ are equivalent.

Proof. Let $\alpha = \alpha_1^{p^k}$ with $\alpha_1 \notin F^{*p}$, $\beta = \beta_1^{p^l}$ with $\beta_1 \notin F^{*p}$. We can assume $l \leqslant n$. Then $\alpha_1^{p^k} \in N_{F(\sqrt[p^{n-l}]{\beta_1})/F} F(\sqrt[p^{n-l}]{\beta_1})^*$. By Corollary (3.2) α_1 belongs to $N_{F(\sqrt[p^{n-l-k}]{\beta_1})/F} F(\sqrt[p^{n-l-k}]{\beta_1})^*$ if $n-l-k \geqslant 0$ (if $n < l+k$ then it is easy to show that $\beta \in N_{F(\sqrt[p^n]{\alpha})/F} F(\sqrt[p^n]{\alpha})^*$). We also get

$$-\beta_1 \in N_{F(\sqrt[p^{n-l-k}]{\beta_1})/F} F(\sqrt[p^{n-l-k}]{\beta_1})^*.$$

Let i be an integer relatively prime to p such that $\alpha_1^i \beta_1 \notin F^{*p}$. Introduce the field $M = F(\sqrt[p^{n-l-k}]{\beta_1}) \cap F(\sqrt[p^{n-l-k}]{\alpha_1^i \beta_1})$, $M_1 = F(\sqrt[p^{n-l-k}]{\beta_1})$, $M_2 = F(\sqrt[p^{n-l-k}]{\alpha_1^i \beta_1})$, $M_3 = F(\sqrt[p^{n-l-k}]{\alpha_1})$. Then $M_3 \supset M$. Let $-\alpha_1^i \beta_1 = N_{M_1/F}\gamma$, $-\alpha_1^i \beta_1 = N_{M_2/F}\delta$. Then

$$N_{M/F}(N_{M_1/M}\gamma N_{M_2/M}\delta^{-1}) = 1$$

and by Proposition (4.1) Ch. III we deduce that $N_{M_1/M}\gamma = N_{M_2/M}\delta^{-1}\varepsilon^{-1}\sigma(\varepsilon)$ for some $\varepsilon \in M$, where σ is a generator of $\mathrm{Gal}(M/F)$. The arguments adduced in the proof of the preceding Proposition show that $\varepsilon^{-1}\sigma(\varepsilon) \in N_{M_2/M} M_2^*$. Therefore, $N_{M_1/M}\gamma \in N_{M_1/M} M_1^* \cap N_{M_2/M} M_2^*$. Now Proposition (3.1) implies $N_{M_1/M}\gamma \in N_{M_3/M} M_3^*$, $-\alpha_1^i \beta_1 = N_{M_3/F}\eta$ for some $\eta \in M_3^*$. Since $-\alpha_1^i \in N_{M_3/F} M_3^*$, we conclude that $\beta_1 \in N_{F(\sqrt[p^{n-l-k}]{\alpha_1})/F} F(\sqrt[p^{n-l-k}]{\alpha_1})^*$ and, by Corollary (3.2), that $\beta \in N_{F(\sqrt[p^n]{\alpha})/F} F(\sqrt[p^n]{\alpha})^*$. □

COROLLARY. *Let $\gamma \in N_{F(\sqrt[p^n]{\alpha})/F} F(\sqrt[p^n]{\alpha})^* \cap N_{F(\sqrt[p^n]{\beta})/F} F(\sqrt[p^n]{\beta})^*$.*

Then $\gamma \in N_{F(\sqrt[p^n]{\alpha\beta})/F} F(\sqrt[p^n]{\alpha\beta})^*$.

Proof. Since $\alpha \in N_{F(\sqrt[p^n]{\gamma})/F} F(\sqrt[p^n]{\gamma})^*$, $\beta \in N_{F(\sqrt[p^n]{\gamma})/F} F(\sqrt[p^n]{\gamma})^*$, we get $\alpha\beta \in N_{F(\sqrt[p^n]{\gamma})/F} F(\sqrt[p^n]{\gamma})^*$ and $\gamma \in N_{F(\sqrt[p^n]{\alpha\beta})/F} F(\sqrt[p^n]{\alpha\beta})^*$. □

(3.4). THEOREM. *Let F be a local number field as in section 2, and let $p > 2$.*

Then $\langle \alpha, \beta \rangle_\pi = 1$ if and only if $\alpha \in N_{F(\sqrt[p^n]{\beta})/F} F(\sqrt[p^n]{\beta})^*$ and if and only if $\beta \in N_{F(\sqrt[p^n]{\alpha})/F} F(\sqrt[p^n]{\alpha})^*$.

Proof. In accordance with the previous Proposition, we must show that

$$\alpha \in N_{F(\sqrt[p^n]{\beta})/F} F(\sqrt[p^n]{\beta})^* \quad \text{or} \quad \beta \in N_{F(\sqrt[p^n]{\alpha})/F} F(\sqrt[p^n]{\alpha})^*.$$

Note that for $p > 2$ $\langle \alpha, \alpha \rangle_\pi = 1$ for $\alpha \in F^*$ and $\alpha \in N_{F(\sqrt[p^n]{\alpha})/F} F(\sqrt[p^n]{\alpha})^*$. In this proof principal units in F^* without p^n-primary part, i.e., $E_X(w(X))|_{X=\pi}$ for $w(X) = \sum_{\substack{1 \leqslant i < pe_1 \\ (i,p)=1}} \alpha_1 X^i$, $\alpha_i \in \mathcal{O}_0$, will be denoted by ε, ρ. Remark (2.4) shows that

$$\langle \pi, \varepsilon \rangle_\pi = 1, \quad \langle \omega_*, \varepsilon \rangle_\pi = 1,$$

where ω_* denotes a p^n-primary element $E_X(as(X))|_{X=\pi}$ with $\operatorname{Tr} a \equiv 1 \mod p^n$. Since $F(\sqrt[p^n]{\omega_*})/F$ is unramified, $\varepsilon \in N_{F(\sqrt[p^n]{\omega_*})/F} F(\sqrt[p^n]{\omega_*})^*$. We can also write $E_X(w(X)) = \prod_{\substack{1 \leqslant i < pe_1 \\ (i,p)=1 \\ 1 \leqslant j \leqslant f}} E(\theta_{ij} X^i)^{a_{ij}}$ with $\theta_{ij} \in \mathcal{R}$, $a_{ij} \in \mathbb{Z}_p$ (see, e.g., the proof of Proposition (2.2) Ch. VI). Now Lemma (9.1) Ch. I shows that

$$E_X(w(X))|_{X=\pi} = \prod_{\substack{1 \leqslant i < pe_1 \\ (i,p)=1 \\ 1 \leqslant j \leqslant f \\ k \geqslant 1 \\ (k,p)=1}} (1 - \theta_{ij}^k \pi^{ki})^{-a_{ij}\mu(k)/k}.$$

Note that $1 - \theta_{ij}^k \pi^{ki} = 1 - \theta_1^{p^n} \pi^{ki}$ with $\theta_1 \in \mathcal{R}$. Hence $1 - \theta_{ij}^k \pi^{ki}$ belongs to $N_{F(\sqrt[p^n]{\pi})/F} F(\sqrt[p^n]{\pi})$ and we obtain that $\varepsilon \in N_{F(\sqrt[p^n]{\pi})/F} F(\sqrt[p^n]{\pi})$. Therefore, ε belongs to $N_{F(\sqrt[p^n]{\beta})/F} F(\sqrt[p^n]{\beta})^*$ for $\beta = \pi^b \eta \omega_*^l$, $\eta \in \mathcal{R}$.

Now we will use the following

Lemma. *Let $\alpha = \pi^a \theta \varepsilon \omega_*^k$, $\beta = \pi^b \eta \omega_*^l$, with $\theta, \eta \in \mathcal{R}^*$. Then*

$$\langle \alpha, \beta \rangle_\pi = 1 \iff a\,l - b\,k \equiv 0 \mod p^n \iff \alpha \in N_{F(\sqrt[p^n]{\beta})/F} F(\sqrt[p^n]{\beta})^*.$$

Proof. We get

$$\langle \alpha, \beta \rangle_\pi = \zeta^{a\,l - b\,k}.$$

Furthermore, $\varepsilon \in N_{F(\sqrt[p^n]{\beta})/F} F(\sqrt[p^n]{\beta})^*$. Let $a = p^m a_1$, $k = p^m k_1$, $b = p^m b_1$, $l = p^m l_1$ and g.c.d.$(a_1, k_1, b_1, l_1, p) = 1$. If $a\,l - b\,k \equiv 0 \mod p^n$, then $a_1 l_1 - b_1 k_1 \equiv 0 \mod p^{n-2m}$ when $n - 2m \geqslant 1$. Suppose that b_1 is relatively prime to p. Then $(\pi^{a_1} \omega_*^{k_1})^{b_1} F^{*p^{n-2m}} = (\pi^{b_1} \omega_*^{l_1})^{a_1} F^{*p^{n-2m}}$. This means that $(\pi^{a_1} \omega_*^{k_1})^{b_1}$ belongs to $N_{F(\sqrt[p^{n-m}]{\beta})/F} F(\sqrt[p^{n-m}]{\beta})^*$, and by the preceding considerations we conclude that $\alpha \in N_{F(\sqrt[p^n]{\beta})/F} F(\sqrt[p^n]{\beta})^*$. Other cases are treated similarly. If $n \geqslant 2m$, then obviously $\alpha \in N_{F(\sqrt[p^n]{\beta})/F} F(\sqrt[p^n]{\beta})^*$.

Conversely, let $\alpha \in N_{F(\sqrt[p^n]{\beta})/F} F(\sqrt[p^n]{\beta})^*$. Suppose that $b \in a\mathbb{Z}_p$. Then $a\,c \equiv b \mod p^n$ for some integer c. Then

$$\beta \equiv \pi^b \omega_*^l = (\pi^a \omega_*^k)^c \omega_*^{l-kc} \mod F^{*p^n}.$$

Therefore, by Corollary (3.3) and the preceding considerations,

$$\alpha \in N_{F(\sqrt[p^n]{\omega_*^{l-kc}})/F} F(\sqrt[p^n]{\omega_*^{l-kc}})^*.$$

Then $\pi^a \in N_{F(\sqrt[p^n]{\omega_*^{l-kc}})/F} F(\sqrt[p^n]{\omega_*^{l-kc}})^*$. Since the quotient group

$$F^*/N_{F(\sqrt[p^n]{\omega_*^{l-kc}})/F} F(\sqrt[p^n]{\omega_*^{l-kc}})^*,$$

which corresponds to an unramified extension, is cyclic and generated by π, we deduce that $a(l-kc) \equiv 0 \mod p^n$. This means $al - kb \equiv 0 \mod p^n$. Assume that $a \in b\mathbb{Z}_p$. Then $bc \equiv a \mod p^n$ for some integer c. Then $\alpha \equiv (\pi^b\omega_*^l)^c\omega_*^{k-lc}\varepsilon \mod F^{*p^n}$. By Corollary (3.3) and the preceding considerations, we deduce that $\omega_*^{k-lc} \in N_{F(\sqrt[p^n]{\pi^b})/F} F(\sqrt[p^n]{\pi^b})^*$ and $b(k-lc) \equiv 0 \mod p^n$, i.e., $bk - la \equiv 0 \mod p^n$. □

To complete the proof of the Theorem, let $\alpha = \pi^a\theta\varepsilon\omega_*^k$ with $\theta \in \mathcal{R}$. Assume first that $k \in a\mathbb{Z}_p$. Then $k \equiv ac \mod p^n$ for some integer c. Then, by the Lemma,

$$\langle \alpha, \omega_*^c\rangle_\pi = \zeta^{ac} = \langle \alpha, \pi\rangle_\pi^{-1}.$$

For a prime element $\tau = \pi\omega_*^c$ we get $\langle \alpha, \tau\rangle_\pi = \langle \alpha, \tau\rangle_\tau = 1$. Therefore, the element α can be written as $\alpha = \tau^a\theta\varepsilon_1$ without p^n-primary part. However, this case has been considered above.

Assume $a \in k\mathbb{Z}_p$. Let $\beta = \pi^b\eta\rho\omega_*^l$ with $\eta \in \mathcal{R}$. Let c be an integer such that $kb - al \equiv ck \mod p^n$. Then, by the Lemma and Corollary (3.3),

$$\langle \alpha, \pi^{b-c}\omega_*^l\rangle_\pi = 1, \quad \pi^{b-c}\omega_*^l \in N_{F(\sqrt[p^n]{\alpha})/F} F(\sqrt[p^n]{\alpha})^*.$$

Also $\langle \alpha, \beta\rangle_\pi = \langle \alpha, \pi^{b-c}\omega_*^l\rangle_\pi \langle \alpha, \pi^c\rho\rangle_\pi$. If $\langle \alpha, \beta\rangle_\pi = 1$, then $\langle \alpha, \pi^c\rho\rangle_\pi = 1$. By the Lemma and Corollary (3.3) we obtain that

$$\pi^{b-c}\omega_*^l \in N_{F(\sqrt[p^n]{\alpha})/F} F(\sqrt[p^n]{\alpha})^*, \quad \pi^c\rho \in N_{F(\sqrt[p^n]{\alpha})/F} F(\sqrt[p^n]{\alpha})^*.$$

Then $\beta \in N_{F(\sqrt[p^n]{\alpha})/F} F(\sqrt[p^n]{\alpha})^*$.

Conversely, if $\beta \in N_{F(\sqrt[p^n]{\alpha})/F} F(\sqrt[p^n]{\alpha})^*$, then $\pi^c\rho \in N_{F(\sqrt[p^n]{\alpha})/F} F(\sqrt[p^n]{\alpha})^*$. Now from the Lemma we get $ck \equiv 0 \mod p$ and $\langle \alpha, \pi^{b-c}\omega_*^l\rangle_\pi = 1$. We conclude that $\langle \alpha, \beta\rangle_\pi = 1$. □

(3.5). PROPOSITION (NONDEGENERACY OF THE PAIRING $\langle \cdot, \cdot\rangle_\pi$). *Let $\alpha \in F^* \setminus F^{*p}$. Then there exists an element $\beta \in F^*$ such that $\langle \alpha, \beta\rangle_\pi = \zeta$.*

Proof. If $\alpha = \pi^a\theta\varepsilon\omega_*^k$ is as in the proof of Theorem (3.4) and a is relatively prime to p, then we can put $\beta = \omega_*^l$ for a suitable integer l. If a is divisible by p and k is relatively prime to p, then we can put $\beta = \pi^b$ for a suitable integer b. If a and k are divisible by p, then let $\varepsilon \equiv 1 + \eta\pi^i \mod \pi^{i+1}$ with $(i,p) = 1$, $1 \leqslant i < pe_1$, $\eta \in \mathcal{O}_F$. For a unit $\rho = 1 + \eta_1\pi^{pe_1 - i}$ with $\eta_1 \in \mathcal{O}_F$ we get

$$\begin{aligned}\langle \varepsilon, \rho\rangle_\pi &= \langle 1 + \eta\pi^i, -\eta\pi^i(1 + \eta_1\pi^{pe_1 - i}\rangle_\pi \\ &= \langle 1 + \eta\eta_1\pi^{pe_1}(1 + \eta\pi^i)^{-1}, -\eta\pi^i(1 + \eta_1\pi^{pe_1-i})\rangle_\pi^{-1},\end{aligned}$$

because $\langle \gamma, 1 - \gamma\rangle_\pi = 1$ by Theorem (3.4) ($1 - \gamma \in N_{F(\sqrt[p^n]{\gamma})/F}F(\sqrt[p^n]{\gamma})^*$). Since $U_{F,pe_1+1} \subset F^{*p}$ from (5.7) Ch. I we deduce that $\langle \varepsilon, \rho\rangle_\pi = \zeta$ for a proper η_1. This completes the proof. □

(3.6). THEOREM. *Let F be as in section 2, $p > 2$. Let A be a subgroup in F^* such that $F^{*p^n} \subset \mathrm{A}$. Let $\mathrm{B} = \mathrm{A}^\perp$ denote its orthogonal complement with respect to the pairing $\langle \cdot, \cdot \rangle_\pi$. Then $\mathrm{A} = N_{L/F}L^*$, where $L = F(\sqrt[p^n]{\mathrm{B}})$ and $\mathrm{B}^\perp = \mathrm{A}$.*

Proof. First, using Proposition (3.5) and the arguments of the last paragraph of the proof of Theorem (5.2) Ch. IV we deduce that $\mathrm{B}^\perp = \mathrm{A}$. Then from Theorem (3.4) we conclude that $N_{L/F}L^* \subset \mathrm{A}$.

In the same way as in the last paragraph of the proof of Proposition (3.2) we deduce that the index of $N_{L/F}L^*$ in F^* isn't greater than the degree of the extension L/F. By Kummer theory the latter is equal to $|\mathrm{B} : F^{*p^n}|$ which is equal to $|F^* : \mathrm{A}|$. Thus, $\mathrm{A} = N_{L/F}L^*$. □

4. Explicit Formulas

In this section following [Vo1] (the case of $p = 2$ [Fe1]) we will verify that the pairing $\langle \cdot, \cdot \rangle_\pi$ coincides with the p^n th Hilbert symbol $(\cdot, \cdot)_{p^n}$, thereby obtaining explicit formulas which compute the values of the Hilbert symbol $(\alpha, \beta)_{p^n}$ are computed by expansions of α, β in series in a prime element π.

THEOREM. *For $\alpha, \beta \in F^*$ and $p > 2$*

$$\boxed{(\alpha, \beta)_{p^n} = \zeta^{\operatorname{Tr}\operatorname{res}\Phi_{\alpha(X),\beta(X)}V(X)}}$$

where $\alpha(X), \beta(X) \in \mathcal{O}_0((X))^*$ are such that $\alpha(\pi) = \alpha, \beta(\pi) = \beta$;

$$\Phi_{\alpha(X),\beta(X)} = \frac{\alpha(X)'}{\alpha(X)}\, l(\beta(X)) - l(\alpha(X))\,\frac{1}{p}\,\frac{(\beta(X)^\Delta)'}{\beta(X)^\Delta}$$

$V(X) = 1/2 + 1/s(X)$, $s(X) = z(X)^{p^n} - 1$ and $z(X) \in 1 + X\mathcal{O}_0[[X]]$ is such that $z(\pi) = \zeta$ is a p^n th primitive root of unity in F.

For $\alpha, \beta \in F^*$ *and* $p = 2$

$$\boxed{(\alpha, \beta)_{p^n} = \zeta^{\operatorname{Tr}\operatorname{res}\left(\Phi_{\alpha(X),\beta(X)} + \Phi^{(1)}_{\alpha(X),\beta(X)} + \Phi^{(2)}_{\alpha(X),\beta(X)}\right) r(X) V(X)}}$$

where $\alpha(X), \beta(X) \in Q$ *with* $\alpha(\pi) = \alpha, \ \ \beta(\pi) = \beta,$

$$Q = \left\{X^m a\psi(X) : \psi(X) \in 1 + X\mathcal{O}[[X]], a \in \mathcal{O}^*, a^{\varphi} \equiv a^2 \mod 4, m \in \mathbb{Z}\right\};$$

$$\Phi^{(1)}_{\alpha(X),\beta(X)} = \left(\frac{\Delta}{2}\left(\frac{\alpha^2 - \alpha^{\Delta}}{2\alpha^{\Delta}} \ \frac{\beta^2 - \beta^{\Delta}}{2\beta^{\Delta}}\right)\right)';$$

$$\Phi^{(2)}_{\alpha(X),\beta(X)} = X^{-1} v_X(\alpha(X)) v_X(\beta(X)) l_X(1 + s_{n-1}(X)),$$

where v_X *is the discrete valuation associated to* X, $s_{n-1}(X) = z(X)^{p^{n-1}} - 1$; $r(X) = 1 + 2^{n-1} \Delta_X \, r_0(X)$ *where* $r_0(X) \in X\mathcal{O}_0[X]$ *satisfies*

$$\Delta^2 r_0 + (1 + (2^{n-1} - 1)s_{n-1}) \Delta r_0 + s_{n-1} r_0 \equiv h \operatorname{modev}(2, \deg 2e),$$

e *is the absolute ramification index of* F *(see* (3.4) *Ch.* VI).

Proof. Let $\mathcal{O}_0$ be the ring of integers in $F_0 = F \cap \mathbb{Q}_p^{\mathrm{ur}}$. Let ε be a principal unit in F. We have its factorization with respect to the Shafarevich basis (see section 5 Ch. VI)

$$\varepsilon = E_X(w(X))|_{X=\pi} \omega(a), \quad w(X) = \sum_{\substack{1 \leqslant i < pe_1 \\ (i,p)=1}} \alpha_i X^i, \quad \alpha_i, a \in \mathcal{O}_0.$$

As we have seen in (1.1)

$$(\pi, \varepsilon)_{p^n} = \zeta^{\operatorname{Tr} a}.$$

On the other hand, Property 2 in (2.3) shows that

$$\langle \pi, \varepsilon \rangle_\pi = \zeta^{\operatorname{Tr} a}$$

for $p > 2$. The same equality can be verified for $p = 2$ (see Exercise 1).

Now let ρ be a principal unit in F. Then $\tau = \pi\rho$ is prime, and the invariance of $\langle \cdot, \cdot \rangle_\pi$ shows that

$$\langle \rho, \varepsilon \rangle_\pi = \langle \pi\rho, \varepsilon \rangle_\pi \langle \pi, \varepsilon \rangle_\pi^{-1} = \langle \tau, \varepsilon \rangle_\tau \langle \pi, \varepsilon \rangle_\pi^{-1}.$$

Then

$$\langle \rho, \varepsilon \rangle_\pi = (\tau, \varepsilon)_{p^n} (\pi, \varepsilon)_{p^n}^{-1} = (\rho, \varepsilon)_{p^n}.$$

Finally, for $\alpha = \pi^i \theta \varepsilon, \ \beta = \pi^j \eta \rho$ with $\theta, \eta \in \mathcal{R}^*, \ \varepsilon, \rho \in U_{1,F}$ we get

$$\begin{aligned} \langle \alpha, \beta \rangle_\pi &= \langle \pi, \eta^i \theta^{-j} \rangle_\pi \langle \pi, (-1)^{ij} \rho^i \varepsilon^{-j} \rangle_\pi \langle \varepsilon, \rho \rangle_\pi \\ &= \langle \pi, (-1)^{ij} \rho^i \varepsilon^{-j} \rangle_\pi \langle \varepsilon, \rho \rangle_\pi = (\pi, (-1)^{ij} \rho^i \varepsilon^{-j})_{p^n} (\varepsilon, \rho)_{p^n} = (\alpha, \beta)_{p^n}, \end{aligned}$$

because $\eta^i \theta^{-j} \in \mathcal{R}^*$. This completes the proof. □

REMARKS.

1. If one does not intend to have an independent pairing $\langle \cdot, \cdot \rangle$, then as in Exercise 2 below one can reduce calculations to the case of $(\pi, \beta)_{p^n}$ and then find an explicit formula in the way similar to (1.1). This is the method of *H. Brückner* [Bru1–2], see also [Henn1–2]. This method does not seem to have a generalization to formal groups.

2. Compare the formulas of this section with the formulas of (5.5), (5.6) Ch. IV for the local functional fields.

Exercises.

1. Let $p = 2$. Using Exercise 2 of section 2 and elements $1 - \theta_j \pi^i$ and $\omega(a)$ as in the proof of Proposition (2.4) show that for $\varepsilon = \prod(1 - \theta_j \pi^i)^{a_{ij}}\ \omega(a)$
$$\langle \pi, \varepsilon \rangle_\pi = (\pi, \varepsilon)_{p^n} = \zeta^{\operatorname{Tr} a}.$$

2. a) (*H. Brückner* [Bru1–2]) Prove the equality $\langle \cdot, \cdot \rangle_\pi = (\cdot, \cdot)_{p^n}$ using (1.1), the Steinberg property for $\langle, \cdot, \cdot \rangle_\pi$ (Proposition (2.1) and Exercise 2 of section 2) and the equalities of Exercise 1f), g) section 1 that hold also for the pairing $\langle \cdot, \cdot \rangle_\pi$.
   b) Prove the equality $\langle \cdot, \cdot \rangle_\pi = (\cdot, \cdot)_{p^n}$ using (1.1), the Steinberg property for $\langle, \cdot, \cdot \rangle_\pi$ and the theory of (4.3) Ch. IX instead of Exercise 1f), g) section 1.

3. Let $p > 2$. Show that
$$(\pi, a)_{p^n} = \zeta_{p^n}^{\log (N_{F_0/\mathbb{Q}_p} a^{p-1})/(2p)}$$
for $a \in \mathcal{O}_0^*$.

4. a) Show that
$$\left(\alpha, E_F(a\, s(X))|_{X=\pi}\right)_{p^n} = \zeta^{v(\alpha) \operatorname{Tr} a}, \quad a \in \mathcal{O}_0,$$
where v is the discrete valuation in F.
   b) Show that for every $i,\ 1 \leqslant i < pe_1,\ (i, p) = 1,\ \theta \in \mathcal{R}^*$, there exists $\eta \in \mathcal{R}^*$ such that
$$(1 + \theta\pi^i, 1 + \eta\pi^{pe_1 - i})_{p^n} = \zeta.$$
(Hint. First prove this for $n = 1$.)
   c) If $i + j > pe_1,\ v(\alpha - 1) = i,\ v(\beta - 1) = j$, then
$$(\alpha, \beta)_p = 1.$$

5. ($\diamond$) (*H. Koch* [Ko1]). Let F be a local number field, L/F a tamely ramified finite Galois extension, $G = \operatorname{Gal}(L/F)$, and let a primitive p^n th root of unity ζ belong to $L,\ p > 2$. Let $(\cdot, \cdot)$ be the p^n th Hilbert symbol in L.
   a) Using Exercise 6, show that there exist elements $\alpha_1, \alpha_2, \dots, \alpha_{r+2} \in L^*,\ k = |L : \mathbb{Q}_p|$, such that $\alpha_i \in U_{1,L}$ for $1 \leqslant i \leqslant r$, and, for $i \leqslant j$, (α_i, α_j) is a primitive p^n th root of unity if $j = i + 1$, i is odd, and $(\alpha_i, \alpha_j) = 1$ otherwise.
   b) Show that $U_{1,L}/U_{1,L}^{p^n}$ as a $\mathbb{Z}/p^n\mathbb{Z}[G]$-module is the sum of two submodules A_1, A_2, each of rank $m/2$, $m = |F : \mathbb{Q}_p|$, such that $(A_1, A_1) = (A_2, A_2) = 1$.

5. Applications and Generalizations

In this section we deduce formulas for the Hilbert symbol in some special cases in (5.1) and (5.2). Then in (5.3)–(5.4) we comment on various aspects of the explicit formulas and their generalizations. In (5.5) we describe a higher dimensional formula in higher dimensional local fields.

(5.1). First we prove formulas of *E. Kummer* and *G. Eisenstein* that had played a central role before the works of *E. Artin* and *H. Hasse*. By use of the formulas of section 4 we will rewrite these formulas in a form that is somewhat different and appears more natural in the context of Ch. VII.

Let ζ be a primitive pth root of unity, $p > 2$. Then $\pi = \zeta - 1$ is prime in $\mathbb{Q}_p(\zeta)$. Let $\varepsilon \in 1 + X\mathbb{Z}_p[[X]]$, $\eta \in 1 + X\mathbb{Z}_p[[X]]$, and $\varepsilon = \varepsilon(\pi)$, $\eta = \eta(\pi)$.

PROPOSITION (KUMMER FORMULA).

$$(\varepsilon, \eta)_p = \zeta^{\operatorname{res} \log \eta(X)(\log \varepsilon(X))' X^{-p}}.$$

Proof. We have $s \equiv X^p \mod p$ (see (1.2)) and

$$1/s \equiv X^{-p} \mod p.$$

Next, for $f \in 1 + X\mathbb{Z}_p[[X]]$

$$l(f(X)) = \left(1 - \frac{\Delta}{p}\right) \log (f(X)) \equiv \log f(X) \mod \deg p.$$

Then

$$\begin{aligned} \Phi_{\varepsilon(X),\eta(X)} &= l(\varepsilon)l(\eta)' - l(\varepsilon)\eta'/\eta + l(\eta)\varepsilon'/\varepsilon \\ &= -l(\varepsilon)\left(\frac{\Delta}{p}\log(\eta)\right)' + l(\eta)\varepsilon'/\varepsilon \equiv \log\eta(X)(\log\varepsilon(X))' \mod \deg p \end{aligned}$$

and

$$\operatorname{res} \Phi_{\varepsilon(X),\eta(X)} V(X) \equiv \log \eta(X)(\log \varepsilon(X))' X^{-p} \mod p,$$

as desired. □

(5.2). PROPOSITION (EISENSTEIN FORMULA). *Let $p > 2$. Let $\beta \in \mathbb{Z}_p[\zeta]$, $\beta = b \mod \pi^2$, $b \in \mathbb{Z}$. Suppose that b is relatively prime to p and an integer a is relatively prime to p. Then*

$$(a, \beta)_p = 1.$$

Proof. It is clear that $(\alpha,\beta)_p = 1$ if and only if $(\alpha^{p-1},\beta^{p-1})_p = 1$. Note that α^{p-1},β^{p-1} are principal units in $\mathbb{Q}_p(\zeta)$. Further, as $\alpha^{p-1} \equiv 1 \mod p$ and $p \equiv -\pi^{p-1} \mod \pi^p$, we get

$$\alpha^{p-1} \equiv 1 - c\pi^{p-1} \mod \pi^p \qquad \text{for some} \quad c \in \mathbb{Z}_p.$$

Hence, if $\alpha^{p-1} = \varepsilon(\pi)$ with $\varepsilon(X) \in 1 + X\mathbb{Z}_p[[X]]$, then we can assume that

$$\log \varepsilon(X) \equiv 1 - cX^{p-1} \mod X^p.$$

Next, as $\beta \equiv b \mod \pi^2$, we deduce $\beta^{p-1} \equiv 1 \mod \pi^2$. Then if $\beta^{p-1} = \eta(\pi)$ with $\eta(X) \in 1 + X\mathbb{Z}_p[[X]]$, we obtain

$$\eta(X) \equiv 1 + dX^2 \mod X^3, \quad d \in \mathbb{Z}_p,$$

and $\log \eta(X) \equiv dX^2 \mod X^3$. Thus, by Proposition (5.1),

$$\Phi_{\varepsilon(X),\eta(X)} \equiv (1 - cX^{p-1})' dX^2 \equiv cd X^p \mod (p, \deg p + 1).$$

This implies $\operatorname{res} \Phi_{\varepsilon(X),\eta(X)} V(X) \equiv 0 \mod p, \quad (\alpha,\beta)_p = 1.$ □

(5.3). REMARKS. 1. Some other formulas for the Hilbert symbol (biquadratic formula, *Kummer–Takagi* formula, *Artin–Hasse–Iwasawa* formulas, *Sen* formulas) can be found in the Exercises. For a review of explicit formulas see [V11].

2. Let A be a local ring of characteristic 0 whose maximal ideal M contains p. Assume that A is p-adically complete (for example, $A = \mathcal{O}_0$ or $A = \mathcal{O}_0\{\{X\}\}$). Suppose that there is a ring homomorphism $\Delta\colon A \to A$ such that for every $a \in A$

$$a^{\Delta} - a^p \in pA.$$

The logarithm map induces an isomorphism

$$\log\colon 1 + 2pA \to 2pA, \quad 1 - a \mapsto -\sum_{i\geqslant 1} a^i/i.$$

So if $p > 2$ then for every $a \in A^*$ the element $\log(a^p/a^{\Delta}) \in pA$ is well defined. The map

$$a \mapsto l(a) = \frac{1}{p}\log(a^p/a^{\Delta})$$

is related to the map

$$a \mapsto \frac{a^{\Delta} - a^p}{p},$$

see the proof of Lemma 2 of (3.6) Ch. VI. When $A = \mathcal{O}_0$ and Δ is the Frobenius automorphism the latter map is sometimes interpreted as a p-adic derivation (of the identity map of A) and the right hand side as the derivative of the p-adic number a, see [Bu1–4].

3. Let A be as above and $p > 2$. Let $\widehat{\Omega}^1_A$ be the M-adic completion of the module of differential forms Ω^1_A. For $a, b \in A^*$ define

$$\Theta_{a,b} = l(b)\frac{da}{a} - l(a)\frac{1}{p}\frac{db^\Delta}{b^\Delta}$$

as an element of $\widehat{\Omega}^1_A$.

Using Milnor K_2-groups of local rings (which are defined similarly to how the K_2-group of a field is introduced in Ch. IX) one can interpret the properties of $\Phi_{\alpha(X),\beta(X)}$ proved in (2.1) as the existence of a well defined homomorphism

$$K_2(\mathcal{O}_0((X))) \to \Omega^1_{\mathcal{O}_0((X))/p^n}/d(\mathcal{O}_0((X))/p^n),$$
$$\{\alpha(X), \beta(X)\} \mapsto \Theta_{\alpha(X),\beta(X)} \in \Omega^1_{\mathcal{O}_0((X))/p^n}/d(\mathcal{O}_0((X))/p^n)$$

so that the diagram

$$\begin{array}{ccc} K_2(\mathcal{O}_0((X))) & \longrightarrow & \Omega^1_{\mathcal{O}_0((X))/p^n}/d(\mathcal{O}_0((X))/p^n) \\ \downarrow & & \downarrow \\ K_2(F) & \xrightarrow{(\cdot,\cdot)_{p^n}} & \mu_{p^n} \end{array}$$

is commutative where the left vertical homomorphism is induced by the substitution

$$\mathcal{O}_0((X))^* \to F, \qquad f(X) \mapsto f(\pi)$$

and the right vertical homomorphism is given by $\omega \mapsto \zeta^{\operatorname{Tr}\operatorname{res}(\omega V)}$.

4. *K. Kato* in [Kat6] gave an interpretation of the pairing $\langle\cdot,\cdot\rangle_\pi$ in terms of syntomic cohomologies: the image of $\{\alpha, \beta\} \in K_2(\mathcal{O}_F)$ with respect to the symbol map

$$K_2(\mathcal{O}_F) \to H^2(\operatorname{Spec}(\mathcal{O}_F), \mathcal{S}_n(2))$$

coincides with the class of

$$\left(\frac{d\alpha(X)}{\alpha(X)} \wedge \frac{d\beta(X)}{\beta(X)}, \Theta_{\alpha(X),\beta(X)}\right)$$

where $\alpha(X), \beta(X) \in \mathcal{O}_0((X))^*$ are such that $\alpha(\pi) = \alpha$, $\beta(\pi) = \beta$. Using the product structure of the syntomic complex [Kat8] one can reduce the proof of independence of this map of the choice of $\alpha(X), \beta(X)$ to the independence in the case of the appropriate map

$$K_1(\mathcal{O}_F) \to H^1(\operatorname{Spec}(\mathcal{O}_F), \mathcal{S}_n(1)), \quad \alpha \mapsto \left(\frac{d\alpha(X)}{\alpha(X)}, l(\alpha(X)\right)$$

which is easier to show.

5. As explained in the work of *M. Kurihara* [Ku3], if in the case of $A = \mathcal{O}_0\{\{X\}\}$ one chooses the action of the map $\Delta: A \to A$ (as in Remark 2 above) on X as

$(1+X)^p - 1$ (and not X^p as in this book), then one can derive formulas for the Hilbert symbol of *R. Coleman*'s type [Col2].

6. Using the theory of fields of norms (see section 5 of Ch. III) for arithmetically profinite extension $E = F(\{\pi_i\})$ over F, where $\pi_i^p = \pi_{i-1}$, $\pi_0 = p$ and a method of *J.-M. Fontaine* to obtain a crystalline interpretation of Witt equations in positive characteristic *V.A. Abrashkin* derived the explicit formula for odd p in [Ab5].

(5.4). Now let F be a complete discrete valuation field of characteristic 0 with perfect residue field of characteristic p. Let a primitive p^n th root of unity ζ belong to F.

Using Exercise 4 in section 2 Ch. VI and Exercise 3 in section 4 Ch. VI we introduce the pairing

$$\langle\cdot,\cdot\rangle_X\colon W(\overline{F})((X))^* \times W(\overline{F})((X))^* \to \mu_{p^n} \otimes W_n(\overline{F})/\wp(W_n(\overline{F})),$$

where $W(\overline{F})$ is the Witt ring of $\overline{F}$, which can be identified with the ring of integers of the absolute inertia subfield F_0 in F, $W_n(\overline{F})$ is the group of Witt vectors of length n, $\wp$ is defined in section 8 Ch. I, by the formula

$$\langle\alpha(X),\beta(X)\rangle_X = \zeta \otimes \operatorname{res}(\Phi_{\alpha(X),\beta(X)}V(X)) \quad \text{where}$$
$$\Phi_{\alpha(X),\beta(X)} = \frac{\alpha(X)'}{\alpha(X)}\, l(\beta(X)) - l(\alpha(X))\, \frac{1}{p}\, \frac{(\beta(X)^{\Delta})'}{\beta(X)^{\Delta}}$$

for $p > 2$ and $V(X)$ is defined as in (2.1). If $\alpha, \beta \in F^*$ and $p > 2$ put

$$\langle\alpha,\beta\rangle_\pi = \langle\alpha(X),\beta(X)\rangle_X,$$

where π is prime in F, $\alpha(X)$, $\beta(X) \in W(\overline{F})((X))^*$, such that $\alpha(\pi) = \alpha$, $\beta(\pi) = \beta$.

Applying Exercise 1 in section 5 Ch. VI and the same arguments as in section 2, one can show that the pairing $\langle\cdot,\cdot\rangle_\pi\colon F^* \times F^* \to \mu_{p^n} \otimes W_n(\overline{F})/\wp(W_n(\overline{F}))$ is well defined, bilinear, symmetric, satisfies the Steinberg property, and invariant with respect to the choice of π.

If $\overline{F}$ is quasi-finite, then the choice of φ in (1.3) Ch. V, when we fixed an isomorphism of $\operatorname{Gal}(\overline{F}^{\mathrm{sep}}/\overline{F})$ onto $\widehat{\mathbb{Z}}$, corresponds, due to Witt theory (see Exercise 6 section 5 Ch. IV), to the choice of a generator of the cyclic group $W_n(\overline{F})/\wp(W_n(\overline{F}))$ of order p^n. We get the corresponding isomorphism

$$\mu_{p^n} \otimes W_n(\overline{F})/\wp(W_n(\overline{F})) \simeq \mu_{p^n}$$

with respect to which $\langle\alpha,\beta\rangle_\pi$ coincides with the Hilbert pairing $(\alpha,\beta)_{p^n}$ defined in (1.3) Ch. V.

Using class field theory of a complete discrete valuation field with perfect residue field $\overline{F}$ of characteristic p with $\overline{F} \neq \wp(\overline{F})$ (see section 4 Ch. V), define the Hilbert symbol as

$$U_{1,F} \times U_{1,F} \to \operatorname{Hom}_{\mathbb{Z}_p}(\operatorname{Gal}(\widetilde{F}/F), \mu_{p^n}), \quad (\varepsilon,\eta)_{p^n}(\varphi) = \rho^{\Psi_F(\varepsilon)(\varphi)-1}$$

where $\rho^{p^n} = \eta$ and Ψ_F is the reciprocity map of (4.8) Ch. V. Note that by Witt theory $\mathrm{Hom}_{\mathbb{Z}_p}(\mathrm{Gal}(\widetilde{F}/F), \mu_{p^n})$ is canonically isomorphic to $\mu_{p^n} \otimes W_n(\overline{F})/\wp(W_n(\overline{F}))$. One can prove that with respect to this isomorphism

$$(\varepsilon, \eta)_{p^n} = \langle \varepsilon, \eta \rangle_\pi \qquad \text{for every } \varepsilon, \eta \in U_{1,F}\,.$$

(5.5). Let K be an n-dimensional field of characteristic 0 as defined in (4.6) Ch. I. Associated to K we have fields $K = K_n, K_{n-1}, \ldots, K_1, K_0$ where K_{i-1} is the residue field of complete discrete valuation field K_i for $i > 0$.

Assume that K_{n-1} is of characteristic p and K_0 is a finite field.

Assume that p is odd and ζ_{p^m} belongs to K.

Let $t_1, \ldots, t_n$ be a lifting of prime elements of $K_1, \ldots, K_{n-1}, K$ to K. Denote by $\mathcal{R}$ the multiplicative representatives of K_0 in K.

For an element

$$\alpha = t_n^{i_n} \ldots t_1^{i_1} \theta \big(1 + \sum a_J t_n^{j_n} \ldots t_1^{j_1}\big), \qquad \theta \in \mathcal{R}^*, a_J \in W(K_0),$$

$(j_1, \ldots, j_n) > (0, \ldots, 0)$ denote by $\underline{\alpha}$ the series

$$X_n^{i_n} \ldots X_1^{i_1} \theta (1 + \sum a_J X_n^{j_n} \ldots X_1^{j_1})$$

in $W(K_0)\{\{X_1\}\} \ldots \{\{X_n\}\}$. Clearly, $\underline{\alpha}$ is not uniquely determined even if the choice of a system of local parameters is fixed.

Define the following explicit pairing [V5]

$$\langle \cdot, \cdot \rangle \colon (K^*)^{n+1} \to \mu_{p^m}$$

by the formula

$$\langle \alpha_1, \ldots, \alpha_{n+1} \rangle = \zeta_{p^m}^{\operatorname{Tr} \operatorname{res} \Phi_{\alpha_1, \ldots, \alpha_{n+1}} / \underline{s}},$$

$$\Phi_{\alpha_1, \ldots, \alpha_{n+1}} = \sum_{i=1}^{n+1} \frac{(-1)^{n-i+1}}{p^{n-i+1}} l\left(\underline{\alpha_i}\right) \frac{d\underline{\alpha_1}}{\underline{\alpha_1}} \wedge \cdots \wedge \frac{d\underline{\alpha_{i-1}}}{\underline{\alpha_{i-1}}} \wedge \frac{d\underline{\alpha_{i+1}}^{\triangle}}{\underline{\alpha_{i+1}}^{\triangle}} \wedge \cdots \wedge \frac{d\underline{\alpha_{n+1}}^{\triangle}}{\underline{\alpha_{n+1}}^{\triangle}}$$

where $\underline{s} = \underline{\zeta_{p^m}}^{p^m} - 1$, $\operatorname{Tr} = \operatorname{Tr}_{W(K_0)/\mathbb{Z}_p}$, $\operatorname{res} = \operatorname{res}_{X_1, \ldots, X_n}$,

$$l\left(\underline{\alpha}\right) = \frac{1}{p} \log\left(\underline{\alpha}^p / \underline{\alpha}^{\triangle}\right), \quad \big(\sum a_J X_n^{j_n} \cdots X_1^{j_i}\big)^{\triangle} = \sum \mathbf{F}(a_J) X_n^{pj_n} \cdots X_1^{pj_1}$$

where $\mathbf{F}$ is defined in section 9 Ch. I.

One can prove that the pairing $\langle \cdot, \cdot \rangle$ is well defined, multilinear and satisfies the Steinberg property.

This pairing plays an important role in the study of (topological) K-groups of higher local fields, see sections of [FK]. Certainly, the pairing coincides with the Hilbert symbol as soon as the latter is defined by higher class field theory (see (4.13) Ch. IX).

Exercises.

1. Let $F = \mathbb{Q}_p(\zeta_p)$, let ζ_p be a primitive pth root of unity, $p > 2$.

a) Show that

$$\frac{1}{p}\operatorname{Tr}(\zeta_p\pi^i) \equiv \begin{cases} 1 \mod p & \text{if} \quad i = p-1 \\ 0 \mod p & \text{if} \quad i \neq p-1, i \geqslant 1, \end{cases}$$

where $\operatorname{Tr} = \operatorname{Tr}_{F/\mathbb{Q}_p}$, $\pi = \zeta_p - 1$.

b) Let $\alpha \equiv 1 \mod \pi^2$, $\beta \equiv 1 \mod \pi$. If $\gamma = \sum a_i\pi^i$, $a_i \in \mathbb{Z}_p$, then let $\operatorname{Dlog}\gamma$ denote the element

$$\gamma^{-1}\left(\sum ia_i\pi^{i-1}\right),$$

depending on the choice of expansion of β in a series in π. Let $\log\beta$ denote the element $(\beta-1) - \dfrac{(\beta-1)^2}{2} + \dfrac{(\beta-1)^3}{3} - \dots$. Prove the *Artin–Hasse* formula

$$(\alpha,\beta)_p = \zeta_p^{\operatorname{Tr}(\zeta_p \log\alpha\cdot \operatorname{Dlog}\beta)/p}$$

c) Using a suitable expansion in a series in π, show that $\operatorname{Dlog}\zeta_p$ can be made equal to $-\zeta_p^{-1}$, $\operatorname{Dlog}\pi$ to π^{-1}. Prove the *Artin–Hasse* formulas

$$(\zeta_p,\beta)_p = \zeta_p^{\operatorname{Tr}(\log\beta)/p} \qquad \text{for} \quad \beta \equiv 1 \mod \pi,$$

$$(\beta,\pi)_p = \zeta_p^{\operatorname{Tr}(\zeta_p\pi^{-1}\log\beta)/p} \qquad \text{for} \quad \beta \equiv 1 \mod \pi.$$

2. Let F be as in Exercise 1.

a) Let τ be a prime element in F such that $\pi \equiv a\tau \mod \tau^2$ for some $a \in \mathbb{Z}_p$. Show that for $\varepsilon, \eta \in U_{1,F}$

$$(\varepsilon,\eta)_p = \zeta_p^{\operatorname{res} a^{-1}\log\eta(X)(\log\varepsilon(X))' X^{-p}},$$

where $\varepsilon(X), \eta(X) \in 1 + X\mathbb{Z}_p[[X]]$, $\varepsilon(\tau) = \varepsilon$, $\eta(\tau) = \eta$.

b) Put

$$E[X] = 1 + X + \frac{X^2}{2!} + \dots + \frac{X^{p-1}}{(p-1)!}.$$

Show that $\zeta_p = E[\tau]$ for some prime element τ in F such that $\tau \equiv \pi = \zeta_p - 1 \mod \tau^2$.

c) Let $\varepsilon, \eta \in U_{1,F}$ and $f(X), g(X) \in \mathbb{Z}[X]$ such that $f(\zeta_p) = \varepsilon$, $g(\zeta_p) = \eta$, $f(1) = g(1) = 1$. Show that $\varepsilon = f(E(X))|_{X=\tau}$, $\eta = g(E(X))|_{X=\tau}$.

d) Put

$$l_i(h(X)) = \left.\frac{d^i L(h(X))}{dX^i}\right|_{X=0},$$

where $L(1-X) = -\left(X + \frac{X^2}{2} + \dots + \frac{X^{p-1}}{p-1}\right)$. Prove the *Kummer–Takagi* formula

$$(\varepsilon, \eta)_p = \zeta_p^{\gamma}, \quad \text{where } \gamma = \sum_{i=1}^{p-1} (-1)^i l_i(g \circ E) l_{p-i}(f \circ E).$$

3. ($\diamond$) (*Biquadratic formula*)
Let $F = \mathbb{Q}_2(i)$, $i^2 = -1$. Show that if $\alpha, \beta \equiv 1 \mod (i-1)^3$, then

$$(\alpha, \beta)_4 = (-1)^{\frac{\alpha - 1}{(i-1)^3} \frac{\beta - 1}{(i-1)^3}}.$$

4. ($\diamond$) Let $F = \mathbb{Q}_p(\zeta_{p^n})$, where ζ_{p^n} is a p^nth primitive root of unity, $p > 2$. Let $\pi_n = \zeta_{p^n} - 1$; then π_n is prime in F, see (1.3) Ch. IV. Put $\mathrm{Tr} = \mathrm{Tr}_{F/\mathbb{Q}_p}$.
   a) Prove the *Artin–Hasse* formulas

$$(\zeta_{p^n}, \beta)_{p^n} = \zeta_{p^n}^{\mathrm{Tr}(\log \beta)/p^n} \qquad \text{for} \quad \beta \equiv 1 \mod \pi_n$$

   and

$$(\beta, \pi_n)_{p^n} = \zeta_{p^n}^{\mathrm{Tr}(\zeta_{p^n} \pi_n^{-1} \log \beta)/p^n} \qquad \text{for} \quad \beta \equiv 1 \mod \pi_n.$$

   b) Prove the *Artin–Hasse–Iwasawa* formula

$$(\alpha, \beta)_{p^n} = \zeta_{p^n}^{\mathrm{Tr}(\zeta_{p^n} \log \alpha \, \mathrm{Dlog}\, \beta)/p^n}$$

   for $\alpha \equiv 1 \mod \pi_n^2$, $\beta \equiv 1 \mod \pi_n$.

5. ($\diamond$) Let a primitive p^nth root of unity ζ_{p^n} belong to F and $n \geqslant 2$ if $p = 2$. Let π be a prime element in F. Put $\mathrm{Tr} = \mathrm{Tr}_{F/\mathbb{Q}_p}$. Prove the *Sen* formulas

$$(\alpha, \pi)_{p^n} = \zeta_{p^n}^{\mathrm{Tr}\left(\frac{\zeta_{p^n}}{f'(\pi)\pi} \log \alpha\right)/p^n} \qquad \text{for} \quad \alpha \equiv 1 \mod (\pi(\zeta_{p^n} - 1)^2)$$

and

$$(\alpha, \beta)_{p^n} = \zeta_{p^n}^{\mathrm{Tr}\left(\frac{\zeta_{p^n}}{f'(\pi)} \frac{g'(\pi)}{g(\pi)} \log \alpha\right)/p^n} \qquad \text{for } \alpha \equiv 1 \bmod ((\zeta_{p^n} - 1)^2), \beta \in U_F,$$

where $f(X), g(X)$ are arbitrary polynomials over the ring of integers $\mathcal{O}_0$ of $F_0 = F \cap \mathbb{Q}_p^{\mathrm{ur}}$ such that $f(\pi) = \zeta_{p^n}$, $g(\pi) = \beta$ (see also [Sen 3]). This formula was deduced by *Sh. Sen* using in particular J. Tate's theory [T2], see (6.5) Ch. IV.

6. ($\diamond$) Let $F = \mathbb{Q}_p(\zeta_p)$, where ζ_p is a pth primitive root of unity, $p > 2$.
   a) Let w be a root of $X^p + pX$ in F such that $w \equiv \pi = \zeta_p - 1 \mod \pi^2$. Let $\pi = f(w)$ for some $f(X) \in X\mathbb{Z}_p[[X]]$. Show that

$$\lambda(X) = \log(1 + f(X)) \equiv X \mod X^p,$$
$$(e^{a\lambda(X)} - 1)^{-1}(e^{a\lambda(X)} - 1)' \equiv (e^{aX} - 1)^{-1}(e^{aX} - 1)' \mod X^{p-1}.$$

   b) Put $\eta_i = 1 + f(w^i)$ for $1 \leqslant i \leqslant p$. Let σ be a generator of $G = \mathrm{Gal}(F/\mathbb{Q}_p)$. Show that (η_i) form a $\mathbb{Z}/p\mathbb{Z}[G]$-basis of $U_{1,F}/U_{1,F}^p$ and $\sigma(\eta_i) = \eta_i^{a^i}$, where the element

a belongs to the set of multiplicative representatives in F and is determined by the condition $\sigma(w) = aw$.

c) Show that

$$(\eta_i, \eta_j)_p = \begin{cases} 1 & \text{if } \ i+j \neq p, \\ \zeta_p^j & \text{if } \ i+j = p. \end{cases}$$

d) Let

$$u = \prod_i (\zeta_p^{a_i} - 1)^{n_i},$$

where a_i is relatively prime to p, $\prod_i a_i^{n_i} \equiv 1 \mod p$, $\sum_i n_i = 0$. The unit u is called cyclotomic. Using *Bernoulli*'s numbers B_k, determined from the equality

$$(e^{aX} - 1)^{-1}(e^{aX} - 1)' = a + \sum_{k \geqslant 0} \frac{1}{k!} B_k a^k X^{k-1},$$

show that

$$(\log u(X))' \equiv \frac{1}{2} \sum_i n_i a_i + \sum_{k \geqslant 2} \frac{B_k}{k!} \left(\sum_i n_i a_i^k \right) X^{k-1} \mod X^{p-1},$$

where $u(X) \in \mathbb{Z}_p[[X]]$ such that $u = u(w)$.

e) Show that if u_1, u_2 are cyclotomic units, then $(u_1, u_2)_p = 1$.

f) Introduce the cyclotomic units

$$u_k = \prod_{a=1}^{p-1} \left(\zeta_p^{a(1-g)/2} \frac{\zeta_p^{ag} - 1}{\zeta_p^g - 1} \right)^{a^{p-1-k}}, \quad 2 \leqslant k \leqslant p-3,$$

where the integer g is such that $g^i \not\equiv 1 \mod p$ for $1 \leqslant i < p-1$. Using d), show that

$$(\log u_k(X))' \equiv -\frac{B_k}{k!} g^k X^{k-1} \mod (p, X^{p-1}),$$

where $u_k(X) \in \mathbb{Z}_p[[X]]$, $u_k(w) = u_k$.

g) Show that if $u_k \in F^{*p}$, then B_k is divisible by p. (Hint: Consider $(u_k, 1+w^{p-k}u)_p$ for $u \in U_F$).

CHAPTER 8

Explicit Formulas for Hilbert Pairings on Formal Groups

The method of the previous chapter possesses a valuable property: it can be relatively easily applied to derive explicit formulas for various generalizations of the Hilbert symbol. This chapter explains how to establish explicit formulas for the generalized Hilbert pairing associated to a formal group of *Lubin–Tate* type or more generally of *Honda* type. Section 1 briefly recalls the theory of Lubin–Tate groups and their applications to local class field theory. In section 2 we discuss for Lubin–Tate formal groups a generalization of the exponential and logarithm maps E_X and l_X of Ch. VI and the arithmetic of the points of formal module. Then we describe explicit formulas for the Hilbert pairing. In section 3 we discuss the arithmetic and explicit formulas in the case of Honda formal groups.

The presentation in this chapter is more concise than in the rest of the book.

1. Formal Groups

(1.1). Let A be a commutative ring with unity. A formal power series $F(X,Y)$ over A is said to determine the commutative *formal group* F over A if

$$\begin{aligned} &F(X,0) = F(0,X) = X, && \\ &F(F(X,Y),Z) = F(X,F(Y,Z)) && \text{(associativity)},\\ &F(X,Y) = F(Y,X) && \text{(commutativity)}. \end{aligned}$$

Natural examples of such formal groups are the additive formal group

$$F_+(X,Y) = X + Y$$

and the multiplicative formal group

$$F_\times(X,Y) = X + Y + XY = (1+X)(1+Y) - 1.$$

Other examples will be exposed below and in Exercises. The definition implies that $F(X,Y) = X + Y + \sum_{i+j\geqslant 2} a_{ij}X^iY^j,\ a_{ij} \in A$.

A formal power series $f(X) \in XA[[X]]$ is called a *homomorphism* from a formal group F to a formal group G if

$$f(F(X,Y)) = G(f(X), f(Y)).$$

f is called an isomorphism if there exists a series $g = f^{-1}$ inverse to it with respect to composition, i.e. such that $(f \circ g)(X) = (g \circ f)(X) = X$. The set $\mathrm{End}_A(F)$ of all homomorphisms of F to F has a structure of a ring:

$$f(X) \oplus_F g(X) = F(f(X), g(X)),$$
$$f(X) \cdot g(X) = f(g(X)).$$

LEMMA. *There exists a uniquely determined homomorphism*

$$\mathbb{Z} \to \mathrm{End}_A(F): \quad n \to [n]_F.$$

Proof. Put $[0]_F(X) = 0$, $[1]_F(X) = X$, $[n+1]_F(X) = F([n]_F(X), X)$ for $n \geqslant 0$. Now we will verify that there exists a formal power series $[-1]_F(X) \in XA[[X]]$ such that $F(X, [-1]_F(X)) = 0$. Put $\varphi_1(X) = -X$ and assume that

$$F(X, \varphi_i(X)) \equiv 0 \mod \deg i+1 \quad \text{for} \quad 1 \leqslant i \leqslant m.$$

Let $F(X, \varphi_m(X)) \equiv c_{m+1}X^{m+1} \mod \deg m+2$, $c_{m+1} \in A$. Then for

$$\varphi_{m+1}(X) = \varphi_m(X) - c_{m+1}X^{m+1}$$

we obtain

$$F(X, \varphi_{m+1}(X)) = X + \varphi_m(X) - c_{m+1}X^{m+1} + \sum_{i+j \geqslant 2} a_{ij}\varphi_m(X)^j$$
$$\equiv F(X, \varphi_m(X)) - c_{m+1}X^{m+1} \equiv 0 \mod \deg m+2.$$

The limit of $\varphi_m(X)$ in $A[[X]]$ is the desired series $[-1]_F(X)$. Finally, we put $[n]_F(X) = F([n+1]_F(X), [-1]_F(X))$ for $n \leqslant -2$. This completes the proof. □

From now on let $A = K$ be a field of characteristic 0.

PROPOSITION. *Any formal group F over K is isomorphic to the additive group F_+, i.e. there exists a formal power series $\lambda(X) \in XK[[X]]$, $\lambda(X) \equiv X \mod \deg 2$ such that*

$$F(X, Y) = \lambda^{-1}(\lambda(X) + \lambda(Y)).$$

Proof. Denote the partial derivative $\frac{\partial F}{\partial Y}(X, Y)$ by $F_2'(X, Y)$. First we show that

$$F_2'(F(X, Y), 0) = F_2'(X, Y)F_2'(Y, 0).$$

To do this, we write

$$F_2'(X, F(Y, Z))F_2'(Y, Z) = \frac{\partial}{\partial Z}F(X, F(Y, Z)) = \frac{\partial}{\partial Z}F(F(X, Y), Z),$$

and put $Z = 0$. Now let $\lambda(X) = X + \sum_{i \geqslant 2} c_i X^i$ be such that

$$\lambda'(X) = 1 + \sum_{n \geqslant 2} nc_n X^{n-1} = \frac{1}{F_2'(X, 0)} = \frac{1}{1 + X + \sum_{i \geqslant 1} a_{i1} X^i}.$$

Then

$$\frac{\partial}{\partial Y}\lambda(F(X,Y)) = \frac{F_2'(X,Y)}{F_2'(F(X,Y),0)} = \frac{1}{F_2'(Y,0)} = \frac{\partial}{\partial Y}\lambda(Y).$$

Therefore,

$$\frac{\partial}{\partial Y}(\lambda(F(X,Y)) - \lambda(Y)) = 0$$

and $\lambda(F(X,Y)) = \lambda(Y)+g(X)$ for some formal power series $g(X) \in K[[X]]$. Setting $Y = 0$, we get $\lambda(X) = \lambda(F(X,0)) = g(X)$. Thus, we conclude that

$$F(X,Y) = \lambda^{-1}(\lambda(X) + \lambda(Y)).\square$$

The series $\lambda(X)$ is called the *logarithm* of the formal group F. We will denote it by $\log_F(X)$. The series inverse to it with respect to composition is denoted by $\exp_F(X)$. Then $F(X,Y) = \exp_F(\log_F(X) + \log_F(Y))$.

The theory of formal groups is presented in [Fr], [Haz3].

(1.2). From now on we assume that K is a local number field. For such a field the Lubin–Tate formal groups play an important role. Let $\mathcal{F}_\pi$ denote the set of formal power series $f(X) \in \mathcal{O}_K[[X]]$ such that $f(X) \equiv \pi X \mod \deg 2$, $f(X) \equiv X^q \mod \pi$, where π is a prime element in K and q is the cardinality of the residue field $\overline{K}$. The following assertion makes it possible to deduce a number of properties of the Lubin–Tate formal groups.

LEMMA. *Let $f(X), g(X) \in \mathcal{F}_\pi$ and $\alpha_i \in \mathcal{O}_K$ for $1 \leqslant i \leqslant m$. Then there exists a formal power series $h(X_1, \dots, X_m) \in K[[X_1, \dots, X_m]]$ uniquely determined by the conditions*:

$$h(X_1, \dots, X_m) \equiv \alpha_1 X_1 + \dots + \alpha_m X_m \mod \deg 2,$$
$$f(h(X_1, \dots, X_m)) = h(g(X_1), \dots, g(X_m)).$$

Proof. It is immediately carried out putting $h_1 = \alpha_1 X_1 + \dots + \alpha_m X_m$ and constructing polynomials $h_i \in K[X_1, \dots, X_m]$ such that

$$h_i \equiv h_{i-1} \mod \deg i,$$
$$f(h_i(X_1, \dots, X_m)) \equiv h_i(g(X_1), \dots, g(X_m)) \mod \deg i + 1.$$

Then $h = \lim h_i$ is the desired series. □

PROPOSITION. *Let $f(X) \in \mathcal{F}_\pi$. Then there exists a unique formal group $F = F_f$ over $\mathcal{O}_K$ such that*

$$F_f(f(X), f(Y)) = f(F_f(X,Y)).$$

For each $\alpha \in \mathcal{O}_K$ there exists a unique $[\alpha]_F \in \operatorname{End}_{\mathcal{O}_K}(F)$ such that

$$[\alpha]_F(X) \equiv \alpha X \mod \deg 2.$$

The map $\mathcal{O}_K \to \mathrm{End}_{\mathcal{O}_K}(F)$: $\alpha \to [\alpha]_F$ *is a ring homomorphism, and* $f = [\pi]_F$. *If* $g(X) \in \mathcal{F}_\pi$ *and* $G = F_g$ *is the corresponding formal group, then* F_f *and* F_g *are isomorphic over* $\mathcal{O}_K$, *i.e., there is a series* $\rho(X) \in K[[X]]$, $\rho(X) \equiv X \mod \deg 2$, *such that*

$$\rho(F_f(X,Y)) = F_g(\rho(X), \rho(Y)).$$

Proof. All assertions follow from the preceding Lemma. For instance, there exists a unique $F_f(X,Y) \in K[[X,Y]]$ such that

$$F_f(X,Y) \equiv X + Y \mod \deg 2, \qquad F_f(f(X), f(Y)) = f(F_f(X,Y)).$$

Then $F_f(X,0) = F_f(0,X) = X$. Both series $F_f(X, F_f(Y,Z))$ and $F_f(F_f(X,Y),Z)$ satisfy the conditions for h:

$$h(X,Y,Z) \equiv X + Y + Z \mod \deg 2, \qquad h(f(X), f(Y), f(Z)) = f(h(X,Y,Z)).$$

Therefore, by the Lemma $F_f(X, F_f(Y,Z)) = F_f(F_f(X,Y),Z)$. In the same way we get $F_f(X,Y) = F_f(Y,X)$. This means that F_f is a formal group. □

The formal group F_f is called a *Lubin–Tate formal group*. Note that the multiplicative formal group $F_\times$ is a Lubin–Tate group for $\pi = p$.

(1.3). Let $F = F_f$, $f \in \mathcal{F}_\pi$, be a Lubin–Tate formal group over $\mathcal{O}_K$, K a local number field. Let L be the completion of an algebraic extension over K. On the set $\mathcal{M}_L$ of elements on which the valuation takes positive values one can define the structure of $\mathcal{O}_K$-module $F(\mathcal{M}_L)$:

$$\alpha +_F \beta = F(\alpha, \beta), \qquad a \cdot \alpha = [a]_F(\alpha), \qquad a \in \mathcal{O}_K, \alpha, \beta \in \mathcal{M}_L.$$

Let κ_n denote the group of π^n-division points:

$$\kappa_n = \{\alpha \in \mathcal{M}_{K^{\mathrm{sep}}} : [\pi^n]_F(\alpha) = 0\}.$$

It can be shown (see Exercise 5) that κ_n is a free $\mathcal{O}_K/\pi^n\mathcal{O}_K$-module of rank 1, $\mathcal{O}_K/\pi^n\mathcal{O}_K$ is isomorphic to $\mathrm{End}_{\mathcal{O}_K}(\kappa_n)$, and $U_K/U_{n,K}$ is isomorphic to $\mathrm{Aut}_{\mathcal{O}_K}(\kappa_n)$. Define the *field of* π^n*-division points* by

$$L_n = K(\kappa_n).$$

Then one can prove (see Exercise 6) that L_n/K is a totally ramified abelian extension of degree $q^{n-1}(q-1)$ and $\mathrm{Gal}(L_n/K)$ is isomorphic to $U_K/U_{n,K}$. Put $K_\pi = \bigcup_{n\geqslant 1} L_n$ and let Ψ_K be the reciprocity map (see section 4 Ch. IV).

The significance of the Lubin–Tate groups for class field theory is expressed by the following

THEOREM. *The field* L_n *is the class field of* $\langle\pi\rangle \times U_{n,K}$ *and the field* K_π *is the class field of* $\langle\pi\rangle$.

The group $\operatorname{Gal}(K^{\mathrm{ab}}/K)$ *is isomorphic to the product* $\operatorname{Gal}(K^{\mathrm{ur}}/K) \times \operatorname{Gal}(K_\pi/K)$ *and*

$$\Psi_K(\pi^a u)(\xi) = [u^{-1}]_F(\xi) \qquad \text{for} \quad \xi \in \bigcup_{n\geqslant 1} \kappa_n,\ a \in \mathbb{Z},\ u \in U_K.$$

See Exercise 7.

Exercises.

1. a) Let $A = \mathbb{F}_p[Z]/(Z^2)$. Show that

$$F(X,Y) = X + Y + ZXY^p$$

determines a noncommutative formal group over A.

b) Let A be a commutative ring with unity and let 2 be invertible in A. Show that

$$F_\alpha(X,Y) = \frac{X\sqrt{(1-Y^2)(1-\alpha^2Y^2)} + Y\sqrt{(1-X^2)(1-\alpha^2X^2)}}{1+\alpha^2X^2Y^2}$$

with $\alpha \in A$, determines a formal group over A (this is the addition formula for the Jacobi functions for elliptic curves).

c) Let $F(X,Y) \in \mathbb{Z}[X,Y]$. Show that F determines a formal group over $\mathbb{Z}$ if and only if $F(X,Y) = X + Y + \alpha XY$ for some $\alpha \in \mathbb{Z}$.

2. a) Show that $\log_F(X) = \sum_{n\geqslant 1} d\frac{a_n}{n} X^n$ and $\exp_F(X) = \sum_{n\geqslant 1} d\frac{b_n}{n!} X^n$ for some $a_n \in \mathcal{O}_K$, $b_n \in \mathcal{O}_K$.

b) Let F be a Lubin–Tate formal group over $\mathcal{O}_K$. Show that $\log_F$ induces an isomorphism of $\mathcal{O}_K$-module $F(\mathcal{M}_K^m)$ onto $\mathcal{O}_K$-module $F_a(\mathcal{M}_K^m)$, where m is an integer, $m > v_K(p)/(p-1)$.

c) Let F be as in b). Let $\mathcal{M}$ be the maximal ideal of the completion of the separable closure of K. Show that the kernel of the homomorphism $F(\mathcal{M}) \to K^{\mathrm{sep}}$ induced by $\log_F$ coincides with $\kappa = \cup \kappa_n$.

3. Show that the homomorphism $\mathcal{O}_K \to \operatorname{End}_{\mathcal{O}_K}(F_f)$ of Proposition (1.2) is an isomorphism.

4. a) Let F be a formal group over $\mathcal{O}_K$ and π a prime element in K. Assume that $\log_F(X) - \pi^{-1}\log_F(X^q) \in \mathcal{O}_K[[X]]$. Put

$$f(X) = \exp_F(\pi \log_F(X)), \quad f_i(X) = f(X)^i - X^{qi}, \quad \text{for} \quad i \geqslant 1.$$

Show that if $\log_F(X) = \sum_{i\geqslant 1} c_i X^i$, $c_1 = 1$, then

$$\sum_{i\geqslant 1} c_i f_i(X) \equiv 0 \mod \pi.$$

Deduce that $f_1(X) \equiv 0 \mod \pi$. Since $f(X) \equiv \pi X \mod \deg 2$, this means that $f \in \mathcal{F}_\pi$ and F is a Lubin–Tate formal group.

b) Show that the series

$$\log_{F_{ah}}(X) = X + \frac{X^q}{\pi} + \frac{X^{q^2}}{\pi^2} + \dots$$

determines the Lubin–Tate formal group

$$F_{ah}(X,Y) = \log_{F_{ah}}^{-1}(\log_{F_{ah}}(X) + \log_{F_{ah}}(Y))$$

over $\mathcal{O}_K$.

c) Using Proposition (1.2), show that if $F = F_f$ is a Lubin–Tate formal group over $\mathcal{O}_K$ and $f \in \mathcal{F}_\pi$, then the series $\mathcal{E}_\pi(X) = \exp_F(\log_{F_{ah}}(X))$ belongs to $\mathcal{O}_K[[X]]$ and determines an isomorphism of F_0 onto F. The series $\mathcal{E}_\pi(X)$ is a generalization of the Artin–Hasse function considered in (9.1) Ch. I.

d) Show using Lemma (7.2) Ch. I that if F is as in c), then $\log_F(X) - \pi^{-1}\log_F(X^q) \in \mathcal{O}_K[[X]]$. Thus, a formal group F over $\mathcal{O}_K$ is a Lubin–Tate formal group over $\mathcal{O}_K$ if and only if $\log_F(X) - \pi^{-1}\log_F(X^q) \in \mathcal{O}_K[[X]]$.

5. a) Let $f, g \in \mathcal{F}_\pi$. Show that κ_n associated to f is isomorphic to κ_n associated to g. Taking $g = \pi X + X^q$ show that $|\kappa_n| = q^n$.

b) Let $\xi \in \kappa_n \setminus \kappa_{n-1}$. Using the map $\mathcal{O}_K \to \kappa_n$, $a \mapsto [a]_F(\xi)$ show that κ_n is isomorphic to $\mathcal{O}_K/\pi^n\mathcal{O}_K$.

c) Using the map $\mathcal{O}_K \to \operatorname{End}_{\mathcal{O}_K}(\kappa_n)$, $a \mapsto (\xi \mapsto [a]_F(\xi))$ show that $\mathcal{O}_K/\pi^n\mathcal{O}_K$ is isomorphic to $\operatorname{End}_{\mathcal{O}_K}(\kappa_n)$ and $U_K/U_{n,K}$ is isomorphic to $\operatorname{Aut}_{\mathcal{O}_K}(\kappa_n)$.

6. Let $\xi \in \kappa_n \setminus \kappa_{n-1}$. Define the field of π^n-division points $L_n = K(\xi)$. Using Exercise 5 show that L_n/K is a totally ramified abelian extension of degree $q^{n-1}(q-1)$, $N_{L_n/L}(-\xi) = \pi$ and $\operatorname{Gal}(L_n/K)$ is isomorphic to $U_K/U_{n,K}$.

7. a) Define a linear operator ϕ acting on power series with coefficients in the completion of the ring of integers $\widehat{\mathcal{O}}$ of the maximal unramified extension of K as $\phi(\sum a_i X^i) = \sum a_i^{\varphi_K} X^i$. Let $u \in U_K$ and $u = v^{\phi-1}$ for some $v \in \widehat{\mathcal{O}}^*$ according to Proposition (1.8) in Ch. IV. Let $f \in \mathcal{F}_\pi$ and $g \in \mathcal{F}_{\pi u}$. Using the method of (1.2) show that there is a unique $h(X) \in \widehat{\mathcal{O}}[[X]]$ such that $h(X) \equiv vX \mod \deg 2$ and $f \circ h = h^\phi \circ g$.

b) Let $u \in U_K$ and let $\sigma \in \operatorname{Gal}(L_n/K)$ be such that $\sigma(\xi) = [u^{-1}]_F(\xi)$. Denote by Σ the fixed field of $\widetilde{\sigma} = \varphi_{L_n}\sigma \in \operatorname{Gal}(L_n^{\mathrm{ur}}/K)$. Show that Σ is the field of π^n-division points of $\mathcal{F}_g$.

c) Let h be as in a). Show that $h(\xi)$ is a prime element of Σ. Deduce using sections 2 and 3 Ch. IV that

$$\Upsilon_{L_n/K}(\sigma) \equiv N_{\Sigma/K}(-h(\xi)) = \pi u \equiv u \mod N_{L_n/K} L_n^*.$$

Thus, $\Psi_{L_n/K}(\pi^a u)(\xi) = [u^{-1}]_F(\xi)$.

d) Deduce that $N_{L_n/K}L_n^* = \langle\pi\rangle \times U_{n,K}$.

e) Show that $K^{\mathrm{ab}} = K^{\mathrm{ur}}K_\pi$ where $K_\pi = \bigcup_{n\geqslant 1} L_n$.

2. Generalized Hilbert Pairing for Lubin–Tate Groups

In this section K is a local number field with residue field $\mathbb{F}_q$, π is a prime element in K, $F = F_f$ is a Lubin–Tate formal group over $\mathcal{O}_K$ for $f \in \mathcal{F}_\pi$. Let L/K be a finite extension such that the $\mathcal{O}_K$-module κ_n of π^n -division points is contained in L. Let $\mathcal{O}_T$ be the ring of integers of $T = L \cap K^{\mathrm{ur}}$ and let $\mathcal{O}_0$ be the ring of integers of $L \cap \mathbb{Q}_p^{\mathrm{ur}}$. Put $e = e(L|\mathbb{Q}_p)$, $e_0 = e(K|\mathbb{Q}_p)$.

(2.1). Define the *generalized Hilbert pairing*

$$(\cdot,\cdot)_F = (\cdot,\cdot)_{F,n}\colon L^* \times F(\mathcal{M}_L) \to \kappa_n$$

by the formula

$$(\alpha,\beta)_F = \Psi_F(\alpha)(\gamma) +_F [-1]_F(\gamma),$$

where $\gamma \in F(\mathcal{M}_{K^{\text{sep}}})$ is such that $[\pi^n]_F(\gamma) = \beta$. If $F = F_\times$, $\pi = p$, then $(\alpha,\beta)_{F,n}$ coincides with the Hilbert symbol $(\alpha, 1+\beta)_{p^n}$.

PROPOSITION. *The generalized Hilbert pairing has the following properties:*

(1) $(\alpha_1\alpha_2,\beta)_F = (\alpha_1,\beta)_{F,n} +_F (\alpha_2,\beta)_F$, $\quad(\alpha,\beta_1 +_F \beta_2)_F = (\alpha,\beta_1)_F +_F (\alpha,\beta_2)_F$;

(2) $(\alpha,\beta)_F = 0$ *if and only if* $\alpha \in N_{L(\gamma)/L}L(\gamma)^*$, *where* $[\pi^n]_F(\gamma) = \beta$;

(3) $(\alpha,\beta)_F = 0$ *for all* $\alpha \in L^*$ *if and only if* $\beta \in [\pi^n]_F\, F(\mathcal{M}_L)$;

(4) $(\alpha,\beta)_F$ *in the field* E *coincides with* $(N_{E/L}(\alpha),\beta)_F$ *in the field* L *for* $\alpha \in E^*$, $\beta \in F(\mathcal{M}_L)$, *where* E *is a finite extension of* L;

(5) $(\sigma\alpha,\sigma\beta)_F$ *in the field* σL *coincides with* $\sigma(\alpha,\beta)_F$, *where* $(\alpha,\beta)_F$ *is considered to be taken in the field* L, $\sigma \in \mathrm{Gal}(K^{\text{sep}}/K)$.

Proof. It is carried out similarly to the proof of Proposition (5.1) Ch. IV. □

Now we shall briefly discuss a generalization of the relevant assertions of Chapters VI, VII to the case of formal groups.

(2.2). For α_i in the completion of the maximal unramified extension K^{ur} of K put

$$\triangle\left(\sum \alpha_i X^i\right) = \sum \varphi_K(\alpha_i) X^{qi},$$

where φ_K is the continuous extension of the Frobenius automorphism of K, and q is the cardinality of $\overline{K}$. Let $\widehat{\mathcal{O}}$ be the ring of integers in the completion of K^{ur}. Let $F(X\widehat{\mathcal{O}}[[X]])$ denote the $\mathcal{O}_K$-module of formal power series in $X\widehat{\mathcal{O}}[[X]]$ with respect to operations

$$f +_F g = F(f,g), \quad a\cdot f = [a]_F(f), \quad a \in \mathcal{O}_K.$$

Analogs of the maps E_X, l_X of section 2 Ch. VI are the following $E_F = E_{F,X}$, $l_F = l_{F,X}$:

$$E_F(f(X)) = \exp_F\left(\left(1 + \frac{\triangle}{\pi} + \frac{\triangle^2}{\pi^2} + \dots\right) f(X)\right), \qquad f(X) \in X\widehat{\mathcal{O}}[[X]]$$

$$l_F(g(X)) = \left(1 - \frac{\triangle}{\pi}\right)\left(\log_F(g(X))\right), \qquad g(X) \in X\widehat{\mathcal{O}}[[X]].$$

Then E_F is a $\mathcal{O}_K$-isomorphism of $X\widehat{\mathcal{O}}[[X]]$ onto $F(X\widehat{\mathcal{O}}[[X]])$ and l_F is the inverse one. This assertion can be proved in the same way as Proposition (2.2) Ch. VI, using the equality

$$E_F(\theta X^m) = \exp_F\bigl(\theta X^m + \frac{(\theta X^m)^q}{\pi} + \dots\bigr) = \exp_F(\log_{F_{ah}}(\theta X^m)),$$

where θ is an l th root of unity, $(l,p)=1$, and $\log_{F_{ah}}$ is the logarithm of the Lubin–Tate formal group F_{ah}, defined in Exercise 4 section 1.

(2.3). Let Π be a prime element in L. Let ξ be a generator of the $\mathcal{O}_K$-module κ_n. To ξ we relate a series $z(X) = c_1X + c_2X^2 + \dots$, $c_i \in \mathcal{O}_T$, such that $z(\Pi) = \xi$. Put $s_m(X) = [\pi^m]_F(z(X))$, $s(X) = s_n(X)$.

An element $\alpha \in \mathcal{M}_L$ is called *π^n-primary* if the extension $L(\gamma)/L$ is unramified where $[\pi^n](\gamma) = \alpha$. As in sections 3 and 4 of Ch. VI one can prove that

$$\omega(a) = E_F(a\, s(X))|_{X=\Pi}, \qquad a \in \mathcal{O}_T,$$

is a π^n-primary element and, moreover,

$$(\pi, \omega(a))_F = [\operatorname{Tr} a](\xi),$$

where $\operatorname{Tr} = \operatorname{Tr}_{T/K}$ (see [V3]). The $\mathcal{O}_K$-module Ω of π^n-primary elements is generated by an element $\omega(a_0)$ with $\operatorname{Tr} a_0 \notin \pi\mathcal{O}_K$, $a_0 \in \mathcal{O}_T$.

An analog of the Shafarevich basis considered in section 5 Ch. VI can be stated as follows: every element $\alpha \in F(\mathcal{M}_L)$ can be expressed as

$$\alpha = \sum_i {}_{(F)} E_F(a_iX^i)|_{X=\Pi} +_F \omega(a), \quad a_i, a \in \mathcal{O}_T,$$

where $1 \leqslant i < qe/(q-1)$, i is not divisible by q. The element α belongs to $[\pi^n]_F F(\mathcal{M}_L)$ if and only if $\operatorname{Tr} a \in \pi^n\mathcal{O}_K$, $a_i \in \pi^n\mathcal{O}_T$. There are also other forms of generalizations of the Shafarevich basis; see [V2–3].

(2.4). To describe formulas for the generalized Hilbert pairing, we introduce the following notions. For $a_i \in \mathcal{O}_0$ put

$$\delta\left(\sum a_iX^i\right) = \sum \varphi(a_i)X^{pi},$$

where $\varphi = \varphi_{\mathbb{Q}_p}$ is the Frobenius automorphism of $\mathbb{Q}_p$.

For the series $\alpha(X) = \theta X^m\varepsilon(X)$, where θ is a l th root of unity, l is relatively prime to p, $\varepsilon(X) \in 1 + X\mathcal{O}_0[[X]]$, put

$$l(\alpha(X)) = l(\varepsilon(X)) = \left(1 - \frac{\Delta}{q}\right)(\log\varepsilon(X)) = \frac{1}{q}\log\left(\frac{\alpha(X)^q}{\alpha(X)^\Delta}\right)$$

and

$$L(\alpha(X)) = (1 + \delta + \delta^2 + \dots)l(\alpha(X)).$$

Let $\alpha \in L^*$, $\beta \in F(\mathcal{M}_L)$. Let $\alpha = \alpha(X)|_{X=\Pi}$, $\beta = \beta(X)|_{X=\Pi}$, where $\alpha(X)$ is as just above, $\beta(X) \in X\mathcal{O}_T[[X]]$. Put

$$\Phi_{\alpha(X),\beta(X)} = \frac{\alpha(X)'}{\alpha(X)} l_F(\beta(X)) - l(\alpha(X))\Big(\frac{\Delta}{q}\log_F(\beta(X))\Big)'$$

and

$$\Phi^{(1)}_{\alpha(X),\beta(X)} = \frac{2}{\pi}\left(\frac{\delta}{q} m l_F(\beta(X)) + \frac{L(\alpha(X))X\varepsilon(X)}{(X\varepsilon(X))'}(\log_F(\beta(X)))'\right)',$$

$$\Phi^{(2)}_{\alpha(X),\beta(X)} = \Big(l(\alpha(X))(\Delta + \Delta^2 + \Delta^3 + \dots) l_F(\beta(X))\Big)',$$

$$\Phi^{(3)}_{\alpha(X),\beta(X)} = \Big(\frac{\Delta}{2}\big(L(\alpha(X))(1 + \Delta + \Delta^2 + \dots) l_F(\beta(X))\big)\Big)'$$

(concerning the form of $\Phi^{(3)}$ see Exercise 2 section 2 Ch. VII keeping in mind the restriction on the series $\alpha(X), \beta(X)$ above).

Similarly to (2.1) Ch. VII we can introduce an appropriate pairing $\langle\cdot,\cdot\rangle_X$ on power series using the series $1/s(X)$ instead of $V(X)$. Similarly to (2.2) Ch. VII we can introduce a pairing $\langle\cdot,\cdot\rangle_\pi$ on $L^* \times F(\mathcal{M}_L)$ and then prove that it coincides with the generalized Hilbert pairing.

Thus, there are the following explicit formulas for the generalized Hilbert pairing. If $p > 2$ then

$$\boxed{(\alpha,\beta)_F = [\operatorname{Tr}\operatorname{res}_X \Phi_{\alpha(X),\beta(X)}/s(X)](\xi_n)}$$

If $p = 2$ and $q > 2$, then

$$\boxed{(\alpha,\beta)_F = [\operatorname{Tr}\operatorname{res}_X(\Phi_{\alpha(X),\beta(X)} + \Phi^{(1)}_{\alpha(X),\beta(X)})/s(X)](\xi_n)}$$

If $p = 2$, $q = 2$, $e_0 > 1$, then

$$\boxed{(\alpha,\beta)_F = [\operatorname{Tr}\operatorname{res}_X(\Phi_{\alpha(X),\beta(X)} + \Phi^{(2)}_{\alpha(X),\beta(X)})r(X)/s(X)](\xi_n)}$$

If $p = 2$, $q = 2$, $e_0 = 1$, then

$$\boxed{(\alpha,\beta)_F = [\operatorname{Tr}\operatorname{res}_X(\Phi_{\alpha(X),\beta(X)} + \Phi^{(3)}_{\alpha(X),\beta(X)})r(X)/s(X)](\xi_n)}$$

see [V2–4] (odd p) and [VF], [Fe2] ($p = 2$).

Here for $q = 2$, $e_0 > 1$, we put $r(X) = 1 + \pi^{n-1} r_0(X)$ and the polynomial $r_0(X)$ is determined by the congruence

$$\Delta^2 r_0 + (1 + (\pi^{n-1} - 1)s)\Delta r_0 + s r_0 \equiv (\Delta^2 s_{n-1} - \Delta s)/\pi^n \operatorname{modev}(\pi, \deg X^{4e})$$

(modev is as in (3.4) Ch. VI).

For $q = 2$, $e_0 = 1$, we put $r(X) = 1 + \pi^{n-1}\Delta r_0(X)$, where the polynomial $r_0(X)$ is determined by the congruence

$$\Delta^2 r_0 + (1 + (\pi^{n-1} - 1)s_{n-1})\Delta r_0 + s_{n-1} r_0 \equiv (\Delta s_{n-1} - s)/\pi^n \operatorname{modev}(\pi, \deg X^{2e}).$$

(2.5). REMARKS.

1. These formulas can be applied to deduce the theory of symbols on Lubin–Tate formal groups; see [V3]. For a review of different types of formulas see [V11].

2. If in the case $p > 2$ the series $\alpha(X)$ is chosen in $\mathcal{O}_0(X))^*$, then the series $1/s(X)$ should be replaced with $V(X) = 1/s(X) + c/(\pi^2 - \pi)$ where c is the coefficient of X^2 in $[\pi](X) = \pi X + cX^2 + \dots$. In particular, if $[\pi](X) = \pi X + X^q$, then $c = 0$ and $V(X) = 1/s(X)$.

3. In connection with Remark 4 in (5.3) Ch. VII we note that no syntomic theory related to formal groups, which could provide an interpretation of explicit formulas discussed in this chapter, is available so far.

3. Generalized Hilbert Pairing for Honda Groups

We assume in this section that $p > 2$. Let K be a local field with residue field of cardinality $q = p^f$ and L be a finite unramified extension of K. Let π be a prime element of K. In this section we put $\varphi = \varphi_K$ which differs from the notation in section 2.

(3.1). Let $\triangle$ be defined in the same way as in (2.2). The set of operators of the form $\sum_{i\geqslant 0} a_i \triangle^i$, where $a_i \in \mathcal{O}_L$, form a noncommutative ring $\mathcal{O}_L[[\triangle]]$ of series in $\triangle$ in which $\triangle\, a = a^{\varphi} \triangle$ for $a \in \mathcal{O}_L$.

DEFINITION. A formal group $F \in \mathcal{O}_L[[X, Y]]$ with logarithm $\log_F(X) \in L[[X]]$ is called a *Honda formal group* if

$$u \circ \log_F \equiv 0 \mod \pi$$

for some operator $u = \pi + a_1 \triangle + \dots \in \mathcal{O}_L[[\triangle]]$. The operator u is called *the type* of the formal group F.

Every 1-dimensional formal group over an unramified extension of $\mathbb{Q}_p$ is a Honda formal group [Hon].

Types u and v of a formal group F are called equivalent if $u = \varepsilon \circ v$ for some $\varepsilon \in \mathcal{O}_L[[\triangle]]$, $\varepsilon(0) = 1$.

Let F be of type u. Then $v = \pi + b_1 \triangle + \dots \in \mathcal{O}_L[[X]]$ is a type of F if and only if v is equivalent to u.

Using the Weierstrass preparation theorem for the ring $\mathcal{O}_L[[\triangle]]$, one can prove [Hon] that for every formal Honda group F there is a unique *canonical type*

$$(*) \qquad u = \pi - a_1 \triangle - \dots - a_h \triangle^h, \quad a_1, \dots, a_{h-1} \in \mathcal{M}_L, a_h \in \mathcal{O}_L^*.$$

This type determines the group F uniquely up to isomorphism. Here h is the *height* of F.

If F and G are Honda formal groups of types u and v respectively, then

$$\operatorname{Hom}_{\mathcal{O}_L}(F, G) = \{a \in \mathcal{O}_L : au = va\}, \quad \operatorname{End}_{\mathcal{O}_L}(F) = \mathcal{O}_K.$$

Along with $(*)$ we can use the following equivalent type

$$\widetilde{u} = \pi - a_h \vartriangle^h - a_{h+1} \vartriangle^{h+1} - \dots,$$

where $\widetilde{u} = C^{-1}u$, $C = 1 - \frac{a_1}{\pi} \vartriangle - \dots - \frac{a_{h-1}}{\pi} \vartriangle^{h-1}$, i.e.,

$$\begin{aligned}\widetilde{u} &= (\pi^{-1}(u + a_h \vartriangle^h))^{-1}u = \pi - (\pi^{-1}(u + a_h \vartriangle^h))^{-1}a_h \vartriangle^h \\ &= \pi - a_h \vartriangle^h - a_{h+1} \vartriangle^{h+1} - \dots .\end{aligned}$$

Now we state classification theorems [De1] that connect Honda formal groups with Lubin–Tate groups.

THEOREM 1. *Let F be a Honda formal group of type*

$$\widetilde{u} = \pi - a_h \vartriangle^h - a_{h+1} \vartriangle^{h+1} - \dots, \quad a_i \in \mathcal{O}_L,$$

where a_h is invertible in $\mathcal{O}_L$. Let $u = \pi - a_1 \vartriangle - \dots - a_{h-1} \vartriangle^{h-1} - a_h \vartriangle^h$ be the canonical type of F, $a_1, \dots, a_{h-1} \in \mathcal{M}_L$. Let $\lambda = \log_F$ be the logarithm of F. Put $\lambda_1 = B_1 \lambda^{\varphi^h}$, where

$$B_1 = 1 + \frac{a_{h+1}}{a_h} \vartriangle + \frac{a_{h+2}}{a_h} \vartriangle^2 + \dots$$

(i.e., $\widetilde{u} = \pi - a_h B_1 \vartriangle^h$). Then

(1) *λ_1 is the logarithm of the Honda formal group F_1 of type $\widetilde{u}_1 = a_h^{-1}\widetilde{u}a_h$ and of canonical type $u_1 = a_h^{-1}ua_h$;*

(2) $f = \left[\dfrac{\pi}{a_h}\right]_{F,F_1} \in \operatorname{Hom}_{\mathcal{O}_L}(F, F_1)$ *and* $f(X) \equiv X^{q^h} \mod \pi$.

EXAMPLES. 1. A formal Lubin–Tate group F has type $u = \pi - \vartriangle$, its height is $h = 1$ and $F_1 = F$.

2. A relative Lubin–Tate group F has type $u = \pi - a_1 \vartriangle$, where $a_1 = \pi/\pi'$, $h = 1$, and $F_1 = F^{\varphi}$.

THEOREM 2 (CONVERSE TO THEOREM 1). *Let $f \in \mathcal{O}_L[[X]]$ be a series satisfying relations*

$$f(X) \equiv X^{q^h} \mod \pi, \quad f(X) \equiv \frac{\pi}{a_h} X \mod \deg 2,$$

where a_h is an invertible element of $\mathcal{O}_L$. Let $u = \pi - a_1 \vartriangle - \dots - a_h \vartriangle^h$, where $a_1, \dots, a_{h-1} \in \mathcal{M}_L$. Let

$$C = 1 - \frac{a_1}{\pi} \vartriangle - \dots - \frac{a_{h-1}}{\pi} \vartriangle^{h-1}$$

and $\widetilde{u} = C^{-1}u = \pi - a_h \bigtriangleup^h - a_{h+1} \bigtriangleup^{h+1} - \dots$. *Then there exists a unique Honda formal group* F *of type* $\widetilde{u}$ *and of canonical type* u *such that* $f = \left[\dfrac{\pi}{a_h}\right]_{F,F_1}$ *is a homomorphism from* F *to the formal group* F_1 *defined and given by Theorem 1.*

REMARKS.

1. If λ and λ_1 are the logarithms of F and F_1 respectively, then

$$f = \left[\frac{\pi}{a_h}\right]_{F,F_1} = \lambda_1^{-1} \circ \left(\frac{\pi}{a_h}\right) \circ \lambda.$$

2. Theorem 2 can be viewed as a generalization of Proposition (1.2).

These theorems allow one to define on the set of Honda formal groups over the ring $\mathcal{O}_L$ the invertible operator $\mathcal{A}: F \to F_1$. Define the sequence of Honda formal groups

$$(**) \qquad F \xrightarrow{f} F_1 \xrightarrow{f_1} \dots \xrightarrow{f_{n-1}} F_n,$$

where $F_m = \mathcal{A}^m F$.

Let $\lambda_m = \log_{F_m}$ be the logarithm of F_m and let u_m be the canonical type of F_m. Put

$$(***) \qquad \begin{aligned} \pi_1 &= \pi/a_h, \pi_m = \pi_1^{\varphi^{h(m-1)}} = \pi/a_h^{\varphi^{h(m-1)}}, \\ \pi_1^{(m)} &= \prod_{i=1}^{m} \pi_i = \pi^m/a_h^{1+\varphi^h+\dots+\varphi^{h(m-1)}}. \end{aligned}$$

Then $u_m \circ \pi_1^{(m)} = \pi_1^{(m)} u$.

Denote $f^{(m)} = f_{m-1} \circ f_{m-2} \circ \dots \circ f_1 \circ f$. From Theorem 1 one can deduce that

$$f_{m-1}(X) \equiv \pi_m X \mod \deg 2, \quad f^{(m)}(X) \equiv \pi_1^{(m)} X \mod \deg 2.$$

(3.2). Define the generalized Hilbert pairing for a Honda formal group.

Let E be a finite extension of L which contains all elements of π^n-division points $\kappa_n = \ker[\pi^n]_F$.

Along with the *generalized Hilbert pairing*

$$(\cdot,\cdot)_F = (\cdot,\cdot)_{F,n}: E^* \times F(\mathcal{M}_E) \to \kappa_n, \quad (\alpha,\beta)_F = \Psi_E(\alpha)(\gamma) -_F \gamma,$$

where Ψ_E is the reciprocity map, γ is such that $[\pi^n]_F(\gamma) = \beta$, we also need another generalization that uses the homomorphism $f^{(n)}$:

$$\{\cdot,\cdot\}_F = \{\cdot,\cdot\}_{F,n}: E^* \times F(\mathcal{M}_E) \to \kappa_n, \quad \{\alpha,\beta\}_F = \Psi_E(\alpha)(\delta) -_F \delta,$$

where δ is such that $f^{(n)}(\delta) = \beta$. Then

$$(\alpha,\beta)_F = \{\alpha, [\pi_1^{(n)}/\pi^n]_{F,F_n}(\beta)\}_F.$$

We get the usual norm property for both $(\cdot,\cdot)_F$ and $\{\cdot,\cdot\}_F$.

(3.3). We introduce a generalization for Honda formal modules of the maps E_F, l_F defined in (2.2).

Let T be the maximal unramified extension of K in E.

Denote by $F(X\mathcal{O}_T[[X]])$ the $\mathcal{O}_K$-module whose underlying set is $X\mathcal{O}_T[[X]]$ and operations are given by

$$f +_F g = F(f,g); \quad a \cdot f = [a]_F(f), \quad a \in \mathcal{O}_K.$$

The class of isomorphic Honda formal groups F contains the canonical group F_{ah} of type

$$u = \pi - a_1 \triangle - \dots - a_h \triangle^h, \quad a_i, \dots, a_{h-1} \in \mathcal{M}_L, \quad a_h \in \mathcal{O}_L^*$$

with Artin–Hasse type logarithm

$$\log_{F_{ah}} = (u^{-1}\pi)(X) = X + \alpha_1 X^q + \alpha_2 X^{q^2} + \dots, \quad \alpha_i \in L.$$

Define the map E_F and its inverse l_F as follows:

$$E_F(g) = \log_F^{-1} \circ (1 + \alpha_1 \triangle + \alpha_2 \triangle^2 + \dots)(g)$$
$$l_F(g) = \left(1 - \frac{a_1}{\pi} \triangle - \dots - \frac{a_h}{\pi} \triangle^h\right)(\log_F \circ g),$$

where $g \in X\mathcal{O}_T[[X]]$.

We also need similar maps for the formal group $F_n = \mathcal{A}^n F$ with logarithm $\lambda_n = \log_{F_n}$ defined in the previous section. Let

$$u_n = \pi - b_1 \triangle - \dots - b_h \triangle^h$$

be the canonical type of F_n. Consider the canonical formal group F_b of type u_n whose logarithm is

$$\lambda_b = (u_n^{-1}\pi)(X) = X + \beta_1 X^q + \beta_2 X^{q^2} + \dots, \quad \beta_i \in L.$$

The groups F_n and F_b are isomorphic because they have the same type u_n. Now we define the functions

$$E_{F_n}(g) = \lambda_n^{-1} \circ (u_n^{-1}\pi)(g) = \lambda_n^{-1} \circ (1 + \beta_1 \triangle + \beta_2 \triangle^2 + \dots)(g)$$

$$l_{F_n}(g) = (u_n \pi^{-1})(\lambda_n \circ \psi) = \left(1 - \frac{b_1}{\pi} \triangle - \dots - \frac{b_h}{\pi} \triangle^h\right)(\lambda_n \circ g).$$

The functions E_F and l_F yield inverse isomorphisms between $X\mathcal{O}_T[[X]]$ and $F(X\mathcal{O}_T[[X]])$, and the functions E_{F_n} and l_{F_n} yield inverse isomorphisms between $X\mathcal{O}_T[[X]]$ and $F_n(X\mathcal{O}_T[[X]])$, see [De2].

(3.4). We discuss an analog of the Shafarevich basis for a Honda formal module.

Let Π be a prime element of E.

First we construct primary elements. An element $\omega \in F(\mathcal{M}_E)$ is called *π^n-primary* if the extension $E(\nu)/E$ is unramified, where $[\pi^n]_F(\nu) = \omega$.

The $\mathcal{O}_K$-module module κ_n has h generators [Hon], [De 2]. Fix a set of generators $\xi_1, \dots, \xi_h$. Let $z_i(X) \in \mathcal{O}_T[[X]]$ be the series corresponding to an expansion of ξ_i into a power series in Π, i.e. $z_i(\Pi) = \xi_i$. Similarly define $z_i(X)$.

Put

$$s^{(i)} = f^{(n)} \circ z_i(X), \quad 1 \leqslant i \leqslant h.$$

Fix an element $b \in \mathcal{O}_T$ and put $\widehat{b} = b + b^{\varphi} + \dots + b^{\varphi^{h-1}}$.

Let Tr be the trace map for the extension K_h/K where K_h is the unramified extension of K of degree h; note that $\widehat{b} \in K_h$.

PROPOSITION. *The element*

$$\omega_i(b) = E_{F_n}(\widehat{b}\lambda_n \circ s^{(i)})\big|_{X=\Pi}$$

is well-defined. It belongs to $F_n(\mathcal{M}_E)$, and it is π^n-primary. Moreover,

$$\{\Pi, \omega_i(b)\}_F = [\mathrm{Tr}\, b]_F(\xi_i).$$

See [De2], [DV2].

Further, let

$$g_0(X) = \pi_{n-1}X + X^{q^h}$$
$$g_{\rho,a}(X) = \pi_{n-1}X + \pi_{n-1}aX^{p^\rho} + X^{q^h}, \quad a \in \mathcal{O}_T, \quad 1 \leqslant \rho < fh.$$

Let u_{n-1} be the type of the formal group F_{n-1} from the sequence $(**)$. By Theorem 2 in (3.1) there exist unique Honda formal groups G_0 and $G_{\rho,a}$ of type u_{n-1} which correspond to $g_0(X)$ and $g_{\rho,a}(X)$ respectively. Then $\mathcal{A}G_0$ and $\mathcal{A}G_{\rho,a}$ are of the same type as F_n. Denote by $\mathcal{E}_n^0 \colon \mathcal{A}G_0 \to F_n$ and $\mathcal{E}_n^{\rho,a} \colon \mathcal{A}G_{\rho,a} \to F_n$ the corresponding isomorphisms.

THEOREM. *Let $\mathcal{R}$ be the set of multiplicative representatives in T. Elements*

$$\{\omega_i(b);\ b \in \mathcal{O}_T,\ 1 \leqslant i \leqslant h\},$$
$$\{\mathcal{E}_n^0(\theta\Pi^i);\ \theta \in \mathcal{R},\ 1 \leqslant i < q^h e/(q^h - 1),\ (i,p) = 1\},$$
$$\{\mathcal{E}_n^{\rho,a}(\theta\Pi^i);\ \theta \in \mathcal{R},\ a \in \mathcal{O}_T^*,\ 1 \leqslant \rho < fh,\ 1 \leqslant i < q^h e/(q^h - 1),\ (i,p) = 1\}$$

form a set of generators of the $\mathcal{O}_K$-module $F_n(\mathcal{M}_E)$. Furthermore,

$$\{\Pi, \mathcal{E}_n^0(\theta\Pi^i)\}_F = \{\Pi, \mathcal{E}_n^{\rho,a}(\theta\Pi^i)\}_F = 0, \quad \{\Pi, \omega_i(b)\}_F = [\mathrm{Tr}\, b]_F(z_i).$$

See [De2], [DV2].

(3.5). Similarly to the case of multiplicative groups discussed Ch. VII and the case of formal Lubin–Tate groups in section 2 one can introduce a pairing on formal power series, check its correctness and various properties and then prove that when X is specialized to Π it gives explicit formulas for the generalized Hilbert pairing.

For a monomial $d_iX^i \in T((X))$ put $\nu(d_iX^i) = v_T(d_i) + i/q^h$ where v_T is the discrete valuation of T. Denote by $\mathcal{L}$ the T-algebra of series

$$\mathcal{L} = \Big\{\sum_{i\in\mathbb{Z}} d_iX^i : d_i \in T, \quad \inf_i \nu(d_iX^i) > -\infty, \quad \lim_{i\to+\infty} \nu(d_iX^i) = +\infty\Big\}.$$

Since the $\mathcal{O}_K$-module κ_n has h generators, we are naturally led to work with $h\times h$ matrices. Denote the ring of integers of the maximal unramified extension of $\mathbb{Q}_p$ in E by $\mathcal{O}_0$.

THEOREM.

For $\alpha \in E^*$ let $\alpha(X)$ be a series in $\{X^i\theta\varepsilon(X) : \theta \in \mathcal{R}^*, \varepsilon \in 1 + X\mathcal{O}_0[[X]]\}$. For $\beta \in F(\mathcal{M}_E)$ let $\beta(X)$ be a series in $X\mathcal{O}_T[[X]]$ such that $\beta(\Pi) = \beta$.

The generalized Hilbert symbol $(\cdot,\cdot)_F$ is given by the following explicit formula:

$$(\alpha,\beta)_F = \sum_{j=1}^{h} {}_{(F)}[\operatorname{Tr}\operatorname{res}\Phi V_j]_F(\xi_j),$$

where $\Phi(X)V_j(X)$ belongs to $\mathcal{L}$, $V_j = A_j/\det A$, $1 \leqslant j \leqslant h$,

$$A = \begin{pmatrix} \pi^n\lambda\circ z_1(X) & \dots & \pi^n\lambda\circ z_h(X) \\ \pi^n\,\Delta\,(\lambda\circ z_1(X)) & \dots & \pi^n\,\Delta\,(\lambda\circ z_h(X)) \\ \dots & \dots & \dots \\ \pi^n\,\Delta^{h-1}\,(\lambda\circ z_1(X)) & \dots & \pi^n\,\Delta^{h-1}\,(\lambda\circ z_h(X)) \end{pmatrix},$$

A_j is the cofactor of the $(j,1)$-element of A,

$$\Phi = \frac{\alpha(X)'}{\alpha(X)} l_F(\beta(X)) - \frac{1}{\pi}\sum_{i=1}^{h} a_i\left(1 - \frac{\Delta^i}{q^i}\right)\big(\log\varepsilon(X)\big)\ \Delta^i\,(\lambda\circ\beta(X)).$$

See [DV2].

REMARKS. 1. The formula above can be simplified in the case of $n = 1$, see [BeV1].

2. The first explicit formula for the generalized Hilbert pairing for formal Honda group and arbitrary n in the case of odd p under some additional assumptions on the field E was obtained by *V A. Abrashkin* [Ab6] using the link between the Hilbert pairing and the Witt pairing via an auxiliary construction of a crystalline symbol as a generalization of his method in [Ab5] (see Remark 6 in (5.3) Ch. VII).

CHAPTER 9

The Milnor K-groups of a Local Field

In this chapter we treat J. Milnor's K-ring of a field and its properties. Milnor K-groups of a field is a sort of a weak generalization of the multiplicative group. The Steinberg property which lies at the heart of Milnor K-groups has already shown itself in the previous chapters in the study of the Hilbert pairing. Section 1 contains basic definitions. The study of K-groups of discrete valuation fields is initiated in section 2. We treat the norm (transfer) map on Milnor K-groups of fields in section 3 using several results from section 2. Finally, in section 4 we describe the structure of Milnor K-groups of local fields with finite residue field by using results of the previous chapters.

1. The Milnor Ring of a Field

In this section we just introduce basic definitions. See Exercises for some simple formulas which hold in K_2-groups.

(1.1). Let F be a field, A an additive abelian group. A map

$$f\colon \underbrace{F^* \times \cdots \times F^*}_{n \text{ times}} \to A$$

is called an *n-symbolic map* on F (a *Steinberg cocycle*) if

1. $f(\dots, \alpha_i\beta_i, \dots) = f(\dots, \alpha_i, \dots) + f(\dots, \beta_i, \dots)$ for $1 \leqslant i \leqslant n$ (multiplicativity).
2. $f(\alpha_1, \dots, \alpha_n) = 0$ if $\alpha_i + \alpha_j = 1$ for some $i \neq j$, $1 \leqslant i, j \leqslant n$ (*Steinberg property*).

Let I_n denote the subgroup in $\underbrace{F^* \otimes_{\mathbb{Z}} \cdots \otimes_{\mathbb{Z}} F^*}_{n \text{ times}}$ generated by the elements $\alpha_1 \otimes \cdots \otimes \alpha_n$ with $\alpha_i + \alpha_j = 1$ for some $i \neq j$. The *nth Milnor K-group* of the field F is the quotient

$$K_n(F) = \underbrace{F^* \otimes_{\mathbb{Z}} \cdots \otimes_{\mathbb{Z}} F^*}_{n \text{ times}} / I_n.$$

The multiplication in $K_n(F)$ will be written additively although for $K_1(F) = F^*$ the multiplicative writing will also be used. The image of $\alpha_1 \otimes \cdots \otimes \alpha_n \in F^* \otimes \cdots \otimes F^*$

in $K_n(F)$ is called a *symbol*. The symbols generate $K_n(F)$ and $\{\dots,\alpha_i,\dots\}+\{\dots,\beta_i,\dots\}=\{\dots,\alpha_i\beta_i,\dots\}$; $\{\alpha_1,\dots,\alpha_n\}=0$ if $\alpha_i+\alpha_j=1$ for $i\neq j$.

It is convenient to put $K_0(F)=\mathbb{Z}$. For natural n, m the images of $I_n\otimes\underbrace{F^*\otimes\dots\otimes F^*}_{m\text{ times}}$, $\underbrace{F^*\otimes\dots\otimes F^*}_{n\text{ times}}\otimes I_m$ in $\underbrace{F^*\otimes\dots\otimes F^*}_{n+m\text{ times}}$ are contained in I_{n+m}; thus, we obtain the homomorphism $K_n(F)\times K_m(F)\to K_{n+m}(F)$:

$$(\{\alpha_1,\dots,\alpha_n\},\{\beta_1,\dots,\beta_m\})\to\{\alpha_1,\dots,\alpha_n,\beta_1,\dots,\beta_m\}.$$

We also have the homomorphisms $K_0(F)\times K_n(F)\to K_n(F)$, $K_n(F)\times K_0(F)\to K_n(F)$ mapping an element $x\in K_n(F)$ to $ax\in K_n(F)$ for $a\in\mathbb{Z}=K_0(F)$.

Thus, we obtain the graded ring

$$K(F)=K_0(F)\oplus K_1(F)\oplus K_2(F)\oplus\dots,$$

which is called *the Milnor ring* of the field F.

LEMMA.
(1) $\{\alpha_1,\dots,\alpha_n\}=0$ *if* $\alpha_i+\alpha_j=0$ *for some* $i\neq j$;
(2) $\{\dots,\alpha_i,\dots,\alpha_j,\dots\}=-\{\dots,\alpha_j,\dots,\alpha_i,\dots\}$; $K(F)$ *is anticommutative.*

Proof. Since $\alpha_j=-\alpha_i=(1-\alpha_i^{-1})^{-1}(1-\alpha_i)$ in (1), we get

$$\{\alpha_i,\alpha_j\}=\{\alpha_i^{-1},1-\alpha_i^{-1}\}+\{\alpha_i,1-\alpha_i\}=0.$$

Now for (2) we obtain that

$$\{\alpha_i,\alpha_j\}+\{\alpha_j,\alpha_i\}+(\{\alpha_i,-\alpha_i\}+\{\alpha_j,-\alpha_j\})=\{\alpha_i\alpha_j,-\alpha_i\alpha_j\}=0.$$

□

The definition of $K_n(F)$ implies that an n-symbolic map f on F can be uniquely extended to a homomorphism $f\colon K_n(F)\to A$. Therefore, for an extension L/F of fields the embedding $F^*\to L^*$ induces the homomorphism

$$j_{F/L}\colon K_n(F)\to K_n(L)$$

(if $n=0$, then $j_{F/L}$ is the identical map).

(1.2). The first information on the Milnor K-groups follows from the following

PROPOSITION. *Let* $F^*=F^{*m}$ *for* m *natural, and let either* $m=\operatorname{char}(F)$ *or the group* μ_m *of mth roots of unity in* F^{sep} *be contained in* F. *Then* $K_n(F)$ *is a uniquely* m*-divisible group for* $n\geqslant 2$.

Proof. Define the map $f_m\colon\underbrace{F^*\times\dots\times F^*}_{n\text{ times}}\to K_n(F)$ by the formula

$$f_m(\alpha_1,\dots,\alpha_n)=\{\beta_1,\alpha_2,\dots,\alpha_n\},$$

where $\beta_1 \in F^*$ is such that $\beta_1^m = \alpha_1$. If $\gamma_1^m = \alpha_1$, then $\beta_1\gamma_1{}^{-1} = \zeta$ for some mth root of unity ζ in F. Since $\alpha_2 = \beta_2^m$ for some $\beta_2 \in F^*$, we obtain

$$\{\beta_1, \alpha_2, \ldots, \alpha_n\} = \{\gamma_1, \alpha_2, \ldots, \alpha_n\} + \{\zeta, \beta_2^m, \ldots, \alpha_n\} = \{\gamma_1, \alpha_2, \ldots, \alpha_n\}.$$

Hence, the map f_m is well defined. Next,

$$f_m(\alpha_1\alpha_1{}', \alpha_2, \ldots, \alpha_n) = f_m(\alpha_1, \alpha_2, \ldots, \alpha_n) + f_m(\alpha_1{}', \alpha_2, \ldots, \alpha_n),$$

and f_m is multiplicative with respect to other arguments as well. If $\alpha_i + \alpha_j = 1$ for some $i \neq j$, $1 < i, j$, then $f_m(\alpha_1, \ldots, \alpha_n) = 0$. If char($F$) $= m$ and $\alpha_1 + \alpha_2 = 1$, $\alpha_1 = \beta_1^m$ for some $\beta_1 \in F^*$, then $\alpha_2 = (1-\beta_1)^m$ and we obtain $f_m(\alpha_1, \ldots, \alpha_n) = 0$. Otherwise $\alpha_2 = \prod_{i=1}^m (1 - \zeta_m^i \beta_1)$, where ζ_m is a generator of μ_m. Then

$$\{\beta_1, 1 - \zeta_m^i \beta_1\} = -\{\zeta_m^i, 1 - \zeta_m^i \beta_1\} = -\{\zeta_m^i, \delta^m\} = 0,$$

where $\delta^m = 1 - \zeta_m^i \beta_1$, $\delta \in F^*$. We conclude, that f_m is an n-symbolic map. Its extension on $K_n(F)$ determines the homomorphism $f_m \colon K_n(F) \to K_n(F)$. Then $mf_m =$ id, because $mf_m\{\alpha_1, \ldots, \alpha_n\} = \{\alpha_1, \ldots, \alpha_n\}$. Therefore, $K_n(F)$ is uniquely m-divisible. □

COROLLARY. *If F is algebraically closed, then $K_n(F)$ is a uniquely divisible group for $n \geqslant 2$.*

(1.3). PROPOSITION. *Let F be a finite field. Then $K_n(F) = 0$ for $n \geqslant 2$.*

Proof. It suffices to show that $\{\alpha, \beta\} = 0$ for $\alpha, \beta \in F^*$. Let θ be a generator of F^*; then $\alpha = \theta^i$, $\beta = \theta^j$ and $\{\alpha, \beta\} = ij\{\theta, \theta\}$. By Lemma (1.1) we get $2\{\theta, \theta\} = 2\{-1, \theta\} = 0$. If char($F$) $= 2$, then F^* is of order $2^m - 1$ for some natural m and $(2^m - 1)\{\theta, \theta\} = \{1, \theta\} = 0$. Hence, $\{\theta, \theta\} = 0$ and $\{\alpha, \beta\} = 0$. If char(F) $= p > 2$, then there are exactly $(p^m - 1)/2$ squares and $(p^m - 1)/2$ non-squares in F^*, where p^m is the order of F. The map $\alpha \to 1 - \alpha$ can not transfer all non-squares into squares, because 1 does not belong to its image. Therefore, for some odd k, l we get $\theta^k = 1 - \theta^l$ and $0 = \{\theta^k, \theta^l\} = kl\{\theta, \theta\}$. Thus, $\{\theta, \theta\} = 0$ and $\{\alpha, \beta\} = 0$. □

Exercises.

1. Show that condition 2 of (1.1) can be replaced with condition 2':

$$f(\alpha_1, \ldots, \alpha_n) = 0$$

   if $\alpha_i + \alpha_{i+1} = 1$ for some $1 \leqslant i \leqslant n-1$.
2. Show that $\{\alpha_1, \ldots, \alpha_n\} = 0$ in $K_n(F)$ if, either $\alpha_1 + \cdots + \alpha_n = 0$ or $\alpha_1 + \cdots + \alpha_n = 1$.
3. Show that $\{\alpha, \beta\} = \{\alpha + \beta, -\alpha^{-1}\beta\}$ in $K_2(F)$.
4. (*R.K. Dennis* and *M.R. Stein* [DS])

a) Let $\alpha,\beta,\gamma\in F^*$, and $\alpha,\beta,\gamma,\alpha\beta,\beta\gamma,\alpha\gamma,\alpha\beta\gamma\neq 1$. Show that

$$\left\{-\frac{1-\beta\gamma}{1-\alpha},\frac{1-\alpha\beta\gamma}{1-\alpha}\right\}+\left\{-\frac{1-\alpha\gamma}{1-\beta},\frac{1-\alpha\beta\gamma}{1-\beta}\right\}+\left\{-\frac{1-\alpha\beta}{1-\gamma},\frac{1-\alpha\beta\gamma}{1-\gamma}\right\}=0.$$

b) Let $\alpha,\beta,\gamma\in F^*$, $\alpha,\beta,\alpha\gamma,\beta\gamma,\alpha\beta\gamma\neq 1$. Show that

$$\{\gamma,1-\alpha\beta\gamma\}=\left\{-\frac{1-\beta\gamma}{1-\alpha},\frac{1-\alpha\beta\gamma}{1-\alpha}\right\}+\left\{-\frac{1-\alpha\gamma}{1-\beta},\frac{1-\alpha\beta\gamma}{1-\beta}\right\}.$$

5. (*A.A. Suslin* [Sus1]) Let $\alpha_i,\beta_i\in F^*$ and $\beta_i\neq\beta_j$ for $i\neq j$. Show that

$$\begin{aligned}&\{\beta_1\alpha_1,\dots,\beta_n\alpha_n\}-\{\alpha_1,\dots,\alpha_n\}\\&=\sum\nolimits_{i=1}^{n}(-1)^{i+n}\{\alpha_1(\beta_1-\beta_i),\dots,\alpha_{i-1}(\beta_{i-1}-\beta_i),\alpha_{i+1}(\beta_{i+1}-\beta_i),\dots,\alpha_n(\beta_n-\beta_i),\beta_i\}.\end{aligned}$$

6. Show that $K_n(F)=0$ for an algebraic extension F of a finite field, $n\geqslant 2$.
7. Let F be a field of characteristic $p>0$. Show that the differential symbol

$$d\colon K_n(F)/pK_n(F)\longrightarrow\Omega_F^n,\quad\{a_1,\dots,a_n\}\mapsto\frac{da_1}{a_1}\wedge\dots\wedge\frac{da_n}{a_n}$$

is well defined. Show that the image $d(K_n(F)/pK_n(F))$ is contained in

$$\nu_n(F)=\ker(\wp\colon\Omega_F^n\longrightarrow\Omega_F^n/d\Omega_F^{n-1})$$

where $\wp(a\frac{db_1}{b_1}\wedge\dots\wedge\frac{db_n}{b_n})=(a^p-a)\frac{db_1}{b_1}\wedge\dots\wedge\frac{db_n}{b_n}$.
A theorem of S. Bloch–K. Kato–O. Gabber asserts that d is an isomorphism between the quotient group $K_n(F)/pK_n(F)$ and $\nu_n(F)$ which allows one to calculate the quotient of the Milnor K-group by using differential forms. For a sketch of the proof see [FK, Append. to sect. 2].

2. The Milnor Ring of a Discrete Valuation Field

In this section we establish a relation between K-groups of a discrete valuation field and K-groups of its residue field. In the case of a field $F(X)$ we obtain a complete description of its K-groups in terms of K-groups of finite extensions of F.

(2.1). Let F be a discrete valuation field, v its valuation, $\mathcal{O}_v$ the ring of integers, U_v the group of units, and $\overline{F}_v$ its residue field. Let $\overline{\alpha}$ denote the image of an element $\alpha\in\mathcal{O}_v$ in $\overline{F}_v$. Let π be a prime element in F with respect to the discrete valuation v.

Now we define the border homomorphism

$$\partial_\pi=(\partial_1,\partial_2)\colon K_n(F)\to K_n(\overline{F}_v)\oplus K_{n-1}(\overline{F}_v).$$

Let $\alpha_i=\pi^{a_i}\varepsilon_i$ with $\varepsilon_i\in U_v$, $a_i=v(\alpha_i)$. For $n\geqslant 1$ introduce the map

$$\partial_1\colon\underbrace{F^*\times\dots\times F^*}_{n\text{ times}}\to K_n(\overline{F}_v),\quad(\alpha_1,\dots,\alpha_n)\mapsto\{\overline{\varepsilon}_1,\dots,\overline{\varepsilon}_n\}.$$

Furthermore, for $n = 1$ put $\partial_2(\alpha_1) = a_1$. For $n > 1$ and indices $k_1, \dots, k_m$ with $1 \leqslant k_1 < \dots < k_m \leqslant n$, $m \leqslant n$, put

$$\partial^{k_1,\dots,k_m}(\alpha_1, \dots, \alpha_n) = a_{k_1} \cdots a_{k_m} xy,$$

where x is equal to the symbol $\{\overline{\varepsilon}_1, \dots, \overline{\varepsilon}_n\} \in K_{n-m}(\overline{F}_v)$ with omitted elements at the k_1 th, ... t, k_m th places if $m < n$, and equal to 1 if $m = n$; y is equal to $\{-1, \dots, -1\} \in K_{m-1}(\overline{F}_v)$ if $m > 1$, and equal to 1 otherwise. The element y takes care of π standing at $k_2, \dots, k_m$ th places: $\{\pi, \dots, \pi\} = \{\pi, -1, \dots, -1\}$.

Define the map

$$\partial_2 \colon \underbrace{F^* \times \dots \times F^*}_{n \text{ times}} \to K_{n-1}(\overline{F}_v)$$

by the formula

$$\partial_2(\alpha_1, \dots, \alpha_n) = \sum_{\substack{1 \leqslant k_1 < \dots < k_m \leqslant n \\ 1 \leqslant m \leqslant n}} (-1)^{n-k_1-\dots-k_m} \partial^{k_1,\dots,k_m}(\alpha_1, \dots, \alpha_n).$$

So, in particular, we have $\partial_2(\alpha_1, \alpha_2) = (-1)^{a_1 a_2} \overline{\alpha_1^{a_2} \alpha_2^{-a_1}} \in K_1(\overline{F}_v)$.

For $n > 1$

$$\partial_\pi(\varepsilon_1, \dots, \varepsilon_n) = (\{\overline{\varepsilon}_1, \dots, \overline{\varepsilon}_n\}, 0), \quad \partial_\pi(\varepsilon_1, \dots, \varepsilon_{n-1}, \pi) = (0, \{\overline{\varepsilon}_1, \dots, \overline{\varepsilon}_{n-1}\}).$$

It is easy to verify that $\partial_\pi(\alpha_1, \dots, \alpha_n) = 0$ if $\alpha_i + \alpha_j = 0$ for some $i \neq j$, and

$$\partial_\pi(\dots, \alpha_i \alpha_i', \dots) = \partial_\pi(\dots, \alpha_i, \dots) + \partial_\pi(\dots, \alpha_i', \dots).$$

Now in the same way as in the proof of Lemma (1.1), one can show that

$$\partial_\pi(\dots, \alpha_i, \dots, \alpha_j, \dots) = -\partial_\pi(\dots, \alpha_j, \dots, \alpha_i, \dots).$$

Moreover, if $\alpha \in U_v$, $1 - \alpha \notin U_v$, then $\alpha \in 1 + \pi \mathcal{O}_v$ and

$$\partial_\pi(\dots, \alpha, \dots, 1 - \alpha, \dots) = (0, 0);$$

the same equality holds if $\alpha, 1 - \alpha \in U_v$. If $\alpha \notin \mathcal{O}_v$, then $\alpha^{-1} \in \mathcal{O}_v$ and $1 - \alpha^{-1} \in 1 + \pi \mathcal{O}_v$, so that

$$\partial_\pi(\dots, \alpha, \cdots, 1 - \alpha, \dots) = -\partial_\pi(\dots, \alpha^{-1}, \dots, 1 - \alpha^{-1}, \dots) = (0, 0).$$

Therefore, the map ∂_π induces the required homomorphism

$$\partial_\pi = (\partial_1, \partial_2) \colon K_n(F) \to K_n(\overline{F}_v) \oplus K_{n-1}(\overline{F}_v).$$

(2.2). PROPOSITION. *Let $U_1 K_n(F)$, $\{\pi\} K_{n-1}(F)$ denote the subgroups in the group $K_n(F)$, generated by the symbols $\{\alpha_1, \alpha_2, \dots, \alpha_n\}$, where $\alpha_2, \dots, \alpha_n \in F^*$ and $\alpha_1 \in 1 + \pi \mathcal{O}_v$, $\alpha_1 = \pi$, respectively. Let $U_n(F)$ denote the subgroup in $K_n(F)$ generated by the symbols $\{\alpha_1, \dots, \alpha_n\}$ with $\alpha_i \in U_v$. Then $\partial_\pi, \partial_1, \partial_2$ are surjective homomorphisms with the kernels*

$$U_1 K_n(F), \quad U_1 K_n(F) + \{\pi\} K_{n-1}(F), \quad U_1 K_n(F) + U_n(F),$$

respectively. The homomorphism ∂_2 does not depend on the choice of a prime element π.

Proof. The surjectivity of ∂_π follows from its definition. Introduce the maps

$$f_1: \underbrace{\overline{F}_v^* \times \cdots \times \overline{F}_v^*}_{n \text{ times}} \to K_n(F)/U_1K_n(F),$$

$$f_2: \underbrace{\overline{F}_v^* \times \cdots \times \overline{F}_v^*}_{n-1 \text{ times}} \to K_n(F)/U_1K_n(F)$$

by the formulas

$$f_1(\overline{\varepsilon}_1, \dots, \overline{\varepsilon}_n) = \{\varepsilon_1, \dots, \varepsilon_n\} \mod U_1K_n(F),$$
$$f_2(\overline{\varepsilon}_1, \dots, \overline{\varepsilon}_{n-1}) = \{\varepsilon_1, \dots, \varepsilon_{n-1}, \pi\} \mod U_1K_n(F).$$

The maps f_1, f_2 are well defined because $\alpha\beta^{-1} \in 1 + \pi\mathcal{O}_v$ for $\alpha, \beta \in \mathcal{O}_v$ if $\overline{\alpha} = \overline{\beta}$. The maps f_1, f_2 are symbolic and induce the homomorphisms

$$f_1: K_n(\overline{F}_v) \to K_n(F)/U_1K_n(F),$$
$$f_2: K_{n-1}(\overline{F}_v) \to K_n(F)/U_1K_n(F),$$
$$f = (f_1, f_2): K_n(\overline{F}_v) \oplus K_{n-1}(\overline{F}_v) \to K_n(F)/U_1K_n(F).$$

We get

$$f\partial_\pi(x) = x \mod U_1K_n(F),$$
$$f_1\partial_1(x) = x \mod U_1K_n(F) + \{\pi\}K_{n-1}(F),$$
$$f_2\partial_2(x) = x \mod U_1K_n(F) + U_n(F)$$

for $x \in K_n(F)$. Hence $\ker \partial_\pi \subset U_1K_n(F)$,

$$\ker \partial_1 \subset U_1K_n(F) + \{\pi\}K_{n-1}(F), \ker \partial_2 \subset U_1K_n(F) + U_n(F).$$

It immediately follows that the inverse inclusions also hold.

Furthermore, let π_1 be a prime element in F with respect to v, and $\pi_1 = \pi\varepsilon$ for some $\varepsilon \in U_v$. Then

$$\partial_\pi(\{\varepsilon_1, \dots, \varepsilon_{n-1}, \pi_1\}) = (\{\overline{\varepsilon}_1, \dots, \overline{\varepsilon}_{n-1}, \overline{\varepsilon}\}, \{\overline{\varepsilon}_1, \dots, \overline{\varepsilon}_{n-1}\}),$$
$$\partial_{\pi_1}(\{\varepsilon_1, \dots, \varepsilon_{n-1}, \pi_1\}) = (0, \{\overline{\varepsilon}_1, \dots, \overline{\varepsilon}_{n-1}\}),$$

and ∂_2 does not depend on the choice of a prime element. □

REMARK. The first component ∂_1 of ∂_π does depend in general on the choice of π. If it were not so, then we would have $U_n(F) \subset U_1K_n(F) + \{\pi\}K_{n-1}(F)$ and $K_n(\overline{F}_v) = 0$, and that is not the case for many fields (see (3.9) below).

(2.3). Denote $\partial_v = \partial_2$.

LEMMA. *If $\alpha \in F^*$ then*

$$\partial_v(\{\alpha\}) = v(\alpha) \in \mathbb{Z} = K_0(F).$$

If $\{\alpha, \beta\} \in K_2(F)$ then

$$\partial_v(\{\alpha, \beta\}) \equiv (-1)^{v(\alpha)v(\beta)} \alpha^{v(\beta)} \beta^{-v(\alpha)} \mod \pi \mathcal{O}_v.$$

Proof. The first assertion follows from the definitions. Let $v(\alpha) = a$, $v(\beta) = b$, $\alpha = \pi^a \varepsilon$, $\beta = \pi^b \eta$ for some $\varepsilon, \eta \in U_v$. Then

$$\{\pi^a \varepsilon, \pi^b \eta\} = \{\varepsilon, \eta\} + \{\varepsilon^b \eta^{-a} (-1)^{ab}, \pi\}$$

and

$$\partial_v(\{\alpha, \beta\}) = (-1)^{ab} \overline{\varepsilon}^b \overline{\eta}^{-a} \equiv (-1)^{ab} \alpha^b \beta^{-a} \mod \pi \mathcal{O}_v. \square$$

REMARK. Compare the formula for $\partial_v \colon K_2(F) \to \overline{F}_v^*$ with that for the Hilbert symbol $(\cdot, \cdot)_{q-1}$ in Theorem (5.3) Ch. IV.

PROPOSITION. *Let L be an algebraic extension of F, v and w discrete valuations of F and L, such that the restriction $w|_F$ is equivalent to v (we write $w|v$). Let $e = e(w|v)$ (see (2.3) Ch. II), $j_{v/w} = j_{\overline{F}_v/\overline{L}_w}$. Then the diagram*

$$\begin{array}{ccc} K_n(F) & \xrightarrow{j_{F/L}} & K_n(L) \\ \partial_v \downarrow & & \partial_w \downarrow \\ K_{n-1}(\overline{F}_v) & \xrightarrow{e j_{v/w}} & K_{n-1}(\overline{L}_w) \end{array}$$

is commutative.

Let L/F be a Galois extension, $\sigma \in \mathrm{Gal}(L/F)$ belong to the decomposition group of $\mathrm{Gal}(L/F)$ (see Remark 2 in (2.7) Ch. II), and let $\overline{\sigma}$ be its image in $\mathrm{Gal}(\overline{L}_w/\overline{F}_v)$. Then the diagram

$$\begin{array}{ccc} K_n(L) & \xrightarrow{\sigma} & K_n(L) \\ \partial_w \downarrow & & \partial_w \downarrow \\ K_{n-1}(\overline{L}_w) & \xrightarrow{\overline{\sigma}} & K_{n-1}(\overline{L}_w) \end{array}$$

is commutative.

Proof. Let $\varepsilon_i \in U_v$. Then

$$\partial_v(\{\varepsilon_1, \dots, \varepsilon_{n-1}, \pi\}) = \{\overline{\varepsilon}_1, \dots, \overline{\varepsilon}_{n-1}\}$$

and

$$\partial_w(j_{F/L}\{\varepsilon_1, \dots, \varepsilon_{n-1}, \pi\}) = e\{\overline{\varepsilon}_1, \dots, \overline{\varepsilon}_{n-1}\}$$

by (2.3) Ch. II. Furthermore, by Proposition (2.2)

$$\partial_w(\{\sigma(\varepsilon_1), \dots, \sigma(\varepsilon_{n-1}), \sigma(\pi_w)\}) = \{\overline{\sigma}(\overline{\varepsilon}_1), \dots, \overline{\sigma}(\overline{\varepsilon}_{n-1})\},$$

where π_w is a prime element with respect to w. □

(2.4). Now let $E = F(X)$ and let v be a nontrivial discrete valuation of E, trivial on F. Such valuations are in one-to-one correspondence with monic irreducible polynomials of positive degree over F and also with $\frac{1}{X}$ The latter corresponds to the valuation v_∞: $v_\infty(f(X)/g(X)) = \deg g(X) - \deg f(X)$ for $f(X), g(X) \in F[X]$.

THEOREM (BASS–TATE). *The sequence*

$$0 \to K_n(F) \xrightarrow{j_{F/E}} K_n(F(X)) \xrightarrow{\oplus \partial_v} \bigoplus_{v \neq v_\infty} K_{n-1}(\overline{E}_v) \to 0$$

is exact and splits, where v runs through all nontrivial discrete valuations of E that are trivial on F, $v \neq v_\infty$.

Proof. If ∂_1 is the first component of the homomorphism

$$\partial_{\frac{1}{X}} : K_n(E) \to K_n(F) \oplus K_{n-1}(F),$$

which corresponds to the valuation v_∞ and the prime element $\frac{1}{X}$ with respect to this valuation, then $\partial_1 \circ j_{F/E}(x) = x$ for $x \in K_n(F)$. This means that $j_{F/E}$ is injective. Let $m \geqslant 0$ and let A_m be the subgroup in $K_n(E)$ generated by the symbols $\{f_1(X), \dots, f_n(X)\}$, where $f_i(X) \in F[X]$, $\deg f_i \leqslant m$. Note that for two monic polynomials $p(X)$, $q(X)$ of the same degree $l > 0$ one can write $p(X) = q(X) + r(X)$ with $\deg r(X) < l$, and

$$0 = \left\{ \frac{r(X)}{p(X)}, \frac{q(X)}{p(X)} \right\} = \{r(X), q(X)\} - \{-r(X), p(X)\} - \{p(X), q(X)\}$$

by Lemma (1.1). Hence, the quotient group A_m/A_{m-1} for $m \geqslant 1$ is generated by the symbols

$$\{\alpha_1, \dots, \alpha_{i-1}, p_i(X), \dots, p_n(X)\},$$

where $\alpha_1, \dots, \alpha_{i-1} \in F$ and the polynomials $p_i(X), \dots, p_n(X)$ are monic irreducible over F, such that $0 < \deg p_i(X) < \dots < \deg p_n(X) = m$. Let v be the discrete valuation on $F(X)$ which corresponds to a monic irreducible polynomial $p_v(X)$ of degree $m > 0$. An element of $\overline{E}_v$ can be written as $\overline{g(X)}$ for some polynomial $g(X)$ over F of degree $< m$. Define the map

$$f_v : \underbrace{\overline{E}_v^* \times \dots \times \overline{E}_v^*}_{n-1 \text{ times}} \to A_m/A_{m-1}$$

by the formula

$$f_v(\overline{g_1(X)}, \dots, \overline{g_{n-1}(X)}) = \{g_1(X), \dots, g_{n-1}(X), p_v(X)\} \mod A_{m-1},$$

where $\deg g_i < m$ for $1 \leqslant i \leqslant n-1$. Denote by B_v the subgroup of A_m generated by symbols $\{g_1(X), \dots, g_{n-1}(X), p_v(X)\}$.

We first show that f_v is multiplicative. Indeed, let $g_1(X), h_1(X), r_1(X)$ be polynomials over F of degree $< m$, such that $\overline{g_1(X)h_1(X)} = \overline{r_1(X)}$, i.e., $g_1(X)h_1(X) = p_v(X)q(X) + r_1(X)$ for some $q(X) \in F[X]$. Then $\deg q(X) < m$ and

$$\{g_1(X)h_1(X)/r_1(X), p_v(X)\} - \{g_1(X)h_1(X)/r_1(X), -q(X)/r_1(X)\} \in A_{m-1}$$

in $K_2(\overline{E}_v)$. Therefore,

$$\begin{aligned}\{r_1(X), \dots, g_{n-1}(X), p_v(X)\} &\equiv \{g_1(X), \dots, g_{n-1}(X), p_v(X)\} \\ &\quad + \{h_1(X), \dots, g_{n-1}(X), p_v(X)\} \mod A_{m-1}\end{aligned}$$

and f_v is multiplicative. Furthermore, if $\overline{g_1(X)} = 1 - \overline{g_2(X)} = \overline{1 - g_2(X)}$, then

$$f_v(\overline{g_1(X)}, \overline{g_2(X)}, \dots) \in A_{m-1}$$

and f_v is a symbolic map. Thus, f_v induces the homomorphism

$$f_v \colon K_{n-1}(\overline{E}_v) \to A_m/A_{m-1}.$$

Now we define

$$f_m = \bigoplus_{\deg p_v = m} f_v \colon \bigoplus_{\deg p_v = m} K_{n-1}(\overline{E}_v) \to A_m/A_{m-1}.$$

This homomorphism is surjective, which follows from the above description of the group A_m/A_{m-1}. The homomorphism f_m is injective, because $\partial_v A_{m-1} = 0$ for any v with $\deg p_v(X) = m$ and $\Big(\bigoplus_{\deg p_v = m} \partial_v\Big) f_m(x) = x$. We obtain that f_m is an isomorphism and that $A_m = A_{m-1} \oplus \bigoplus_{\deg p_v = m} B_v$.

Hence, we get an isomorphism $K_n(F) \bigoplus_{v \neq v_\infty} K_{n-1}(\overline{E}_v) \xrightarrow{\sim} K_n(E)$. □

(2.5). COROLLARY 1. *Let v be the discrete valuation on E which corresponds to a monic irreducible polynomial $p_v(X)$ of degree $m > 0$. Then $K_n(\overline{E}_v)$ is generated by the symbols $\{\beta_1, \dots, \beta_{i-1}, p_i(\alpha_v), \dots, p_n(\alpha_v)\}$, where α_v is the image of X in $\overline{E}_v$ (and hence $\overline{E}_v = F(\alpha_v)$), $\beta_1, \dots, \beta_{i-1} \in F$, and $p_i(X), \dots, p_n(X)$ are monic irreducible polynomials over F, $0 < \deg p_i(X) < \dots < \deg p_n(X) < m$.*

Proof. It follows from the description of the quotient groups A_i/A_{i-1} in the proof of the Theorem. □

COROLLARY 2. *Let L/F be an extension of prime degree m and let there be no extensions over F of degree $l < m, l > 1$. Then the group $K_n(L)$ is generated by the symbols $\{\alpha_1, \dots, \alpha_{n-1}, \alpha_n\}$ with $\alpha_1, \dots, \alpha_{n-1} \in F^*, \alpha_n \in L^*$.*

Proof. In this case any polynomial of degree l over F is reducible. □

COROLLARY 3. *Let L/F be an extension of prime degree m, $x \in K_n(L)$. Then there is a finite extension F_1/F of degree relatively prime to m, such that $j_{L/LF_1}(x)$ is a sum of symbols $\{\alpha_1, \dots, \alpha_{n-1}, \alpha_n\}$ with $\alpha_1, \dots, \alpha_{n-1} \in F_1, \alpha_n \in LF_1$.*

Proof. Let $L = F(\alpha)$. Without loss of generality one may assume that $x = \{\beta_1, \dots, \beta_n\}$. Let $\beta_i = f_i(\alpha)$ with polynomials $f_i(X)$ of degree $< m$ over F. Let F_1/F be an extension of degree relatively prime to m, such that all polynomials $f_i(X)$ split into linear factors over F_1. Then $j_{L/LF_1}(x)$ is a sum of symbols $\{\gamma_1, \dots, \gamma_k, \alpha - \delta_1, \dots, \alpha - \delta_{n-k}\}$ with $\gamma_i, \delta_j \in F_1$. Now the required assertion follows from the relation

$$\{\alpha - \delta_1, \alpha - \delta_2\} = \{-1, \alpha - \delta_1\} + \{\delta_2 - \delta_1, \alpha - \delta_2\} - \{\delta_2 - \delta_1, \alpha - \delta_1\}$$

for $\delta_1 \neq \delta_2$. □

Exercises.

1. Let F be a complete discrete valuation field with a residue field $\overline{F}$ of characteristic $p > 0$. Show that if $(m, p) = 1$, then $U_1 K_n(F) \subset m K_n(F)$ and
$$K_n(F)/mK_n(F) \xrightarrow{\sim} K_n(\overline{F})/mK_n(\overline{F}) \oplus K_{n-1}(\overline{F})/mK_{n-1}(\overline{F}).$$
2. Show that $K_n(\mathbb{F}_q(X)) = 0$ for $n \geqslant 3$ and that $K_2(\mathbb{F}_q(X))$ is a nontrivial torsion group.
3. Let A_m denote the subgroup in $K_n(\mathbb{Q})$ generated by $\{a_1, \dots, a_n\}$, where the integers a_i satisfy the condition $|a_i| \leqslant m$ for $1 \leqslant i \leqslant n$. Show in the same way as in the proof of Theorem (2.4) that $A_m = A_{m-1}$, if $m > 1$ is not prime, and
$$\partial_{v_p}: A_p/A_{p-1} \xrightarrow{\sim} K_{n-1}(\mathbb{F}_p),$$
where v_p is the p-adic valuation of $\mathbb{Q}$.
4. Define the map
$$f: \underbrace{\mathbb{Q}^* \times \dots \times \mathbb{Q}^*}_{n \text{ times}} \to \{\pm 1\}$$
setting $f(\alpha_1, \dots, \alpha_n) = -1$ if $\alpha_1 < 0, \dots, \alpha_n < 0$ and $f(\alpha_1, \dots, \alpha_n) = 1$ otherwise. Show that f is a symbolic map. Thus, we have a homomorphism $f: K_n(\mathbb{Q}) \to \mu_2$ and $\{-1, \dots, -1\}$ is of order 2 in $K_n(\mathbb{Q})$. The subgroup $A_1 \subset K_n(\mathbb{Q})$ is mapped isomorphically onto μ_2.
5. Using Exercises 3 and 4, show that
$$K_2(\mathbb{Q}) \xrightarrow{\sim} \mu_2 \oplus \mathbb{F}_3^* \oplus \mathbb{F}_5^* \oplus \mathbb{F}_7^* \oplus \mathbb{F}_{11}^* \oplus \dots$$
and $K_n(\mathbb{Q}) \xrightarrow{\sim} \mu_2$ for $n \geqslant 3$.
In general, a theorem of *H. Garland* asserts that for a finite extension F over $\mathbb{Q}$ the kernel of $K_2(F) \to \bigoplus_v K_1(\overline{F}_v)$ is of finite order.

3. The Norm Map

The norm map in the Milnor ring of fields allows one to calculate it. In this section we define the norm map for Milnor K-groups and study its properties. For algebraic extensions generated by one element we define the norm map in subsection (3.1). Propositions (3.2) and (3.3) demonstrate first properties of this norm map. Subsection (3.4) introduces the norm map for an arbitrary finite extension and its correctness and properties are established by Theorem (3.8) after auxiliary results are described in subsections (3.5)–(3.7).

Let $E = F(X)$ and let v be a nontrivial discrete valuation on E trivial on F. In this section the residue field $\overline{E}_v$ will be denoted by $F(v)$. Then $|F(v) : F| = \deg p_v(X)$, where $p_v(X)$ is the monic irreducible polynomial over F corresponding to v. We get $F(v_\infty) = F$.

(3.1). The homomorphisms $j_{F/E}\colon K(F) \to K(E)$, $j_{F/F(v)}K(F) \to K(F(v))$ induce the structure of $K(F)$-modules on $K(E)$, $K(F(v))$:

$$x \cdot y = j_{F/E}(x) \cdot y \qquad \text{or} \qquad x \cdot z = j_{F/F(v)}(x) \cdot z, \qquad x \in F.$$

The homomorphism $\partial_v\colon K(E) \to K(F(v))$ is a homomorphism of $K(F)$-modules. Instead of the sequence of Theorem (2.4), one can consider the sequence

$$0 \to K_n(F) \xrightarrow{j_{F/E}} K_n(E) \xrightarrow{\oplus} \oplus K_{n-1}(F(v)) \to 0,$$

where v runs through all discrete valuations on E trivial on F. This sequence is not exact in the term $\oplus K_{n-1}(F(v))$ but it is exact in the terms $K_n(F)$, $K_n(E)$, because $\partial_{v_\infty} j_{F/E}(K_n(F)) = 0$. Introduce the homomorphism

$$N = \oplus N_v\colon \oplus K_{n-1}(F(v)) \to K_{n-1}(F),$$

where N_{v_∞} is the identity automorphism of $K_n(F(v_\infty)) = K_n(F)$ so that the sequence

$$0 \to K_n(F) \xrightarrow{j_{F/E}} K_n(E) \xrightarrow{\oplus\partial_v} \oplus K_{n-1}(F(v)) \xrightarrow{\oplus N_v} K_{n-1}(F) \to 0$$

is exact. The exactness in the term $K_{n-1}(F)$ follows from the definition of N_{v_∞}. As for the exactness in the term $\oplus K_{n-1}(F(v))$, we must take N_v for $v \neq v_\infty$ such that the composition

$$K_n(E)/j_{F/E}K_n(F) \xrightarrow{\sim} \bigoplus_{v \neq v_\infty} K_{n-1}(F(v)) \xrightarrow{\oplus N_v} K_{n-1}(F)$$

coincides with

$$-\partial_{v_\infty}\colon K_n(E)/j_{F/E}K_n(F) \to K_{n-1}(F).$$

Such homomorphisms N_v, $v \neq v_\infty$, do exist and are uniquely determined. Then $N_v\colon K(F(v)) \to K(F)$ is a homomorphism of $K(F)$-modules, i.e.,

$$N_v(j_{F/F(v)}(x) \cdot y) = x \cdot N_v(y) \qquad \text{for} \quad x \in K_n(F), y \in K_m(F(v)).$$

(3.2). PROPOSITION.

(1) *The composition* $N_v \circ j_{F/F(v)}: K_n(F) \to K_n(F)$ *coincides with multiplication by* $N_v(1) = |F(v) : F| \in \mathbb{Z}$.

(2) *The homomorphism* $N_v: K_1(F(v)) \to K_1(F)$ *coincides with the norm map*

$$N_{F(v)/F}: F(v)^* \to F^*.$$

Proof. For $f(X) \in F(X)$ we get

$$\sum_{v \neq v_\infty} \deg p_v(X) \cdot v(f(X)) + v_\infty(f(X)) = 0.$$

By Lemma (2.3) v coincides with $\partial_v: F(X)^* \to \mathbb{Z}$, consequently

$$N_v \circ j_{F/F(v)}(x) = N_v(1)x = \deg p_v(X)x \qquad \text{for} \quad x \in K_n(F).$$

To verify (2) it suffices to show, by the uniqueness of (N_v), that for polynomials $f(X), g(X)$ over F

$$\langle f(X), g(X) \rangle = \prod_v N_{F(v)/F} \partial_v \{f(X), g(X)\} = 1.$$

Lemma (2.3) implies that

$$\partial_v \{f(X), g(X)\} = (-1)^{v(f(X))v(g(X))} f(\alpha_v)^{v(g(X))} g(\alpha_v)^{-v(f(X))} \in F(v)^*,$$

where α_v is the image of X in the field $F(v)$. Taking into account the multiplicativity of $\langle \cdot, \cdot \rangle$, and the relation $\langle f(X), f(X) \rangle = \langle -1, f(X) \rangle$, we may assume that $f(X) = \beta p_{v_1}(X)$, $g(X) = \gamma p_{v_2}(X)$ with monic irreducible polynomials p_{v_1}, p_{v_2} of positive degree, $\beta, \gamma \in F^*$. Then $\langle f(X), g(X) \rangle$ is equal to

$$\partial_{v_\infty} \{f(X), g(X)\} \cdot N_{F(v_1)/F} \partial_{v_1} \{f(X), g(X)\} N_{F(v_2)/F} \partial_{v_2} \{f(X), g(X)\}$$

and

$$\partial_{v_\infty} \{f(X), g(X)\} = (-1)^{\deg f(X) \deg g(X)} \beta^{-\deg g(X)} \gamma^{\deg f(X)},$$
$$\partial_{v_1} \{f(X), g(X)\} = g(\alpha_{v_1})^{-1}, \quad \partial_{v_2} \{f(X), g(X)\} = f(\alpha_{v_2}).$$

Let

$$f(X) = \beta(X - \beta_1) \dots (X - \beta_n), \quad g(X) = \gamma(X - \gamma_1) \dots (X - \gamma_m)$$

be the decompositions of $f(X), g(X)$ over F^{alg}. Then

$$N_{F(v_1)/F} g(\alpha_{v_1})^{-1} = \gamma^{-n} \prod (\beta_i - \gamma_j)^{-1}, \qquad 1 \leqslant i \leqslant n, 1 \leqslant j \leqslant m,$$
$$N_{F(v_2)/F} f(\alpha_{v_2}) = \beta^m \prod (\gamma_j - \beta_i), \qquad 1 \leqslant j \leqslant m, 1 \leqslant i \leqslant n.$$

Thus, we deduce that $\langle f(X), g(X) \rangle = 1$. □

(3.3). Now let F_1 be an extension of F, $E = F(X)$, $E_1 = F_1(X)$. Let $v \neq v_\infty$ be a discrete valuation on E trivial on F, and $p_v(X) \in F[X]$ the corresponding monic irreducible polynomial. Let $p_v(X) = \prod_{w|v} p_w(X)^{e(w|v)}$ be the decomposition over F_1 (see Example in (2.7) Ch. II), where $p_w(X)$ are monic irreducible over F_1 polynomials corresponding to the discrete valuations w on F_1, $w|v$. Then $F(v)$ is embedded in $F_1(w)$. There is exactly one discrete valuation w_∞ over v_∞ and $e(w_\infty|v_\infty) = 1$.

PROPOSITION. *Let $j_{v/w} = j_{F(v)/F_1(w)}$. Then the diagram*

$$\begin{array}{ccccccccc}
0 \longrightarrow & K_n(F_1) & \xrightarrow{j_{F_1/E_1}} & K_n(E_1) & \xrightarrow{\oplus \partial_w} & \bigoplus\limits_v \bigoplus\limits_{w|v} K_{n-1}(F_1(w)) & \xrightarrow{\oplus N_w} & K_{n-1}(F_1) & \longrightarrow 0 \\
& \uparrow j_{F/F_1} & & \uparrow j_{E/E_1} & & \uparrow \bigoplus\limits_v \bigoplus\limits_{w|v} e(w|v) j_{v/w} & & \uparrow j_{F/F_1} & \\
0 \longrightarrow & K_n(F) & \xrightarrow{j_{F/E}} & K_n(E) & \xrightarrow{\oplus \partial_v} & \bigoplus\limits_v K_{n-1}(F(v)) & \xrightarrow{\oplus N_v} & K_{n-1}(F) & \longrightarrow 0
\end{array}$$

is commutative.

Proof. The commutativity of the left square follows immediately. The commutativity of the middle square follows from Proposition (2.3). Next, there is exactly one homomorphism

$$g: K_{n-1}(F) \to K_{n-1}(F_1)$$

instead of j_{F/F_1}, which makes the right square commutative. Indeed, for $x = \sum N_v(y_v)$ with $y_v \in K_{n-1}(F(v))$ we are to get

$$g(x) = \sum_v \sum_{w|v} N_w(e(w|v) j_{v/w}(y_v)).$$

If $x = \sum N_v(z_v)$, $z_v \in K_{n-1}(F(v))$, then the exactness of the sequences and the middle square shows that

$$\sum_v \sum_{w|v} N_w(e(w|v) j_{v/w}(z_v)) = g(x).$$

In particular,

$$g(N_{v_\infty}(x)) = N_{w_\infty}(j_{v_\infty/w_\infty}(x)) \qquad \text{for} \quad x \in K_{n-1}(F(v_\infty)) = K_{n-1}(F).$$

Thus, $g(x) = j_{F/F_1}(x)$ and the right square is commutative. □

COROLLARY 1. *Let $F_1 = F(\alpha)$ be an algebraic extension of F, and v the discrete valuation of $F(X)$ which corresponds to the monic irreducible polynomial $p(X)$ of α over F. Let F_2/F be a normal extension and F_1/F a subextension in F_2/F. Let σ_i be distinct embeddings of F_1 in F_2 over F, m the degree of inseparability of F_1/F. Then the composition*

$$j_{F/F_2} \circ N_v: K_n(F_1) \to K_n(F_2)$$

coincides with $m\sum\sigma_i\colon K_n(F_1)\to K_n(F_2)$, *where the maps* $\sigma_i\colon K_n(F_1)\to K_n(F_2)$ *are induced by* $\sigma_i\colon F_1\to F_2$.

Proof. We have the decomposition

$$p(X)=\prod(X-\alpha_i)^m$$

over F_2, where $\alpha_i=\sigma_i(\alpha)$. Now the right commutative square of the diagram with F_2 instead of F_1 implies the desired assertion. □

COROLLARY 2. *Let* $F(v)\cap F_1=F$, $F(v)F_1=F_1(w)$. *Then the diagram*

$$\begin{array}{ccc} K_{n-1}(F_1(w)) & \xrightarrow{N_w} & K_{n-1}(F_1) \\ \uparrow{\scriptstyle j_{v/w}} & & \uparrow{\scriptstyle j_{F/F_1}} \\ K_{n-1}(F(v)) & \xrightarrow{N_v} & K_{n-1}(F) \end{array}$$

is commutative.

Proof. In this case $p_v(X)=p_w(X)$. □

(3.4). Let L/F be a finite extension and $L=F(\alpha_1,\dots,\alpha_l)$, where α_i are algebraic over F. Put $F_0=F$, $F_i=F_{i-1}(\alpha_i)$. Then there is the homomorphism $N_{v_i}\colon K_n(F_i)\to K_n(F_{i-1})$, where v_i is the discrete valuation of the field $F_{i-1}(X)$ which corresponds to α_i. We shall denote this homomorphism by N_{α_i} or $N_{\alpha_i/F_{i-1}}$. Put

$$N_{\alpha_1,\dots,\alpha_l}=N_{\alpha_1}\circ\dots\circ N_{\alpha_l}\colon K_n(L)\to K_n(F).$$

Our first goal is to verify that the homomorphism $N_{\alpha_1,\dots,\alpha_l}$ does not depend on the choice of $\alpha_1,\dots,\alpha_l$. Then we obtain the norm map $N_{L/F}\colon K_n(L)\to K_n(F)$.

From (3.1), (3.2) we deduce that

$$N_{\alpha_1,\dots,\alpha_l}(j_{F/L}(x)\cdot y)=xN_{\alpha_1,\dots,\alpha_l}(y)\qquad\text{for}\quad x\in K_n(F),y\in K_m(L).$$

The composition $N_{\alpha_1,\dots,\alpha_l}\circ j_{F/L}\colon K_n(F)\to K_n(F)$ coincides with multiplication by $|L:F|$, the action of $N_{\alpha_1,\dots,\alpha_l}$ coincides on $K_0(L)$ with multiplication by $|L:F|$ and on $K_1(L)$ with the norm map $N_{L/F}\colon L^*\to F^*$. Similarly to Corollary 1 of (3.3), Proposition (3.3) implies that for a normal extension L_1/F with $L_1\supset L$ the composition $j_{F/L_1}\circ N_{\alpha_1,\dots,\alpha_l}$ coincides with $m\sum\sigma_i$, where m is the degree of inseparability of L/F and $\sigma_i\colon K_n(L)\to K_n(L_1)$ are induced by $\sigma_i\colon L\to L_1$ over F.

LEMMA. *Let* L/F *be a finite extension. Then the kernel of the homomorphism*

$$j_{F/L}\colon K_n(F)\to K_n(L)$$

is contained in the subgroup of $|L:F|$*-torsion in* $K_n(F)$. *For an algebraic extension* L/F *the kernel of* $j_{F/L}$ *is contained in the torsion subgroup of* $K_n(F)$.

Proof. If $j_{F/L}x = 0$, then $N_{\alpha_1,\dots,\alpha_l} j_{F/L}x = |L : F|x = 0$. If L/F is algebraic and $j_{F/L}x = 0$, then $j_{F/M}x = 0$ for some finite subextension M/F in L/F. □

(3.5). For subsequent considerations, it is convenient to have the following results at hand

PROPOSITION. *For a field F and a prime p there exists an algebraic extension F' of F with the following properties:*

(1) *for any finite subextension L/F in F'/F the degree $|L : F|$ is relatively prime to p;*
(2) *any finite extension F''/F' is of degree p^m for some $m \geqslant 0$;*
(3) *if $F'(\alpha)/F'$ is an extension of degree p, then $K_n(F'(\alpha))$ is generated by symbols $\{\alpha_1, \dots, \alpha_{n-1}, \alpha_n\}$ with $\alpha_1, \dots, \alpha_{n-1} \in F'$, $\alpha_n \in F'(\alpha)$;*
(4) *if $p^m x = 0$ for some $x \in K_n(F)$, $m \geqslant 0$ and $j_{F/F'}(x) = 0$, then $x = 0$.*

Proof. Consider the set of all algebraic extensions $\widetilde{F}/F$ with the property: any finite subextension L/F in $\tilde{F}/F$ is of degree prime to p. This set is not empty. Let F'/F be an extension from this set, maximal with respect to embedding of fields. Then property (1) holds for F'.

Let α be a root of an irreducible polynomial $f(X)$ over F'. Then $f(X) \in L[X]$ for some finite extension L/F. Assume that $\deg f(X)$ is prime to p. Then $|L(\alpha) : L|$ is relatively prime to p and so is $|L(\alpha) : F|$. Let F_1/F be a finite subextension in $F'(\alpha)/F$; then $|F_1L(\alpha) : F| = |F_1L(\alpha) : L(\alpha)| \cdot |L(\alpha) : F|$ is relatively prime to p because $|F_1L(\alpha) : L(\alpha)|$ is relatively prime to p. Therefore, $|F_1 : F|$ is relatively prime to p and $F'(\alpha) = F'$. We obtain that any finite extension of F' of degree relatively prime to p coincides with F'.

Now let F''/F' be a finite extension and let F'''/F' be a normal finite extension with $F'' \subset F'''$. If $\operatorname{char}(F) \neq p$ and G is the group of automorphisms of F''' over F', then the fixed field M of G is purely inseparable over F' of degree relatively prime to p. Hence, $M = F'$ and F'''/F' is Galois. Let G_p be a Sylow p-subgroup in G and let M_1 be the fixed field of G. Then $M_1 = F'$ and F''/F' is of degree p^m for some $m \geqslant 0$. If $\operatorname{char}(F) = p$, then let L'/F' be the maximal separable subextension in F''/F'. In the same way as just above, we deduce that L'/F' is of degree p^m and, consequently, F''/F' is of degree p^k, $k \geqslant 0$. Thus, property (2) holds for F'.

Since a polynomial $p(X) \in F'[X]$ of degree $1 < \deg p(X) < p$ is not irreducible over F', Corollary 2 of (2.5) implies property (3).

Finally, if $j_{F/F'}x = 0$, then $j_{F/L}x = 0$ for some finite subextension L/F in F'/F. Lemma (3.4) shows that $|L : F|x = 0$. Since $p^m x = 0$ and $|L : F|$ is relatively prime to p^m, we deduce that $x = 0$. □

(3.6). PROPOSITION. *Let L/F be a normal extension of prime degree p. Then the homomorphism $N_{\alpha/F}\colon K_n(L) \to K_n(F)$ does not depend on the choice of $\alpha \in L$ and* *determines the norm map*

$$N_{L/F}\colon K_n(L) \to K_n(F).$$

Proof. Let $L = F(\alpha) = F(\beta)$, and let v_1, v_2 be the discrete valuations of $F(X)$ which correspond to the monic irreducible polynomials of α, β over F. Corollary 1 of (3.3) shows that $j_{F/L} \circ N_\alpha = j_{F/L} \circ N_\beta$. Hence, by Lemma (3.4), $p(N_\alpha(x) - N_\beta(x)) = 0$ for any $x \in K_n(L)$. Let F' be as in the preceding Proposition. Then $L' = F'(\alpha) = F'(\beta)$ is of degree p and $K_n(L')$ is generated by symbols $\{\alpha_1, \dots, \alpha_{n-1}, \alpha_n\}$ with $\alpha_1, \dots, \alpha_{n-1} \in F'$, $\alpha_n \in L'$. We deduce that

$$N_{\alpha/F'} \circ \{\alpha_1, \dots, \alpha_n\} = \{\alpha_1, \dots, \alpha_{n-1}, N_{L'/F'}(\alpha_n)\} = N_{\beta/F'} \circ \{\alpha_1, \dots, \alpha_n\}.$$

Therefore, $N_{\alpha/F'} = N_{\beta/F'}$. Corollary 2 of (3.3) implies now that

$$j_{F/F'}(N_{\alpha/F}(x) - N_{\beta/F}(x)) = N_{\alpha/F'}(j_{L/L'}(x)) - N_{\beta/F'}(j_{L/L'}(x)) = 0.$$

The property (4) of the preceding Proposition implies $N_{\alpha/F}(x) = N_{\beta/F}(x)$, as desired. □

(3.7). PROPOSITION. *Let L/F be a normal extension of prime degree p. Let v be a nontrivial discrete valuation of $F(X)$ trivial on F. Then the composition*

$$\partial_v \circ N_{L(X)/F(X)}\colon K_n(L(X)) \to K_{n-1}(F(v))$$

coincides with

$$\sum_{w/v} N_{L(w)/F(v)} \circ \partial_w\colon K_n(L(X)) \to K_{n-1}(F(v)),$$

where w runs through all discrete valuations of $L(X)$ trivial on L, $w|v$.

Proof. Let $\widetilde{v}$ be the discrete valuation on $F(X)(Y)$ which corresponds to the irreducible monic polynomial $p(Y)$ of α over $F(X)$, where $L = F(\alpha)$. Then, by Proposition (3.3), the following diagram is commutative:

$$\begin{array}{ccc} K_n(L(X)) & \xrightarrow{\oplus e(\tilde{w}_i|\tilde{v}) j_{\tilde{v}/\tilde{w}_i}} & \oplus K_n(\widehat{F(X)}_v(\tilde{w}_i)) \\ {\scriptstyle N_{\tilde{v}}}\downarrow & & \downarrow{\scriptstyle \oplus N_{\tilde{w}_i}} \\ K_n(F(X)) & \xrightarrow{j_{F(X)/\widehat{F(X)}_v}} & K_n(\widehat{F(X)}_v) \end{array}$$

where $\widetilde{w}_i$ are discrete valuations on $\widehat{F(X)}_v(Y), \widetilde{w}_i|\widetilde{v}$. According to Example (2.7) Ch. II, the valuations $\widetilde{w}_i$ correspond to the irreducible monic polynomials in the decomposition $p(Y) = \prod p_{\tilde{w}_i}(Y)^{e_i}$ over $\widehat{F(X)}_v$, and $e_i = e(\widetilde{w}_i|\widetilde{v})$. We get also $\widehat{F(X)}_v(\widetilde{w}_i) = \widehat{F(X)}_v(\alpha_i)$, where α_i is a root of $p_{\tilde{w}_i}(Y)$. On the other hand, Proposition (2.6) Ch. II shows that $\widehat{L(X)}_{w_i} = \widehat{F(X)}_v(\alpha_i)$, where $w_i|v$ are discrete valuations on $L(X)$.

We will next verify that $e(\widetilde{w}_i|\widetilde{v}) = 1$. Indeed, if $e(\widetilde{w}|\widetilde{v}) = p$, then the polynomial $p(Y)$ can be decomposed into linear factors over $\widehat{F(X)}_v$. This means that there is a unique extension w of v on $L(X)$ and $L(w) = F(v)$, $e(w|v) = 1$. Applying Example (2.7) Ch. II for $w|v$, we obtain that the irreducible monic polynomial $p_v(X)$ over F corresponding to v is irreducible over L (if $v \neq v_\infty$). Then $|F(v) : F| = \deg p_v(X) = \deg p_w(X) = |L(w) : L|$, which is impossible. Thus, $e(\widetilde{w}_i|\widetilde{v}) = 1$.

Consequently, the following diagram is commutative:

$$\begin{array}{ccc} K_n(L(X)) & \xrightarrow{\oplus_{w|v} j_{L(X)/\widehat{L(X)}_w}} & \bigoplus_{w|v} K_n(\widehat{L(X)}_w) \\ {\scriptstyle N_{L(X)/F(X)}}\big\downarrow & & \big\downarrow{\scriptstyle \oplus_{w|v} N_{\widehat{L(X)}_w/\widehat{F(X)}_v}} \\ K_n(F(X)) & \xrightarrow{j_{F(X)/\widehat{F(X)}_v}} & K_n(\widehat{F(X)}_v) \end{array}$$

The definition of ∂_v implies that it coincides with the composition

$$K_n(F(X)) \xrightarrow{j_{F(X)/\widehat{F(X)}_v}} K_n(\widehat{F(X)}_v) \to K_{n-1}(F(v)).$$

A similar assertion holds for L. Thus, it suffices to show that the diagram

$$\begin{array}{ccc} K_n(L) & \xrightarrow{\partial_w} & K_{n-1}(\overline{L}_w) \\ {\scriptstyle N_{L/F}}\big\downarrow & & \big\downarrow{\scriptstyle N_{\overline{L}_w/\overline{F}_v}} \\ K_n(F) & \xrightarrow{\partial_v} & K_{n-1}(\overline{F}_v) \end{array}$$

is commutative, where F is a complete discrete valuation field with respect to v, and L/F is an extension of degree p.

By Proposition (2.3) we get $e(w|v)j_{\overline{F}_v/\overline{L}_w}(y) = 0$ for $y = \partial_v \circ N_{L/F}(x) - N_{\overline{L}_w/\overline{F}_v} \circ \partial_w(x)$, $x \in K_n(L)$. Hence, $py = 0$. Let F_1/F be a finite extension of degree relatively prime to p such that $j_{L/LF_1}(x)$ is a sum of symbols $\{\alpha_1, \dots, \alpha_n\}$ with $\alpha_1, \dots, \alpha_{n-1} \in F_1$, $\alpha_n \in LF_1$, according to Corollary 3 of (2.5). Proposition (2.3), Corollary 2 of (3.3) and Lemma (3.4) show that one may assume that $x = \{\alpha_1, \dots, \alpha_n\}$. We get

$$\partial_v \circ N_{L/F}(\{\alpha_1, \dots, \alpha_n\}) = \begin{cases} f(w|v)\{\overline{\alpha}_1, \dots, \overline{\alpha}_{n-1}\}v_L(\alpha_n), & \text{if } \alpha_1, \dots, \alpha_{n-1} \in U_F, \\ -\{\overline{N_{L/F}(\alpha_n)}, \overline{\alpha}_2, \dots, \overline{\alpha}_{n-1}\}v_F(\alpha_1), & \text{if } \alpha_2, \dots, \alpha_{n-1} \in U_F,\ \alpha_n \in U_L. \end{cases}$$

The same expression holds for $N_{\overline{L}_w/\overline{F}_v} \circ \partial_w\{\alpha_1, \dots, \alpha_n\}$. □

COROLLARY. *Let L/F be a normal extension of prime degree p and let $F_1 = F(\alpha)$ be an algebraic extension of F. Let $L_1 = L(\alpha)$. Then*

$$N_{\alpha/F} \circ N_{L_1/F_1} = N_{L/F} \circ N_{\alpha/L}: K_n(L_1) \to K_n(F).$$

Proof. Let v be the discrete valuation of $F(X)$ corresponding to α. Then $F_1 = F(v)$, $L_1 = L(w)$ for some discrete valuation w of $L(X)$, $w|v$. Let $x \in K_n(L_1)$ and $x = \partial_w(y)$ for some $y \in K_n(L(X))$, such that $\partial_{w'}(y) = 0$ for all $w' \neq w, w_\infty$ (such an element y exists by Theorem (2.4)). Then

$$N_{\alpha/F} \circ N_{L(w)/F(v)}(x) = N_{\alpha/F} \circ \partial_v \circ N_{L(X)/F(X)}(y)$$

by the Proposition, and

$$N_{v'} \circ \partial_{v'}\left(N_{L(X)/F(X)}(y)\right) = 0 \qquad \text{for} \quad v' \neq v, v_\infty.$$

Hence, we deduce from the definition of N_v that

$$N_{\alpha/F} \circ N_{L(w)/F(v)}(x) = -\partial_{v_\infty} \circ N_{L(X)/F(X)}(y).$$

On the other hand, $N_{\alpha/L} \circ \partial_w(y) = -\partial_{w_\infty}(y)$ and

$$N_{L/F} \circ N_{\alpha/L} \circ \partial_w(y) = -N_{L/F} \circ \partial_{w_\infty}(y) = -\partial_{v_\infty} \circ N_{L(X)/F(X)}(y)$$

by Corollary 2 of (3.3). This completes the proof. □

(3.8). THEOREM (BASS–TATE–KATO). *Let L/F be a finite extension. Then there is the norm map $N_{L/F}\colon K(L) \to K(F)$ which is a homomorphism of $K(F)$-modules, with the properties*:

(1) $N_{L/F}$ *coincides with* $N_{\alpha_1,\dots,\alpha_l}$ *for any* $\alpha_1, \dots, \alpha_l \in L$ *with* $L = F(\alpha_1, \dots, \alpha_l)$.

(2) *For every subextension* M/F *in* L/F

$$N_{L/F} = N_{M/F} \circ N_{L/M}.$$

(3) *The map* $N_{L/F}$ *acts on* $K_0(L)$ *as the multiplication by* $|L : F|$ *and on* $K_1(L)$ *as the norm map of fields* $N_{L/F}\colon L^* \to F^*$.

(4) *The composition* $N_{L/F} \circ j_{F/L}$ *coincides with the multiplication by* $|L : F|$.

(5) *If* L_1/F *is a normal finite extension with* $L_1 \supset L$, *then the composition* $j_{F/L_1} \circ N_{L/F}$ *coincides with* $m \sum \sigma_i$, *where* m *is the degree of inseparability of* L/F *and* $\sigma_i\colon K(L) \to K(L_1)$ *are induced by distinct embeddings of* L *in* L_1 *over* F.

(6) *If* σ *is an automorphism of* L *over* F, *then* $N_{L/F} \circ \sigma = N_{L/F}$, *where* $\sigma\colon K(L) \to K(L)$ *is induced by* σ.

Proof. Let $L = F(\alpha_1, \dots, \alpha_l) = F(\beta_1, \dots, \beta_k)$. Let L_1/F be a normal finite extension with $L_1 \supset L$. By (3.4) we get

$$j_{F/L_1}(N_{\alpha_1,\dots,\alpha_l} - N_{\beta_1,\dots,\beta_k}) = m \sum \sigma_i - m \sum \sigma_i = 0.$$

Lemma (3.4) implies $|L_1 : F| y = 0$ for the element $y = N_{\alpha_1,\dots,\alpha_l}(x) - N_{\beta_1,\dots,\beta_k}(x)$, where $x \in K(L)$. Let $|L_1 : F| = p^r q$ with $(q, p) = 1$ and a prime p. Let F' be as in Proposition (3.5), and let $L' = LF'$, $L_1' = L_1F'$. Then L_1'/F' is of degree p^r. Let L''/F' be the maximal separable subextension in L'/F', and let L'''/F' be the minimal normal extension with $L''' \supset L''$. Then Gal(L'''/F') is a finite p-group. Therefore, there exists a chain of subgroups

$$\mathrm{Gal}(L'''/L'') = G_0 \leqslant G_1 \leqslant \ldots \leqslant G_s = \mathrm{Gal}(L'''/F'),$$

such that G_i is normal in G_{i+1} of index p. We obtain a tower of fields

$$F' = F_0' \subset F_1' \subset \cdots \subset L'' \subset \cdots \subset F_k' = L',$$

such that F_j'/F_{j-1}' is a normal extension of degree p.

Let $p_i(X)$ be the monic irreducible polynomial of α_i over

$$F_{i-1} = F(\alpha_1, \ldots, \alpha_{i-1}), \quad F_1 = F,$$

and v_i the corresponding discrete valuation of $F_{i-1}(X)$. Then $N_{\alpha_1,\ldots,\alpha_l} = N_{v_1} \circ \cdots \circ N_{v_l}$. By Proposition (3.3),

$$j_{F/F'} \circ N_{\alpha_1,\ldots,\alpha_l} = \sum_{w_{1i}|v_1,\ldots,w_{li}|v_l} e(w_{1i}|v_1)\ldots e(w_{li}|v_l) N_{w_{1i}} \circ \cdots \circ N_{w_{li}} \circ j_{L/L'},$$

where w_{ji} are the discrete valuation of $F'F_{r-1}(X)$ with $w_{ji}|v_r$. Therefore, if we show that $N_{w_1} \circ \cdots \circ N_{w_l} \colon K(L') \to K(F')$ does not depend on the choice of generating algebraic elements in L' over F', then we shall obtain $j_{F/F'}(y) = 0$, $j_{F/F'}(qy) = 0$. Since $p^r(qy) = 0$, Proposition (3.5) implies $qy = 0$. Continuing in this way for qy we finally deduce $y = 0$, as required.

Now let $N_{w_1} \circ \cdots \circ N_{w_l} = N_{\gamma_1,\ldots,\gamma_l}$ and $F_i'' = F'(\gamma_1, \ldots, \gamma_i)$, $F_0'' = F'$. Put $F_{i,j}' = F_i''F_j'$ for $0 \leqslant i \leqslant l$, $0 \leqslant j \leqslant k$. Then $F_{i,j-1}' = F_{i-1,j-1}'(\gamma_i)$, and $F_{i-1,j}'/F_{i-1,j-1}'$ is a normal extension of degree 1 or p. Applying Corollary (3.7), we get $N_{\gamma_i/F_{i-1,j-1}'} \circ N_{F_{i,j}'/F_{i,j-1}'} = N_{F_{i-1,j}'/F_{i-1,j-1}'} \circ N_{\gamma_i/F_{i-1,j}'}$. Therefore,

$$N_{w_1} \circ \cdots \circ N_{w_l} = N_{F_1'/F'} \circ \cdots \circ N_{L'/F_{k-1}'}$$

and $N_{w_1} \circ \cdots \circ N_{w_l}$ does not depend on the choice of generating elements.

Furthermore, if σ is an automorphism of L over F, then

$$L = F(\alpha_1, \ldots, \alpha_l) = F(\sigma\alpha_1, \ldots, \sigma\alpha_l) \quad \text{and} \quad N_{L/F} \circ \sigma = N_{L/F}.$$

Other properties of $N_{L/F}$ follow from the corresponding properties of the homomorphism $N_{\alpha_1,\ldots,\alpha_l}$ discussed in (3.4). □

REMARK. For the properties of $N_{L/F}$ see also Exercises 2–5.

(3.9). One application of the norm map $N_{L/F}$ is the following. Let T_n be the torsion group of $K_n(F)$. The cardinality of $\mathbb{Z}$-module $K_n(F)/T_n$ is said to be the rank of $K_n(F)$.

PROPOSITION. *Let $\delta(F)$ be the Kronecker dimension of F, i.e., the degree of transcendence of F over $\mathbb{F}_p$ in the case of* char$(F) = p$, *and* 1+ (*the degree of transcendence* *of F over $\mathbb{Q}$), in the case of* $\operatorname{char}(F) = 0$. *Then the rank of $K_n(F)$ is equal to the cardinality of F if $1 \leqslant n \leqslant \delta(F)$.*

Proof. Let $n = 1$, $\delta(F) \geqslant 1$. Then the cardinality of F is equal to the cardinality of F^*/T_1. Let $\delta(F) \neq 1$. Let E be a subfield in F such that F is an algebraic extension of $E(X)$ for some element X in F transcendental over E. The cardinality of F is equal to the cardinalities of $E(X)$ and E. By Theorem (2.4) there is a surjective homomorphism

$$K_n(E(X)) \to \bigoplus_v K_{n-1}(\overline{E(X)_v}).$$

If $2 \leqslant n \leqslant \delta(F)$, then $1 \leqslant n-1 \leqslant \delta(\overline{E(X)_v})$. By induction we can assume that the rank of $K_{n-1}(\overline{E(X)_v})$ is equal to the cardinality of E. Therefore, the rank of $K_n(E(X)) \geqslant$ the cardinality of E. Lemma (3.4) implies now that the rank of $K_n(F) \geqslant$ the cardinality of F. The inverse inequality follows from the definition of $K_n(F)$. □

COROLLARY. *$K_n(\mathbb{C})$ is an uncountable uniquely divisible group for $n \geqslant 2$.*

Exercises.

1. Show that the field F' in Proposition (3.5) is not uniquely determined and is not a normal extension of F, in general. Show that for an extension L/F of degree p and the field $L' = LF'$ the pair (L', L) does not possess, in general, all the properties formulated in Proposition (3.5) with respect to L.
2. Let F be a complete discrete valuation field, and let L be a normal extension of F of finite degree. Show that the diagram

$$\begin{array}{ccc} K_n(L) & \xrightarrow{\partial} & K_{n-1}(\overline{L}) \\ {\scriptstyle N_{L/F}}\downarrow & & \downarrow{\scriptstyle N_{\overline{L}/\overline{F}}} \\ K_n(F) & \xrightarrow{\partial} & K_{n-1}(\overline{F}) \end{array}$$

   is commutative.
3. Let L be a finite extension of F, σ an automorphism of L. Show that the diagram

$$\begin{array}{ccc} K_n(L) & \xrightarrow{\sigma} & K_n(\sigma L) \\ {\scriptstyle N_{L/F}}\downarrow & & \downarrow{\scriptstyle N_{\sigma L/\sigma F}} \\ K_n(F) & \xrightarrow{\sigma} & K_n(\sigma F) \end{array}$$

   is commutative, where $\sigma\colon K_n(L) \to K_n(\sigma L)$ is induced by $\sigma\colon L^* \to (\sigma L)^*$.
4. Let L, M be finite separable extensions of F and

$$L \otimes_F M = \bigoplus_\sigma L\sigma(M),$$

where σ runs through embeddings of M in F^{sep} over $L \cap M$. Show that the diagram

$$\begin{array}{ccc} K_n(M) & \xrightarrow{\oplus j_{\sigma M/L\sigma(M)} \circ \sigma} & \oplus K_n(L\sigma(M)) \\ {\scriptstyle N_{M/F}} \downarrow & & \downarrow {\scriptstyle \oplus N_{L\sigma(M)/L}} \\ K_n(F) & \xrightarrow{j_{F/L}} & K_n(L) \end{array}$$

is commutative.

5. ($\diamond$) (*S. Rosset* and *J. Tate* [RT])

a) Let $f(X),\ g(X)$ be relatively prime polynomials over F. If g is a monic irreducible polynomial of positive degree, $g(X) \neq X$, then put

$$\left(\frac{f}{g}\right) = N_{F(\alpha)/F}\{\alpha, f(\alpha)\},$$

where α is a root of $g(X)$. If $g(X)$ is a constant or $g(X) = X$, then put $\left(\frac{f}{g}\right) = 0$. If $g = g_1 g_2$ and $g_1,\ g_2$ are relatively prime to f then put

$$\left(\frac{f}{g_1 g_2}\right) = \left(\frac{f}{g_1}\right) + \left(\frac{f}{g_2}\right).$$

Show that $\left(\frac{f}{g}\right) = N_{E/F}\{\alpha, \beta\}$, where $\alpha, \beta \in E^*$, $g(X) \in F[X]$ is the monic irreducible polynomial with the root α, and $f(X) \in F[X]$ is the polynomial of minimal degree such that $N_{E/F(\alpha)}\beta = f(\alpha)$. For a polynomial $p(X) = \alpha_n X^n + \cdots + \alpha_m X^m$ with $\alpha_n \alpha_m \neq 0$, $n \geqslant m$, put $p^*(X) = \alpha_m^{-1} X^{-m} p(X)$, $c(p) = (-1)^n \alpha_n$. Prove the reciprocity law

$$\left(\frac{f}{g}\right) = \left(\frac{g^*}{f}\right) - (c(g^*), c(f))$$

b) Let for $\alpha, \beta \in E^*$ the polynomials $f(X), g(X) \in F[X]$ be as in a). Put $g_0 = g$, $g_1 = f$, and let g_{i+1} be the remainder of the division of g_{i-1}^* by g_i if $g_i \neq 0$ for $i \geqslant 1$. Show that $g_m \neq 0, g_{m+1} = 0$ for some $m \leqslant |E : F|$, and that

$$N_{E/F}\{\alpha, \beta\} = -\sum_{i=1}^{m} \left\{c(g_{i-1}^*), c(g_i)\right\}.$$

In particular, $N_{E/F}\{\alpha, \beta\}$ is a sum of at most $|E : F|$ symbols.

6. Let the group μ_m of all mth roots of unity in F^{sep} be contained in F. Let $\alpha \in F^*$, $x \in K_n(F)$ and $x \in N_{F(\sqrt[m]{\alpha})/F} K_n(F(\sqrt[m]{\alpha}))$. Show that the element $\{\alpha\} \cdot x \in K_{n+1}(F)$ is m-divisible in $K_{n+1}(F)$. (The converse assertion for $n = 1$ and arbitrary field (μ_m is not necessarily contained in F) is true if m is square-free and $\neq 0$ in F; see [Mil1, sect. 15]).

7. Let $\operatorname{char}(F) = p$ and $|F : F^p| = p^d$. Put $E = F^{1/p}$. Then the homomorphism $g(\alpha) = \alpha^{1/p}$ is an isomorphism of F^* onto E^* and $j_{F/E}(\alpha) = g(\alpha^p)$. Show that the homomorphism $j_{F/E}\colon K_n(F) \to K_n(E)$ coincides with $p^n g\colon K_n(F) \to K_n(E)$. Show that the group $p^d K_n(F)$ is uniquely p-divisible for $n > d$ and $p^{d-1} K_n(F) = p^d K_n(F)$.

8. a) Let the map $f: \underbrace{\mathbb{R}^* \times \cdots \times \mathbb{R}^*}_{n \text{ times}} \to \mu_2 = \{\pm 1\}$ be determined in the same manner as in Exercise 4 section 2. Show that f is n-symbolic and

$$K_n(\mathbb{R})/2K_n(\mathbb{R}) \simeq \mu_2.$$

b) Let the map $g: \mathbb{R}^* \times \mathbb{R}^* \to K_2(\mathbb{R})/T$, where T is the subgroup generated by the symbol $\{-1, -1\}$, be defined by the formula $g(\alpha, \beta) \equiv \{\sqrt{|\alpha|}, \beta\} \mod T$. Show that g is 2-symbolic and the subgroup of 2-torsion of $K_2(\mathbb{R})$ is contained in T.

c) Show that if $x \in 2K_2(\mathbb{R})$, then $x \in N_{\mathbb{C}/\mathbb{R}} K_2(\mathbb{C})$, and hence $2K_2(\mathbb{R})$ is a divisible group. Deduce that

$$K_2(\mathbb{R}) \simeq \mu_2 \oplus 2K_2(\mathbb{R}),$$

where μ_2 corresponds to T. Show that $2K_2(\mathbb{R})$ is an uncountable uniquely divisible group.

d) Show that $K_n(\mathbb{R}) \simeq \mu_2 \oplus 2K_n(\mathbb{R})$ and $2K_n(\mathbb{R})$ is an uncountable uniquely divisible group, μ_2 corresponds to $\{-1, \dots, -1\}$, $n \geqslant 2$.

9. Let F be a Brauer field (see Exercise 4 in section 1 Ch. V). Show that $K_n(F)$ is a uniquely divisible group for $n \geqslant 2$, if $d(2) \neq 1$ (in terms of the same Exercise).

4. The Milnor Ring of a Local Field

In this section F is a local field with finite residue field $\mathbb{F}_q$ of characteristic p. We shall describe the Milnor groups of F using the Hilbert symbol. The main results are Theorems (4.3), (4.7) which together give a complete description of $K_2(F)$ and (4.11) which describes higher Milnor $K_n(F)$. The role of the Hilbert symbol is demonstrated in Theorem (4.3) and its Corollary, and in the end of the proof of Theorem (4.7). In (4.13) we briefly discuss how Milnor K-groups are involved in higher local class field theory.

(4.1). LEMMA. *Let $\theta_1, \theta_2 \in \mu_{q-1} \subset F$, $\varepsilon \in U_{1,F}$. Then*

$$\{\theta_1, \theta_2\} = \{\theta_1, \varepsilon\} = 0.$$

Proof. By (5.5) Ch. I the group $U_{1,F}$ is uniquely $(q-1)$-divisible. Then $\{\theta_1, \varepsilon\} = \{\theta_1^{q-1}, \eta\} = 0$, where $\eta^{q-1} = \varepsilon$, $\eta \in U_{1,F}$.

section 7 Repeating the arguments of the proof of Proposition (1.3) using the relation $\theta^k + \theta^l - 1 \in U_{1,F}$ instead of the equality $\theta^k = 1 - \theta^l$, we deduce that $\{\theta_1, \theta_2\} = 0$ for $\theta_1, \theta_2 \in \mu_{q-1}$. □

PROPOSITION. *Let θ be a generator of μ_{q-1} and $(m, p) = 1$. Then the quotient group $K_2(F)/mK_2(F)$ is a cyclic group of order $d = (m, q-1)$ and generated by $\{\theta, \pi\}$ mod $mK_2(F)$, where π is a prime element in F. The group $(q-1)K_2(F)$ coincides with $l(q-1)K_2(F)$ for $l \geqslant 1$ relatively prime to p.*

Proof. The m-divisibility of $U_{1,F}$ implies that $\{\alpha, \beta\} \in mK_2(F)$ for $\alpha \in U_{1,F}, \beta \in F^*$. Since $\{\pi, \pi\} = \{-1, \pi\}$, applying Proposition (5.4) Ch. I we deduce, that $\{\theta, \pi\}$ mod $mK_2(F)$ generates $K_2(F)/mK_2(F)$. The Hilbert symbol

$$(\cdot, \cdot)_{q-1}: F^* \times F^* \to \mu_{q-1}$$

is 2-symbolic by Proposition (5.1) Ch. IV, and hence determines the surjective homomorphism $H_{q-1}: K_2(F) \to \mu_{q-1}$. Therefore, $K_2(F)/(q-1)K_2(F)$ is cyclic of order $q-1$. □

(4.2). PROPOSITION. *If there are no nontrivial pth roots of unity in F, then $K_2(F) = pK_2(F)$. Otherwise $K_2(F)/pK_2(F)$ is of order p.*

Proof. Since μ_{q-1} is p-divisible, we obtain that $K_2(F)/pK_2(F)$ is generated by symbols $\{\varepsilon_1, \varepsilon_2\}$, $\{\pi, \varepsilon_2\}$ mod $pK_2(F)$ with $\varepsilon_1, \varepsilon_2 \in U_{1,F}$. Put $\varepsilon_1 = 1 + \beta_1$, $\varepsilon_2 = 1 + \beta_2$ with $\beta_1 = \pi^i \gamma_1$, $\beta_2 = \pi^j \gamma_2$, $\gamma_1, \gamma_2 \in U_F$. Then

$$\begin{aligned}\{\varepsilon_1, \varepsilon_2\} &= \{1 + \beta_1, -\beta_1(1 + \beta_2)\} = -\{1 + \beta_1\beta_2(1 + \beta_1)^{-1}, -\beta_1(1 + \beta_2)\} \\ &= -i\left\{\left(1 + \frac{\beta_1\beta_2}{1 + \beta_1}\right), \pi\right\} - \left\{1 + \frac{\beta_1\beta_2}{1 + \beta_1}, -\gamma_1(1 + \beta_2)\right\}.\end{aligned}$$

Continue this calculation for $\varepsilon_1' = 1 + \beta_1\beta_2(1 + \beta_1)^{-1}$, $\varepsilon_2' = -\gamma_1(1 + \beta_2)$ with the help of Lemma (4.1). We deduce that $\{\varepsilon_1, \varepsilon_2\} = \{\varepsilon, \eta\} + \{\varepsilon_3, \pi\}$ with $\varepsilon \in U_{k,F}$, $k > pe/(p-1)$, where $e = e(F|\mathbb{Q}_p)$.

Let $\operatorname{char}(F) = 0$. Then $\varepsilon \in F^{*p}$ by (5.8) Ch. I, and

$$\{\varepsilon_1, \varepsilon_2\} \equiv \{\varepsilon_3, \pi\} \mod pK_2(F).$$

Thus, $K_2(F)/pK_2(F)$ is generated by symbols $\{\varepsilon, \pi\}$, $\varepsilon \in U_{1,F}$. Assume that $e/(p-1)$ is an integer. Let $t = t(F)$ be the maximal integer such that $pe/(p-1)$ is divisible by p^t, $e_* = pe/(p^t(p-1))$. Using (6.5) Ch. I take the following generators of the quotient group $U_{1,F}/U_{1,F}^p$:

- $(1 - \theta\pi^i)^i$ for $1 \leqslant i < pe/(p-1), (i, p) = 1, \theta \in R_0$, where R_0 is a subset in μ_{q-1} such that the residues of its elements form a basis of $\mathbb{F}_q$ over $\mathbb{F}_p$;
- $\omega_* = \left(1 - \theta_*\pi^{pe/(p-1)}\right)^{e_*}$ if $\mu_p \subset F^*$ and $\omega_* = 1$ if $\mu_p \not\subset F^*$, where θ_* is an element of μ_{q-1} such that $1 - \theta_*\pi^{pe/(p-1)} \notin U_{1,F}^p$.

Then by Lemma (4.1)

$$\begin{aligned}\{\pi, (1 - \theta\pi^i)^i\} &= \{\pi^i, 1 - \theta\pi^i\} = \{\pi^i\theta, 1 - \theta\pi^i\} = 0, \\ p^t\{\pi, \omega_*\} &= \{\theta_*\pi^{pe/(p-1)}, 1 - \theta_*\pi^{pe/(p-1)}\} = 0.\end{aligned}$$

This means that $K_2(F) = pK_2(F)$ if $\mu_p \not\subset F^*$. If $\mu_p \subset F^*$, then the Hilbert symbol $(\cdot, \cdot)_p: F^* \times F^* \to \mu_p$ induces the surjective homomorphism

$$H_p: K_2(F) \to \mu_p.$$

Therefore, $K_2(F)/pK_2(F)$ is a cyclic group of order p and generated by $\{\omega_*, \pi\}$ mod $pK_2(F)$.

Now let $\operatorname{char}(F) = p$. Then the field $F^{1/p}$ is an inseparable extension of F of degree p. The norm map

$$N_{F^{1/p}/F}: F^{1/p*} \to F^*, \qquad N_{F^{1/p}/F}(\alpha^{1/p}) = \alpha$$

is surjective and

$$\{\alpha, \beta\} = \{N_{F^{1/p}/F}(\alpha^{1/p}), \beta\} = pN_{F^{1/p}/F}\{\alpha^{1/p}, \beta^{1/p}\}$$

in $K_2(F)$. Thus, $K_2(F) = pK_2(F)$ in this case. □

COROLLARY. *The quotient group $K_2(F)/p^sK_2(F)$ is cyclic and generated by $\{\omega_*, \pi\}$ mod $p^sK_2(F)$ for $s \geqslant 1$ if* $\operatorname{char}(F) = 0$, $\mu_p \subset F^*$. *Otherwise $K_2(F) = p^sK_2(F)$.*

(4.3). THEOREM (C. Moore). *Let m be the cardinality of the torsion group in F^*. Then the mth Hilbert symbol $(\cdot,\cdot)_m$ induces the exact sequence*

$$0 \to mK_2(F) \to K_2(F) \to \mu_m \to 1$$

which splits: $K_2(F) \simeq \mu_m \oplus mK_2(F)$. The group $mK_2(F)$ is divisible.

Proof. Let $m = p^r(q-1)$, $r \geqslant 0$, and let ζ_m be a primitive mth root of unity in F. Assume that $r \geqslant 1$. Then ζ_m^{q-1} is a primitive p^rth root of unity. By property (6) of Proposition (5.1) Ch. IV there is an element $\alpha \in F^*$ such that $(\zeta_m^{q-1}, \alpha)_p \neq 1$. Then $\{\zeta_m^{q-1}, \alpha\}$ mod $pK_2(F)$ generates $K_2(F)/pK_2(F)$ and so $\{\zeta_m^{q-1}, \alpha\}$ mod $p^sK_2(F)$ generates the quotient group $K_2(F)/p^sK_2(F)$. Since $p^r\{\zeta_m^{q-1}, \alpha\} = \{1, \alpha\} = 0$, we obtain that

$$p^rK_2(F) = p^{r+1}K_2(F),$$

and $K_2(F)/p^rK_2(F)$ is a cyclic group of order $\leqslant p^r$. On the other hand, the Hilbert symbol $(\cdot,\cdot)_{p^r}$ induces the surjective homomorphism

$$H_{p^r}: K_2(F) \to \mu_{p^r}.$$

Therefore, $K_2(F)/p^rK_2(F)$ is a cyclic group of order p^r if $r \geqslant 1$.

Now Proposition (4.1) implies that $K_2(F)/mK_2(F)$ is a cyclic group of order m and generated by $\{\zeta_m, \beta\}$ mod $mK_2(F)$ for some $\beta \in F^*$. We also deduce that $mK_2(F) = lmK_2(F)$ for $l \geqslant 1$. This means that $mK_2(F)$ is divisible and the exact sequence of the Theorem splits. □

COROLLARY. *Let A be a finite group, and let $f: K_2(F) \to A$ be a homomorphism. Then $f(mK_2(F)) = 1$ and there is a homomorphism $g: \mu_m \to A$ such that $f = g \circ H_m$.*

Proof. Let n be the order of A. Then $nK_2(F) \subset \ker(f)$ and

$$nK_2(F) \simeq \mu_m^n \oplus mK_2(F).$$

Therefore, the order of $K_2(F)/nK_2(F)$ is a divisor of m. Let $x \mod nK_2(F)$ generate $K_2(F)/nK_2(F)$ and $g\colon \mu_m \to A$ be a homomorphism such that $f(x) = g(H_m(x))$. Then $f = g \circ H_m$. □

(4.4). Our nearest purpose is to verify that $mK_2(F)$ is in fact a uniquely divisible group. The following assertion will be useful in the study of l-torsion in $K_2(F)$ for l relatively prime to p.

LEMMA. *Let l be a prime, $q-1$ divisible by l. Let $\theta \in \mu_{q-1}$, $\varepsilon \in U_{1,F}$. Then $\{1-\theta\varepsilon^l, \varepsilon\} = 0$ in $K_2(F)$.*

Proof. Put $L = F(\sqrt[l]{\theta})$. Then L/F is a cyclic extension of degree l or $L = F$. We get

$$1 - \theta\varepsilon^l = \prod_{i=1}^{l}(1 - \zeta_l^i\theta_1\varepsilon),$$

where $\theta_1 \in L$, $\theta_1^l = \theta$, and ζ_l is a primitive lth root of unity in F. If $L = F$, then

$$\{1 - \theta\varepsilon^l, \varepsilon\} = -\sum_{i=1}^{l}\{1 - \zeta_l^i\theta_1\varepsilon, \zeta_l^i\theta_1\},$$

and $\zeta_l^i\theta_1 \in \mu_{q-1}$, $1-\zeta_l^i\theta_1\varepsilon \in \mu_{q-1}U_{1,F}$. Lemma (4.1) implies now that $\{1-\theta\varepsilon^l, \varepsilon\} = 0$. If $L \neq F$, then $1 - \theta\varepsilon^l = N_{L/F}(1-\theta_1\varepsilon)$ and

$$\{1 - \theta\varepsilon^l, \varepsilon\} = N_{L/F}\{1 - \theta_1\varepsilon, \varepsilon\} = -N_{L/F}\{1 - \theta_1\varepsilon, \theta_1\}. \square$$

Let μ denote the group $\mu_{2(q-1)}$ for $p = 2$ and the group μ_{q-1} for $p > 2$. Then $-1 \in \mu$.

PROPOSITION (CARROLL). *Let l be prime, $q-1$ divisible by l. Let $lx = 0$ for some $x \in K_2(F)$. Then $x = \{\gamma, \alpha\}$ for some $\gamma \in \mu$, $\alpha \in F^*$.*

Proof. Introduce the map $f\colon F^* \times F^* \to K_2(F)/C$, where C is the subgroup in $K_2(F)$ generated by $\{\gamma, \alpha\}$ with $\gamma \in \mu$, $\alpha \in F^*$, by the formula

$$f(\alpha_1, \alpha_2) \equiv \{\pi, (\varepsilon_2^{a_1}\varepsilon_1^{-a_2})^{1/l}\} + \{\varepsilon_1, \varepsilon_2^{1/l}\} \mod C,$$

where $\alpha_1 = \pi^{a_1}\theta_1\varepsilon_1$, $\alpha_2 = \pi^{a_2}\theta_2\varepsilon_2$, $\theta_1, \theta_2 \in \mu_{q-1}$, $\varepsilon_1, \varepsilon_2 \in U_{1,F}$. Note that f is well defined, because the element $\varepsilon^{1/l} \in U_{1,F}$ for $\varepsilon \in U_{1,F}$ is uniquely determined. First we verify that f is 2-symbolic. Indeed, f is multiplicative, because the expression $(\varepsilon_2^{a_1}\varepsilon_1^{-a_2})^{1/l}$ depends multiplicatively on α_1, α_2. Next, if $p > 2$, then $-1 \in \mu_{q-1}$

and $f(\alpha_1,-\alpha_1)\in C$. If $p=2$ and $\alpha_2=-\alpha_1$, then $\varepsilon_2=-\varepsilon_1$, $\varepsilon_2^{1/l}=-\varepsilon_1^{1/l}$, and $f(\alpha_1,-\alpha_1)\in C$. Now let $\alpha_2=1-\alpha_1$. If $a_1>0$, then $1-\alpha_1=\varepsilon_2$ and

$$\begin{aligned}f(\alpha_1,1-\alpha_1)&\equiv\{\pi,(1-\pi^{a_1}\theta_1\varepsilon_1)^{a_1/l}\}+\{\varepsilon_1,(1-\pi^{a_1}\theta_1\varepsilon_1)^{1/l}\}\\&=\{\pi^{a_1}\varepsilon_1,(1-\pi^{a_1}\theta_1\varepsilon_1)^{1/l}\}\mod C.\end{aligned}$$

But $\{\pi^{a_1}\varepsilon_1,(1-\pi^{a_1}\theta_1\varepsilon_1)^{1/l}\}=\{\pi^{a_1}\varepsilon_1\theta_1,(1-\pi^{a_1}\theta_1\varepsilon_1)^{1/l}\}=\{1-\varepsilon_2,\varepsilon_2^{1/l}\}$ by Lemma (4.1). Then by the preceding Lemma we deduce that $f(\alpha_1,1-\alpha_1)\in C$. If $a_1=0$ and $a_2=0$, then $\alpha_1=\theta_1\varepsilon_1$, $1-\alpha_1=\theta_2\varepsilon_2$ and, likewise,

$$f(\alpha_1,1-\alpha_1)\equiv\{\varepsilon_1,\varepsilon_2^{1/l}\}=\{\varepsilon_1\theta_1,\varepsilon_2^{1/l}\}=\{1-\theta_2\varepsilon_2,\varepsilon_2^{1/l}\}=0\mod C.$$

If $a_1<0$ then

$$\begin{aligned}f(\alpha_1,1-\alpha_1)\equiv-f(\alpha_1^{-1},1-\alpha_1)&\equiv-f(\alpha_1^{-1},-\alpha_1^{-1}(1-\alpha_1))\\&=-f(\alpha_1^{-1},1-\alpha_1^{-1})\equiv0\mod C.\end{aligned}$$

Thus, f induces the homomorphism $f\colon K_2(F)\to K_2(F)/C$. We observe that

$$\{\alpha_1,\alpha_2\}\equiv\{\pi,\varepsilon_2^{a_1}\varepsilon_1^{-a_2}\}+\{\varepsilon_1,\varepsilon_2\}\mod C$$

by Lemma (4.1). Therefore, $lf(x)\equiv f(lx)\equiv x\mod C$ for $x\in K_2(F)$. This means that the condition $lx=0$ implies $x\in C$. □

COROLLARY. *Let $q-1$ be divisible by l. Let $lx=0$ for $x\in K_2(F)$. Then $x=\{\zeta_l,\pi\}$ for some l th root of unity ζ_l, where π is prime in F.*

Proof. First assume that l is prime. If $p>2$, then $\mu=\mu_{q-1}$ and the Proposition (4.1) and Lemma (4.1) imply $x=\{\theta,\pi^a\}$ for a generator θ of μ_{q-1} and an integer a. Proposition (4.1) shows that la is divisible by $q-1$, therefore $x=\{\zeta_l,\pi\}$ for $\zeta_l=\theta^a\in\mu_l$. If $p=2$ then by the same arguments we obtain $x=\{\theta,\pi^a\}+\{-1,\alpha\}$ for some $\alpha\in F^*$. Then $0=\{\theta^l,\pi^a\}+\{-1,\alpha\}$. As the order of $\{\theta,\pi^a\}$ is relatively prime to the order of $\{-1,\alpha\}$, we conclude that $\{-1,\alpha\}=al\{\theta,\pi\}=0$ and $x=\{\zeta_l,\pi\}$ for $\zeta_l=\theta^a\in\mu_l$.

Now let $l=l_1l_2$, $l_1>1$, $l_2>1$. We may suppose by induction that $l_1x=\{\zeta_{l_2},\pi\}$ for some $\zeta_{l_2}\in\mu_{l_2}$. Then $l_1(x-\{\zeta,\pi\})=0$ for $\zeta\in\mu_l$ with $\zeta^{l_1}=\zeta_{l_2}$. Hence, $x=\{\zeta,\pi\}+\{\zeta_{l_1},\pi\}=\{\zeta_l,\pi\}$ for a suitable root ζ_l. □

(4.5). Now we formulate without proof the following assertion due to *J.Tate* ([T6]):

Let char$(F)=0$, *$\mu_p\subset F^*$ and $px=0$ for $x\in K_2(F)$. Then $x=\{\zeta_p,\alpha\}$ for some $\zeta_p\in\mu_p$, $\alpha\in F^*$.*

A discussion of Tate's proof of this assertion would have involved introducing theories out of the scope of this book. It would be interesting to verify this assertion in an elementary way in the context of this book. Note that a theorem of *A.A. Suslin* asserts that

For an arbitrary field F and l relatively prime to char(F), $\mu_l \subset F^*$, *the equality* $lx = 0$ *in* $K_2(F)$ *implies* $x = \{\zeta_l, \alpha\}$ *for some* $\zeta_l \in \mu_l$, $\alpha \in F^*$ ([Sus 3]).

(4.6). For a Theorem of *A.S. Merkur'ev* to follow it is convenient first to prove

PROPOSITION. *Let $m = p^r(q-1)$ be the cardinality of the torsion group in F^*, $r \geqslant 1$. Let $2r-1 \geqslant t$, where t is the maximal integer such that $pe/(p-1)$ is divisible by p^t, $e = e(F/\mathbb{Q}_p)$. Let ζ_{p^r} be a primitive p^r th root of unity and $\zeta_{p^r} \notin U^p U_{e+1,F}$. Then the condition $\{\zeta_{p^r}, \pi\} \in p^r K_2(F)$ for a prime π in F implies $\{\zeta_p, \pi\} = 0$ for $\zeta_p \in \mu_p$.*

Proof. Take the generators of $U_{1,F}/U_{1,F}^p$ as in the proof of the Proposition (4.2). Let

$$\zeta_{p^r} = \prod (1 - \theta_{ij}\pi^i)^{ia_{ij}} (1 - \theta_* \pi^{pe/(p-1)})^{e_* a}$$

with $1 \leqslant i < pe/(p-1)$, where i is relatively prime to p, a_{ij}, $a \in \mathbb{Z}_p$. Then

$$\{\zeta_{p^r}, \pi\} = \{(1 - \theta_* \pi^{pe/(p-1)})^{e_* a}, \pi\},$$

and using Corollary (4.2) we obtain that $a \in p^r \mathbb{Z}_p$.

Our goal is to show that *there exists a prime element π_1 in F such that $\pi_1 \pi^{-1} \in U_{1,F}^p$ and*

$$\zeta_{p^r} = \prod (1 - \theta_{ij}\pi_1^i)^{ib_{ij}} (1 - \theta_* \pi_1^{pe/(p-1)})^{e_* b}$$

with $1 \leqslant i < pe/(p-1)$, where i is relatively prime to p, b_{ij}, $b \in \mathbb{Z}$ and b is divisible by p^r. From this assertion we deduce that

$$\{\zeta_p, \pi\} = \{\zeta_p, \pi_1\} = e_* b p^{r-1} \{1 - \theta_* \pi_1^{pe/(p-1)}, \pi_1\} = 0,$$

because bp^{r-1} is divisible by p^t and $p^t\{1 - \theta_* \pi_1^{pe/(p-1)},\ \pi_1\} = 0$.

To prove the assertion note that, since $pe/(p-1) \geqslant e+1$ and $\zeta_{p^r} \notin U_{e+1,F} U_{1,F}^p$, there is $i \leqslant e$ such that a_{ij} is relatively prime to p. Let $i_0 \leqslant e$ be the minimal number among all i. Consider the element

$$\alpha = \sum \frac{i^2 a_{ij} \theta_{ij} \pi^i}{1 - \theta_{ij}\pi^i} + \frac{e_* a \theta_* \pi^{pe/(p-1)}}{1 - \theta_* \pi^{pe/(p-1)}} \cdot pe/(p-1).$$

For $i < i_0$ we get $ia_i \in p\mathbb{Z}_p$ and hence

$$\alpha \equiv \sum_j \frac{i_0^2 a_{i_0 j} \theta_{i_0 j} \pi^{i_0}}{1 - \theta_{i_0 j}\pi^{i_0}} \equiv \left(\sum_j i_0^2 a_{i_0 j}\theta_{i_0 j}\right)\pi^{i_0} \mod \pi^{i_0+1}.$$

Recall that $\theta_{ij} \in R_0$ (see the proof of Proposition (4.2)). Then we obtain that

$$\sum i_0 a_{i_0 j}\theta_{i_0 j} \not\equiv 0 \mod \pi \qquad \text{and } v_F(\alpha) = i_0 .$$

Put

$$f(X) = \zeta_{p^r} - \prod (1 - \theta_{ij} X^i)^{ib_{ij}} (1 - \theta_* X^{pe/(p-1)})^{e_* b}$$

with b_{ij}, $b \in \mathbb{Z}$. If $v_p(b_{ij} - a_{ij}) \to +\infty$, $v_p(b - a) \to +\infty$, where v_p is the p-adic valuation, then $v_p(f(\pi)) \to +\infty$ and $v_p(f'(\pi) - \alpha\pi^{-1}\zeta_{p^r}) \to +\infty$. In particular, for $n \geqslant pe/(p-1) + 1$ there are integer b_{ij}, b such that $v_F(f(\pi)) \geqslant 2n + 1$ and $v_F(f'(\pi)) \leqslant n$. Corollary 3 of (1.3) Ch. II implies now that there exists an element $\pi_1 \in F$ such that $\pi_1\pi^{-1} \in U_{n,F} \subset U_{1,F}^p$ and $f(\pi_1) = 0$. Then π_1 is prime and is the desired element. □

(4.7). THEOREM (MERKUR'EV). *The group $mK_2(F)$ of Theorem* (4.3) *is an uncountable uniquely divisible group.*

Proof. $K_2(F)$ is uncountable by Proposition (3.9), because F and $\delta(F)$ are uncountable.

First we verify that the group $mK_2(F)$ has no nontrivial l-torsion for a prime $l \neq p$. If $\mu_l \subset F^*$ and $lx = 0$ for $x \in mK_2(F)$, then by Corollary (4.4) we get $x = \{\zeta_l, \pi\}$ for some $\zeta_l \in \mu_l$ and a prime π. Then Proposition (4.1) shows $x = 0$. If $\mu_l \not\subset F^*$, then put $F_1 = F(\mu_l)$. Assume that $lx = 0$ for $x \in mK_2(F)$. The divisibility of $mK_2(F)$ implies the existence of $y \in K_2(F)$ such that $x = m_L y$, where m_L is the cardinality of the torsion group of F_1^*. Then $lm_L j_{F/F_1}(y) = 0$ and $m_L j_{F/F_1}(y) = 0$. By Lemma (3.4) we obtain $m_L|F_1 : F|y = 0$. Thus, the order of x divides l and $|F_1 : F| < l$, and hence $x = 0$.

It remains to verify that there is no nontrivial p-torsion in $mK_2(F)$. If char$(F) = p$, then $\{\alpha, \beta\} = pN_{F^{1/p}/F}\{\alpha^{1/p}, \beta^{1/p}\}$. The map

$$f\colon F^* \times F^* \to K_2(F), \quad (\alpha, \beta) \mapsto N_{F^{1/p}/F}\{\alpha^{1/p}, \beta^{1/p}\}$$

is multiplicative, and $f(\alpha, 1 - \alpha) = N_{F^{1/p}/F}\{\alpha^{1/p}, 1 - \alpha^{1/p}\} = 0$. Therefore, f induces the homomorphism $f\colon K_2(F) \to K_2(F)$ such that $pf(x) = f(px) = x$ for $x \in K_2(F)$. This means that $K_2(F)$ has no nontrivial p-torsion.

Now we treat the most difficult case of char$(F) = 0$. Suppose that for some finite extension L/F the group $m_L K_2(L)$ is uniquely divisible, where m_L is the cardinality of the torsion group of L^*. Then if $x \in mK_2(F)$ and $px = 0$ we obtain $x = |L : F|m_L y$ for some $y \in K_2(F)$ and $p|L : F|j_{F/L}(m_L y) = 0$. Hence, $x = N_{L/F} j_{F/L}(m_L y) = 0$ and $mK_2(F)$ is uniquely divisible. Therefore, we can replace the field F by its proper finite extension. More specifically, we take the field $L = F^{(k)}$ of the following

PROPOSITION. *Put $F^{(m)} = F(\mu_{p^m})$. Let r_m denote the maximal integer such that $\mu_{p^{r_m}}$ is contained in $F^{(m)*}$, and let t_m be the maximal integer for which $pe(F^{(m)}|\mathbb{Q}_p)$ is divisible by p^{t_m}. Then there exists a natural $k \geqslant 1$, such that extensions $F^{(k+m)}/F^{(k)}$ are totally ramified of degree p^m, $m \geqslant 1$ and $2r_k \geqslant t_k + 1$, $\mu_{p^k} \not\subset U_{F^{(k)}}^p U_{e+1,F^{(k)}}$, where $e = e(F^{(k)}|\mathbb{Q}_p)$.*

Proof. Since $F^{(m)} \supset \mathbb{Q}_p^{(r_m)}$, we get $e(F^{(m)}|\mathbb{Q}_p) \in (p-1)p^{r_m-1}\mathbb{Z}$ by (1.3) Ch. IV. Hence, $t_m \geqslant r_m - 1$. If $r_{m+1} = r_m$, then $F^{(m+1)} = F^{(m)}$ and $t_{m+1} = t_m$; if $r_{m+1} > r_m$ then $F^{(m+1)}/F^{(m)}$ is of degree p and $t_{m+1} \leqslant t_m + 1$. Therefore, in any case $t_m - r_m \geqslant -1$, and $t_m - r_m$ does not increase when m increases. This means that there is a natural n such that $r_{n+m} = r_{n+m-1} + 1$, $t_{n+m} = t_{n+m-1} + 1$ for $m \geqslant 1$, $r_n = n$ and $2r_{n+m} \geqslant t_{n+m} + 1$ for $m \geqslant 0$. We obtain that $F^{(n+m)}/F^{(n)}$ is a totally ramified extension of degree p^m.

We next show that, for a sufficiently large m,

$$(*) \qquad N_{F^{(n+m)}/F^{(n)}} U_{e_{n+m}+1, F^{(n+m)}} \subset U_{F^{(n)}}^p,$$

where $e_{n+m} = e(F^{(n+m)}|\mathbb{Q}_p)$.

Then if $\mu_{p^{n+m}} \subset U_{F^{(n+m)}}^p U_{e_{n+m}+1, F^{(n+m)}}$ we would have

$$\mu_{p^n} = N_{F^{(n+m)}/F^{(n)}} \mu_{p^{n+m}} \subset U_{F^{(n)}}^p,$$

which is impossible in view of the choice of n. Thus, we deduce that the assertion of the Proposition holds for $k = n + m$.

To verify $(*)$ write $\varepsilon \in U_{e_{n+m}+1, F^{(n+m)}}$ as $\varepsilon = 1 + p\alpha$ with $\alpha \in \mathcal{M}_{F^{(n+m)}}$. Then the formula of Lemma (1.1) Ch. III implies the congruence

$$N_{F^{(n+m)}/F^{(n)}} \varepsilon \equiv 1 + p \operatorname{Tr}_{F^{(n+m)}/F^{(n)}}(\alpha) \mod \pi_n^{pe_n/(p-1)+1},$$

where π_n is prime in $F^{(n)}$ because $p^2 \in \pi_n^{pe_n/(p-1)} \mathcal{O}_{F^{(n)}}$. Therefore, it suffices to verify that $\operatorname{Tr}_{F^{(n+m)}/F^{(n)}}(\mathcal{O}_{F^{(n+m)}}) \to 0$ as $m \to +\infty$.

Put $\beta = \operatorname{Tr}_{F^{(n+m)}/F^{(n)}}(\alpha) \in \mathcal{O}_{F^{(n)}}$. Let $i = [e_n^{-1} v_{F^{(n)}}(\beta)]$. Then there exists $\delta \in \mathbb{Z}_p$ with $v_{F^{(n)}}(\delta) = (i+1)e_n$. For $\gamma = \delta\beta^{-1}$ we get $0 < v_{F^{(n)}}(\gamma) \leqslant e_n$, $\gamma\beta \in \mathbb{Z}_p$. Put $d_n = |F^{(n)} : \mathbb{Q}_p|$. Then $e_n \leqslant d_n$ and

$$\operatorname{Tr}_{F^{(n+m)}/\mathbb{Q}_p}(\gamma\alpha) = \operatorname{Tr}_{F^{(n)}/\mathbb{Q}_p}(\gamma \operatorname{Tr}_{F^{(n+m)}/F^{(n)}}(\alpha)) = d_n\gamma\beta.$$

Hence

$$\operatorname{Tr}_{F^{(n+m)}/F^{(n)}}(\alpha) = d_n^{-1}\gamma^{-1} \operatorname{Tr}_{\mathbb{Q}_p^{(n+m)}/\mathbb{Q}_p}(\operatorname{Tr}_{F^{(n+m)}/\mathbb{Q}_p^{(n+m)}}(\gamma\alpha)).$$

Therefore, it remains to show that

$$\operatorname{Tr}_{\mathbb{Q}_p^{(m)}/\mathbb{Q}_p}\left(\mathcal{O}_{\mathbb{Q}_p^{(m)}}\right) \to 0 \quad \text{as} \quad m \to +\infty.$$

By (1.3) Ch. IV $\mathcal{O}_{\mathbb{Q}_p^{(m)}} = \mathcal{O}_{\mathbb{Q}_p}[\zeta_{p^m}]$. Now, since

$$\operatorname{Tr}_{\mathbb{Q}_p^{(m)}/\mathbb{Q}_p}(\mu_{p^m}) \to 0 \qquad \text{as} \quad m \to +\infty$$

(it is straightforward to compute the coefficient of $X^{p^{m-1}(p-1)-1}$ in the polynomial $f_m(X)$ of (1.3) Ch. IV), we obtain the required assertion $(*)$ and complete the proof. □

Returning to the proof of the Theorem, we set $L = F^{(k)}$. Let $px = 0$ for $x \in m_L K_2(L)$. By (4.5) we get $x = \{\zeta_p, \alpha_0\}$ for some $\zeta_p \in \mu_p$, $\alpha_0 \in L^*$. Let $m_L = p^r(q-1)$, $r \geqslant 1$. As $\{\zeta_p, \alpha_0\} \in p^r K_2(L)$, we deduce with the help of Corollary (4.2) that $\{\zeta_{p^r}, \alpha_0\} \in pK_2(L)$ for an element $\zeta_{p^r} \in \mu_{p^r}$ with $\zeta_{p^r}^{p^{r-1}} = \zeta_p$. Since $\zeta_{p^r} \notin L^{*p}$, we conclude, by the same arguments as in the proof of Theorem (4.3), that $\{\zeta_{p^r}, \alpha\} \mod p^r K_2(L)$ generates $K_2(L)/p^r K_2(L)$ for some $\alpha \in L^*$. This means that $\{\zeta_{p^r}, \alpha_0\alpha^{-c}\} \in p^r K_2(L)$ for some $c \in p\mathbb{Z}$. If $\alpha_0\alpha^{-c}$ is prime in L, then Proposition (4.6) implies $x = \{\zeta_p, \alpha_0\} = \{\zeta_p, \alpha_0\alpha^{-c}\} = 0$. If this is not the case, then let π_0 be a prime element in L belonging to the norm subgroup $N_{L^{(2r)}/L} L^{(2r)*}$. Property (5) of Proposition (5.1) Ch. IV shows that $(\zeta_{p^r}, \pi_0)_{p^r} = 1$. Then by Theorem (4.3) we deduce $\{\zeta_{p^r}, \pi_0\} \in p^r K_2(L)$ and by Proposition (4.6) $\{\zeta_p, \pi_0\} = 0$. Let s be an integer such that $\pi_0^s \alpha_0 \alpha^{-c}$ is prime in L. Then $\{\zeta_{p^r}, \pi_0^s\alpha_0\alpha^{-c}\} \in p^r K_2(L)$ and by Proposition (4.6) $\{\zeta_p, \pi_0^s\alpha_0\alpha^{-c}\} = 0$. Thus, $x = \{\zeta_p, \alpha_0\} = \{\zeta_p, \pi_0^s\alpha_0\} - s\{\zeta_p, \pi_0\} = 0$. This completes the proof. □

(4.8). The rest of this section is concerned with the Milnor K_n-groups of F for $n \geqslant 3$.

PROPOSITION. *Let L/F be an abelian extension of finite degree. Then the norm map $N_{L/F}\colon K_2(L) \to K_2(F)$ is surjective.*

Proof. Let $d = |L : F|$. It suffices to prove the assertion for a prime d. If d is relatively prime to m, where m is the cardinality of the torsion group in F^*, then by Theorem (4.3)

$$dK_2(F) \simeq mK_2(F) \oplus \mu_m^d \simeq mK_2(F) \oplus \mu_m \simeq K_2(F),$$

and $K_2(F) = N_{L/F}(j_{F/L} K_2(F))$ by Theorem (3.8). Let m be divisible by d. The norm subgroup $N_{L/F} L^*$ is of index d in F^*, according to (1.5) Ch. IV. Since the index of F^{*d} in F^* is $> d$, there exists an element $\alpha \in N_{L/F} L^*$, $\alpha \notin F^{*d}$. Property (6) of Proposition (5.1) Ch. IV shows that $(\alpha, \beta)_d \neq 1$ for some $\beta \in F^*$. Therefore, $\{\alpha, \beta\}$ mod $dK_2(F)$ generates the cyclic group $K_2(F)/dK_2(F)$ and $\{\alpha, \beta\} \in N_{L/F} K_2(L)$. Since $dK_2(F) = N_{L/F}(j_{F/L} K_2(F))$, we deduce $K_2(F) = N_{L/F} K_2(L)$. □

(4.9). PROPOSITION. *Let l be relatively prime to* char$(\overline{F})$, *$\mu_l \subset F^*$. Let L/F be a cyclic extension of degree l, σ a generator of* Gal(L/F), *and $\sigma\colon K_2(L) \to K_2(L)$ the homomorphism induced by σ. Then the sequence*

$$K_2(L) \xrightarrow{1-\sigma} K_2(L) \xrightarrow{N_{L/F}} K_2(F)$$

is exact.

Proof. By Theorem (4.3) the groups $K_2(L)/lK_2(L)$, $K_2(F)/lK_2(F)$ are cyclic of order l. The preceding Proposition implies now that the homomorphism

$$K_2(L)/lK_2(L) \to K_2(F)/lK_2(F)$$

induced by the norm map $N_{L/F}$ is an isomorphism. Therefore, if $N_{L/F}(x) = 0$ for $x \in K_2(L)$, then $x = ly$ for some $y \in K_2(L)$ and $lN_{L/F}(y) = 0$. By subsections (4.4) and (4.5) $N_{L/F}(y) = \{\zeta_l, \alpha\}$, for some $\alpha \in F^*$ and a primitive lth root ζ_l of unity. By Theorem (3.8) we get

$$j_{F/L}N_{L/F}(y) = (1 + \cdots + \sigma^{l-1})y = j_{F/L}\{\zeta_l, \alpha\}.$$

Let $L = F(\sqrt[l]{\beta})$. Then we may assume $\zeta_l = \beta/\sigma(\beta)$. We obtain

$$\begin{aligned} x = ly &= (l - 1 - \cdots - \sigma^{l-1})y + (1 - \sigma)\{\beta, \alpha\} \\ &= (1 - \sigma)((\sigma^{l-2} + 2\sigma^{l-3} + \cdots + l - 1)y + \{\beta, \alpha\}). \end{aligned}$$

Conversely, $N_{L/F}(1 - \sigma)x = 0$ for $x \in K_2(L)$ by Theorem (3.8). □

REMARK. The assertion of the Proposition, so-called "Satz 90" for K_2-groups, holds for arbitrary fields (Merkur'ev-Suslin, [MS]).

(4.10). PROPOSITION. *Let l be prime, $\mu_l \subset F^*$. Then $\{\zeta_l\} \cdot x = 0$ in $K_3(F)$ for $x \in K_2(F)$, $\zeta_l \in \mu_l$.*

Proof. If $q - 1$ is divisible by l, then Proposition (4.1) implies that $\{\zeta_l\} \cdot x = \{\zeta_l, \zeta_{q-1}, \alpha\}$ for some $\alpha \in F^*$ and Lemma (4.1) shows that $\{\zeta_l\} \cdot x = 0$. If $l = p$ and $p \neq 2$, $m = p^r(q - 1)$, then $\{-\zeta_{p^r}, \alpha\}$ mod $pK_2(F)$ generates the quotient group $K_2(F)/pK_2(F)$ for a primitive p^rth root of unity ζ_{p^r} and some $\alpha \in F^*$. Therefore, $\{\zeta_p\} \cdot x = \{\zeta_p, -\zeta_{p^r}, \alpha^c\}$ for some integer c and $\{\zeta_p\} \cdot x = p^{r-1}\{\zeta_{p^r}, -\zeta_{p^r}, \alpha^c\} = 0$. Finally, if $l = p = 2$, then $\{-1\} \cdot x = \{-1, \zeta_{2^r}, \alpha\}$ for some $\alpha \in F^*$. If $r > 1$, then $\{-1, \zeta_{2^r}\} = 2^{r-1}\{\zeta_{2^r}, \zeta_{2^r}\} = 2^{r-1}\{-1, \zeta_{2^r}\} = 0$. If $r = 1$, then $x = N_{F(\sqrt{-1})/F}y$ for some $y \in K_2(F(\sqrt{-1}))$, by Proposition (4.8). Then $\{-1\} \cdot x = N_{F(\sqrt{-1})/F}(\{-1\} \cdot y)$ and $\{-1\} \cdot y = 0$ in $K_2(F(\sqrt{-1}))$. □

(4.11). THEOREM (SIVITSKII). *The group $K_n(F)$ is an uncountable uniquely divisible group for $n \geqslant 3$.*

Proof. First we assume that $\mu_l \subset F^*$, and that l is relatively prime to char(F). Define the map $f\colon \underbrace{F^* \times \cdots \times F^*}_{n \text{ times}} \to K_n(F)$ by the formula

$$f(\alpha_1, \ldots \alpha_n) = N_{F(\sqrt[l]{\alpha_1})/F}(\{\alpha\} \cdot x \cdot \{\alpha_4, \ldots, \alpha_n\}),$$

where $x \in K_2(F(\sqrt[l]{\alpha_1}))$ with $N_{F(\sqrt[l]{\alpha_1})/F}(x) = \{\alpha_2, \alpha_3\}$ (x exists by Proposition (4.8)), $\alpha \in F(\sqrt[l]{\alpha})$ with $\alpha^l = \alpha_1$. By Proposition (4.10) f does not depend on the choice of α. Moreover, if $N_{F(\sqrt[l]{\alpha_1})/F}x = N_{F(\sqrt[l]{\alpha_1})/F}y$, then $x-y=\sigma(z)-z$ for some $z \in K_2(F(\sqrt[l]{\alpha_1}))$ by Proposition (4.9), where σ is a generator of $\mathrm{Gal}(F(\sqrt[l]{\alpha_1})/F)$. Then

$$\{\alpha\}\cdot(x-y)=(\sigma-1)(\{\alpha\}\cdot z)+\{\alpha\sigma(\alpha^{-1})\}\cdot\sigma(z),$$

and

$$N_{F(\sqrt[l]{\alpha_1})/F}(\{\alpha\}\cdot(x-y))=\{\zeta_l\}\cdot N_{F(\sqrt[l]{\alpha_1})/F}(z)=0,$$

by Theorem (3.8) and Proposition (4.10), where $\zeta_l=\alpha\cdot\sigma(\alpha^{-1})\in\mu_l$. Thus, the map f is well defined.

Furthermore, the map f is multiplicative on $\alpha_4,\dots,\alpha_n$. It is also multiplicative on α_2,α_3, because if $\alpha_2=\alpha_2'\alpha_2''$, then $x=x'+x''$ and $f(\alpha_1,\alpha_2'\alpha_2'',\dots)=f(\alpha_1,\alpha_2',\dots)$ $+f(\alpha_1,\alpha_2'',\dots)$. Let $\alpha_1=\alpha_1'\alpha_1''$ and $L=F(\sqrt[l]{\alpha_1'},\sqrt[l]{\alpha_1''})$. Then $\{\alpha_2,\alpha_3\}=N_{L/F}y$ for some $y\in K_2(L)$ by Proposition (4.8). Therefore, for $\alpha'^l=\alpha_1'$, $\alpha''^l=\alpha_1''$, $\alpha^l=\alpha_1$ we get

$$\begin{aligned}
&f(\alpha_1',\alpha_2,\dots)=N_{F(\sqrt[l]{\alpha_1'})/F}(\{\alpha'\}\cdot N_{L/F(\sqrt[l]{\alpha_1'})}y\cdot\dots)=N_{L/F}(\{\alpha'\}\cdot y\cdot\dots),\\
&f(\alpha_1'',\alpha_2,\dots)=N_{L/F}(\{\alpha''\}\cdot y\cdot\dots),\\
&f(\alpha_1'\alpha_1'',\alpha_2,\dots)=N_{L/F}(\{\alpha'\alpha''\}\cdot y\cdot\dots).
\end{aligned}$$

Thus, f is multiplicative on α_1.

Let $\alpha_1+\alpha_n=1$, then $f(\alpha_1,\dots,\alpha_n)=N_{F(\sqrt[l]{\alpha_1})/F}(\{\alpha\}\cdot x\cdot\{\dots,1-\alpha_1\})$. Since $1-\alpha_1=\prod_{i=1}^{l}(1-\zeta_l^i\alpha)$, we deduce

$$\{\alpha,1-\alpha_1\}=\sum_{i=1}^{l}\{\zeta_l^{-i},1-\zeta_l^i\alpha\}.$$

Now Proposition (4.10) implies that $f(\alpha_1,\dots,\alpha_n)=0$. Let $\alpha_1+\alpha_2=1$ and $\alpha\notin F$. Then $\alpha_2=N_{F(\sqrt[l]{\alpha_1})/F}(1-\alpha)$ and

$$f(\alpha_1,\dots,\alpha_n)=N_{F(\sqrt[l]{\alpha_1})/F}\{\alpha,1-\alpha,\alpha_3,\dots\}=0.$$

Let $\alpha_1+\alpha_2=1$ and $\alpha\in F$. Then $\alpha_2=\prod_{i=1}^{l}(1-\zeta_l^i\alpha)$ and

$$f(\alpha_1,\dots,\alpha_n)=\sum_{i=1}^{l}\{\alpha,1-\zeta_l^i\alpha,\dots\}=\sum_{i=1}^{l}\{\zeta_l^{-i},1-\zeta_l^i\alpha,\dots\}=0.$$

Let $\alpha_1+\alpha_n=1$. Then $f(\alpha_1,\dots,\alpha_n)=-f(\alpha_2,\alpha_1,\dots,\alpha_n)=0$. Thus, f is n-symbolic. It induces the homomorphism $f\colon K_n(F)\to K_n(F)$, $n\geqslant 3$. We get $lf(x)=f(lx)=x$ for $x\in K_n(F)$. Hence, $K_n(F)$ is uniquely l-divisible.

Suppose now that l is relatively prime to char(F), $\mu_l\not\subset F^*$. Put $F_1=F(\mu_l)$. Then for an element $x\in K_n(F)$ with $lx=0$ we get $j_{F/F_1}(x)=0$. Therefore, $|F_1:F|x=0$ by Lemma (3.4). As $|F_1:F|$ is relatively prime to l, we conclude that

$x = 0$. By Proposition (4.8) for $x \in K_n(F)$ there is $y \in K_n(F_1)$ with $N_{F_1/F}(y) = x$. Then $y = lz$ for some $z \in K_n(F_1)$ and $x = lN_{F_1/F}(z)$. Thus, $K_n(F)$ is uniquely l-divisible.

Assume finally that $l = p = \operatorname{char}(F)$. Then the map

$$f\colon \underbrace{F^* \times \cdots \times F^*}_{n \text{ times}} \to K_n(F), \quad (\alpha_1, \dots, \alpha_n) \mapsto N_{F^{1/p}/F}\{\alpha_1^{1/p}, \alpha_2^{1/p}, \alpha_3, \dots, \alpha_n\},$$

is n-symbolic (see the proof of Theorem (4.7)). We conclude, that it induces the map $f\colon K_n(F) \to K_n(F)$ with $pf(x) = f(px) = x$. Thus, $K_n(F)$ is uniquely p-divisible. □

(4.12). REMARKS.

1. For further information on the Milnor ring of a complete discrete valuation field with a perfect residue field see [Kah], and also [Bog]. A computation of Quillen's K-groups of a local field can be found in [Sus2].

2. Differential forms are important for the study of quotients of the Milnor K-groups annihilated by a power of p, see Exercise 7 section 1; they are useful in the proof of a theorem of *S. Bloch–K.Kato* which claims that for every l not divisible by char(F) the symbol map

$$K_m(F)/lK_m(F) \to H^m(F, \mu_l^{\otimes m})$$

is an isomorphism for Henselian discrete valuation fields. For an arbitrary field F and $m = 2$ the symbol map is an isomorphism according to the famous *Merkur'ev–Suslin* theorem [MS].

When the residue field $\overline{F}$ of a complete discrete valuation fields of characteristic zero is imperfect an effective tool for the description of quotients of $K_m(F)$ is *Kurihara*'s exponential map [Ku4]

$$\exp\colon \varprojlim_n \Omega^m_{\mathcal{O}_F} \otimes \mathbb{Z}/p^n\mathbb{Z} \to \varprojlim_n K_m(F) \otimes \mathbb{Z}/p^n\mathbb{Z}.$$

(4.13). Theorems (4.3) and (4.11) demonstrate that the most interesting group in the list of Milnor K-groups of a local field F with finite residue field is the group $K_1(F)$ (infinitely divisible parts are not of great arithmetical interest). The latter group is related via the local reciprocity map described in Ch. IV and V to the maximal abelian extension of F.

One can interpret the injective homomorphism $\mathbb{Z} \to \operatorname{Gal}(K^{\mathrm{sep}}/K)$ for a finite field K as the 0-dimensional local reciprocity map

$$K_0(K) \to \operatorname{Gal}(K^{\mathrm{ab}}/K).$$

It is then natural to expect that for an n-dimensional local field F, as in (5.5) Ch. VII, its nth Milnor K-group $K_n(F)$ should be related to abelian extensions of F. And indeed, there is a higher dimensional local class field theory first developed by *A.N. Parshin* in characteristic p [Pa1–5], *K. Kato* in the general case [Kat1–3]. We briefly describe here how one can generalize the theory of Ch. IV and V to obtain another approach [Fe3–5] to a *higher dimensional local reciprocity map*

$$\Psi_F: K_n(F) \to \mathrm{Gal}(F^{\mathrm{ab}}/F).$$

Let L/F be a finite Galois extension and $\sigma \in \mathrm{Gal}(L/F)$. Denote by F' the maximal unramified extension of F corresponding the maximal separable extension of its last residue field $\mathbb{F}_q$ (see (5.5) Ch. VII). Then there is $\widetilde{\sigma} \in \mathrm{Gal}(LF'/F)$ such that $\widetilde{\sigma}|_L = \sigma$ and $\widetilde{\sigma}_{F'}$ is a positive power of the lifting of the Frobenius automorphism of $G_{\mathbb{F}_q}$. The fixed field Σ of $\widetilde{\sigma}$ is a finite extension of F. Let $t_1, \dots, t_n$ be a lifting of prime elements of residue fields $\Sigma_1, \dots, \Sigma_{n-1}, \Sigma$ of Σ to Σ. A generalization of the Neukirch map is then defined as

$$\sigma \mapsto N_{\Sigma/F}\{t_1, \dots, t_n\} \mod N_{L/F}K_n(L).$$

A specific feature of higher dimensional local fields is that in general for an arbitrary finite Galois extension L/F linearly disjoint with F'/F a generalization of the Hazewinkel homomorphism does not exist. This is due to the fact that the map $i_{F/F'}: K_n(F) \to K_n(F')$ is not injective for $n > 1$. Still one can define a generalization of the Hazewinkel map for extensions which are composed of Artin–Schreier extensions, and this is enough to prove that the Neukirch map induces an isomorphism

$$\mathrm{Gal}(L/F)^{\mathrm{ab}} \xrightarrow{\sim} K_n(F)/N_{L/F}K_n(L)$$

[Fe7].

Contrary to the case of $n = 1$ the kernel of the map Ψ_F is nontrivial for $n > 1$; it is equal to $\cap_{l \geqslant 1} lK_n(F)$. The quotient of $K_n(F)$ by the kernel can be described in terms of topological generators as a generalization of results of section 6 Ch. I.

For more details and various approaches to higher local class field theory see papers in [FK].

Exercises.

1. Let A be a topological Hausdorff group, and let $f: F^* \times F^* \to A$ be a continuous symbolic map. Show that if m is the cardinality of the torsion group of F^*, then $mf = 0$. Deduce that there is a homomorphism $\psi: \mu_m \to A$ such that $f = \psi \circ (\cdot,\cdot)_m$, where $(\cdot,\cdot)_m$ is the m th Hilbert symbol.
2. Show that for a finite extension L/F of local number fields the norm homomorphism
$$N_{L/F}: K_2(L) \to K_2(F)$$
is surjective.
3. Show that the cokernel of the homomorphism $N_{\mathbb{C}/\mathbb{R}}: K_n(\mathbb{C}) \to K_n(\mathbb{R})$ is a cyclic group of order 2.
4. (*J. Tate*)
   a) Let F be a field, $\alpha, \beta \in F^*$, and $\beta^{m+1} - \beta^m - \beta + 1 = \alpha$, $m \geqslant 2$. Show that $m\{\alpha, \beta\} = 0$ in $K_2(F)$.

b) Let $F = \mathbb{Q}_p(\zeta_{p^n})$, where ζ_{p^n} is a primitive p^n th root of unity, $n \geqslant 1$. Show that the polynomial $X^{m+1} - X^m - X + 1 - \zeta_{p^n}$ has a unique root β_m such that $v_F(\beta_m) = v_F(\zeta_{p^n} - 1) = 1$.

c) Put $\varepsilon_m = \beta_m(1 - \zeta_{p^n})^{-1}$, $m \geqslant 2$. Show that the elements ε_m, $2 \leqslant m \leqslant p^n + 1$ generate the group $U_{1,F}/U_{1,F}^p$.

d) Show that $\{\zeta_{p^n}, \varepsilon_m\} = 0$ for $2 \leqslant m < p^n$, $m = p^n + 1$, and $\{\zeta_{p^n}, \varepsilon_{p^n}\}$ generates the p-torsion group of $K_2(F)$.

5. a) Let L/F be a cyclic extension, and let σ be a generator of $\mathrm{Gal}(L/F)$. Show that the sequence

$$0 \to K_n(F) \xrightarrow{j_{F/L}} K_n(L) \xrightarrow{1-\sigma} K_n(L) \xrightarrow{N_{L/F}} K_n(F)$$

is exact for $n \geqslant 3$. (*O.T. Izhboldin* proved that if $n \geqslant 2$, then this sequence is exact for an arbitrary field F of characteristic p when L/F is of degree p^m, see [Izh]).

b) Let L/F be a cyclic unramified extension, and let σ be a generator of $\mathrm{Gal}(L/F)$. Show that the sequence

$$0 \to K_2(F) \xrightarrow{j_{F/L}} K_2(L) \xrightarrow{1-\sigma} K_2(L)$$

is exact if $\mathrm{char}(F)$ is positive.

c) Let L/F be a finite extension. Show that the cardinality of the kernel of the homomorphism $j_{F/L}\colon K_2(F) \to K_2(L)$ is equal to $|t_p(F)/N_{L/F}t_p(L)|$ where $t_p(K)$ for a field K stands for the group of roots of unity in K^* of order a power of p.

6. ($\diamond$) Let F be a local number field and $g(X) \in 1 + X\mathbb{Z}_p[[X]], \notin 1 + X^2\mathbb{Z}_p[[X]]$. Let A be a Hausdorff topological group. A continuous multiplicative pairing $c\colon F^* \times U_F \to A$ is called g-symbolic if $c(\alpha, g(\alpha)) = 1$ for all $\alpha \in \mathcal{M}_F$.

a) Show that $c(F^*, U_F)$ is generated by $c(\pi, \omega_*)$ and $c(\pi, \theta)$ for prime elements π in F, $\theta \in \mu_{q-1}, \omega_*$ as in (1.6).

b) Let $E(X)$ be the Artin–Hasse function (see (9.1) Ch. I). Show that the Hilbert symbol $H_{p^n}\colon K_2(F) \to \mu_{p^n}$ is E-symbolic.

c) Let $p > 2$. Show that the Hilbert symbol H_{p^n} is g-symbolic if and only if $v_F(c_m) \geqslant v_{\mathbb{Q}_p}(m)$ for the series $\sum_{m\geqslant 1} c_m X^m = l_X(g(X))$ (for the definition of l_X see in section 2 Ch. VI).

Bibliography

Comments

Introductory sources on related subjects.
local fields [Cas];
algebraic number theory [KKS], [M], [N5], [NSchW], [CF], [FT], [BSh], [Iya], [La2], [IR], [Ko6], [W];
cyclotomic fields [Wa], [La3];
valuation theory [E], [Rib];
formally p-adic fields [PR];
non-Archimedean analysis [Kob1–2], [vR], [Schf], [T4], [BGR];
embedding problems [ILF];
formal groups [Fr], [CF], [Iw6], [Haz3];
elliptic curves over number fields [Silv];
local zeta function and Fourier analysis [T1], [RV], [Ig], [Den];
p-adic L-functions [Wa], [Iw7], [Hi];
local Langlands correspondence [T7], [Bum], [Kudl], [BaK], [Rit2];
pro-p-groups [DdSMS], [Wi], [dSSS];
p-adic Hodge theory [T2], [Fo2], [Sen4,7–9];
p-adic periods [A];
p-adic differential equations [RC];
non-Archimedean analytic geometry [Ber];
field arithmetic [FJ], [Jar], [Ef4];
characteristic p [Gos];
Milnor K-theory [Bas], [Ro], [Silr], [Gr];
higher local fields and higher local class field theory [FK];
power series over local fields, formal groups, and dynamics [Lu1–2], [Li1–4];
non-Archimedean physics [VVZ], [BF], [Chr], [RTVW], [HS], [Kh].

Symbols and explicit formulas (perfect residue field case). [AH1–2], [Has1–11], [Sha2], [Kn], [Rot], [Bru1–2], [Henn1–2], [Iw3], [Col1,3], [Wil], [CW1], [dSh1–3], [Sen3], [Hel], [Des], [Shi], [Sue], [ShI], [V1–7,9], [Fe1–2], [BeV1–2], [Ab5–6], [Kol], [Kuz], [Kat6–7], [Ku3–4], [GK], [VG], [DV1–2], [Ben1–2].

Ramification theory of local fields (perfect residue field case). [Kaw1], [Sa], [Tam], [Hei], [Mar1], [Mau1–5], [Mik5–6], [T2], [Wy], [Sen1–2], [ST], [Ep], [KZ], [Fo4], [Win1–4], [Lau1–6], [LS], [CG], [Ab2–4,7–8], [Fe8,11–12].

Bibliography

[A] *Périodes p-adiques, Astérisque*, vol. 223, SMF, 1994.

[Ab1] V. A. Abrashkin, *Galois modules of group schemes of period p over the ring of Witt vectors*, Izv. Akad. Nauk SSSR Ser. Mat. **51** (1987), no. 4, 691–736; English transl. in Math. USSR-Izv. **31** (1988).

[Ab2] ———, *Ramification filtration of the Galois group of a local field*, Tr. St-Peterbg. Mat. Obsch., vol. 3, 1993; English transl. in Amer. Math. Soc. Transl. Ser. 2, vol. 166, AMS, 1995, pp. 35–100.

[Ab3] ———, *A ramification filtration of the Galois group of a local field. II*, Proc. Steklov Inst. Math., vol. 208, 1995, pp. 18–69; English transl. in Proc. Steklov Inst. Math., vol. 208.

[Ab4] ———, *A ramification filtration of the Galois group of a local field. III*, Izv. Ross. Akad. Nauk Ser. Mat. **62 no. 6** (1998), 3–48; English transl. in Izv. Math. **62** (1998), 857–900.

[Ab5] ———, *The field of norms functor and the Brückner-Vostokov formula*, Math. Ann. **308** (1997), 5–19.

[Ab6] ———, *Explicit formulas for the Hilbert symbol of a formal group over Witt vectors*, Izv. Ross. Akad. Nauk Ser. Mat. **61 no. 3** (1997), 3–56; English transl. in Izv. Math. **61** (1997), 463–515.

[Ab7] ———, *A group-theoretic property of ramification filtration*, Izv. Ross. Akad. Nauk Ser. Mat. **62 no. 6** (1997), 3–26; English transl. in Izv. Math. **62** (1998), 1073–1094.

[Ab8] ———, *On a local analogue of the Grothendieck conjecture*, Internat. J. Math **11** (2000), 133–175.

[AdCK] E. Arbarello, C. de Concini, V.G. Kac, *The infinite wedge representation and the reciprocity law for algebraic curves*, Proc. Symp. Pure Math., vol. 49 Part I, 1989, pp. 171–190.

[AH1] Emil Artin and H. Hasse, *Über den zweiten Ergänzungssatz zum Reziprozitätsgesetz der l-ten Potenzreste im Körper k_ζ der l-ten Einheitswurzeln und Oberkörpern von k_ζ*, J. Reine Angew. Math. **154** (1925), 143–148.

[AH2] ———, *Die beiden Ergänzungssatz zum Reziprzitätsgesetz der l^n-ten Potenzreste im Körper der l^n-ten Einheitswurzeln*, Abh. Math. Sem. Univ. Hamburg **6** (1928), 146–162.

[Am] Shigeru Amano, *Eisenstein equations of degree p in a $\mathfrak{p}$-adic field*, J. Fac. Sci. Univ. Tokyo Sect. IA Math. **18** (1971), 1–22.

[Ar] Emil Artin, *Algebraic numbers and algebraic functions*, Gordon and Breach, New York, London, and Paris, 1967.

[ASch] Emil Artin and O. Schreier, *Eine Kennzeichnung der reell abgeschlossen Körper*, Abh. Math. Sem. Univ. Hamburg **5** (1927), 225–231.

[AT] Emil Artin and John T. Tate, *Class field theory. Second Edition*, Addison–Wesley, 1990.

[AV] David K. Arrowsmith and Franco Vivaldi, *Goemetry of p-adic Siegel discs*, Phys. D. **71** (1994), 222–236.

[Ax] James Ax, *Zeros of polynomials over local fields*, J. Algebra **15** (1970), 417–428.

[AxK] James Ax and Simon Kochen, *Diophantine problems over local fields.* I, Amer. J. Math. **87** (1965), 605–630; II, Amer. J. Math. **87** (1965), 631–648; III, Ann. of Math. (2) **83** (1966), 437–456.

[Bah] George Bachman, *Introduction to p-adic numbers and valuation theory*, Academic Press, New York and London, 1964.

[BaK] T. N. Bailey and A. W. Knapp (eds.), *Representation theory and automorphic forms. Proc. Symp. Pure Math.*, vol. 61, 1997.

[Bas] Hyman Bass, *Algebraic K-theory*, Benjamin, New York and Amsterdam, 1968.

[Ben1] Denis Benois, *Périodes p-adiques et lois de réciprocité explicites*, J. Reine Angew. Math. **493** (1997), 115–151.

[Ben2] ______, *On Iwasawa theory of cristalline representations*, Duke Math. J. **104** (2000), 211–267.

[Ber] Vladimir G. Berkovich, *Spectral theory and analytic geometry over non-Archimedean fields, Math. Surveys Monogr.*, vol. 33, AMS, 1990.

[BeV1] D. G. Benois and S. V. Vostokov, *Norm pairing in formal groups and Galois representations*, Algebra i Analiz **2** (1990), no. 6, 69–97; English transl. in Leningrad Math. J. **2** (1991).

[BeV2] D. G. Benois and S. V. Vostokov, *Galois representaions in Honda's formal groups. Arithmetic of the group of points*, Tr. St-Peterbg. mat. Obsch., vol. 2, 1993, pp. 3–23; English transl. in Amer. Math. Soc. Transl. Ser. 2, vol. 159, AMS, 1994, pp. 1–14.

[BF] Lee Brekke and Peter G. O. Freund, *p-adic numbers in physics*, Phys. Rep. **233** (1993), 1–66.

[BGR] S. Bosch, U. Güntzer, and Reinhold Remmert, *Non-Archimedean analysis*, Grundlehren Math. Wiss., vol. . 261, Springer-Verlag, Berlin and New York, 1984.

[BK1] Spencer Bloch and Kazuya Kato, *p-adic etale cohomology*, Inst. Hautes Etudes Sci. Publ. Math. **63** (1986), 107–152.

[BK2] Spencer Bloch and Kazuya Kato, *L-functions and Tamagawa numbers of motives*, The Grothendieck Festschrift, vol. 1, Birkhäuser, 1990, pp. 333–400.

[BNW] E. Binz, Jürgen Neukirch, and G. H. Wenzel, *A subgroup theorem for free products of pro-finite groups*, J. Algebra **19** (1971), 104–109.

[Bog] R. A. Bogomolov, *Two theorems on divisibility and torsion in the Milnor K-groups*, Mat. Sb. **130** (1986), no. 3, 404–412; English transl. in Math. USSR-Sb. **58** (1987).

[Bor1] Z. I. Borevich, *The multiplicative group of cyclic p-extensions of a local field*, Trudy Mat. Inst. Steklov. **80** (1965), 16–29; English transl. in Proc. Steklov Inst. Math. **1968**, no. 80.

[Bor2] ______, *Groups of principal units of p-extension of a local field*, Dokl. Akad. Nauk SSSR **173** (1967), no. 2, 253–255; English transl. in Soviet Math. Dokl. **8** (1967).

[Bou] N. Bourbaki, *Algébre commutative*, Hermann, Paris, 1965.

[BoV] Z. I. Borevich and S. V. Vostokov, *The ring of integral elements of an extension of prime degree of a local field as a Galois module*, Zap. Nauchn. Sem. Leningrad.

Otdel. Mat. Inst. Steklov (LOMI) **31** (1973), 24–37; English transl. in J. Soviet Math. **6** (1976), no. 3.

[Br] R. Brauer, *Über die Konstruktion der Schiefkörper, die von endlichem Rang in bezug auf ein gegebenes Zentrum sind*, J. Reine Angew. Math. **168** (1932), 44–64.

[Bru1] Helmut Brückner, *Eine explizite Formel zum Reziprozitätsgesetz für Primzahlexponenten p*, Algebraische Zahlentheorie (Ber. Tag. Math. Forschungsinst. Oberwolfach, 1964), Bibliographisches Institut, Mannheim, 1967, pp. 31–39.

[Bru2] ———, *Hilbertsymbole zum Exponenten p^n und Pfaffische Formen*, Preprint, Hamburg, 1979.

[BSh] Z. I. Borevich and I. R. Shafarevich, *Number theory*, Nauka, Moscow, 1964; English transl., Pure Appl. Math., vol. 20, Academic Press, New York and London, 1966.

[BSk] Z. I. Borevich and A. I. Skopin, *Extensions of a local field with normal basis for principal units*, Trudy Mat. Inst. Steklov. **80** (1965), 45–50; English transl. in Proc. Steklov Inst. Math. **1968**, no. 80.

[BT] Hyman Bass and John T. Tate, *The Milnor ring of a global field*, Algebraic K-theory, II: "Classical" algebraic K-theory and connections with arithmetic (Proc. Conf., Seattle, WA, 1972), Lecture Notes in Math., vol. 342, Springer-Verlag, Berlin, 1973, pp. 349–446.

[Bu1] Alexandru Buium, *Differential charaters of abelian varieties over p-fields*, Invent. Math. **122** (1995), 309–340.

[Bu2] ———, *Geometry of p-jets*, Duke Math. J. **82** (1996), 349–367.

[Bu3] ———, *Arithmetic analogues of derivations*, J. Algebra **198** (1997), 290–299.

[Bu4] ———, *Continuous π-adic functions and π-derivations*, J. Number Theory **84** (2000), 34–39.

[Bu5] ———, *Differential algebraic geometry and Diophantine geometry: an overview*, Sympos. Math. **37** (1997), Cambridge Univ. Press, 87–98.

[Bum] Daniel Bump, *Automorphic forms and representations*, Cambridge University Press, 1998.

[Bur] D. J. Burns, *Factorisability and the arithmetic of wildly ramified Galois extensions*, Sém. Théor. Nombres Bordeaux (2) **1** (1989), 59–65.

[Cam] Rachel Camina, *Subgroups of the Nottingham group*, J. Algebra **196** (1997), 101–113.

[Car] Joseph E. Carroll, *On the torsion in K_2 of local fields*, Algebraic K-theory, II: "Classical" algebraic K-theory and connections with arithmetic (Proc. Conf., Seattle, WA, 1972), Lecture Notes in Math., vol. 342, Springer-Verlag, Berlin, 1973, pp. 464–473.

[Cas] J. W. S. Cassels, *Local fields*, Cambridge Univ. Press, London, 1986.

[CF] J. W. S. Cassels and A. Fröhlich (eds.), *Algebraic number theory*, Academic Press, London, and Thompson Book, Washington DC, 1967.

[CG] J. Coates, R. Greenberg, *Kummer theory for abelian varieties over local fields*, Invent. Math. **124** (1996), 129–174.

[Ch1] C. Chevalley, *Sur la théorie du corps de classes dans les corps finis et les corps locaux*, J. Fac. Sci. Tokyo Imp. Univ. Ser. Math. **2** (1933), 363–476.

[Ch2] ———, *Class field theory*, Nagoya University, Nagoya, 1954.

[Chr] Gilles Christol, *p-adic numbers and ultrametricity*, From number theory to physics (Les Houches, 1989), Springer, 1992, pp. 440–475.

[Coh] I. S. Cohen, *On the structure and ideal theory of complete local rings*, Trans. Amer. Math. Soc. **59** (1946), 54–106.

[Col1] Robert F. Coleman, *Division values in local fields*, Invent. Math. **53** (1979), 91–116.

[Col2] ———, *The dilogarithm and the norm residue symbol*, Bull. Soc. Math. France **109** (1981), 373–402.

[Col3] ———, *Arithmetic of Lubin–Tate division towers*, Duke Math. J. **48** (1981), 449–466.

[Colm] Pierre Colmez, *Représentations p-adiques d'un corps local*, Doc. Math. **2** (1998), 153–162.

[Cr] G.-Martin Cram, *The multiplicative group of a local sekw field as Galois group*, J. Reine Angew. Math. **381** (1987), 51–60.

[CW1] J. Coates and A. Wiles, *Explicit reciprocity laws*, Journées Arithmét. de Caen (Caen, 1976), Astérisque, No. 41–42, Soc. Math. France, Paris, 1977, pp. 7–17.

[CW2] ———, *On the conjecture of Birch and Swinnerton–Dyer*, Invent. Math. **39** (1977), 223–251.

[DdSMS] J. D. Dixon, M. du Sautoy, A. Mann, D. Segal, *Analytic pro-p groups. Second ed.*, Cambridge Univ. Press, 1999.

[Del] P. Deligne, *Les corps locaux des caractéristique p, limites de corps locaux de caractéristique* 0, Representations des groupes réductifs sur un corps local, Hermann, Paris, 1984, pp. 119–157.

[De1] O. V. Demchenko, *New relationships between formal Lubin-Tate groups and formal Honda groups*, Algebra i Analiz **10** (1998), no. 5, 77–84; English transl. in St. Petersburg Math. J. **10** (1999), 785–789.

[De2] ———, *Formal Honda groups: the arithmetic of the group of points*, Algebra i Analiz **12** (2000), no. 1, 132–149; English transl. in St. Petersburg Math. J. **12** (2001), 101–115.

[Dem1] S. P. Demushkin, *The group of the maximal p-extension of a local field*, Dokl. Akad. Nauk SSSR **128** (1959), 657–660; English transl. in Amer. Math. Soc. Transl. Ser. 2 **46** (1965).

[Dem2] ———, *The group of a maximal p-extension of a local field*, Izv. Akad. Nauk SSSR Ser. Mat. **25** (1961), 329–346. (Russian)

[Den] Jan Denef, *Report on Igusa's local zeta function*, Séminaire Bourbaki, 1990/ 1991, Astérisque, vol. 201–203, 1992, pp. 359–386.

[Des1] Francois Destrempes, *Generalization of a result of Schankar Sen: integral representations associated with local field extensions*, Acta Arithm. **63** (1993), 267–286.

[Des2] ———, *Explicit reciprocity law for Lubin–Tate modules*, J. Reine Angew. Math. **463** (1995), 27–47.

[DG] Michel Demazure and Pierre Gabriel, *Groupes algébriques. Tome* I: *Geometrie algébrique, generalités, groupes commutatifs*, Masson & Cie, Paris, and North Holland, Amsterdam, 1970.

[DGS] Bernard M. Dwork, G. Gerotto, F. J. Sullivan, *An introduction to G-functions, Ann. Math. Stud.*, vol. 133, Princeton Univ. Press, 1994.

[Di] Volker Diekert, *Über die absolute Galoisgruppe dyadischer Zahlkörper*, J. Reine Angew. Math. **350** (1984), 152–172.

[Dr1] V. G. Drinfeld, *Elliptic modules*, Mat. Sb. **94** (1974), no. 4, 594–627; English transl. in Math. USSR-Sb. **23** (1974).

[**Dr2**] V. G. Drinfeld, *Coverings of p-adic symmetric regions*, Funct. Analiz i Ego Prilozheniya **10** (1976), no. 2, 29–40; English transl. in Funct. Analysis and its Applications **10** (1976), 107–115.

[**DS**] R. Keith Dennis and Michael R. Stein, *K_2 of discrete valuation rings*, Adv. Math. **18** (1975), 182–238.

[**dSF**] Marcus du Sautoy and Ivan Fesenko, *Where the wild things are: ramification groups and the Nottingham group*, in [dSSS], pp. 287–328.

[**dSh1**] E. de Shalit, *The explicit reciprocity law in local field theory*, Duke Math. J. **53** (1986), 163–176.

[**dSh2**] ———, *Iwasawa theory of elliptic curves with complex multiplication*, Academic Press, 1987.

[**dSh3**] ———, *Making class field theory explicit*, CMS Conf. Proc., vol. 7, AMS, 1987, pp. 55–58.

[**dSm1**] Bart de Smit, *Ramification groups of local fields with imperfect residue class fields*, J. Number Theory **44** (1993), 229–236.

[**dSm2**] ———, *The different and differential of local fields with imperfect residue field*, Proc. Edinb. Math. Soc. **40** (1997), 353–365.

[**dSSS**] Marcus du Sautoy, Dan Segal and Aner Shalev (eds), *New horizons in pro-p-groups*, Birkhäuser, 2000.

[**DV1**] O.V. Demchenko and S.V.Vostokov, *Explicit form of Hilbert pairing for relative Lubin-Tate formal groups*, Zap. Nauchn. Sem. POMI **227** (1995), 41–44; English transl. in J. Math.Sci. Ser. 2 **89**, 1105–1107.

[**DV2**] ———, *Explicit formula of the Hilbert symbol for Honda formal group*, Zap. Nauchn. Sem. POMI **272** (2000), 86–128.

[**Dw**] Bernard Dwork, *Norm residue symbol in local number fields*, Abh. Math. Sem. Univ. Hamburg **22** (1958), 180–190.

[**E**] Otto Endler, *Valuation theory*, Springer-Verlag, New York and Heidelberg, 1972.

[**EF**] Ido Efrat and Ivan Fesenko, *Fields Galois-equivalent to a local field of positive characteristic*, Math. Res. Letters **6** (1999), 345–356.

[**Ef1**] Ido Efrat, *A Galois-theoretic characterization of p-adically closed fields*, Israel J. Math. **91** (1995), 273–284.

[**Ef2**] ———, *Construction of valuations from K-theory*, Math. Res. Letters **6** (1999), 335–343.

[**Ef3**] ———, *Finitely generated pro-p Galois groups of p-Henselian fields*, J. Pure Applied Algebra **138** (1999), 215–228.

[**Ef4**] ———, *Recovering higher global and local fields from Galois groups — an algebraic approach*, in **[FK]**, pp. 273–279.

[**Ep**] H. Epp, *Eliminating wild ramification*, Invent. Math. **19** (1973), 235–249.

[**Er1**] Yu. L. Ershov, *On the elementary theory of maximal normed fields*, Soviet Math. Dokl. **6** (1965), 1390–1393.

[**Er2**] ———, *On elementary theories of local fields*, Algebra i logika **4** (1965), no. 2, 5–30. (Russian)

[**Er3**] ———, *On the elementary theory of maximal normed fields.* I, Algebra i Logika **4** (1965), no. 3, 31–70; II, Algebra i Logika **5** (1966), no. 1, 5–40; III, Algebra i Logika **6** (1967), no. 3, 31–38. (Russian)

[Fa1] Gerd Faltings, *Hodge–Tate structures and modular forms*, Math. Ann. **278** (1987), 133–149.

[Fa2] Gerd Faltings, *p-adic Hodge theory*, J. AMS **1** (1988), 255–299.

[Fe1] Ivan B. Fesenko, *The generalized Hilbert symbol in the 2-adic case*, Vestnik Leningrad. Univ. Mat. Mekh. Astronom. **1985**, no. 4, 112–114; English transl. in Vestnik Leningrad Univ. Math. **18** (1985).

[Fe2] ______, *Explicit constructions in local class field theory*, Thesis, Leningrad. Univ., Leningrad, 1987.

[Fe3] ______, *Class field theory of multidimensional local fields of characteristic zero, with residue field of positive characteristic*, Algebra i Analiz **3** (1991), no. 3, 165–196; English transl. in St. Petersburg Math. J. **3, N3** (1992).

[Fe4] ______, *On class field theory of multidimensional local fields of positive characteristic*, Algebraic K-theory, Adv. Soviet Math., vol. 4, Amer. Math. Soc., Providence, RI, 1991, pp. 103–127.

[Fe5] ______, *Multidimensional local class field theory*, Dokl. Akad. Nauk SSSR **318** (1991), no. 1, 47–50; English transl. in Soviet Math. Dokl. **43** (1991).

[Fe6] ______, *Local class field theory: perfect residue field case*, Izvestija Russ. Acad. Nauk. Ser. Mat. **57** (1993), no. 4, 72–91; English transl. in Russ. Acad. Scienc. Izvest. Math. **43** (1994), 65–81; Amer. Math. Soc.

[Fe7] ______, *Abelian local p-class field theory*, Math. Annal. **301** (1995), 561–586.

[Fe8] ______, *Hasse-Arf property and abelian extensions*, Math. Nachr **174** (1995), 81–87.

[Fe9] ______, *On general local reciprocity maps*, J. Reine Angew. Math. **473** (1996), 207–222.

[Fe10] ______, *Abelian extensions of complete discrete valuation fields*, Number Theory Paris 1993–94, Cambridge Univ. Press, 1996.

[Fe11] ______, *On deeply ramified extensions*, Journal of the LMS (2) **57** (1998), 325–335.

[Fe12] ______, *On just infinite pro-p-groups and arithmetically profinite extensions*, J. Reine Angew. Math. **517** (1999), 61–80.

[Fe13] ______, *Class Field Theory – Its Centenary and Prospect, Advanced Studies in Pure Mathematics, Math. Soc. Japan*, vol. 30 (K. Miyake, eds.), Tokyo, 2001, pp. 63–78,.

[Fe14] ______, *On the image of noncommutative reciprocity map*, www.maths.nott.ac.uk/personal/ibf/prepr.html, 2002.

[FI] Jean-Marc Fontaine and L. Illusie, *p-adic periods: a survey*, Proc. Indo-French Conf. Geometry (Bombay), Hindustan Book Agency, Delhi, 1993, pp. 57–93.

[FJ] Michael D. Fried and Moshe Jarden, *Field arithmetic*, Springer, 1986.

[FK] Ivan Fesenko and Masato Kurihara (eds.), *Invitation to higher local fields,* Geometry and Topology Monographs, vol. 3, Geometry and Topology Publications, International Press, 2000; free electronic copy is available from www.maths.warwick.ac.uk/gt/gtmcontents3.html.

[FM] Jean-Marc Fontaine and William Messing, *p-adic periods and p-etale cohomology*, Current trends in arithmetical algebraic geometry (Arcata, CA, 1985), Contemp. Math., vol. 67, Amer. Math. Soc., Providence, RI, 1987, pp. 179–207.

[Fo1] Jean-Marc Fontaine, *Corps de series formelles et extensions galoisiennes des corps locaux*, Séminaire Théorie des Nombres, Grenoble, 1971–72, pp. 28–38.

[Fo2] ______, *Groupes p-divisibles sur les corps locaux*, Astérisque, No. 47–48, Soc. Math. France, Paris, 1977.

[Fo3] ______, *Formes différentiels et modules de Tate des variétés abéliennes sur les corps locaux*, Invent. Math. **65** (1982), 379–409.

[Fo4] ______, *Groupes de ramifications et representations d'Artin*, Ann. Scient. École Norm. Sup. **4** (1971), 337–392.

[Fo5] ______, *Représentations p-adiques des corps locaux,*, The Grothendieck Festschrift, vol. 2, Birkhäuser, 1994, pp. 59–111.

[Fr] A. Fröhlich, *Formal groups, Lecture Notes Math.*, vol. 74, Springer-Verlag, 1968.

[FT] A. Froöhlich and M. J. Taylor, *Algebraic number theory*, Cambridge Univ. Press, 1991.

[Fu] Yasushi Fujiwara, *On Galois actions on p-power torsion points of some one-dimensional formal groups over $\mathbb{F}_p[[t]]$*, J. Algebra **113** (1988), 491–510.

[FVZ] Ivan B. Fesenko, S. V. Vostokov, and I. B. Zhukov, *On the theory of multidimensional local fields. Methods and constructions*, Algebra i Analiz **2** (1990), no. 4, 91–118; English transl. in Leningrad Math. J. **2** (1991).

[FW] Jean-Marc Fontaine and J.-P. Wintenberger, *Le "corps des normes" de certaines extensions algébriques de corps locaux*, C. R. Acad. Sci. Paris Sér. A **288** (1979), 367–370.

[Gi] David Gilbarg, *The structure of the group of $\mathbb{Z}_p$ 1-units*, Duke Math. J. **9** (1942), 262–271.

[GK] M. Gross, M. Kurihara, *Régulateurs syntomigues et valuers de fonctions L p-adiques I* (by M. Gross), *with appendix* by M. Kurihara, Invent. math. **99** (1990), 293–320.

[GMW] K.H.M. Glass, R.A. Moore, G. Whaples, *On extending a norm residue symbol*, J. Reine Angew. Math. **245** (1970), 124–132.

[Gol] Larry Joel Goldstein, *Analytic number theory*, Prentice Hall, Englewood Cliffs, NJ, 1971.

[Gold] Robert Gold, *Local class field theory via Lubin–Tate groups*, Indiana Univ. Math. J. **30** (1981), 795–798.

[Gor] N. L. Gordeev, *Infiniteness of the number of relations in a Galois group of maximal p-extensions with a bounded ramification of a local field*, Izv. Akad. Nauk SSSR Ser. Mat. **45** (1981), no. 3, 592–607; English transl. in Math. USSR-Izv. **18** (1982).

[Gos] David Goss, *Basic structures of function field arithmetic*, Springer, 1996.

[Gou] Fernando Q. Gouvêa, *p-adic numbers*, Springer, 1993.

[Gr] Daniel R. Grayson, *On the K-theory of fields*, Algebraic K-theory and algebraic number theory (Honolulu, HI, 1987), Contemp. Math., vol. 83, Amer. Math. Soc., Providence, RI, 1989, pp. 31–55.

[GR] Hans Grauert and Reinhold Remmert, *Über die Methode der diskret bewerteten Ringe in der nicht-archimedischen Analysis*, Invent. Math. **2** (1966), 87–133.

[Has1] Helmut Hasse, *Über die Normenreste eines relativzyklischen Körpers vom Primzahlgrad l nach einem Primteiler $\mathfrak{l}$ von l*, Math. Ann. **90** (1923), 262–278.

[Has2] ______, *Das allgemeine Reziprozitätsgesetz und seine Ergänzungssätze in beliebigen algebraischen Zahlkörpern für gewisse nicht-primäre Zahlen*, J. Reine Angew. Math. **153** (1924), 192–207.

[Has3] ———, *Direkter Beweis des Zerleguns-und Vertauschungs-satzes für das Hilbertische Normenrestesymbol in einem algebraischen Zahlkörper im Falle eines Primteiler* $\mathfrak{l}$ *des Relativgrades* l, J. Reine Angew. Math. **154** (1925), 20–35.

[Has4] ———, *Über das allgemeine Reziprozitätsgesetz der* l*-ten Potenzreste im Körper* k_ζ *der* l*-ten Einheitswurzeln und in Oberkörpern von* k_ζ, J. Reine Angew. Math. **154** (1925), 96–109.

[Has5] ———, *Das allgemeine Reziprozitätsgesetz der* l*-ten Potenzreste für beliebege, zu* l *prime Zahlen in gewissen Oberkörpern des Körpers der* l*-ten Einheitswurzeln*, J. Reine Angew. Math. **154** (1925), 199–214.

[Has6] ———, *Zum expliziten Reziprozitätsgesetz*, Abh. Math. Sem. Univ. Hamburg **7** (1929), 52–63.

[Has7] ———, *Normenresttheorie galoischer Zahlkörpern mit Anwendungen auf Führer und Discriminante abelscher Zahlkörper*, J. Fac. Sci. Tokyo Imp. Univ. Ser. Math. **2** (1934), 477–498.

[Has8] ———, *Die Gruppe der* p^n*-primären Zahlen für einen Primteiler* $\mathfrak{p}$ *von* p, J. Reine Angew. Math. **176** (1936), 174–183.

[Has9] ———, *Zur Arbeit von I. R. Šafarevič über das allgemeine Reziprozitätsgesetz*, Math. Nachr. **5** (1951), 302–327.

[Has10] ———, *Der* 2^n*-te Potenzcharakter von* 2 *im Körper der* 2^n*-ten Einheitswurzeln*, Rend. Circ. Mat. Palermo (2) **7** (1958), 185–244.

[Has11] ———, *Zum expliziten Reiprozitätsgesetz*, Arch. Math. (Basel) **13** (1962), 479–485.

[Has12] ———, *Zahlentheorie*, Akademie-Verlag, Berlin, 1949.

[Haz1] Michiel Hazewinkel, *Abelian extensions of local fields*, Doctoral Dissertation, Universiteit van Amsterdam, Amsterdam, 1969.

[Haz2] ———, *Local class field theory is easy*, Adv. Math. **18** (1975), 148–181.

[Haz3] ———, *Formal groups and application*, Academic Press, New York, 1978.

[Haz4] ———, *Twisted Lubin–Tate formal group laws, ramified Witt vectors and (ramified) Artin–Hasse exponentials*, Trans. Amer. Math. Soc. **259** (1980), 47–63.

[Hei] Volker Heiermann, *De nouveaux invaraints numériques pour les extensions totalement ramifiées de corps locaux*, J. Number Theory **59** (1996), 159–202.

[Hel] Charles Helou, *An explicit* 2^n *th reciprocity law*, J. Reine Angew. Math. **389** (1988), 64–89.

[Henn1] Guy Henniart, *Lois de reciprocité explicites*, Séminaire Théorie des Nombres, Paris, 1979–80, Birkhäuser, Boston, 1981, pp. 135–149.

[Henn2] ———, *Sur les lois de reciprocité explicites.* I, J. Reine Angew. Math. **329** (1981), 177–203.

[Henn3] ———, *Une preuve simple des conjectures de Langlands pour GL(n) sur un corps* p*-adiques*, Invent. Math. **139** (2000), 439–455.

[Her] J. Herbrand, *Sur la théorie des groups de decomposition, d'inertie et de ramification*, J. Math. Pures Appl. **10** (1931), 481–498.

[Herr1] Laurent Herr, *Une approche nouvelle de la dualité locale de Tate*, Math. Annalen (2001).

[Herr2] ———, $\Phi-\Gamma$*-modules and Galois cohomology*, in [FK], pp. 263-272.

[Hi] Haruzo Hida, *Elementary theory of* L*-functions and Eisenstein series. LMS student texts*, vol. 26, Cambridge Univ. Press, 1993.

[HJ] Dan Haran and Moshe Jarden, *The absolute Galois group of a preudo p-adically closed field*, J. Reine Angew. Math. **383** (1988), 147–206.

[Ho] G. Hochschild, *Local class field theory*, Ann. of Math. (2) **51** (1950), 331–347.

[Hon] Taira Honda, *On the theory of commutative formal groups*, J. Math. Soc. Japan **22** (1970), 213–246.

[HS] Zvonimir Hlousek and Donald Spector, *p-adic string theories*, Ann. Physics **189** (1989), 370–431.

[HT] M. Harris and R. Taylor, *The geometry and cohomology of some simple Shimura varieties*, Princeton Univ. Press, 2002.

[HSch] Helmut Hasse and F. K. Schmidt, *Die Struktur discret bewerteten Körper*, J. Reine Angew. Math. **170** (1934), 4–63.

[Hy1] Osamu Hyodo, *Wild ramification in the imperfect residue field case*, Galois representations and arithmetic algebraic geometry (Kyoto 1985, Tokyo 1986), Adv. Stud. Pure Math., vol. 12, North-Holland, Amsterdam and New York, 1987, pp. 287–314.

[Hy2] Osamu Hyodo, *On the Hodge–Tate decomposition in the imprefect residue field case*, J. Reine Angew. Math. **365** (1986), 97–113.

[Ig] Jun-ichi Igusa, *An introduction to the theory of local zeta functions*, AMS, 2000.

[ILF] V.V. Ishkhanov, B.B. Lur'e, D.K. Faddeev, *The embedding problem in Galois theory*, Nauka, 1990; English transl. in; Transl. Math. Monographs, vol. 165, AMS, 1997.

[IR] Kenneth F. Ireland and Michael I. Rosen, *A classical introduction to modern number theory*, Springer, Berlin and New York, 1982.

[Izh] O. T. Izhboldin, *On the torsion subgroup of Milnor K-groups*, Dokl. Akad. Nauk SSSR **294** (1987), no. 1, 30–33; English transl. in Soviet Math. Dokl. **37** (1987).

[Iw1] Kenkichi Iwasawa, *On Galois groups of local fields*, Trans. Amer. Math. Soc. **80** (1955), 448–469.

[Iw2] ______, *On local cyclotomic fields*, J. Math. Soc. Japan **12** (1960), 16–21.

[Iw3] ______, *On explicit formulas for the norm residue symbol*, J. Math. Soc. Japan **20** (1968), 151–165.

[Iw4] ______, *On $\mathbb{Z}_l$-extensions of algebraic number fields*, Ann. of Math. (2) **98** (1973), 246–326.

[Iw5] ______, *Local class field theory*, Iwanami-Shoten, Tokyo, 1980.

[Iw6] ______, *Local class field theory*, Oxford Univ. Press, New York, and Clarendon Press, New York, 1986.

[Iw7] ______, *Lectures on p-adic L-function.*, Princeton Univ. Press, 1972.

[Iya] S. Iyanaga (eds.), *The theory of numbers*, North–Holland, Amsterdam, 1975.

[Jan] Uwe Jannsen, *Über Galoisgruppen lokaler Körper*, Invent. Math. **70** (1982), 53–69.

[Jar] Moshe Jarden, *Infinite Galois theory*, Handbook of Algebra, vol. 1, North-Holland, 1996, pp. 269–319.

[JR1] Moshe Jarden and Jürgen Ritter, *On the characterization of local fields by their absolute Galois groups*, J. Number Theory **11** (1979), 1–13.

[JR2] ______, *Normal automorphisms of absolute Galois group of $\mathfrak{p}$-adic fields*, Duke Math. J. **47** (1980), 47–56.

[JR3] ______, *The Frattini subgroup of the absolute Galois group of a local field*, Israel J. Math. **74** (1991), 81–90.

[JW] Uwe Jannsen and Kay Wingberg, *Die struktur der absoluten Galoisgruppe $\mathfrak{p}$-adischer Zahlkörpers*, Invent. Math. **70** (1982), 71–98.

[**Kah**] Bruno Kahn, *L'anneau de Milnor d'un corps local à corps résiduel parfait*, Ann. Inst. Fourier (Grenoble) **34** (1984), 19–65.

[**KaSh**] Tsuneo Kanno and Takeo Shirason, *Value groups of Henselian valuations*, Proc. Japan Acad. Ser. A Math. Sci. **69** (1993), 268–271.

[**Kat1**] Kazuya Kato, *A generalization of local class field theory by using K-groups.* I, J. Fac. Sci. Univ. Tokyo Sect. IA Math. **26** (1979), 303–376.

[**Kat2**] ———, *A generalization of local class field theory by using K-groups.* II, J. Fac. Sci. Univ. Tokyo Sect. IA Math. **27** (1980), 603–683.

[**Kat3**] ———, *Galois cohomology of complete discrete valuation fields*, Algebraic K-theory, Part II (Oberwolfach, 1980), Lecture Notes in Math., vol. 967, Springer, Berlin and New York, 1982, pp. 215–238.

[**Kat4**] ———, *Swan conductors with differential values*, Galois representations and arithmetic algebraic geometry (Kyoto, 1985/Tokyo, 1986), Adv. Stud. Pure Math., vol. 12, North-Holland, Amsterdam, 1987, pp. 315–342.

[**Kat5**] ———, *Swan conductors for characters of degree one in the imperfect residue field case*, Algebraic K-theory and algebraic number theory (Honolulu, HI, 1987), Contemp. Math., vol. 83, Amer. Math. Soc., Providence, RI, 1989, pp. 101–131.

[**Kat6**] ———, *The explicit reciprocity law and the cohomology of Fontaine–Messing*, Bull. Soc. Math. France **119** (1991), 397–441.

[**Kat7**] ———, *Lectures on the approach to Iwasawa theory for Hasse–Weil L-functions via B_{dR}*, Arithmetic Algebraic Geometry, Springer-Verlag, Berlin etc.; Lect. Notes in Math., vol. 1553, 1993, pp. 50–163.

[**Kat8**] ———, *On p-adic vanishing cycles (applications of ideas of Fontaine–Messing)*, Adv. Stud. Pure Math., vol. 10, 1987, pp. 207–251.

[**Kaw1**] Yukiyosi Kawada, *On the ramification theory of infinite algebraic extensions*, Ann. of Math. (2) **58** (1953), 24–47.

[**Kaw2**] ———, *Class formations*, Duke Math. J. **22** (1955), 165–177; IV, J. Math. Soc. Japan **9** (1957), 395–405; V, J. Math. Soc. Japan **12** (1960), 34–64.

[**KdS**] Helmut Koch and Ehud de Shalit, *Metabelian local class field theory*, J. Reine Angew. Math. **478** (1996), 85–106.

[**Ke**] Kevin Keating, *Galois extensions associated to deformations of formal A-modules*, J. Fac. Sci. Univ. Tokyo Sect. IA Math. **37** (1990), 151–170.

[**Kh**] Andrei Khrennikov, *The theory of non-Archimedean generalized functions and its applications to quantum mechanics and field theory*, J. Math. Sci. **73** (1995), 243–298.

[**Kha1**] Sudesh K. Khanduja, *On a result of James Ax*, J. Algebra **172** (1995), 147–151.

[**Kha2**] ———, *On Krasner's constant*, J. Algebra **213** (1999), 225–230.

[**KhaS**] Sudesh K. Khanduja and Jayanti Saha, *Generalized Hensel's lemma*, Proc. Edinb. Math. Soc. **42** (1999), 469–480.

[**KKS**] Kazuya Kato, Nobushige Kurokawa, Takeshi Saito, *Number theory I*, AMS, 2000.

[**Kn**] M. Kneser, *Zum expliziten Reziprozitätsgesetz von I. R. Šafarevič*, Math. Nachr. **6** (1951), 89–96.

[**KnS**] Kiyomi Kanesaka and Koji Sekiguchi, *Representation of Witt vectors by formal power series and its applications*, Tokyo J. Math. **2** (1979), 349–370.

[**Ko1**] Helmut Koch, *Über Darstellungsräume und die Struktur der multiplikativen Gruppe eines p-adischen Zahlkörpers*, Math. Nachr. **26** (1963), 67–100.

[Ko2] ———, *Über Galoissche Gruppen von $\mathfrak{p}$-adischen Zahlkörpern*, Math. Nachr. **29** (1965), 77–111.

[Ko3] ———, *Über die Galoissche Gruppe der algebraischen Abschließung eines Potenzreihenkörpers mit endlichem Konstantenkörper*, Math. Nachr. **35** (1967), 323–327.

[Ko4] ———, *Galoissche Theorie der p-Erweiterungen*, Deutscher Verlag Wissenschaften, Berlin, and Springer-Verlag, Berlin and New York,, 1970.

[Ko5] ———, *The Galois group of a p-closed extension of a local field*, Soviet Math. Dokl. **19** (1978), 10–13.

[Ko6] ———, *Algebraic number theory*, Springer-Verlag, 1997.

[Ko7] ———, *Local class field theory for metabelian extensions*, Proc. 2nd Gauss Symposium. Confer. A: Math. and Theor. Physics (Munich, 1993), de Gruyter, Berlin, 1995, pp. 287-300.

[Kob1] Neal Koblitz, *p-adic analysis: A short course on recent works*, London Math. Soc. Lecture Note Ser., vol. 46, Cambridge Univ. Press, Cambridge and New York, 1980.

[Kob2] ———, *p-adic numbers, p-adic analysis and zeta-functions*, 2nd ed., Springer-Verlag, Berlin and New York, 1984.

[Kö] P. Kölcze, *Die Brückner–Vostokov–Formel für das Hilbersymbol unf ihre Geltung im Fall $p = 2$*, Manuscr. math. **88** (1995), 335–355.

[Koen1] J. Koenigsmann, *p-henselian fields*, manuscr. math. **87** (1995), 89–99.

[Koen2] ———, *From p-rigid elements to valuations (with a Galois-characterization of p-adic fields)*, J. Reine Angew. Math. **465** (1995), 165–182.

[Kol] V. A. Kolyvagin, *Formal groups and the norm residue symbol*, Math. USSR-Izv. **15** (1980), 289–348.

[Kom] K. Komatsu, *On the absolute Galois groups of local fields.* II, Galois groups and their representations (Nagoya, 1981), Adv. Stud. Pure Math., vol. 2, North-Holland, Amsterdam and New York, 1983, pp. 63–68.

[KPR] Franz-Viktor Kuhlmann, Mathias Pank, and Peter Roquette, *Immediate and purely wild extensions of valued fields*, Manuscripta Math. **55** (1986), 39–67.

[Kr1] M. Krasner, *Sur la representation exponentielle dans les corps relativement galoisiens de nombres $\mathfrak{p}$-adiques*, Acta Arith. **3** (1939), 133–173.

[Kr2] ———, *Rapport sur le prolongement analytique dans les corps values complets par la méthode des éléments analytiques quasi-connexes*, Table Ronde d'Analyse nonarchimedienne (Paris, 1972), Bull. Soc. France, Mem. No. 39–40, Soc. Math. France, Paris, 1974, pp. 131–254.

[KtS] Kazuya Kato and S. Saito, *Two-dimensional class field theory*, Galois groups and their representations (Nagoya, 1981), Adv. Stud. Pure Math., vol. 2, North-Holland, Amsterdam, 1983, pp. 103–152.

[Ku1] M. Kurihara, *On two types of complete discrete valuation fields*, Compos. Math. **63** (1987), 237–257.

[Ku2] ———, *Abelian extensions of an absolutely unramified local field with general residue field*, Invent. Math. **93** (1988), 451–480.

[Ku3] ———, *Computation of the syntomic regulator in the cyclotomic case, Appendix to M. Gross paper*, Invent. Math. **99** (1990), 313–320.

[Ku4] ———, *The exponential homomorphism for the Milnor K-groups and an explicit reciprocity law*, J. Reine Angew. Math. **498** (1998), 201–221.

[**Kub**] Tomio Kubota, *Geometry of numbers and class field theory*, Japan J. Math. **13** (1987), 235–275.

[**Kudo**] Aichi Kudo, *On Iwasawa's explicit formula for the norm residue symbol*, Mem. Fac. Sci. Kyushu Univ. Ser. A **26** (1972), 139–148.

[**Kudl**] Stephen S. Kudla, *The local Langlands correspondence: the non-Archimedean case*, Motives (Seattle, WA, 1991), Proc. Symp. Pure Math., vol. 55, part 2, AMS, 1994, pp. 365–391.

[**Kuh**] Franz-Viktor Kuhlmann, *Henselian function fields and tame fields*, Preprint.

[**Kuz**] L. V. Kuzmin, *New explicit formulas for the norm residue symbol and their applications*, Izv. Akad. Nauk SSSR Ser. Mat. **54** (1990), no. 6, 1196–1228; English transl. in Math. USSR-Izv. **37** (1991).

[**KwS**] Yukiyosi Kawada and Ichiro Satake, *Class formations.* II, J. Fac. Sci. Univ. Tokyo Sect. IA Math. **7** (1956), 353–389.

[**KZ**] M. V. Koroteev and I. B. Zhukov, *Elimination of wild ramification*, Algebra i Analiz **11** (1999), no. 6, 153–177; English transl. in St. Petersburg Math. J. **11** (2000), no. 6, 1063–1083.

[**L**] Laurent Lafforgue, *Chtoukas de Drinfeld et correspondance de Langlands*, Invent. Math. **147** (2002), 1–241.

[**La1**] Serge Lang, *Algebra*, Addison-Wesley, Reading, MA, 1965.

[**La2**] ______, *Algebraic number theory*, Springer-Verlag, Berlin and New York, 1986.

[**La3**] ______, *Cyclotomic fields*, 2nd ed., Springer-Verlag, Berlin and New York, 1986.

[**Lab**] J.-P. Labute, *Classification of Demushkin groups*, Canad. J. Math. **19** (1967), 106–132.

[**Lau1**] Francois Laubie, *Groupes de ramification et corps residuels*, Bull. Sci. Math. (2) **105** (1981), 309–320.

[**Lau2**] ______, *Sur la ramification des extensions de Lie*, Compositio Math. **55** (1985), 253–262.

[**Lau3**] ______, *Sur la ramification des extensions infinies des corps locaux*, Séminaire de Théorie des Nombres, Paris, 1985–86, Birkhäuser, Boston, 1987, pp. 97–117.

[**Lau4**] ______, *Extensions de Lie et groups d'automorphismes de corps locaux*, Compositio Math. **67** (1988), 165–189.

[**Lau5**] ______, *Sur la ramification des groupes de Weil*, C. R. Acad. Sci. Paris Sér. I Math. **308** (1989), 333–336.

[**Lau6**] ______, *La ramification des extensions galoisiennes est déterminée par les discriminants de certaines sous-extensions*, Acta Arithm. **65** (1993), 283–291.

[**Le**] Heinrich W. Leopoldt, *Zur Approximation des p-adischen Logarithmus*, Abh. Math. Sem. Univ. Hamburg **25** (1961), 77–81.

[**Li1**] Hua-Chieh Li, *p-adic dynamical systems and formal groups*, Compos. Math. **104** (1996), 41–54.

[**Li2**] ______, *Counting periodic points of p-adic power series*, Compos. Math. **100** (1996), 351–364.

[**Li3**] ______, *p-adic periodic points and Sen's theorem*, J. Number Theory **56** (1996), 309–318.

[**Li4**] ______, *When a p-adic power series an endomorphism of a formal group*, Proc. Amer. Math. Soc. **124** (1996), 2325–2329.

[**LR**] Jonathan Lubin and Michael Rosen, *The norm map for ordinary abelian varieties*, J. Algebra **52** (1978), 236–240.

[**LRS**] G. Laumon, M. Rapoport and U. Stuhler, *D-elliptic sheaves and the Langlands correspondence*, Invent. Math. **113** (1993), 217–338 F.

[**LS**] F. Laubie and M. Saine, *Ramification of some automorphisms of local fields*, J. Number Theory **72** (1998), 174–182.

[**LT**] Jonathan Lubin and John T. Tate, *Formal complex multiplication in local fields*, Ann. of Math. (2) **81** (1965), 380–387.

[**Lu1**] Jonathan Lubin, *Non-Archimedean dynamical systems*, Compositio Math. **94** (1994), 321–346.

[**Lu2**] ______, *Sen's theorem on iteration of power series*, Proc. Amer. Math. Soc. **123** (1995), 63–66.

[**M**] D.A. Marcus, *Number fields*, Springer-Verlag, 1977.

[**Mah**] K. Mahler, *Introduction to p-adic numbers and their functions*, Cambridge Univ. Press, London and New York, 1973.

[**Man**] Ju. I. Manin, *Cyclotomic fields and modular curves*, Uspekhi Mat. Nauk **26** (1971), no. 6, 7–71; English transl. in Russian Math. Surveys **26** (1971).

[**Mar1**] Murray A. Marshall, *Ramification groups of abelian local field extensions*, Canad. J. Math. **23** (1971), 271–281.

[**Mar2**] ______, *The maximal p-extension of a local field*, Canad. J. Math. **23** (1971), 398–402.

[**Mau1**] Eckart Maus, *Arithmetisch disjunkte Körper*, J. Reine Angew. Math. **226** (1967), 184–203.

[**Mau2**] ______, *Die gruppentheoretische Struktur der Verzweigungsgruppenreihen*, J. Reine Angew. Math. **230** (1968), 1–28.

[**Mau3**] ______, *On the jumps in the series of ramification groups*, Colloque de Théorie des Nombres (Bordeaux, 1969), Bull. Soc. Math. France, Mem. No. 25, Soc. Math. France, Paris, 1971, pp. 127–133.

[**Mau4**] ______, *Über die Verteilung der Grundverzweigungszahlen von wild verzweigten Erweiterungen $\mathfrak{p}$-adischer Zahlkörper*, J. Reine Angew. Math. **257** (1972), 47–79.

[**Mau5**] ______, *Relationen in Verzweigungsgruppen*, J. Reine Angew. Math. **258** (1973), 23–50.

[**Maz**] Barry Mazur, *Rational points of abelian varieties with values in towers of number fields*, Invent. Math. **18** (1972), 183–266.

[**McC**] William G. McCallum, *Tate duality and wild ramification*, Math. Ann. **288** (1990), 553–558.

[**McL**] S. MacLane, *Subfields and automorphism groups of p-adic fields*, Ann. of Math. (2) **40** (1939), 423–442.

[**Me**] A. S. Merkurjev, *On the torsion in K_2 of local fields*, Ann. of Math. (2) **118** (1983), 375–381.

[**Mi**] J. S. Milne, *Arithmetic duality theorems*, Academic Press, 1986.

[**Mik1**] Hiroo Miki, *On $\mathbb{Z}_p$-extensions of complete p-adic power series fields and function fields*, J. Fac. Sci. Univ. Tokyo Sect. IA Math. **21** (1974), 377–393.

[**Mik2**] ______, *On some Galois cohomology groups of a local field and its application to the maximal p-extension*, J. Math. Soc. Japan **28** (1976), 114–122.

[Mik3] ———, *On the absolute Galois group of local fields.* I, Galois groups and their representations (Nagoya, 1981), Adv. Stud. Pure Math., vol. 2, North-Holland, Amsterdam and New York, 1983, pp. 55–61.

[Mik4] ———, *On unramified abelian extensions of a complete field under a discrete valuation with arbitrary residue field of characteristic $p \neq 0$ and its application to wildly ramified $\mathbb{Z}_p$-extensions*, J. Math. Soc. Japan **29** (1977), no. 2, 363–371.

[Mik5] ———, *A note on Maus' theorem on ramification groups*, Tohoku Math. J. (2) **29** (1977), 61–68.

[Mik6] ———, *On the ramification numbers of cyclic p-extensions over local fields*, J. Reine Angew. Math. **328** (1981), 99–115.

[Mil1] John Milnor, *Introducion to algebraic K-theory*, Princeton Univ. Press, Princeton, NJ, and Univ. Tokyo Press, Tokyo, 1971.

[Mil2] ———, *Algebraic K-theory and quadratic forms*, Invent. Math. **9** (1970), 318–344.

[Miy] Katsuya Miyake, *A fundamental theorem on p-extensions of algebraic number fields*, Japan J. Math. **16** (1990), 307–315.

[Mo1] Mikao Moriya, *Einige Eigenschaften der endlichen separablen algebraischen Erzweiterungen über perfekten Körpern*, Proc. Imp. Acad. Tokyo **17** (1941), 405–410.

[Mo2] ———, *Die Theorie der Klassenkörper im Kleinen über diskret perfekten Körpern.* I, Proc. Imp. Acad. Tokyo **18** (1942), 39–44; II, Proc. Imp. Acad. Tokyo **18** (1942), 452–459.

[Mo3] ———, *Zur theorie der Klassenkörper im Kleinen*, J. Math. Soc. Japan **3** (1951), 195–203.

[Moc1] Shinichi Mochizuki, *A version of the Grothendieck conjecture for p-adic fields*, Int. J. Math. **8** (1997), 499–506.

[Moc2] ———, *The local pro-p anabelian geometry of curves*, Invent. Math. 138(1999), 319–423.

[Moo] Calvin C. Moore, *Group extensions of p-adic and adelic linear groups*, Inst. Hautes Études Sci. Publ. Math. **1968**, no. 35, 157–222.

[MS] A. S. Merkurjev and A. A. Suslin, *K-cohomology of Severi–Brauer varieties and the norm residue homomorphism*, Math. USSR-Izv. **21** (1983), 307–340.

[MSh] O. V. Melnikov and A. A. Sharomet, *The Galois group of a multidimensional local field of positive characteristic*, Mat. Sb. **180** (1989), no. 8, 1132–1147; English transl. in Math. USSR-Sb. **67** (1990).

[MW] R. E. MacKenzie and George Whaples, *Artin–Schreier equations in characteristic zero*, Amer. J. Math. **78** (1956), 473–485.

[MZh] A. I. Madunts, I. B. Zhukov, *Multidimensional complete fields: topology and other basic constructions*, Trudy S.-Peterb. Mat. Obsch. **3** (1995), 4–46; English transl. in Amer. Math. Soc. Transl. Ser. 2, vol. 166, AMS, 1995, pp. 1–34.

[N1] Jürgen Neukirch, *Kennzeichnung der $\mathfrak{p}$-adischen und der endlichen algebraischen Zählkörper*, Invent. Math. **6** (1969), 296–314.

[N2] ———, *Freie Produkte pro-endlicher Gruppen und ihre Kohomologie*, Arch. Math. (Basel) **22** (1971), 337–357.

[N3] ———, *Neubegründung der Klassenkörpertheorie*, Math. Z. **186** (1984), 557–574.

[N4] ———, *Class field theory*, Springer-Verlag, Berlin and New York, 1986.

[N5] ———, *Algebraic number theory*, Springer-Verlag, Berlin and New York, 1999.

[Na] Masayoshi Nagata, *Local rings*, Interscience, New York, 1962.

[NSchW] Jürgen Neukirch, Alexander Schmidt, Kay Wingberg, *Cohomology of number fields*, Springer-Verlag, 2000.

[Pa1] A. N. Parshin, *Class fields and algebraic K-theory*, Uspekhi Mat. Nauk **30** (1975), no. 1, 253–254. (Russian)

[Pa2] ———, *On the arithmetic of two dimensional schemes. I. Distributions and residues*, Izv. Akad. Nauk SSSR Ser. Mat. **40** (1976), no. 4, 736–773; English transl. in Math. USSR-Izv. **10** (1976).

[Pa3] ———, *Abelian coverings of arithmetic schemes*, Dokl. Akad. Nauk SSSR **243** (1978), no. 4, 855–858; English transl. in Soviet Math. Dokl. **19** (1978).

[Pa4] ———, *Local class field theory*, Trudy Mat. Inst. Steklov. **165** (1985), 143–170; English transl. in Proc. Steklov Inst. Math. **1985**, no. 3.

[Pa5] ———, *Galois cohomology and Brauer group of local fields*, Trudy Mat. Inst. Steklov. **183** (1990), 159–169; English transl. in Proc. Steklov Inst. Math. **1991**, no. 4.

[Po1] Florian Pop, *Galoissche Kennzeichnung p-adisch abgeschlossener Körper*, J. Reine Angew. Math. **392** (1988), 145–175.

[Po2] ———, *On Grothendieck's conjecture of birational anabelian geometry*, Ann. Math. **139** (1994), 145–182.

[Pr] Gopal Prasad, *On the wild norm residue symbol in an abelian extension*, Math. Ann. **274** (1986), 419-422.

[PR] Alexander Prestel and Peter Roquette, *Formally p-adic fields*, Lecture Notes in Math., vol. 1050, Springer-Verlag, Berlin and New York, 1984.

[Ra] Michele Raynaud, *Anneaux locaux henseliens*, Lecture Notes in Math., 169, Springer-Verlag, Berlin and New York, 1970.

[RC] Philippe Robba and Gilles Christol, *Équations différentielles p-adiques. Applications aux sommes exponentielles*, Hermann, 1994.

[Rib] P. Ribenboim, *Theorie des Valuations*, 2nd ed., Les Presses Univ. de Montreal, Montreal, 1968.

[Rie] C. Riehm, *The Schur subgroup of the Brauer group of a local field*, Enseign. Math. (2) **34** (1988), 1–11.

[Rim] Dock Sang Rim, *Relatively complete fields*, Duke Math. J. **24** (1957), 197–200.

[Rit1] Jürgen Ritter, $\mathfrak{p}$*-adic fields having the same type of algebraic extensions*, Math. Ann. **238** (1978), 281–288.

[Rit2] Jürgen Ritter (eds.), *Representation theory and number theory in connection with the local Langlands conjecture, Contemporary Math.*, vol. 86, AMS, 1989.

[Ro] Jonathan Rosenberg, *Algebraic K-theory and its applications*, Springer, 1996.

[Roq1] Peter Roquette, *Abspaltung des Radikals in vollständigen lokalen Ringen*, Abh. Math. Sem. Univ. Hamburg **23** (1959), 75–113.

[Roq2] ———, *Some tendencies in contemporary algebra*, Anniv. Oberwolfach 1984, Birkhäuser, Basel and Boston, 1984, pp. 393–422.

[Ros] Michael Rosen, *An elementary proof of the local Kronecker–Weber theorem*, Trans. Amer. Math. Soc. **265** (1981), 599–605.

[Rot] H. Rothgiesser, *Zum Reziprotitätsgesetz für p^n*, Abh. Math. Sem. Univ. Hamburg **11** (1934).

[RT] Samuel Rosset and John T. Tate, *A reciprocity law for K_2-traces*, Comment. Math. Helv. **58** (1983), 38–47.

[**RTVW**] Ph. Ruelle, E. Thiran, D. Verstegen, J. Weyers, *Quantum mechanics on p-adic fields*, J. Math. Phys. **30** (1989), 2854–2874.

[**RV**] Dinakar Ramakrishnan and Robert J. Valenza, *Fourier analysis on number fields*, Springer, 1999.

[**Sa**] Ichiro Satake, *On a generalization of Hilbert's theory of ramifications*, Sci. Papers College Gen. Ed. Univ. Tokyo **2** (1952), 25–39.

[**Sch**] O. F. G. Schilling, *The theory of valuations*, Math. Surveys, No. 4, Amer Math. Soc., New York, 1950.

[**Schm**] F. K. Schmidt, *Zur Klassenkörpertheorie in Kleinen*, J. Reine Angew. Math. **162** (1930), 155–168.

[**Schf**] W. H. Schikhof, *Ultrametric calculus*, Cambr. Univ. Press, 1984.

[**Schi**] A. Schinzel, *The number of zeros of polynomials in valuation rings of complete discretely valued fields*, Fund. Math. **124** (1984), 41–97.

[**Se1**] Jean-Pierre Serre, *Groupes algébriques et corps de classes*, Hermann, Paris, 1959.

[**Se2**] ———, *Sur les corps locaux á corps residuel algébriquement clos*, Bull. Soc. Math. France **89** (1961), 105–154.

[**Se3**] ———, *Local fields*, Springer-Verlag, 1979.

[**Se4**] ———, *Cohomologie galoissiene*, Lecture Notes in Math., 4th ed., vol. 5, Springer-Verlag, Berlin and New York, 1973.

[**Se5**] ———, *A course in arithmetic*, 2nd ed., Springer-Verlag, Berlin and New York, 1978.

[**Se6**] ———, *Abelian l-adic representationa and elliptic curves*, Benjamin, 1968.

[**Sek1**] Koji Sekiguchi, *Class field theory of p-extensions over a formal power series field with a p-quasifinite coefficient field*, Tokyo J. Math. **6** (1983), 167–190.

[**Sek2**] ———, *The Lubin–Tate theory for formal power series fields with finite coefficient fields*, J. Number Theory **18** (1984), 360–370.

[**Sen1**] Shankar Sen, *On automorphisms of local fields*, Ann. of Math. (2) **90** (1969), 33–46.

[**Sen2**] ———, *Ramification in p-adic Lie extensions*, Invent. Math. **17** (1972), 44–50.

[**Sen3**] ———, *On explicit reciprocity laws.* I, J. Reine Angew. Math. **313** (1980), 1–26; II, J. Reine Angew. Math. **323** (1981), 68–87.

[**Sen4**] ———, *Lie algebras of Galois groups arising from Hodge–Tate modules*, Ann. Math. (2) **97** (1973), 160–170.

[**Sen5**] ———, *Continuous cohomology and p-adic Galois representations*, Invent. Math. **62** (1980/81), 89–116.

[**Sen6**] ———, *Integral representations associated with p-adic field extensions*, Invent. Math. **94** (1988), 1–12.

[**Sen7**] ———, *The analytic variation of p-adic Hodge structure*, Ann. Math. (2) **127** (1988), 647–661.

[**Sen8**] ———, *An infinite-dimensional Hodge–Tate theory*, Bull. Soc. Math. France **121** (1993), 13–34.

[**Sen9**] ———, *Galois cohomology and Galois representations*, Invent. MAth. **112** (1993), 639–656.

[**Sha1**] I. R. Shafarevich, *On p-extensions*, Amer. Math. Soc. Transl. Ser. 2 **4** (1956), 59–72.

[**Sha2**] ———, *A general reciprocity law*, Amer. Math. Soc. Transl. Ser. 2 **4** (1956), 73–106.

[**Shi**] Katsumi Shiratani, *Note on the Kummer–Hilbert reciprocity law*, J. Math. Soc. Japan **12** (1960), 412–421.

[**ShI**] Katsumi Shiratani and Makoto Ishibashi, *On explicit formulas for the norm residue symbol in prime cyclotomic fields*, Mem. Fac. Sci. Kyushu Univ. Ser. A **38** (1984), 203–231.

[**Si**] I. Ya. Sivitskiĭ, *On torsion in Milnor's K-groups for a local field*, Mat. Sb. **126** (1985), no. 4, 576–583; English transl. in Math. USSR-Sb. **54** (1985).

[**Siln**] Joseph H. Silverman, *Advanced topics in the arithmetic of elliptic curves*, Springer, 1994.

[**Silr**] John R. Silvester, *Introduction to algebraic K-theory*, Chapman and Hall, London, 1981.

[**Sim**] Lloyd Simons, *The Hilbert symbol for tamely ramified abelian extensions of 2-adic fields*, Manuscripta Math. **58** (1987), 345–362.

[**ST**] Shankar Sen and John T. Tate, *Ramification groups of local fields*, J. Indian Math. Soc. **27** (1963), 197–202.

[**Sue**] Y. Sueyoshi, *Explicit reciprocity laws on relative Lubin–Tate groups*, Acta Arith. **55** (1990), 291–299.

[**Sus1**] A. A. Suslin, *Homology of GL_n, characteristic classes and Milnor K-theory*, Algebraic K-theory, number theory, geometry and analysis (Bielefeld, 1982), Lecture Notes in Math., vol. 1046, Springer-Verlag, Berlin and New York, 1984, pp. 357–375.

[**Sus2**] ______, *On the K-theory of local fields*, J. Pure Appl. Algebra **34** (1984), 301–318.

[**Sus3**] ______, *Torsion in K_2 of fields*, K-Theory **1** (1987), 5–29.

[**T1**] John T. Tate, *Fourier analysis in number fields, and Hecke's zeta-function*, [**CF**], pp. 305–347.

[**T2**] ______, *p-divisible groups*, Proc. Conference Local Fields (Driebergen, 1966), Springer-Verlag, Berlin, 1967, pp. 158–183.

[**T3**] ______, *Symbols in arithmetic*, Actes du Congrés International des Mathématiciens (Nice, 1970), Tome 1, Gauthier-Villars, Paris, 1971, pp. 201–211.

[**T4**] ______, *Rigid analytic spaces*, Invent. Math. **12** (1971), 257–289.

[**T5**] ______, *Relations between K_2 and Galois cohomology*, Invent. Math. **36** (1976), 257–274.

[**T6**] ______, *On the torsion in K_2 of fields*, Algebraic Number Theory (Kyoto, 1976), Japan Soc. Promotion Sci., Tokyo, 1977, pp. 243–261.

[**T7**] ______, *Number theoretic background*, Automorphic forms, representations and L-functions, Proc. Symp. Pure Math., vol. 33, 1979, pp. 3–26.

[**T8**] ______, *Residues of differentials on curves*, Ann. Sci. École Norm. Sup. **1** (1968), 149–159.

[**Tam**] Tsuneo Tamagawa, *On the theory of ramification groups and conductors*, Japan J. Math. **21** (1951), 197–215.

[**Tay1**] Martin Taylor, *Formal groups and the Galois module structure of local rings of integers*, J. Reine Angew. Math. **358** (1985), 97–103.

[**Tay2**] Martin Taylor, *Hopf structure and the Kummer theory of formal groups*, J. Reine Angew. Math. **375/376** (1987), 1–11.

[**Te**] O. Teichmüller, *Diskret bewertete perfekte Körper mit unvollkommen Restklassenkörper*, J. Reine Angew. Math. **176** (1936), 141–152.

[TV] John Tate and Jose Filipe Voloch, *Linear forms in p-adic roots of unity*, Intern. Math. Res. Notices **12** (1996), 589–601.

[VF] S. V. Vostokov and Ivan B. Fesenko, *The Hilbert symbol for Lubin–Tate formal groups*. II, Zap. Nauchn. Sem. Leningrad. Otdel. Mat. Inst. Steklov (LOMI) **132** (1983), 85–96; English transl. in J. Soviet Math. **30** (1985), no. 1.

[VG] S. V. Vostokov and A. N. Gurevich, *Relation between Hilbert symbol and Witt symbol*, Zap. Nauchn. Sem. POMI **227** (1995), 45–51; English transl. in J. Math. Sci. **89** (1998).

[V1] S. V. Vostokov, *Explicit form of the law of reciprocity*, Izv. Akad. Nauk SSSR Ser. Mat. **42** (1978), no. 6, 1288–1321; English transl. in Math. USSR-Izv. **13** (1979).

[V2] ———, *A norm pairing in formal modules*, Izv. Akad. Nauk SSSR Ser. Mat. **43** (1979), no. 4, 765–794; English transl. in Math. USSR-Izv. **15** (1980).

[V3] ———, *Symbols on formal groups*, Izv. Akad. Nauk SSSR Ser. Mat. **45** (1981), no. 5, 985–1014; English transl. in Math. USSR-Izv. **19** (1982).

[V4] ———, *The Hilbert symbol for Lubin–Tate formal groups*. I, Zap. Nauchn. Sem. Leningrad. Otdel. Mat. Inst. Steklov. (LOMI) **114** (1982), 77–95; English transl. in J. Soviet Math. **27** (1984), no. 4.

[V5] ———, *Explicit construction of class field theory for a multidimensional local field*, Izv. Akad. Nauk SSSR Ser. Mat. **49** (1985), no. 2, 238–308; English transl. in Math. USSR-Izv. **26** (1986).

[V6] ———, *The Lutz filtration as a Galois module in an extension without higher ramification*, Zap. Nauchn. Sem. Leningrad. Otdel. Mat. Inst. Steklov. (LOMI) **160** (1987), 182–192; English transl. in J. Soviet Math. **52** (1990), no. 3.

[V7] ———, *A remark on the space of cyclotomic units*, Vestnik Leningrad. Univ. Mat. Mekh. Astronom. **1988**, no. 1, 14–17; English transl. in Vestnik Leningrad Univ. Math. **21** (1988), no. 1.

[V8] ———, *Decomposablity of ideals in splitting p-extensions of local fields.*, Vestnik Leningrad. Univ. Mat. Mekh. Astronom. **26** (1993), no. 2, 15–21; English transl. in Vestnik St.Petersburg University: Mathematics **26** (1993), 10–16.

[V9] ———, *The pairing on K-groups in fields of valuation of rank n*, Trudy Sankt-Peterb. Mat. Obschestva **3** (1995), 140–184; English transl. in Amer. Math. Soc. Transl. Ser. 2, vol. 166, AMS, 1995.

[V10] ———, *Artin-Hasse exponentials and Bernoulli numbers*, Trudy Sankt-Peterb. Mat. Obschestva **3** (1995); English transl. in Amer. Math. Soc. Transl. Ser. 2, vol. 166, AMS, 1995.

[V11] ———, *Explicit formulas for the Hilbert symbol*, in **[FK]**, pp. 81–89.

[Vo] David A. Vogan, *The local Langlands conjecture*, Contemp. Math. **155** (1993), AMS, 305–379.

[vR] A. C. M. van Rooij, *Non–Archimedean functional analysis*, Marcel Dekker, 1978.

[VVZ] V. S. Vladimirov, I. V. Volovich, E. I. Zelenov, *p-adic analysis and mathematical physics*, World Sci., River Edge, 1994.

[VZh1] S. V. Vostokov, I. B. Zhukov, *Abelian semiramified extensions of a two-dimensional local field*, Rings and modules. Limit theorems of probability theory, vol. 2, Lenigrad Univ., pp. 39–50; Russian, 1988.

[VZh2] S. V. Vostokov, I. B. Zhukov, *Some approaches to the construction of abelian extensions for* $\mathfrak{p}$*-adic fields*, Tr. St-Peterbg. mat. Obsch., vol. 3, 1995; English transl. in Amer. Math. Soc. Transl. Ser. 2, vol. 166, AMS, 1995, pp. 157–174.

[W] Andre Weil, *Basic number theory*, 3rd ed., Springer-Verlag, Berlin and New York, 1974.

[Wa] Lawrence C. Washington, *Introduction to cyclotomic fields*, Springer-Verlag, Berlin and New York, 1982.

[Wd] Adrian R. Wadsworth, *p-Henselian field: K-theory, Galois cohomology and graded Witt rings*, Pacific J. Math. **105** (1983), 473–496.

[Wen] G.H. Wenzel, *Note on G. Whaples' paper "Algebraic extensions of arbitrary fields"*, Duke Math. J. **35** (1968), 47–47.

[Wes] Edwin Weiss, *Algebraic number theory*, McGraw-Hill, New York, 1963.

[Wh1] George Whaples, *Generalized local class field theory.* I, Duke Math. J. **19** (1952), 505–517; II, Duke Math. J. **21** (1954), 247–255; III, Duke Math. J. **21** (1954), 575–581; IV, Duke Math. J. **21** (1954), 583–586.

[Wh2] ______, *Additive polynomials*, Duke Math. J. **21** (1954), 55–66.

[Wh3] ______, *Galois cohomology of additive polynomials and nth power mapping of fields*, Duke Math. J. **24** (1957), 143–150.

[Wh4] ______, *Algebraic extensions of arbitrary fields*, Duke Math. J. **24** (1957), 201–204.

[Wh5] ______, *The generality of local class field theory* (*Generalized local class field theory.* V), Proc. Amer. Math. Soc. **8** (1957), 137–140.

[Wig1] Kay Wingberg, *Der Eindeutigkeitssatz für Demuškinformationen*, Invent. Math. **70** (1982), 99–113.

[Wig2] ______, *Galois groups of Poincare-type over algebraic number fields*, Galois groups over $\mathbb{Q}$ (Berkeley, CA, 1987), Math. Sci. Res. Inst. Publ., 16, Springer-Verlag, Berlin and New York, 1989, pp. 439–449.

[Wi] John S. Wilson, *Profinite groups*, Clarendon Press, Oxford, 1998.

[Wil] A. Wiles, *Higher explicit reciprocity laws*, Ann. of Math. (2) **107** (1978), 235–254.

[Win1] J.-P. Wintenberger, *Extensions de Lie et groupes d'automorphismes des corps locaux de caractéristique p*, C. R. Acad. Sci. Paris Sér. A **288** (1979), 477–479.

[Win2] ______, *Extensions abeliennes et groupes d'automorphismes der corps locaux*, C. R. Acad. Sci. Paris Sér. A **290** (1980), 201–203.

[Win3] ______, *Le corps des normes de certaines extensions infinies des corps locaux; applications*, Ann. Sci. École Norm. Sup. (4) **16** (1983), 59–89.

[Win4] ______, *Automorphismes des corps locaux de caractéristique p*, preprint, Strasbourg (2000).

[Wit1] E. Witt, *Der Existenzsatz für abelsche Functionenkörper*, J. Reine Angew. Math. **173** (1935), 43–51.

[Win4] ______, *Une généralisation d'un théorème de Tate–Sen–Ax*, C. R. Acad. Sci. Paris Sér. I Math. **307** (1988), 63–65.

[Wit2] ______, *Zyklische Körper und Algebren der Characteristik p vom grade p^n*, J. Reine Angew. Math. **176** (1936), 126–140.

[Wit3] ______, *Schiefkörper über diskret bewerteten Körpern*, J. Reine Angew. Math. **176** (1936), 153–156.

[Wy] Bostick F. Wyman, *Wildly ramified gamma extensions*, Amer. J. Math. **91** (1969), 135–152.

[Ya1] Koichi Yamamoto, *Isomorphism theorem in the local class field theory*, Mem. Fac. Sci. Kyushu Univ. Ser. A **12** (1958), 67–103.

[Ya2] ______, *On the Kummer–Hilbert reciprocity law*, Mem. Fac. Sci. Kyushu Univ. Ser. A **13** (1959), 85–95.

[Yak1] A. V. Yakovlev, *The Galois group of the algebraic closure of a local field*, Izv. Akad. Nauk SSSR Ser. Mat. **32** (1968), no. 6, 1283–1322; English transl. in Math. USSR-Izv. **2** (1968).

[Yak2] ______, *Remarks on my paper "The Galois group of the algebraic closure of a local field"*, Izv. Akad. Nauk SSSR Ser. Mat. **42** (1978), no. 1, 212; English transl. in Math. USSR-Izv. **12** (1978).

[Yak3] ______, *Symplectic spaces with operators over commutative rings*, Vestnik Leningrad. Univ. **25** (1970), no. 19, 58–64; English transl. in Vestnik Leningrad univ. Math. **3** (1976), 339–346.

[Yak4] ______, *Abstract characterization of the Galois group of the algebraic closure of a local field*, Zap. Nauchn. Sem. Leningrad. Otdel. Mat. Inst. Steklov. (LOMI) **75** (1978), 179–193; English transl. in J. Soviet Math. **37** (1987), no. 2.

[Yak5] ______, *Structure of the multiplicative group of a simply ramified extension of a local field of odd degree*, Mat. Sb. **107** (1978), no. 2, 304–316; English transl. in Math. USSR-Sb. **35** (1979).

[Yam] Shuji Yamagata, *A remark on integral representations associated with p-adic field extensions*, Proc. Japan Acad. Ser. A Math. Sci **71** (1995), no. 9, 215-217.

[Ze] I. G. Zel'venskiĭ, *The algebraic closure of a local field when $p = 2$*, Izv. Akad. Nauk SSSR Ser. Mat. **36** (1972), no. 5, 933–946; English transl. in Math. USSR-Izv. **6** (1972).

[Zh1] Igor B. Zhukov, *Structure theorems for complete fields*, Tr. St-Peterbg. mat. Obsch., vol. 3, 1995, pp. 215–234; English transl. in Amer. Math. Soc. Transl. Ser. 2, vol. 166, AMS, 1995, pp. 175–192.

[Zh2] ______, *On ramificaion theory in the imperfect residue field case*, Preprint no. 98-02, Nottingham Univ., Nottingham, 1998, to appear in the Proc. of the Luminy conference on Ramification theory for arithmetic schemes.

List of Notations

Index

Titles in This Series

210 **Toshitake Kohno,** Conformal field theory and topology, 2002
209 **Yasumasa Nishiura,** Far-from-equilibrium dynamics, 2002
208 **Yukio Matsumoto,** An introduction to Morse theory, 2002
207 **Ken'ichi Ohshika,** Discrete groups, 2002
206 **Yuji Shimizu and Kenji Ueno,** Advances in moduli theory, 2002
205 **Seiki Nishikawa,** Variational problems in geometry, 2001
204 **A. M. Vinogradov,** Cohomological analysis of partial differential equations and Secondary Calculus, 2001
203 **Te Sun Han and Kingo Kobayashi,** Mathematics of information and coding, 2002
202 **V. P. Maslov and G. A. Omel′yanov,** Geometric asymptotics for nonlinear PDE. I, 2001
201 **Shigeyuki Morita,** Geometry of differential forms, 2001
200 **V. V. Prasolov and V. M. Tikhomirov,** Geometry, 2001
199 **Shigeyuki Morita,** Geometry of characteristic classes, 2001
198 **V. A. Smirnov,** Simplicial and operad methods in algebraic topology, 2001
197 **Kenji Ueno,** Algebraic geometry 2: Sheaves and cohomology, 2001
196 **Yu. N. Lin′kov,** Asymptotic statistical methods for stochastic processes, 2001
195 **Minoru Wakimoto,** Infinite-dimensional Lie algebras, 2001
194 **Valery B. Nevzorov,** Records: Mathematical theory, 2001
193 **Toshio Nishino,** Function theory in several complex variables, 2001
192 **Yu. P. Solovyov and E. V. Troitsky,** C^*-algebras and elliptic operators in differential topology, 2001
191 **Shun-ichi Amari and Hiroshi Nagaoka,** Methods of information geometry, 2000
190 **Alexander N. Starkov,** Dynamical systems on homogeneous spaces, 2000
189 **Mitsuru Ikawa,** Hyperbolic partial differential equations and wave phenomena, 2000
188 **V. V. Buldygin and Yu. V. Kozachenko,** Metric characterization of random variables and random processes, 2000
187 **A. V. Fursikov,** Optimal control of distributed systems. Theory and applications, 2000
186 **Kazuya Kato, Nobushige Kurokawa, and Takeshi Saito,** Number theory 1: Fermat's dream, 2000
185 **Kenji Ueno,** Algebraic Geometry 1: From algebraic varieties to schemes, 1999
184 **A. V. Mel′nikov,** Financial markets, 1999
183 **Hajime Sato,** Algebraic topology: an intuitive approach, 1999
182 **I. S. Krasil′shchik and A. M. Vinogradov, Editors,** Symmetries and conservation laws for differential equations of mathematical physics, 1999
181 **Ya. G. Berkovich and E. M. Zhmud′,** Characters of finite groups. Part 2, 1999
180 **A. A. Milyutin and N. P. Osmolovskii,** Calculus of variations and optimal control, 1998
179 **V. E. Voskresenskiĭ,** Algebraic groups and their birational invariants, 1998
178 **Mitsuo Morimoto,** Analytic functionals on the sphere, 1998
177 **Satoru Igari,** Real analysis—with an introduction to wavelet theory, 1998
176 **L. M. Lerman and Ya. L. Umanskiy,** Four-dimensional integrable Hamiltonian systems with simple singular points (topological aspects), 1998
175 **S. K. Godunov,** Modern aspects of linear algebra, 1998
174 **Ya-Zhe Chen and Lan-Cheng Wu,** Second order elliptic equations and elliptic systems, 1998
173 **Yu. A. Davydov, M. A. Lifshits, and N. V. Smorodina,** Local properties of distributions of stochastic functionals, 1998

172 **Ya. G. Berkovich and E. M. Zhmud′,** Characters of finite groups. Part 1, 1998
171 **E. M. Landis,** Second order equations of elliptic and parabolic type, 1998
170 **Viktor Prasolov and Yuri Solovyev,** Elliptic functions and elliptic integrals, 1997
169 **S. K. Godunov,** Ordinary differential equations with constant coefficient, 1997
168 **Junjiro Noguchi,** Introduction to complex analysis, 1998
167 **Masaya Yamaguti, Masayoshi Hata, and Jun Kigami,** Mathematics of fractals, 1997
166 **Kenji Ueno,** An introduction to algebraic geometry, 1997
165 **V. V. Ishkhanov, B. B. Lur′e, and D. K. Faddeev,** The embedding problem in Galois theory, 1997
164 **E. I. Gordon,** Nonstandard methods in commutative harmonic analysis, 1997
163 **A. Ya. Dorogovtsev, D. S. Silvestrov, A. V. Skorokhod, and M. I. Yadrenko,** Probability theory: Collection of problems, 1997
162 **M. V. Boldin, G. I. Simonova, and Yu. N. Tyurin,** Sign-based methods in linear statistical models, 1997
161 **Michael Blank,** Discreteness and continuity in problems of chaotic dynamics, 1997
160 **V. G. Osmolovskiĭ,** Linear and nonlinear perturbations of the operator div, 1997
159 **S. Ya. Khavinson,** Best approximation by linear superpositions (approximate nomography), 1997
158 **Hideki Omori,** Infinite-dimensional Lie groups, 1997
157 **V. B. Kolmanovskiĭ and L. E. Shaĭkhet,** Control of systems with aftereffect, 1996
156 **V. N. Shevchenko,** Qualitative topics in integer linear programming, 1997
155 **Yu. Safarov and D. Vassiliev,** The asymptotic distribution of eigenvalues of partial differential operators, 1997
154 **V. V. Prasolov and A. B. Sossinsky,** Knots, links, braids and 3-manifolds. An introduction to the new invariants in low-dimensional topology, 1997
153 **S. Kh. Aranson, G. R. Belitsky, and E. V. Zhuzhoma,** Introduction to the qualitative theory of dynamical systems on surfaces, 1996
152 **R. S. Ismagilov,** Representations of infinite-dimensional groups, 1996
151 **S. Yu. Slavyanov,** Asymptotic solutions of the one-dimensional Schrödinger equation, 1996
150 **B. Ya. Levin,** Lectures on entire functions, 1996
149 **Takashi Sakai,** Riemannian geometry, 1996
148 **Vladimir I. Piterbarg,** Asymptotic methods in the theory of Gaussian processes and fields, 1996
147 **S. G. Gindikin and L. R. Volevich,** Mixed problem for partial differential equations with quasihomogeneous principal part, 1996
146 **L. Ya. Adrianova,** Introduction to linear systems of differential equations, 1995
145 **A. N. Andrianov and V. G. Zhuravlev,** Modular forms and Hecke operators, 1995
144 **O. V. Troshkin,** Nontraditional methods in mathematical hydrodynamics, 1995
143 **V. A. Malyshev and R. A. Minlos,** Linear infinite-particle operators, 1995
142 **N. V. Krylov,** Introduction to the theory of diffusion processes, 1995
141 **A. A. Davydov,** Qualitative theory of control systems, 1994
140 **Aizik I. Volpert, Vitaly A. Volpert, and Vladimir A. Volpert,** Traveling wave solutions of parabolic systems, 1994
139 **I. V. Skrypnik,** Methods for analysis of nonlinear elliptic boundary value problems, 1994

For a complete list of titles in this series, visit the AMS Bookstore at **www.ams.org/bookstore/**.